Regulation of Gene Expression in Animal Viruses

NATO ASI Series

Advanced Science Institutes Series

A series presenting the results of activities sponsored by the NATO Science Committee, which aims at the dissemination of advanced scientific and technological knowledge, with a view to strengthening links between scientific communities.

The series is published by an international board of publishers in conjunction with the NATO Scientific Affairs Division

A	**Life Sciences**	Plenum Publishing Corporation
B	**Physics**	New York and London
C	**Mathematical and Physical Sciences**	Kluwer Academic Publishers
D	**Behavioral and Social Sciences**	Dordrecht, Boston, and London
E	**Applied Sciences**	
F	**Computer and Systems Sciences**	Springer-Verlag
G	**Ecological Sciences**	Berlin, Heidelberg, New York, London,
H	**Cell Biology**	Paris, Tokyo, Hong Kong, and Barcelona
I	**Global Environmental Change**	

Recent Volumes in this Series

Volume 234—Development of the Central Nervous System in Vertebrates
edited by S. C. Sharma and A. M. Goffinet

Volume 235—Advances in Cardiovascular Engineering
edited by Ned H. C. Hwang, Vincent T. Turitto, and Michael R. T. Yen

Volume 236—Rhythms in Fishes
edited by M. A. Ali

Volume 237—The Photosynthetic Bacterial Reaction Center II: Structure, Spectroscopy, and Dynamics
edited by Jacques Breton and André Verméglio

Volume 238—Targeting of Drugs 3: The Challenge of Peptides and Proteins
edited by Gregory Gregoriadis, Alexander T. Florence, and George Poste

Volume 239—The Mammalian Cochlear Nuclei: Organization and Function
edited by Miguel A. Merchán, José M. Juiz, Donald A. Godfrey, and Enrico Mugnaini

Volume 240—Regulation of Gene Expression in Animal Viruses
edited by Luis Carrasco, Nahum Sonenberg, and Eckard Wimmer

Series A: Life Sciences

Regulation of Gene Expression in Animal Viruses

Edited by

Luis Carrasco
Universidad Autónoma de Madrid
Madrid, Spain

Nahum Sonenberg
McGill University
Montréal, Canada

and

Eckard Wimmer
State University of New York at Stony Brook
Stony Brook, New York

Springer Science+Business Media, LLC

Proceedings of a NATO Advanced Study Institute on
Regulation of Gene Expression in Animal Viruses,
held May 30–June 8, 1992,
in Mallorca, Spain

NATO-PCO-DATA BASE

The electronic index to the NATO ASI Series provides full bibliographical references (with keywords and/or abstracts) to more than 30,000 contributions from international scientists published in all sections of the NATO ASI Series. Access to the NATO-PCO-DATA BASE is possible in two ways:

—via online FILE 128 (NATO-PCO-DATA BASE) hosted by ESRIN, Via Galileo Galilei, I-00044 Frascati, Italy

—via CD-ROM "NATO-PCO-DATA BASE" with user-friendly retrieval software in English, French, and German (©WTV GmbH and DATAWARE Technologies, Inc. 1989)

The CD-ROM can be ordered through any member of the Board of Publishers or through NATO-PCO, Overijse, Belgium.

Library of Congress Cataloging in Publication Data

Regulation of gene expression in animal viruses / edited by Luis Carrasco, Nahum Sonenberg, and Eckard Wimmer.
p. cm.—(NATO ASI series. Series A, Life sciences; v. 240)
"Proceedings of a NATO Advanced Study Institute on Regulation of Gene Expression in Animal Viruses, held May 30–June 8, 1992, in Mallorca, Spain"—T.p. verso.
Includes bibliographical reference and index.
ISBN 978-0-306-44408-1 ISBN 978-1-4615-2928-6 (eBook)
DOI 10.1007/978-1-4615-2928-6
1. Viral genetics—Congresses. 2. Genetic regulation—Congresses. I. Carrasco, Luis, 1949– II. Sonenberg, Nahum. III. Wimmer, Eckard. IV. North Atlantic Treaty Organization. Scientific Affairs Division. V. NATO Advanced Study Institute on Regulation of Gene Expression in Animal Viruses (1992: Palma, Spain) VI. Series.
[DNLM: 1. Gene Expression Regulation, Viral—congresses. 2. Genes, Viral—congresses. QW 160 R344 1992]
QR456.R44 1993
576'.6484—dc20
DNLM/DLC 92-48456
for Library of Congress CIP

ISBN 978-0-306-44408-1

PREFACE

Viruses, being obligatory parasites of their host cells, rely on a vast supply of cellular components for their replication, regardless of whether infection leads to cell death or to the state of persistence. Animal viruses are providing scientists with relatively simple models to study the molecular biology of genome replication and gene expression. Whereas viruses use, in general, pathways of macromolecular biosynthesis common to the host cell, they have a cunning ability to adopt unusual mechanisms of gene expression and gene replication, provided these special pathways offer an advantage in competition for cellular resources. Any study of viral gene expression and replication is likely to lead also to new insights in cellular metabolism. The discoveries of cis-acting regulatory elements in transcription, the phenomenon of splicing of pre-mRNA, and cap-dependent and cap-independent initiation of translation may be cited as examples. In addition, animal virus genomes contain elements and encode proteins that are very useful for the design of vectors for gene cloning and expression in mammalian cells.

Apart from the basic interest in their biology, viruses have gained notoriety, of course, because they are pathogens. Human animal viruses may cause diseases ranging from the deadly (AIDS) to the benign (common cold). All studies on animal viruses potentially lead to the development of tools for their control, be it through prevention by immunization or treatment with antiviral drugs. Finally, viruses have yielded invaluable reagents in molecular biology as, for example, the vaccinia virus vector for the expression of foreign genes.

This volume contains contrubutions from the faculty of a NATO Advanced Study Institute. The articles summarize the subjects presented at a meeting entitled "The Regulation of Gene Expression in Animal Viruses" that took place in Mallorca, Spain, in May/June, 1992; they convey, in the opinion of the editors, an unabated excitement about basic research in animal virology.

The aim of the Mallorca meeting was to gather scientists and students interested in different animal viruses. Many aspects of molecular virology and concepts of viral pathogenesis were discussed: virus structure at the atomic level, gene expression of DNA viruses, molecular biology and gene variation of viruses containing reverse transcriptase, gene expression and pathogenicity of RNA viruses with double- and single-stranded RNA genomes. This overview is far from complete but it summarizes important topics of a rapidly moving field in biological sciences that has had, and will continue to have, an enormous impact on diverse disciplines such as basic cell biology and human health care.

June, 1992

Luis Carrasco

Nahum Sonenberg

Eckard Wimmer

CONTENTS

STRUCTURE OF A HUMAN RHINOVIRUS COMPLEXED WITH ITS RECEPTOR MOLECULE

Norman H. Olson[1], Prasanna R. Kolatkar[1], Marcos A. Oliveira[1], R. Holland Cheng[1], Jeffrey M. Greve[2], Alan McClelland[2,3], Timothy S. Baker[1] and Michael G. Rossmann[1,4]

[1]Department of Biological Sciences
Purdue University
West Lafayette, Indiana 47907-1392, USA

[2]Institute for Molecular Biologicals
Miles INC.
400 Morgan Lane
West Haven, Connecticut 06516-4175, USA

[3]Present address: Genetic Therapy INC.
19 Firstfield Road
Gaithersburg, Maryland, 20878, USA

[4]To whom correspondence should be addressed.

INTRODUCTION

Human rhinoviruses are one of the major causes of the common cold. They, like other picornaviruses, are icosahedral assemblies of 60 protomers that envelope a single, positive-sense strand of RNA. Each protomer consists of four polypeptides, VP1 - VP4. The three external viral proteins (VP1 - VP3) each have an approximate molecular weight of 30,000 and a similar folding topology (Rossmann *et al.*, 1985; Hogle *et al.*, 1985). The external viral radius is ~150 Å and the total molecular weight is roughly 8.5 x 10^6. A surface depression, or canyon, that is about 12 Å deep and 12 - 15 Å wide, encircles each pentagonal vertex (Fig. 1C). Residues lining the canyon are more conserved than other surface residues among rhinovirus serotypes[3]. The most variable surface residues are at the sites of attachment of neutralizing antibodies (Rossmann *et al.*, 1985; Sherry and Ruecker, 1985; Sherry *et al.*, 1986). It has been proposed that the cellular receptor molecule recognized by the virus binds to conserved residues in the canyon, thus escaping neutralization by host antibodies that are too big to penetrate into that region. This hypothesis (Rossmann *et al.*, 1985; Rossmann, 1989) is supported by site-directed mutagenesis of residues lining the canyon which alters the ability of the virus to attach to HeLa cell membranes (Colonno *et al.*, 1988).

Regulation of Gene Expression in Animal Viruses, Edited by
L. Carrasco *et al.*, Plenum Press, New York, 1993

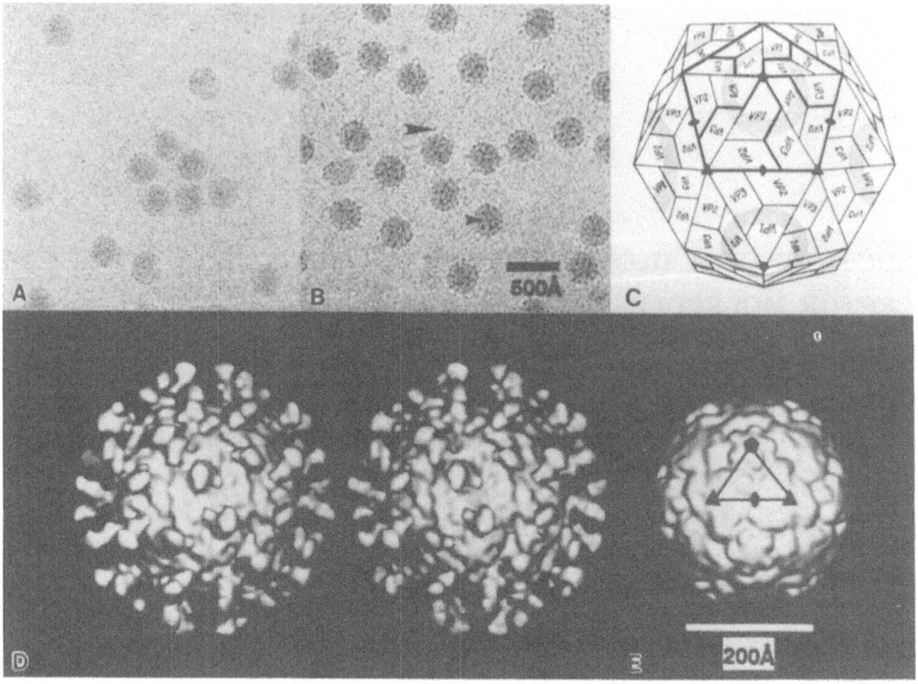

Figure 1. Cryoelectron microscopy of HRV16 particles and their complex with D1D2. (A) Native HRV16. (B) HRV16:D1D2 complex. D1D2 molecules (the two amino terminal domains of ICAM-1) are seen edge-on at the periphery of the virions (large arrow), or end-on in projection (small arrow). Cryoelectron microscopy was performed essentially as described by Cheng *et al.* (1992) with images recorded at a nominal magnification of 49,000X and with an electron dose of ~20e⁻/Å^2. (C) Schematic diagram of HRV showing the icosahedral symmetry, subunit organization and canyon (shaded). Thick lines encircle five protomers of VP1, VP2, and VP3. The fourth viral protein, VP4, is inside the capsid. (D) Stereoview of the reconstruction of the HRV16:D1D2 complex, viewed along an icosahedral twofold axis in approximately the same orientation as in (C). Sixty D1D2 molecules are bound to symmetry-equivalent position at the twelve canyon regions on the virion. The reconstruction was modified to correct for defocus and amplitude contrast effects present in the original micrographs (R.H. Cheng, manuscript submitted). (E) Shaded-surface view of HRV14, computed from the atomic structure (Rossmann *et al.*, 1985), truncated to 20 Å resolution. The triangular outline of one icosahedral asymmetric unit corresponding to that in (C) is indicated.

Also, conformational changes in the floor of the canyon, produced by certain antiviral agents that bind into a pocket beneath the canyon floor, inhibit viral attachment to cellular membranes (Pevear *et al.*, 1985). Conservation of the viral attachment site inside a surface depression has been observed for Mengo (Kim *et al.*, 1990) and influenza virus (Weis *et al.*, 1988; Colman *et al.*, 1983).

There are well over 100 rhinovirus serotypes, which can be divided into roughly two groups according to the cellular receptor they recognize (Abraham and Colonno, 1984; Uncapher *et al.*, 1991). The structures of human rhinovirus 14 (HRV14) (Rossmann *et al.*, 1985), which belongs to the major group of serotypes, and of HRV1A (Kim *et al.*, 1989), which belongs to the minor group of serotypes, have been determined. The structure of HRV16, another major group rhinovirus, is currently being investigated (M.A. Oliveira, R. R. Rueckert & M. G. Rossmann, unpublished results). There are at least 78 serotypes (Tomassini *et al.*, 1989) that bind to intercellular adhesion molecule-1 (ICAM-1), the major group rhinovirus receptor (Greve *et al.*, 1989; Staunton *et al.*, 1989). The ICAM-1 has five immunoglobulin-like domains (D1 to D5 numbered sequentially from the amino end), a transmembrane portion, and a small cytoplasmic domain (Simmons *et al.*, 1988; Staunton *et al.*, 1988). Domains D2, D3 and D4 are glycosylated. Unlike immunoglobulins, ICAM-1 appears to be monomeric (Staunton *et al.*, 1989). Mutational analysis of ICAM-1 has shown that domain D1 contains the primary binding site for rhinoviruses as well as the binding site for its natural ligand, lymphocyte function-associated antigen-1 (LFA-1) (Staunton *et al.*, 1990; Lineberger *et al.*, 1990). Other surface antigens within the immunoglobulin superfamily that are utilized by viruses as receptors include CD4 for human immunodeficiency virus-1 (Robey and Axel, 1990; Maddon *et al.*, 1986), the poliovirus receptor (Mendelsohn *et al.*, 1989), and the mouse coronavirus receptor (Williams *et al.*, 1991). As in ICAM-1, domain D1 of the poliovirus receptor (Koike *et al.*, 1991; Freistadt and Racaniello, 1991) and of CD4 (Arthos *et al.*, 1989) is the primary receptor-virus binding site. The structures of the two amino-terminal domains of CD4 have been determined to atomic resolution (Wang *et al.*, 1990; Ryu *et al.*, 1990). Truncated proteins corresponding to the two amino-terminal domains of ICAM-1 (tICAM-1(185)) as well as the intact extracellular portion of ICAM-1 (tICAM-1(453) or domains D1 to D5) have been expressed in CHO cells (Greve *et al.*, 1991). The desialated form of tICAM-1(185), which will be referred to hereafter as molecule D1D2, has recently been crystallized (Kolatkar *et al.*, 1992).

The attachment of rhinovirus to the receptor molecule at the cell surface is only the first step of virus uncoating. Subsequent to binding receptor, virus is apparently internalized by receptor-mediated endocytosis and enters the endosomal compartment. Productive rhinovirus uncoating and infection requires an intracellular low pH step (Madshus *et al.*, 1984). *In vitro*, low pH treatment will convert rhinovirus to both 135S (missing VP4) and 80S (missing VP4 and RNA) subviral particles (Korant *et al.*, 1975). A number of studies have shown that poliovirus can be conformationally altered to a 135S form upon interaction with its receptor (Gromeier and Wetz, 1990), and rhinovirus can be converted to an 80S empty capsid by incubation in the presence of soluble ICAM-1 (Greve *et al.*, 1991). Thus, both virus-receptor binding and low pH (presumably in the endosomal compartment) appear to play active roles in the controlled disassembly of virus during uncoating, although the relative contributions of these two factors and their temporal relationship *in vivo* are unclear.

A model of the amino-terminal domain D1 of ICAM-1, based on its homology to known structures of the constant domains of immunoglobulins, was reported by Giranda *et al.* (1990) Guided by mutational studies of HRV14 and ICAM-1, they were able to fit this model into the known canyon structure of HRV14. We have utilized cryoelectron microscopy and image analysis techniques to calculate a three-dimensional reconstruction of the complex of HRV16 and D1D2 to ~28 Å resolution. The reconstruction clearly shows that the receptor binds into the canyon of rhinovirus as predicted (Rossmann *et al.*, 1985; Rossmann, 1989). In addition, we use the known structures of HRV14 and CD4 and the predicted structure of D1 of ICAM-1 to identify atomic interactions.

STRUCTURE OF THE VIRUS: RECEPTOR MODEL

Initial attempts at observing complexes of ICAM-1 with HRV14 failed due to the instability of HRV14 in the presence of bound receptor (Greve *et al.*, 1991). Electron micrographs of such specimens revealed severely disrupted particles in a background of protein. However, analysis of several rhinovirus serotypes indicated that HRV14 rapidly uncoated to 80S empty capsids in the presence of soluble ICAM-1, whereas HRV3 and HRV16 formed stable virus-receptor complexes under the same conditions (H. Hoover-Litty & J. M. Greeve, in preparation). HRV16 complexes with the D1D2 molecule or with the complete D1 to D5 extracellular fragment were both used in the investigation. We present here only the results obtained on the HRV16:D1D2 complex.

Unstained vitrified HRV16 and HRV16:D1D2 (Figs. 1A, 1B) have very low, inherent contrast, and the recorded micrographs were very noisy because of the required levels of defocus (~0.8μm) and irradiation (~20e$^-$/Å^2). The only readily visible details on the complexes (Fig. 1B) are the D1D2 molecules that are either seen edge-on at the periphery of the virions or end-on in projection. Forty-four images of the complex were combined to compute a three-dimensional reconstruction (Fig. 1D) with an effective resolution of ~28Å (ref. 40). Although each of the images could be aligned with respect to a consistent choice of enantiomorph, in the absence of additional information there was no way to determine the absolute hand of the reconstruction. However, the asymmetric distribution of density features about the three- and fivefold axes in both the reconstruction and the known HRV14 structure (Fig, 1E) was clearly evident and unambiguously established that the reconstruction had been computed with the correct hand. The excellent correspondence between the asymmetric features provided added confirmation that the reconstruction was accurate. Furthermore, the correlation coefficient between the EM and X-ray maps for densities between radii of 125 to 150 Å was 0.67 for the correct hand versus 0.53 for the opposite hand. The density value of the D1D2 feature in the reconstruction was roughly the same as the density of the virion capsid, thus indicating that the D1D2 molecules nearly saturated the 60 available sites on the virion. Each D1D2 molecule has an approximate dumbbell shape, consistent with the presence of a two-domain structure.

A difference map between the EM model and 20 Å HRV14 model was computed after the radial scale of the EM map was adjusted to correct for magnification errors and after the EM data were compensated to correct for the effects of the microscope contrast transfer function (Toyoshima and Unwin, 1988) (R. H. Cheng, manuscript submitted). The difference map showed that the D1D2 molecule binds to the central portion of the canyon in a manner roughly as predicted by Giranda *et al.* (1990), confirming the predictions inherent in the canyon hypothesis (Rossmann, 1989). The D1D2 molecule is closely associated with the "southern" (see Fig. 2) wall and rim of the canyon, extending 10 Å to 12 Å into the canyon. The binding site is near the center of the triangle formed between a fivelold and two adjacent threefold axes (Fig. 2). The ICAM fragment is oriented roughly perpendicular to the viral surface and extends to a radius of ~205 Å. Its total length is ~75 Å as measured in the difference map. Studies with interspecies chimeras and site-directed mutagenesis of ICAM-1 aimed at identifying the regions necessary for rhinovirus binding indicated that only the first domain is essential, and that the residues most involved in virus binding were concentrated on the outside end of domain D1 (Staunton *et al.*, 1990; McClelland *et al.*, 1991). However, it has proven difficult to produce an active form of domain D1 solution, domains D1 + D2 being the minimal soluble

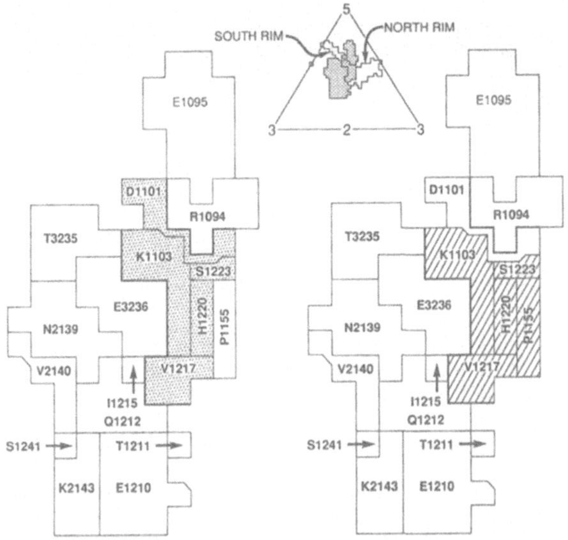

Figure 2. Top: View of the icosahedral asymmetric unit bounded by adjacent five- and threefold axes, outlining residues on the viral surface. Shown are the limits of the canyon, arbitrarily demarcated by a 138 Å radial distance from the viral center[3], and the ICAM-1 footprint (stippled). Left and right: Enlarged view of the residues in the ICAM-1 footprint showing (right) the residues which, when mutated, affect viral attachment (Colonno *et al.*, 1988), and (left) the residues altered in structure by the binding of antiviral compounds that inhibit attachment and uncoating (Smith *et al.*, 1986).

virus-binding species (Greve *et al.*, 1991). The inability to produce domain D1 in isolation and the sequence alignment between ICAM-1 and CD4 suggested that domains D1 and D2 of ICAM-1 are intimately associated through a common, extended β-strand as is seen in the structure of CD4 (Wang *et al.*, 1990; Ryu *et al.*, 1990). Thus, it seemed reasonable to use the known structures of CD4 for fitting the reconstructed density map, although there was slightly too little density for domain D1 and too much density for D2. A better assessment of the fit of domain D1 to the density was obtained by taking the predicted D1 structure of ICAM-1, including all side chains, and superimposing it onto the fitted C_α backbone of CD4. The resultant sequence alignment is shown in Figure 3. One major difference is that although domain D1 of CD4 resembles a variable, immunoglobulin-like domain with extra β-strands, the ICAM-1 prediction is based on a more likely analogy to an immunoglobulin constant domain. This gives domain D1 of ICAM-1 a sleeker appearance, consistent with the observed difference density. The extra density in D2 compared to D2 of CD4 is probably due to the associated carbohydrate groups.

The atomic structure of HRV14 closely matched the reconstructed density that was not occupied by the D1D2 fragment. The only exception occurred in the BC loop of VP1 of HRV14 (see Fig. 4), which extended about 3Å outside the reconstructed density on the "northern" rim of the canyon. However, this region of the polypeptide chain is a highly variable structure and is the site of one of the two largest conformational differences between HRV14 and HRV1A
(Kim *et al.*, 1989). Furthermore, it is also the site of major differences in the structures of homologous poliovirus serotypes 1 and 3 (Filman *et al.*, 1989).

```
                                     10                            20                    30
                  * *                                                          * * * *
Human ICAM-1      Q T S V S P S K V I L P R G G S V L V T C S T S C D Q P K L
Mouse ICAM-1      Q V S I H P R E A F L P Q G G S V Q V N C S S S C K E D L S
                                                  D T V E L T C T A|K S I Q F
CD4, D1                                           10              18 22
                                                    |<----B---->|        |<-C

                                  39                       50                    60
                                  * * * *       * * *
Human ICAM-1      L G I E T P L P K K E L L L P G N N R K V Y E L S N V Q E D
Mouse ICAM-1      L G L E T Q W L K D E L E S - G P N W K L F E L S E T G E D
                  H W         N D R A D S R R|W D Q G N F P L I I K N
CD4, D1           28          52            59 62             70    73
                  C->|          |<----D---->|   |<------E------>|

                                  69                       80                    90
                              *   * * *       * * *
Human ICAM-1      S Q P M C Y S N C P D G Q S T A K T F L T V Y . . . . . V E L A
Mouse ICAM-1      S S P C A L E N C G I V Q S S A S A T I T V V Y P F P E S V E L K
                  S D T Y I C E V E D
CD4, D1           79                88
                    |<----F---->|       |<----G---->|
```

Figure 3. Sequence alignment of domain D1 for human and mouse ICAM-1 (Siu *et al.*, 1989). Also shown is the alignment of ICAM-1 with CD4 based on the predicted structure of the former (Giranda *et al.*, 1990) and the known structure of the latter (Wang *et al.*, 1990; Ryu *et al.*, 1990). Residues marked with asterisks indicate positions implicated in virus binding by site-directed mutagenesis of ICAM-1. The criteria for involvement in virus binding is a reduction of approximately 90% or more in virus binding to ICAM-1 mutants in the absence of evidence for gross structural defects (Staunton *et al.*, 1990; McClelland *et al.*, 1991; Register *et al.*, 1991). Positions identified by Staunton *et al.* (1990) are: 1 - 2, 26 - 29 and 46 - 48; positions identified by McClelland *et al.* (1991) are: 40 - 43, 70 - 72 and 75 - 77; positions identified by Register *et al.* (1991) are: 26 - 30, 67 and 70. An inconsistency was observed at residue 30 (an effect was seen by Register *et al.* (1991) but not by McClelland *et al.* (1991) and is therefore not indicated. Residues of human ICAM-1 that are likely to approach atoms of HRV14 to within 4.5 Å are underlined; based on fitting the CD4 structure into the electron density and then superimposing the predicted ICAM-1, D1 structure (derived from its homology to an immunoglobulin constant domain (Giranda *et al.*, 1990) onto CD4-D1. β-sheet regions in CD4 are marked B, C, D, E and F.

The BC loop has been used extensively for generating poliovirus chimera, permitting the alteration of one serotype to another (Minor *et al.*, 1991; Altermeyer *et al.*, 1991). In the absence of the HRV16 sequence or structure, it is uncertain how far the BC loop of HRV16 would extend towards the ICAM density.

MUTATION DATA

Colonno *et al.* (1988) showed that HRV14 residues H1220*, K1103, P1155, and S1223 all affect the binding of the virus to cellular membranes. Changing P1155 to glycine resulted in enhanced binding, while other alterations of these residues were almost neutral or reduced the binding ability. All these residues are part of the canyon floor and lie centrally within the footprint of the D1D2 molecule binding site (Fig. 2).

---

* The first digit of the residue identification signifies the viral protein, while the last three digits give the amino acid sequence number within the protein.

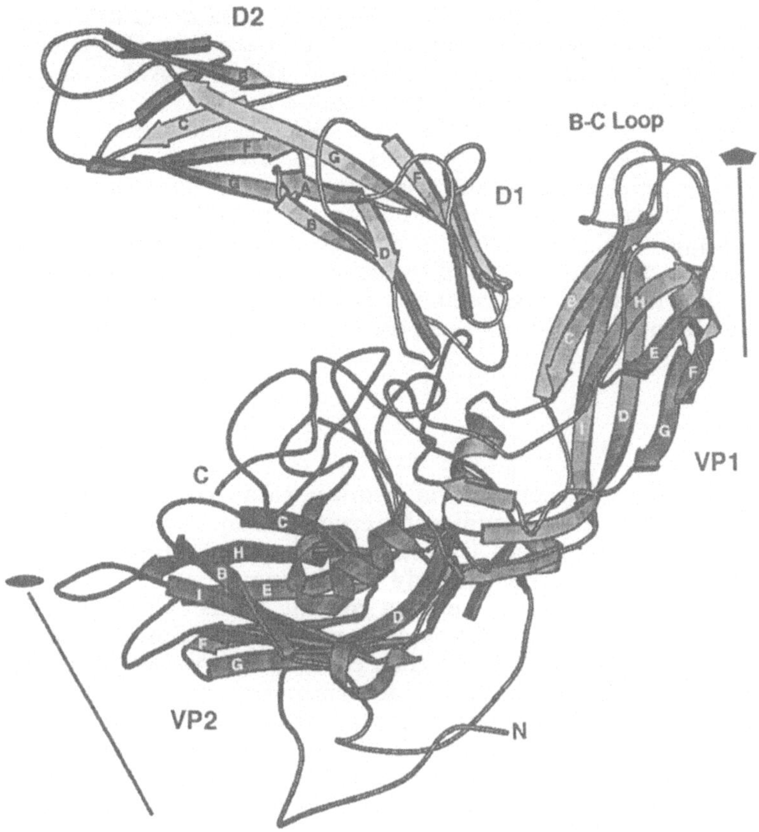

Figure 4. A diagrammatic drawing (Kraulis, 1992) showing the CD4 structure bound to HRV16 as seen in the difference density. Secondary structural elements of the CD4 fragment and HRV14 (homologous structures used to represent ICAM-1 and HRV16, respectively) are identified by the standard nomenclature. The N and C termini of VP1 are also marked.

Certain antiviral agents to rhino-and enteroviruses (Diana *et al.*, 1985) inhibit uncoating and attachment. These agents bind to a pocket beneath the canyon and, in HRV14, significantly alter the structure of the canyon floor (Smith *et al.*, 1986; Badger *et al.*, 1988). These conformational changes inhibitviral attachment (Pevear *et al.*, 1989; Heinz *et al.*, 1989) and are now shown to be exactly at the site of ICAM-1 attachment (Fig. 3).

Heinz *et al.* (1989) have examined a series of escape mutants to the antiviral uncoating inhibitors. Mutants selected in high drug concentrations have larger residues in the binding pocket that inhibit drug entry (Heinz *et al.*, 1989; Badger *et al.*, 1989). However, in lower drug concentrations, resistant mutations are found near the surface residues of the canyon. These mutations (N1100→S, N1105→S, V1153→I, N1219→S, S1223→G) permit binding of drugs into the pocket which induce the usual conformational changes, but do not inhibit viral attachment. The mutated residues are within the region of conformational changes and, hence, might allow some additional flexibility in the floor of the canyon and receptor attachment.

The parts of the predicted ICAM-1 structure that make contact with HRV14 are the amino terminal four residues and loops BC (residues 24 - 26), DE (residues 45 - 49) and FG (residues 71 - 72; see Figs. 3 and 4 for nomenclature). Staunton *et al.* (1990), McClelland *et al.* (1991) and Register *et al.* (1991) have examined the effects of a number of site-directed mutations and mouse-human substitutions in domain D1 of ICAM-1 on rhinovirus binding. Based on these reports seven regions in D1 (Fig. 3), corresponding to the N-terminus (residues 1 - 2), loop BC (residues 26 - 29), strand D (residue 40, 43), loop DE (residues 46 - 48), strand F (residue 67), FG (residues 70 - 72), and the G strand (residues 75 - 77), have been implicated in virus binding. There is correspondence to, or significant overlap between, the four regions of ICAM-1 seen here to be in contact with rhinovirus and four of the seven regions identified by site-directed mutagenesis. Thus, there appears to be reasonable agreement between the mutational studies of ICAM-1 and the observed virus-receptor contacts of the complex.

DISCUSSION

The structure of a complex of Simian rotavirus with a neutralizing antibody Fab fragment was studied by cryoelectron microscopy in a manner similar to that reported here (Prassad *et al.*, 1990), however neither the structure of the virus nor the antibody was known in atomic detail. Recently, the structure of a complex of cowpea mosaic virus (CPMV) and a bound monoclonal antibody Fab fragment was determined with electron microscopy (Wang *et al.*, 1992). In that case, an atomic resolution structure of CPMV was known, permitting the determination of the antibody footprint on the viral surface. Even more recently, the structure of a neutralating antibody Fab fragment, complexed with HRV14, has been determined (Smith *et al.*, 1992), which suggests the mode of bivalent attachment required for neutralization. Here these cryoelectron microscopy techniques are applied to a virus-receptor complex.

Weis *et al.* (1988) and Sauter *et al.* (1992) have explored the interaction of a carbohydrate moiety on the surface of erythrocytes to which influenza virus can attach. We describe here first structure of a virus-receptor complex in which the receptor is a membrane-bound glycoprotein molecule that is used by a virus for recognition of a specific host tissue for attachment and subsequent entry. This receptor molecule belongs to the immunoglobulin superfamily, a class of molecule frequently employed on cell surfaces for the recognition of other molecules (or recognized by viruses) that are subsequently transferred across the membrane. Although the structure of CD4 is known (Wang *et al.*, 1990; Ryu *et al.*, 1990), the structure of the HIV-CD4 complex is not. Mutational studies suggest that the structure recognized by HIV is a ridge made up of β-strands C" and D. Whereas strand C" probably does not exist in domain D1 of ICAM-1 (it is characteristic of immunoglobulin variable type domains), strand D in the D1D2 fragment of ICAM-1 does make particularly close contact, as does also the amino-terminal βA strand. The structure of a complex of human growth factor (hGF) and its immunoglobulin-like receptor has been determined (de Vos *et al.*, 1992). In this complex the primary contact regions between the two receptor molecules and hGF are the AB and EF loops and the βG strand, on the opposite end of the immunoglobulin molecule as those identified for making contact with the rhinovirus canyon.

The structure of a virus-receptor complex might give some indication of the subsequent steps which permit the virus to enter and infect the host cell. Furthermore, since the general nature of the complex described here had been predicted on the basis of the strategy used by HRV to hide its receptor attachment site, perhaps many other viruses use a similar strategy.

Poliovirus is clearly homologous to HRV, and both poliovirus (Mendelsohn *et al.*, 1989) and the major rhinovirus group use an immunoglobulin-like molecule as receptor. Thus, it would be expected that the poliovirus receptor binds into the poliovirus canyon in a manner similar to that of the complex formed for rhinoviruses (Freistadt and Racaniello, 1991). The structure of a mouse-adapted chimera of human poliovirus 2 has been determined (Yeates *et al.*, 1991). The major structural change occurs in the chimera in the BC loop, not in the canyon floor. In this instance, therefore, the BC loop might modulate the virus-receptor interaction.

The determination of residues involved in receptor binding should make it possible to ascertain the origin of specificity of the major rhinovirus serotypes for ICAM-1. Comparison of the amino acid sequences of six rhinoviruses belonging to the major receptor group against four of the minor receptor group did not reveal any clear differentiation. However, structural and binding investigations tentatively suggest that major rhinoviruses do not bind a lipid or fatty acid (Kim *et al.*, 1989; Filman *et al.*, 1989; Kim *et al.*, 1992) within the pocket utilized by the attachment-uncoating inhibitors, in contrast to the minor receptor group of rhinoviruses. Therefore, specificity for the minor receptor may reside in the virus' tendency to bind tightly a cellular fatty acid (Kim *et al.*, 1989; Kim *et al.*, 1992) and thus alter the shape of the canyon floor, rather than the identity of the canyon surface residues themselves. Nevertheless, kwowledge of the virus-receptor interaction will illuminate various strategies currently being developed to interfere with early stages of rhinoviral and other viral infections (Greve *et al.*, 1991; McKinlay *et al.*, 1992; Marlin *et al.*, 1990).

ACKNOWLEDGMENTS

We are grateful for many helpful discussions with Roland Rueckert (University of Wisconsin) and Mark McKinlay, Frank Dutko, Guy Diana and Dan Pevear (Sterling Winthrop Pharmaceuticals Research Division). We are grateful to Marcia Kremer, Carla Forte, and Cindy Music for technical assistance. We thank Wayne Hendrickson and Steve Harrison for sharing coordinate information. We also thank Helene Prongay and Sharon Wilder for help in preparation of this manuscript. The work was supported by National Institutes of Health grants to M.G.R. and T.S.B., a National Science Foundation grant to T.S.B. and Lucille P. Markey Foundation Award.

This article is a reproduction of a paper submitted to Nature in May 1992.

REFERENCES

Abraham, G. and Colonno, R.J., 1984, Many rhinovirus serotypes share the same cellular receptor, J Virol. 51: 340-345.

Altermeyer, R., Murdin, A.D., Harber, J.J. and Wimmer, E., 1991, Construction and characterization of a poliovirus/rhinovirus antigenic hybrid, Virology. 184: 636-644.

Arthos, J., Deen, K.C., Chaikin, M.A., Fornwald, J.A., Sathe, G., Sattentau, Q.J., Clapham, P.R., Weiss, R.A., McDougal, J.S., Pietropaolo, C., Axel, R., Truneh, A., Maddon, P.J. and Sweet, R.W., 1989, Identification of the residues in human CD4 critical for the binding of HIV, Cell. 57:469-481.

Badger, J., Krishnaswamy, S., Kremer, M.J., Oliveira, M.A., Rossmann, M.G., Heinz, B.A., Rueckert, R.R., Dutko, F.J. and McKinlay, M.A., 1989, The three-dimensional structures of drug-resistant mutants of human rhinovirus 14, J Mol Biol. 207: 163-174.

Badger, J., Minor, I., Kremer, M.J., Oliveira, M.A., Smith, T.J., Griffith, J.P., Guerin, D.M.A., Krishnaswamy, S., Luo, M., Rossmann, M.G., McKinlay, M.A., Diana,

G.D., Dutko, F.J., Fancher, M., Rueckert, R.R. and Heinz, B.A., 1988, Structural analysis of a series of antiviral agents complexed with human rhinovirus 14, Proc Natl Acad Sci USA. 85: 3304-3308.

Baker, T.S., Newcomb, W.W., Olson, N.H., Cowsert, L.M., Olson, C. and Brown, J.C., 1991, Structures of bovine and human papillomaviruses: analysis cryoelectron microscopy and three-dimensional image reconstruction, Biophys J. 60: 1445-1456.

Colman,P.M., Varghese, J.N., and Laver, W.G., 1983, Structure of the catalytic and antigenic sites in influenza virus neuraminidase, Nature. 303: 41-44.

Colonno, R.J., Condra, J.H., Mizutani, S., Callahan, P.L., Davies, M.E. and Murcko, M.A., 1988, Evidence for the direct involvement of the rhinovirus canyon in receptor binding, Proc Natl Acad Sci USA. 85:5449-5453.

Cheng, R.H., Olson, N.H. and Baker, T.S., 1992, Cauliflower mosaic virus: a 420 subunit (T = 7), multilayer structure, Virology. 186: 655-668.

Dalgleish, A.G., Beverly, P.C.L., Clapham, P.R., Crawford, D.H., Greaves, M.F. and Weiss, R.A., 1984, The CD4 (T4) antigen is an essential component of the receptor for the AIDS retrovirus, Nature (London). 312: 763-767.

de Vos, A.M., Ultsch, M. and Kossiakoff, A.A., 1992, Human growth hormone and extracellular domain of its receptor: crystal structure of the complex, Science. 255: 306-312.

Diana, G.D., McKinlay, M.A., Otto, M.J., Alullian, V. and Oglesby, C., 1985, [[(4,5-Dihydro-2-oxazolyl)phenoxy]alkyl]isoxazoles. Inhibitors of viral uncoating, J Med Chem. 28: 1906-1910.

Filman, D.J., Syed, R., Chow, M., Macadam, A.J., Minor, P.D. and Hogle, J.M., 1989, Structural factors that control conformational transitions and serotype specificity in type 3 poliovirus, EMBO J. 8: 1567-1579.

Freistadt, M.S. and Racaniello, V.R., 1991, Mutational analysis of the cellular receptors for poliovirus, J Virol. 65: 3873-3876.

Giranda, V.L., Chapman, M.S. and Rossmann, M.G., 1990, Modeling of the human intercellular adhesion molecule-1, the human rhinovirus major group receptor, Proteins. 7: 227-233.

Greve, J.M., Davis, G., Meyer, A.M., Forte, C.P., Yost, S.C., Marlor, C.W., Kamarck, M.E. and McClelland, A., 1989, The major human rhinovirus receptor is ICAM-1, Cell. 56: 839-847.

Greve, J.M., Forte, C.P., Marlor, C.W., Meyer, A.M., Hoover-Litty, H., Wunderlich, D. and McClelland, A., 1991, Mechanisms of receptor-mediated rhinovirus neutralization defined by two soluble forms of ICAM-1, J Virol. 65: 6015-6023.

Gromeier, M. and Wetz, K.J., 1990, Kinetics of poliovirus uncoating in HeLa cells in a nonacidic environment, J Virol. 64: 3590-3597.

Heinz, B.A., Rueckert, R.R., Shepard, D.A., Dutko, F.J., McKinlay, M.A., Fancher, M., Rossmann, M.G., Badger, J. and Smith, T.J., 1989, NGenetic and molecular analyses of spontaneous mutants of human rhinovirus 14 are resistant to an antiviral compound, J Virol. 63: 2476-2485.

Hogle, J.M., Chow, M. and Filman, D.J., 1985, Three-dimensional structure of poliovirus at 2.9 Å resolution, Science. 229: 1358-1365.

Kim, K.H., Willingmann, P., Gong, Z.X., Kremer, M.J., Chapman, M.S., Minor, I., Oliveira, M., Rossmann, M.G., Andries, K., Diana, G.D., Dutko, F.J., McKinlay, M.A. and Pevear, D.C., 1992, A Comparison of the Anti-rhinoviral Drug Binding Pocket in HRV14 and HRV1A, manuscript in preparation.

Kim, S. , Boege, U., Krishnaswamy, S., Minor, I., Smith, T.J., Luo, M., Scraba, D.G. and Rossmann, M.G., 1990, Conformational variability of a picornavirus capsid: pH-dependent structural changes of Mengo virus related to its host receptor attachment site and disassembly, Virology. 175: 176-190.

Kim, S., Smith, T.J., Chapman, M.S., Rossmann, M.G., Pevear, D.C., Dutko, F.J., Felock, P.J., Diana, G.D. and McKinlay, M.A., 1989, The crystal structure of human rhinoviruses serotype 1A (HRV1A), J Mol Biol. 210: 91-111.

Klatzmann, D., Champagne, E., Chamaret, S., Gruest, J., Guetard, D., Hercend, T., Gluckman, J.-C. and Montagnier, L., 1984, T-lymphocyte T4 molecule behaves as the receptor for human retrovirus LAV, Nature (London). 312: 767-768.

Koike, S., Ise, I. and Nomoto, A., 1991, Functional domains of the poliovirus receptor, Proc Natl Acad Sci USA. 88: 4104-4108.

Kolatkar, P.R., Oliveira, M.A., Rossmann, M.G., Robbins, A.H., Katti, S.K., Hooer-Litty, H., Forte, C., Greve, J.M., McClelland, A. and Olson, N.H., 1992, Preliminary X-ray crystallographic analysis of intercellular adhesion molecule-1, J Mol Biol., in press.

Korant, B.D., Lonberg-Holm, K., Yin, F.H. and Noble-Harvey, J., 1975, Fractionation of biologically active and inactive populations of human rhinovirus type 2, Virology. 63: 384-394.

Kraulis, P.J.,1992, J Appl Crystallogr. 24: 946-950.

Lineberger, D.W., Graham, D.J., Tomassini, J.E. and Colonno, R.J., 1990, Antibodies that block rhinovirus attachment map to domain 1 of the major group receptor, J Virol. 64: 2582-2587.

Maddon, P.J., Dalgleish, A.G., McDougal, J.S., Clapham, P.R., Weiss, R.A. and Axel, R., 1986, The T4 gene encodes the AIDS virus receptor and is expressed in the immune system and the brain, Cell. 47: 333-348.

Madshus, I.H., Olsnes, S. and Sandvig, K., 1984, Different pH requirements for entry of the two picornaviruses, human rhinovirus 2 and murine encephalomyocarditis virus, Virology. 139: 346-357.

Marlin, S.D., Staunton, D.E., Springer, T.A., Stratowa, C., Sommergruber, W. and Merluzzi, V.J., 1990, A soluble form of intercellular adhesion molecule-1 inhibits rhinovirus infection, Nature (London). 344: 70-72.

McClelland, A., deBear, J., Yost, S.C., Meyer, A.M., Marlor, C.W. and Greve, J.M., 1991, Identification of monoclonal antibody epitopes and critical residues for rhinovirus binding in domain 1 of intercellular adhesion molecule 1, Proc Natl Acad Sci USA. 88: 7993-7997.

McKinlay, M.A., Pevear, D.C. and Rossmann, M.G., 1992, Treatment of the picornavirus common cold by inhibitors of viral uncoating and attachment, Ann Rev Microbiol. manuscript in press.

Mendelsohn, C.L., Wimmerm, E. and Racaniello, V.R., 1989, Cellular receptors for poliovirus: molecular cloning, nucleotide sequence, and expression of a new member of the immunoglobulin superfamily, Cell. 56: 855-865.

Minor, P.D., Ferguson, M., Katrak, K., Wood, D., John, A., Howlett, J., Dunn, G., Burke, K. and Almond, J.W., 1991, Antigenic structure of chimeras of type 1 and type 3 polioviruses involving antigenic sites 2, 3 and 4, J Gen Virol. 72: 2475-2481.

Pevear, D.C., Fancher, M.J., Felock, P.J., Rossmann, M.G., Miller, M.S., Diana, G., Treasurywala, A.M., McKinlay, M.A. and Dutko, F.J., 1989, Conformation change in the floor of the human rhinovirus canyon blocks adsorption to HeLa cell receptors, J Virol. 63:2002-2007.

Prasad, B.V.V., Brurns, J.W., Marietta, E., Estes, M.K. and Chiu, W., 1990, Localization of VP4 neutralization sites in rotavirus by three-dimensional cryo-electron microscopy, Nature (London). 343: 476-479.

Register, R.B., Uncapher, C.R., Naylor, A.M., Lineberger, D.W. and Colonno, R.J., 1991, Human-murine chimeras of ICAM-1 identify amino acid residues critical for rhinovirus and antibody binding, J Virol. 65: 6589-6596.

Robey, E. and Axel, R., 1990, CD4: collaborator in immune recognition and HIV infection, Cell. 60: 697-700.

Rossmann, M.G., 1989, Hiding the host cell receptor attachment site on a viral surface from immune surveillance, J Biol Chem. 263: 14587-14590.

Rossmann, M.G. and Palmenberg, A.C., 1988, Conservation of the putative receptor attachment site in picornaviruses, Virology. 164: 373-382.

Rossmann, M.G., Arnold, E., Erickson, J.W., Frankenberger, E.A., Griffith, J.P., Hecht, H.J., Johnson, J.E., Kamer, G., Luo, M., Ruekert, A.G., Sherry, B. and Vriend, G., 1985, Structure of a human common cold virus and functional relationship to other picornaviruses, Nature (London). 317:145-153.

Ryu, S.E., Kwong, P.D., Truneh, A., Porter, T.G., Arthos, J., Rosenberg, M., Dai, X., Xuong, N., Axel, R., Sweet, R.W. and Hendrickson, W.A., 1990, Crystal structure of an HIV-binding recomginant fragment of human CD4, Nature (London). 348: 419-426.

Sauter, N.K., Glick, G.D., Crowther, R.L., Park, S.J., Eisen, M.B., Skehel, J.J., Knowles, J.R. and Wiley, D.C., 1992, Crystallographic detection of a second ligand binding site in influenza virus hemagglutinin, Proc Natl Acad Sci USA. 89: 324-328.

Sherry, B. and Rueckert, R., 1985, Evidence for at least two dominant neutralization antigens on human rhinovirus 14, J Virol. 53: 137-143.

Sherry, B., Mosser, A.G., Colonno, R.J. and Rueckert, R.R., 1986, Use of monoclonal antibodies to identify four neutralization immunogens on a common cold picornavirus, human rhinovirus 14, J Virol. 57: 246-257.

Simmons, D., Makgoba, M.W. and Seed, B., 1988, ICAM, an adhesion ligand of LFA-1, is homologous to the neural cell adhesion molecule NCAM, Nature (London). 331: 624-627.

Siu, G., Hedrick, S.M. and Bian, A.A., 1989, Isolation of the murine intercellular adhesion molecule 1 (ICAM-1) gene, J Immunol. 143: 3813-3820.

Smith, T.J., Kremer, M.J., Luo, M., Vriend, G., Arnold, E., Kamer, G., Rossmann, M.G., McKinlay, M.A., Diana, G.D. and Otto, M.J.,1986, The site of attachment in human rhinovirus 14 for antiviral agents that inhibit uncoating, Science. 233: 1286-1293.

Smith, T.J., Olson, N.H., Cheng, R.H., Chase, E., Lee, W.M., Leippe, D., Mosser, A., Rueckert, R.R. and Baker, T.S., 1992, Structure of human rhinovirus complexed with Fab fragments from a neutralizing antibody, Nature (London). Manuscript submitted for publication.

Staunton, D.E., Dustin, M.L., Erickson, H.P. and Springer, T.A., 1990, The arrangement of the immunoglobulin-like domains of ICAM-1 and the binding site for LFA-1 and rhinovirus, Cell, 61: 243-254.

Staunton, D.E., Marlin, S.D., Stratowa, C., Dustin, M.L., and Springer, T.A., 1988, Primary structure of ICAM-1 demonstrates interaction between members of the immunoglobulin and integrin supergene families, Cell. 52: 925-933.

Staunton, D.E., Merluzzi, V.J., Rothlein, R., Barton, R., Marlin, S.D. and Springer, T.A., 1989, A cell adhesion molecule, ICAM-1, is the major suface receptor for rhinoviruses, Cell. 56: 849-853.

Tomassini, J.E., Maxson, T.R. and Colonno, R.J., 1989, Biochemical characterization of a glycoprotein required for rhinovirus attachment, J Biol Chem. 264:1656-1662.

Toyoshima, C. and Unwin, N., 1988, Contrast transfer for frozen-hydrated specimens: determination from pairs of defocused images, Ultramicroscopy. 25: 279-282.

Uncapher, C.R., DeWitt, C.M. and Colonno, R.J., 1991, The major and minor group receptor families contain all but one human rhinovirus serotype, Virology. 180: 814-817.

Wang, G., Porta, C., Chen, Z., Baker, T.S. and Johnson, J.E., 1992, Identification of a Fab interaction footprint site on an icosahedral virus by cryoelectron microscopy and X-ray crystallography, Nature (London). 355: 275-278.

Wang, J., Yan, Y., Garret, T.P.J., Liu, J., Rodgers, D.W., Garlick, R.L., Tarr, G.E., Husain, Y., Reinherz, E.L. and Harrison, S.C., 1990, Atomic structure of a fragment of human CD4 containing two immunoglobulin-like domains, Nature (London). 348: 411-418.

Weis, W., Brown, J.H., Cusack, S., Paulson, J.C., Skehel, J.J. and Wiley, D.C., 1988, Structure of the influenza virus haemagglutinin complexed with its receptor, sialic acid, Nature (London), 333: 426-431.

Williams, R.K., Jiang, G.S. and Holmes, K.V., 1991, Receptor for mouse hepatitis virus is a member of the carcinoembryonic antigen family of glycoproteins, Proc Natl Acad Sci USA. 88: 5533-5536.

Yeates, T.O., Jacobson, D.H., Margin, A., Wychowki, C., Girard, M., Filman, D.J. and Hogle, J. M., 1991, Three-dimensional structure of a mouse-adapted type 2/type 1 poliovirus chimera, EMBO J. 10: 2331-2341.

CASCADE REGULATION OF VACCINIA VIRUS GENE EXPRESSION

Bernard Moss

Laboratory of Viral Diseases
National Institute of Allergy and Infectious Diseases
National Institutes of Health
Bethesda MD 20892, USA

INTRODUCTION

The poxviruses, unlike most other DNA viruses, propagate entirely within the cytoplasm of eukaryotic cells and encode specific enzymes and factors needed for transcription and replication (Moss, 1990; Moss et al., 1991). The complex developmental cycle of vaccinia virus, the best characterized member of the poxvirus family, is represented in Figure 1. The infectious virion is composed of a core structure enveloped by lipoprotein membranes. The linear, double-stranded, 200,000 base pair (bp) DNA genome, as well as the protein components of the early transcription system, reside within the core. After fusion of the virion and cell membranes, the core is released into the cytoplasm where transcription of the early class of genes proceeds. There are about 100 early genes which encode proteins with diverse functions that stimulate proliferation of neighboring cells, protect against the host immune system, replicate the viral genome, and transcribe the intermediate class of genes. Expression of the intermediate genes, however, is delayed until DNA replication occurs. Although the number of genes that belong to the intermediate class is still uncertain, at least three are required to activate expression of the late class of genes. The many late genes encode the principal virion proteins as well as the enzymes and factors needed for transcription of early genes during the next cycle of infection. Thus, there are three temporal classes of vaccinia virus genes - early, intermediate, and late - which are regulated in a cascade fashion.

Regulation of Gene Expression in Animal Viruses, Edited by
L. Carrasco *et al.*, Plenum Press, New York, 1993

TRANSCRIPTION OF EARLY GENES

Early Promoters

Transcriptional regulatory signals have been analyzed by *in vivo* and *in vitro* methods. Promoter containing DNA sequences have been ligated to reporter genes, such as β-galactosidase or chloramphenicol acetyltransferase, and then integrated into the vaccinia virus genome by homologous recombination. The expression of the reporter genes was then measured by infecting cells with the recombinant viruses. In this manner, it was found that the promoters for early genes extend only about 30 bp upstream of the RNA start sites (Cochran et al., 1985; Weir and Moss, 1987). Within one early promoter, the effects of all single nucleotide substitutions were determined (Davison and Moss, 1989a). Using these data, the promoter could be divided into 3 regions relative to the RNA start site at +1: a 15 bp A-rich region (-14 to -28) in which many single nucleotide substitutions have a major effect, separated by a T-rich spacer sequence in which single substitutions are of little consequence, from a 7 bp region within which initiation with a purine occurs. Within the critical region, A-residues are essential at several places and optimal at some others; a G-residue is needed at one position and Ts at one or two places (Figure 2). As with TATA boxes of higher eukaryotic RNA polymerase II promoters, the critical region specifies the distance to the transcription initiation site. Results obtained by analysis of reporter gene expression have been confirmed by direct analysis of RNAs in infected cells and by *in vitro* transcription studies. In addition, the optimal sequence determined by mutational analysis corresponds closely to the consensus sequence obtained by lining up a large number of early promoters.

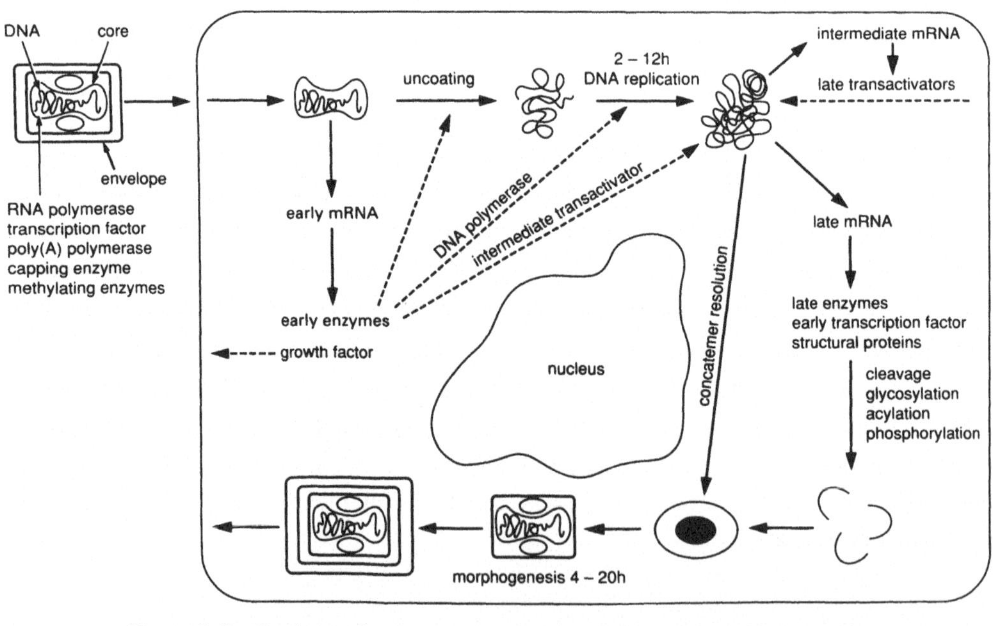

Figure 1. Replication cycle of vaccinia virus. From Moss (1991) with permission.

	-30 -20 -10	+1
Early	AAA TTGAAA A	A/G
Intermediate	T TT AAA AA	TAAA
Late	---A/T-rich region---	TAAAT

Figure 2. Comparison of nucleotide sequences of early, intermediate and late promoters. Nucleotide substitutions of the indicated nucleotides between -14 and -27 of the early and intermediate promoters and at the initiator regions (+1) of intermediate and late promoters resulted in drastic decreases in promoter activities (Baldick et al., 1992; Davison and Moss, 1989a,b).

Termination Signal

The DNA sequence AATAAA, which appears about 20 bp before the polyadenylation site of higher eukaryotic mRNAs and is required for processing and polyadenylation, is not present near the 3' ends of most poxvirus genes. Instead, runs of T residues are frequently present between 20 and 50 bp before the 3 ends of vaccinia virus early mRNAs (Yuen and Moss, 1986). Mutagenesis studies demonstrated that the sequence TTTTTNT, in which N can be any nucleotide, is required for 3' end formation both *in vitro* (Yuen and Moss, 1987) and *in vivo* (Earl et al., 1990). Nucleotide substitutions of any of the T residues completely prevents formation of the 3' end of the RNA. A termination mechanism appears to be operative since long RNAs are not detected as intermediates nor can added long RNAs be processed in *in vitro* systems (Rohrmann et al., 1986). The termination signal is recognized in the RNA as UUUUUNU (Shuman and Moss, 1989) and incorporation of the signal sequence into a hairpin structure can suppress termination (Luo and Shuman, 1991).

CORE-ASSOCIATED ENZYMES AND TRANSCRIPTION FACTORS

DNA-Dependent RNA Polymerase

The presence of a transcriptase activity within vaccinia virus cores was demonstrated by Kates and McAuslan (1967) and Munyon and co-workers (1967). The purified enzyme contains two large and at least six small virus-encoded subunits (Ahn et al., 1990a,b; Ahn et al., 1992; Amegadzie et al., 1991a,b; Amegadzie et al., 1992a; Baroudy and Moss, 1980; Broyles and Moss, 1986; Broyles and Pennington, 1990; Jones et al., 1987; Patel and Pickup, 1989; Quick and Broyles, 1990). The genes, which have been given the prefix *rpo* followed by the predicted size of the polypeptide in kiloDaltons (kDa), are listed in Table 1. The large subunits are homologous to the corresponding subunits of eukaryotic RNA polymerases although the one encoded by *rpo147* lacks the phosphorylated heptapeptide C-terminal repeats of the related RNA polymerase II subunit. Except for *rpo7*, the genes encoding viral small subunits are not closely related to known eukaryotic RNA polymerase

Table 1. Vaccinia Virus RNA Polymerase Genes

Gene	Homolog	Sequence motif
rpo147	RPB1 (yeast)	Zinc finger
rpo132	RPB2 (yeast)	Zinc finger, Leucine zipper
rap94	?RAP30 (man)	-
rpo35	-	-
rpo30	SII (mouse)	Zinc finger
rpo22	-	-
rpo19	-	Acidic N-terminus
rpo18	-	-
rpo7	RPB10 (yeast)	Metal binding

small subunit genes. Interestingly, however, the *rpo30* product has some sequence similarity with eukaryotic transcription elongation factor SII, suggesting that it may have a related function.

RNA Polymerase-Associated Transcription Specificity Factor

The RNA polymerase isolated from virions has been separated by column chromatography into two forms - one that can carry out only non-specific transcription of single-stranded DNA templates and another that can transcribe early vaccinia virus genes in conjunction with the vaccinia virus early transcription factor (to be described in the next section). The RNA polymerase capable of specific transcription was shown to contain an additional virus-encoded polypeptide named RAP94 (Table 1) because of its association with RNA polymerase and size in kDa (Ahn and Moss, 1992). Unassociated RAP94 has not been found in virus particles. However, transfection and immunodepletion experiments indicated that RAP94 is required for specific early transcription.

Early Transcription Factor

In vitro assays containing vaccinia virus RNA polymerase and a specific early gene template were used to detect a vaccinia virus early transcription factor (VETF) in virion extracts (Broyles et al., 1988). The purified factor is a heterodimeric protein that has a DNA-dependent ATPase activity (Broyles and Moss, 1988). VETF binds to the critical region of early promoters and to sequences downstream of the RNA start site and causes DNA bending (Broyles et al., 1991). The ATPase activity of VETF may explain the known requirement for ATP hydrolysis in initiation of transcription (Broyles, 1991; Gershowitz et al., 1978; Watananbi et al., 1974). The viral genes encoding the 70- and 82-kDa subunits of VETF have been identified and are expressed late in infection (Broyles and Fesler, 1990; Gershon and Moss, 1990). The protein sequence predicted for the 70-kDa subunit contains a motif associated with ATP binding or ATPase activity and the sequence for the 82-kDa subunit contains variations of the cannonical zinc finger and leucine zipper motifs (Gershon and Moss, 1990).

RNA polymerase containing RAP94, VETF and a DNA template with an early promoter comprise the minimal macromolecular components necessary for accurate initiation and elongation of RNA molecules.

CAPPING AND METHYLATING ENZYMES

The 5' ends of vaccinia virus mRNAs, like of those of eukaryotes, contain a cap structure that facilitates translation and contributes to stability (Muthukrishnan et al., 1978; Wei and Moss, 1975). Two enzymes are involved in the capping and methylation of vaccinia virus mRNAs. One is a multifunctional enzyme complex composed of 97- and 33-kDa subunits that catalyze (i) removal of the terminal or γ-phosphate from an RNA chain, (ii) transfer of the GMP moiety of GTP to the diphosphate-ended RNA, and (iii) transfer of the methyl group from S-adenosylmethionine to the 7-position of the added guanosine (Martin and Moss, 1976; Venkatesan et al., 1980). The large subunit forms a covalent lysine-GMP intermediate (Roth and Hurwitz, 1984; Shuman and Hurwitz, 1981). The capping enzyme binds to RNA polymerase apparently facilitating its role in capping nascent transcripts (Broyles and Moss, 1987; Hagler and Shuman, 1991). The genes encoding both subunits of capping enzyme have been identified (Morgan et al., 1984; Niles et al., 1989) and the large one has been expressed by itself and with the small subunit as a heterodimer in Escherichia coli (Guo and Moss, 1990; Shuman, 1990; Shuman and Morham, 1990). The large subunit alone has RNA triphosphatase and guanylyltransferase activities. Both subunits, however, appear to be required for methylation (M.A. Higman, N. Bourgeois and E. G. Niles, submitted for publication).

A separate enzyme, of approximately 38 kDa, catalyzes the transfer of a methyl group from S-adenosylmethionine to the ribose of the first encoded nucleotide, which is separated by a triphosphate from the terminal 7-methylguanosine (Barbosa and Moss, 1978). The viral gene encoding this cap-specific RNA (nucleoside-$O^{2'}$-)-methyltransferase was recently identified (Schnierle et al., 1992) and surprisingly was the same one that encodes the stimulatory subunit of the viral poly(A) polymerase (see later section).

Termination Factor

Transcription catalyzed by purified vaccinia virus RNA polymerase and VETF proceeds past termination signals. Using an *in vitro* transcription assay, based on the ability to restore termination, a complementing factor was isolated from vaccinia virus cores. Unexpectedly, the termination factor was identified as the capping enzyme (Shuman, 1990; Shuman et al., 1987). The precise role of the capping enzyme in termination is still not understood since capping *per se* is not required for this activity. However, the association of capping enzyme with RNA polymerase suggests that it functions as a part of the transcription complex (Broyles and Moss, 1987; Hagler and Shuman, 1991).

Poly(A) Polymerase

A primer-dependent poly(A) polymerase, with 55- and 33-kDa polypeptide subunits, has been purified from vaccinia virus cores (Moss et al., 1975) and infected cells (Nevins and Joklik, 1977). Recently, the genes encoding the two subunits were identified (Gershon et al., 1991). The large subunit contains the catalytic activity and is able to add approximately 35 adenylate residues to the 3' end of an RNA primer in a rapid and highly processive manner (Gershon et al., 1991; Gershon and Moss, 1992). The small subunit, stimulates the addition of more adenylate residues to make long poly(A) tails. Both the heterodimeric poly(A) polymerase and the small subunit possess cap-specific RNA (nucleoside-$O^{2'}$-)-methyltransferase activity (Schnierle et al., 1992). It is not known whether the actions of the capping enzyme and the RNA (nucleoside-$O^{2'}$-)-methyltransferase at both ends of the mRNA molecule have special significance with regard to the structure of the transcription complex.

TRANSCRIPTION OF INTERMEDIATE GENES

Intermediate Promoters

Genes belonging to the intermediate temporal class have only recently been identified (Keck et al., 1990; Vos and Stunnenberg, 1988). Intermediate promoters are distinct from early promoters although they resemble them in length and general A/T-richness. Deletion and linker scanning mutations of one intermediate promoter indicated sensitive regions both upstream and downstream of the RNA start site (Hirschmann et al., 1990). A more detailed analysis involving numerous single base substitutions of other intermediate promoters defined 3 regions: an A/T-rich core element located from -26 to -13, an 11 bp spacer element, and an initiator element from -1 to +3 (Baldick et al., 1992). The initiator consists of the sequence TAAA. The spacer does not exhibit sequence specificity but must be 10 or 11 nucleotides long, consistent with a single turn of the double helix. The critical residues within early and late promoters are compared in Figure 2.

Intermediate Transcription Factors

Transfection studies indicated that the factors required for transcription of intermediate genes are synthesized early in infection and in the presence of inhibitors of DNA replication (Keck et al., 1990; Vos and Stunnenberg, 1988). *In vitro* studies by Stunnenberg and co-workers indicated a requirement for the viral RNA polymerase, a transcription factor, and capping enzyme (Vos et al., 1991a,b). Further studies indicated a requirement for at least two transcription factors in addition to capping enzyme (R.

Rosales, N. Harris, and B. Moss, unpublished). The genes encoding the transcription factors have not yet been identified. The capping enzyme seems to have different roles in early and intermediate transcription since it is required for termination in the former case and initiation in the latter.

TRANSCRIPTION OF LATE GENES

Late Promoters

Late promoters also may be considered in terms of 3 regions: an upstream element of about 20 bp with some consecutive T or A residues, separated by a spacer of about 6 bp from a highly conserved TAAAT element within which transcription intitiates (Davison and Moss, 1989b). Mutations of the A residues in the initiator element (Figure 2) drastically decreased transcription (Bertholet et al., 1986; Davison and Moss, 1989b; Hanggi et al., 1986) and substitution of the T residues also had a negative effect (Davison and Moss, 1989b). The late initiator element TAAAT closely resembles the intermediate element TAAA. A very powerful synthetic late promoter was constructed with runs of 18 to 20 T residues starting 7 bp upstream of the TAAAT sequence (Davison and Moss, 1989b).

Late mRNA Structure

Late mRNAs have two curious features. The first is their extremely heterogeneous length (Cooper et al., 1981; Mahr and Roberts, 1984) which results in large amounts of complementary RNAs formed by divergent transcription of opposite DNA strands (Boone et al., 1979; Colby et al., 1971). The RNA length heterogeneity appears to be related to the inability of the late transcription apparatus to recognize and use the early termination signal. The discrete 3' end of a major cowpox virus transcript (Patel and Pickup, 1987) and its vaccinia homolog (Amegadzie et al., 1992b) provide an exception to the rule of length heterogeneity of late RNA. The mechanism of formation of this 3' end, however, has not yet been reported.

The second unusual feature of late mRNAs is the presence of a capped 5' poly(A) leader of about 30 nucleotides (Ahn and Moss, 1989; Bertholet et al., 1987; Patel and Pickup, 1987; Schwer et al., 1987). Only three A residues are encoded within the DNA and therefore the poly(A) leader is probably produced by backward slippage of the RNA polymerase during initiation. The length of the leader was diminished when the T residues of TAAAT were altered and the leader was absent or limited to a few nucleotides when any of the three A residues themselves were changed (Davison and Moss, 1989b). A decreased length poly(A) leader also was noted when mutations downstream of TAAAT were made (de Magistris and Stunnenberg, 1988).

Late Transcription Factors

Biochemical experiments indicated the presence of viral late transcription factors in extracts of infected cells (Schwer and Stunnenberg, 1988; Wright and Moss, 1987, 1989).

Progress in identifying late transcription factors was dramatically accelerated, however, by the use of a novel transfection approach (Keck et al., 1990). Initial experiments demonstrated that transfected vaccinia virus DNA was required for the expression of a late promoter regulated reporter gene from a plasmid co-transfected into cells infected with vaccinia virus in the presence of an inhibitor of DNA replication. By using cloned vaccinia virus DNA fragments, the transactivation activity was shown to be dependent on three genes belonging to the intermediate class and encoding polypeptides of 17-, 26-, and 30-kDa. Analyses of conditional lethal mutant viruses have confirmed roles for the 17- and 30-kDa proteins in late gene expression (Carpenter and DeLange, 1992; Y. Zhang, J. G. Keck and B. Moss, submitted for publication). Further studies revealed that the 30-kDa protein corresponds to VLTF-1, a factor required for in vitro transcription of late genes (Wright et al., 1991). Experiments to determine whether the 17- and 26-kDa polypeptides are transcription factors are in progress.

SUMMARY

Vaccinia virus provides a facile system for combined biochemical and genetic analyses. The three regulatory classes of vaccinia virus genes - early, intermediate and late - have distinctive promoters and stage specific virus-encoded transcription factors. The expression of vaccinia virus early genes occurs soon after infection of the cell and entry of the virus particle into the cytoplasm. The core associated RNA polymerase, VETF, capping and methylating enzymes, and poly(A) polymerase produce viral transcripts that resemble eukaryotic mRNAs in structure and are translated efficiently. The products of the early mRNAs include RNA polymerase, capping enzyme, and poly(A) polymerase subunits, DNA polymerase and factors for specific transcription of intermediate genes. The virion DNA cannot serve as a template for intermediate gene expression, probably because core proteins restrict access by the newly synthesized transcription complex. Following DNA replication, however, intermediate genes are transcribed. The products of the intermediate genes include three transactivators of late gene expression. In the third phase of the cycle, the late genes are expressed and virion enzymes, early transcription factor and structural proteins are assembled into infectious progeny viral particles to begin the cycle again.

ACKNOWLEDGEMENTS

I thank past and present members of my laboratory for carrying out many of the experiments that were discussed in this review.

REFERENCES

Ahn, B.-Y., Gershon, P. D., Jones, E. V. and Moss, B., 1990a, Identification of *rpo30*, a vaccinia virus RNA polymerase gene with structural similarity to a eukaryotic transcription factor. Mol. Cell. Biol. 10: 5433.

Ahn, B.-Y., Jones, E. V. and Moss, B., 1990b, Identification of the vaccinia virus gene encoding an 18-kilodalton subunit of RNA polymerase and demonstration of a 5' poly(A) leader on its early transcript. J. Virol. 64: 3019.

Ahn, B.-Y. and Moss, B., 1989, Capped poly(A) leader of variable lengths at the 5' ends of vaccinia virus late mRNAs. J. Virol. 63: 226.

Ahn, B.-Y. and Moss, B., 1992, RNA polymerase-associated transcription specificity factor encoded by vaccinia virus. Proc. Natl. Acad. Sci. USA. 89: 3536.

Ahn, B.-Y., Rosel, J., Cole, N. B. and Moss, B., 1992, Identification and expression of *rpo19*, a vaccinia virus gene encoding a 19-kiloDalton DNA-dependent RNA polymerase subunit. J. Virol. 66: 971.

Amegadzie, B. Y., Cole, N., Ahn, B. Y. and Moss, B., 1991a, Identification, sequence and expression of the gene encoding a Mr 35,000 subunit of the vaccinia virus DNA-dependent RNA polymerase. J. Biol. Chem. 266: 13712.

Amegadzie, B. Y., Holmes, M., Cole, N. B., Jones, E. V., Earl, P. L. and Moss, B., 1991b, Identification, sequence, and expression of the gene encoding the second-largest subunit of the vaccinia virus DNA polymerase. Virology. 180: 88.

Amegadzie, B. Y., Ahn, B. Y. and Moss, B., 1992a, Characterization of a 7-kilodalton subunit of vaccinia virus DNA-dependent RNA polymerase with structural similarities to the smallest subunit of eukaryotic RNA polymerase II. J. Virol. 66: 3003.

Amegadzie, B. Y., Sisler, J. R. and Moss, B., 1992b, Frame-shift mutations within the vaccinia virus A-type inclusion protein gene. Virology. 186: 777.

Baldick, C. J., Keck, J. G. and Moss, B., 1992, Mutational analysis of the core, spacer and initiator regions of vaccinia virus intermediate class promoters. J. Virol. 66: in press.

Barbosa, E. and Moss, B., 1978, mRNA(nucleoside-2'-)-methyltransferase from vaccinia virus. Purification and physical properties. J. Biol. Chem. 253: 7692.

Baroudy, B. M. and Moss, B., 1980, Purification and characterization of a DNA-dependent RNA polymerase from vaccinia virions. J. Biol. Chem. 255: 4372.

Bertholet, C., Stocco, P., Van Meir, E. and Wittek, R., 1986, Functional analysis of the 5' flanking sequence of a vaccinia virus late gene. EMBO J. 5: 1951.

Bertholet, C., Van Meir, E., ten Heggeler-Bordier, B. and Wittek, R., 1987, Vaccinia virus produces late mRNAs by discontinuous synthesis. Cell. 50: 153.

Boone, R. F., Parr, R. P. and Moss, B., 1979, Intermolecular duplexes formed from polyadenylated vaccinia virus RNA. J. Virol. 30: 365.

Broyles, S. S., 1991, A role for ATP hydrolysis in vaccinia virus early gene transcription. J. Biol. Chem. 266: 15545.

Broyles, S. S. and Fesler, B. S., 1990, Vaccinia virus gene encoding a component of the viral early transcription factor. J. Virol. 64: 1523.

Broyles, S. S. and Moss, B., 1986, Homology between RNA polymerases of poxviruses, prokaryotes, and eukaryotes: Nucleotide sequence and transcriptional analysis of vaccinia virus genes encoding 147-kDa and 22-kDa subunits. Proc. Natl. Acad. Sci. USA. 83: 3141.

Broyles, S. S. and Moss, B., 1987, Sedimentation of an RNA polymerase complex from vaccinia virus that specifically initiates and terminates transcription. Mol. Cell. Biol. 7: 7.

Broyles, S. S. and Moss, B., 1988, DNA-dependent ATPase activity associated with vaccinia virus early transcription factor. J. Biol. Chem. 263: 10761.

Broyles, S. S. and Pennington, M. J., 1990, Vaccinia virus gene encoding a 30-kilodalton subunit of the viral DNA-dependent RNA polymerase. J. Virol. 64: 5376.

Broyles, S. S., Li, J., and Moss, B. 1991, Promoter DNA contacts made by the vaccinia virus early transcription factor. J. Biol. Chem. 266: 15539.

Broyles, S. S., Yuen, L., Shuman, S. and Moss, B., 1988, Purification of a factor required for transcription of vaccinia virus early genes. J. Biol. Chem. 263: 10754.

Carpenter, M. S. and DeLange, A. M., 1992, Identification of a temperature-sensitive mutant of vaccinia virus defective in late but not intermediate gene expression. Virology. 188: 233.

Cochran, M. A., Puckett, C. and Moss, B., 1985, In vitro mutagenesis of the promoter region for a vaccinia virus gene: evidence for tandem early and late regulatory signals. J. Virol. 54: 30.

Colby, C., Jurale, C. and Kates, J. R., 1971, Mechanism of synthesis of vaccinia virus double-stranded ribonucleic acid in vivo and in vitro. J. Virol. 7: 71.

Cooper, J. A., Wittek, R. and Moss, B., 1981, Extension of the transcriptional and translational map of the left end of the vaccinia virus genome to 21 kilobase pairs. J. Virol. 39: 733.

Davison, A. J. and Moss, B., 1989a, The structure of vaccinia virus early promoters. J. Mol. Biol. 210: 749.

Davison, A. J. and Moss, B., 1989b, The structure of vaccinia virus late promoters. J. Mol. Biol. 210: 771.

de Magistris, L. and Stunnenberg, H. G., 1988, Cis-acting sequences affecting the length of the poly(A) head of vaccinia virus late transcripts. Nucleic Acids Res. 16: 3141.

Earl, P. L., Hugen, A. W. and Moss, B., 1990, Removal of cryptic poxvirus transcription termination signals from the human immunodeficiency virus type 1 envelope gene enhances expression and immunogenicity of a recombinant vaccinia virus. J. Virol. 64: 2448.

Gershon, P. D., Ahn, B. Y., Garfield, M. and Moss, B., 1991, Poly(A) polymerase and a dissociable polyadenylation stimulatory factor encoded by vaccinia virus. Cell. 66: 1269.

Gershon, P. D. and Moss, B., 1990, Early transcription factor subunits are encoded by vaccinia virus late genes. Proc. Natl. Acad. Sci USA. 87: 4401.

Gershon, P. D. and Moss, B., 1992, Transition from rapid processive to slow non-processive polyadenylation by vaccinia virus poly(A) polymerase catalytic subunit is regulated by the net length of the poly(A) tail. Genes and Develop. in press.

Gershowitz, A., Boone, R. F. and Moss, B., 1978, Multiple roles for ATP in the synthesis and processing of mRNA by vaccinia virus: specific inhibitory effects of adenosine (β,γ-imido)triphosphate. J. Virol. 27: 399.

Guo, P. and Moss, B., 1990, Interaction and mutual stabilization of the two subunits of vaccinia virus mRNA capping enzyme coexpressed in Escherichia coli. Proc. Natl. Acad. Sci. USA. 87: 4023.

Hagler, J. and Shuman, S., 1991, A freeze-frame view of eukaryotic transcription during elongation and capping of nascent mRNA. Science. 255: 9833

Hanggi, M., Bannwarth, W. and Stunnenberg, H. G., 1986, Conserved TAAAT motif in vaccinia virus late promoters: overlapping TATA box and site of transcription initiation. EMBO J. 5: 1071.

Hirschmann, P., Vos, J. C. and Stunnenberg, H. G., 1990, Mutational analysis of a vaccinia virus intermediate promoter in vivo and in vitro. J. Virol. 64: 6063.

Jones, E. V., Puckett, C. and Moss, B., 1987, DNA-dependent RNA polymerase subunits encoded within the vaccinia virus genome. J. Virol. 61: 1765.

Kates, J. R. and McAuslan, B. R., 1967, Poxvirus DNA-dependent RNA polymerase. Proc. Natl. Acad. Sci. USA. 58: 134.

Keck, J. G., Baldick, C. J. and Moss, B., 1990, Role of DNA replication in vaccinia virus gene expression: A naked template is required for transcription of three late transactivator genes. Cell. 61: 801.

Luo, Y. and Shuman, S., 1991, Antitermination of vaccinia virus early transcription: possible role of RNA secondary structure. Virology. 185: 432.

Mahr, A. and Roberts, B. E., 1984, Arrangement of late RNAs transcribed from a 7.1 kilobase *Eco R1* vaccinia virus DNA fragment. J. Virol. 49: 510.

Martin, S. A. and Moss, B., 1976, mRNA guanylyltransferase and mRNA (guanine-7)methyltransferase from vaccinia virions. Donor and acceptor substrate activities. J. Biol. Chem. 251: 7313.

Morgan, J. R., Cohen, L. K. and Roberts, B. E., 1984, Identification of the DNA sequences encoding the large subunit of the mRNA capping enzyme of vaccinia virus. J. Virol. 52: 206.

Moss, B., 1990, Regulation of vaccinia virus transcription. Ann. Rev. Biochem. 59: 661.

Moss, B., 1991, Vaccinia virus: a tool for research and vaccine development. Science 252: 1662.

Moss, B., Ahn, B.-Y., Amegadzie, B., Gershon, P. D. and Keck, J. G., 1991, Cytoplasmic transcription system encoded by vaccinia virus. J. Biol. Chem. 266: 1355.

Moss, B., Rosenblum, E. N. and Gershowitz, A., 1975, Characterization of a polyriboadenylate polymerase from vaccinia virions. J. Biol. Chem. 250: 4722.

Munyon, W. E., Paoletti, E. and Grace, J. T., Jr., 1967, RNA polymerase activity in purified infectious vaccinia virus. Proc. Natl. Acad. Sci. USA. 58: 2280.

Muthukrishnan, S., Moss, B., Cooper, J. A. and Maxwell, E. S., 1978, Influence of 5' terminal cap structure on the initiation of translation of vaccinia virus mRNA. J. Biol. Chem. 253: 1710.

Nevins, J. R. and Joklik, W. K., 1977, Isolation and partial charactrerization of the poly(A) polymerases from HeLa cells infected with vaccinia virus. J. Biol. Chem. 252: 6939.

Niles, E. G., Lee-Chen, G.-J., Shuman, S., Moss, B. and Broyles, S. S., 1989, Vaccinia virus gene D12L encodes the small subunit of the viral mRNA capping enzyme. Virology. 172: 513.

Patel, D. D. and Pickup, D. J., 1987, Messenger RNAs of a strongly-expressed late gene of cowpox virus contains a 5'-terminal poly(A) leader. EMBO J. 6: 3787.

Patel, D. D. and Pickup, D. J., 1989, The second-largest subunit of the poxvirus RNA polymerase is similar to the corresponding subunits of procaryotic and eucaryotic RNA polymerases. J. Virol. 63: 1076.

Quick, S. D. and Broyles, S. S., 1990, Vaccinia virus gene D7R encodes a 20,000-Dalton subunit of the viral DNA-dependent RNA polymerase. Virology. 178: 603.

Rohrmann, G., Yuen, L. and Moss, B., 1986, Transcription of vaccinia virus early genes by enzymes isolated from vaccinia virions terminates downstream of a regulatory sequence. Cell. 46: 1029.

Roth, M. J. and Hurwitz, J., 1984, RNA capping by the vaccinia virus guanylytransferase. Structure of enzyme-guanylate intermediate. J. Biol. Chem. 259: 13488.

Schnierle, B. S., Gershon, P. D. and Moss, B., 1992, Cap-specific mRNA (nucleoside-$O^{2'}$-)-methyltransferase and poly(A) polymerase stimulatory activities of vaccinia virus are mediated by a single protein. Proc. Natl. Acad. Sci. USA. 89: 2897.

Schwer, B. and Stunnenberg, H. G., 1988, Vaccinia virus late transcripts generated in vitro have a poly(A) head. EMBO J. 7: 1183.

Schwer, B., Visca, P., Vos, J. C. and Stunnenberg, H. G., 1987, Discontinuous transcription or RNA processing of vaccinia virus late messengers results in a 5' poly(A) leader. Cell. 50: 163.

Shuman, S., 1990, Catalytic activity of vaccinia messenger RNA capping enzyme subunits coexpressed in Escherichia Coli. J. Biol. Chem. 265: 11960.

Shuman, S., Broyles, S. S. and Moss, B., 1987, Purification and characterization of a transcription termination factor from vaccinia virions. J. Biol. Chem. 262: 12372.

Shuman, S. and Hurwitz, J., 1981, Mechanism of mRNA capping by vaccinia virus guanylyltransferase: Characterization of an enzyme-guanylate intermediate. Proc. Natl. Acad. Sci. USA. 78: 187.

Shuman, S. and Morham, S. G., 1990, Domain structure of vaccinia virus messenger RNA capping enzyme - activity of the MR 95,000 subunit expressed in Escherichia coli. J. Biol. Chem. 265: 11967.

Shuman, S. and Moss, B., 1989, Bromouridine triphosphate inhibits transcription termination and mRNA release by vaccinia virions. J. Biol. Chem. 264: 21356.

Venkatesan, S., Gershowitz, A. and Moss, B., 1980, Modification of the 5'-end of mRNA: Association of RNA triphosphatase with the RNA guanylyltransferase-RNA (guanine-7)methyltransferase complex from vaccinia virus. J. Biol. Chem. 255: 903.

Vos, J. C., Saskar, M. and Stunnenberg, H. G., 1991a, Promoter melting by a stage-specific vaccinia virus transcription factor is independent of the presence of RNA polymerase. Cell. 65: 105.

Vos, J. C., Sasker, M. and Stunnenberg, H. G., 1991b, Vaccinia virus capping enzyme is a transcription initiation factor. EMBO J. 10: 2553.

Vos, J. C. and Stunnenberg, H. G., 1988, Derepression of a novel class of vaccinia virus genes upon DNA replication. EMBO J. 7: 3487.

Watananbi, Y., Sakuma, S. and Tanaka, S., 1974, A possible biological function of the protein kinase associated with vaccina and vesicular stomatitis viruses. FEBS Lett. 4: 331.

Wei, C. M. and Moss, B., 1975, Methylated nucleotides block 5'-terminus of vaccinia virus mRNA. Proc. Natl. Acad. Sci. USA. 72: 318.

Weir, J. P. and Moss, B., 1987, Determination of the promoter region of an early vaccinia virus gene encoding thymidine kinase. Virology. 158: 206.

Wright, C. F., Keck, J. G. and Moss, B., 1991, A transcription factor for expression of vaccinia virus late genes is encoded by an intermediate gene. J. Virol. 65: 3715.

Wright, C. F. and Moss, B., 1987, In vitro synthesis of vaccinia virus late mRNA containing a 5' poly(A) leader sequence. Proc. Natl. Acad. Sci. USA. 84: 8883.

Wright, C. F. and Moss, B., 1989, Identification of factors specific for transcription of the late class of vaccinia virus genes. J. Virol. 63: 4224.

Yuen, L. and Moss, B., 1986, Multiple 3' ends of mRNA encoding vaccinia virus growth factor occur within a series of repeated sequences downstream of T clusters. J. Virol. 60: 320.

Yuen, L. and Moss, B., 1987, Oligonucleotide sequence signaling transcriptional termination of vaccinia virus early genes. Proc. Natl. Acad. Sci. USA. 84: 6417.

REGULATION OF α AND γ GENE EXPRESSION IN CELLS INFECTED WITH HERPES SIMPLEX VIRUSES

D. Spector, F.C. Purves, R.W. King and B.Roizman

The Marjorie B. Kovler Viral Oncology Laboratories, The University of Chicago, Chicago IL 60637 USA

Herpes simplex virus 1 (HSV-1) genes form three groups designated as α, β and γ, whose expression is coordinately regulated and sequentially ordered in a cascade fashion (Honess and Roizman, 1974, 1975). A unique feature of α gene expression is that these genes are induced by a protein previously described as virion protein No. 16 (VP16). The unique feature of the highly heterogeneous genes comprising the γ group is that the response elements in some of the genes examined to date appears to be in the 5' transcribed non coding domain. In this report we examine the regulation of α and γ genes of HSV-1.

The Transactivation of α Genes by VP16

I. Background

The transactivation of HSV-1 α genes by VP16 is one of the best characterized events in eukaryotic gene expression. A unique feature of this system is that the VP16 protein is located in the tegument, a virion compartment located between the capsid containing the viral DNA and the superficial envelope of the virion. Following the fusion of the envelope with the plasma membrane, the first step in the infection of the cell, VP16 appears to be transported to the nucleus whereas the capsid is transported to the nuclear pore. VP16 specifically transactivates the five α genes (α0, α4, α22, α27, α47) by interacting with specific cis-acting sites located in one to several copies upstream of α gene promoters. To form stable complexes with its cis-acting sites, VP16 requires at least two cellular factors. Of these, the transcription factor Oct-1 is the best characterized to date. The Oct-1 binding site, however, constitutes only a portion of the conserved domain and presumed cis-acting site of VP16. Once bound, the mechanism by which VP16 transactivates gene expression is poorly understood, but the C-terminal acidic activating domain of the protein is the focus of much investigation and evidence for its interaction with several proteins in the basal transcriptional complex located at the TATAA box of eukaryotic promoters is accumulating. Besides providing an ideal system to study protein-protein and protein-DNA interactions during the induction of gene expression, VP16 is also the focus of studies on its role in both lytic and latent infections.

The trans-activation of α genes by a putative structural component of the virus was first reported by Post et al. in experiments designed to map the domains of the α gene promoters (Post *et al.,* 1981). In these experiments, α4 transcribed noncoding and 5' untranscribed

Regulation of Gene Expression in Animal Viruses, Edited by
L. Carrasco *et al.,* Plenum Press, New York, 1993

domains were fused to the transcribed noncoding and coding domains of the thymidine kinase (*tk*) gene. The chimeric *tk* gene was functional in transfected cells and biochemically transformed cell lines were readily obtained. The noteworthy observation was that the *tk* other α gene promoters, suggested the possibility that the inducing factor was brought into the infected cells with virion (Mackem and Roizman, 1982a, 1982b, 1982c,). The compelling evidence that the trans-activator is a component of the virion came from two sources. First Batterson and Roizman reported that UV-inactivated virus was capable of inducing *α-tk* chimeric genes (Batterson and Roizman, 1983). They concluded that the trans-activator was a component of the virion tegument inasmuch as HSV-1(HFEM)*ts*B7, a *ts* mutant, at the nonpermissive temperature induced the *α-tk* chimeric genes even though the capsids accumulated at the nuclear pore and failed to release viral DNA into the nucleus (Batterson *et al.,* 1983). In parallel studies, Mackem and Roizman reported that all DNA fragments derived from *α* promoters that were capable of being induced contained a consensus sequence 5' **GyATGnTAATGArATTCyTTGnGGG** (noncoding) common to the regulatory domains of all *α* genes (Mackem and Roizman, 1982a). They also reported that some promoters contained G+C rich, perfect inverted repeats; some of these stretches were later shown to contain SP1 binding sites. Kristie and Roizman showed that homologs of the consensus sequence imparted on chimeric constructs the capacity to be induced by the virion component, whereas the G+C rich domains affected the basal level of expression of chimeric genes (Kristie and Roizman, 1984).

The second line of evidence concerning the source of the transactivator emerged from co-transfections of fragments of HSV DNA with an *α-tk* chimeric indicator gene. These studies led to the identification of an open reading frame whose product was capable of inducing the indicator gene (Campbell *et al.,* 1984). The product of the gene was identified as the VP16, a relatively abundant tegument protein estimated to be present in 500-1000 molecules per virion (Spear and Roizman, 1972; Heine *et al.,* 1974). The sequence of the HSV-1(F) and HSV(17)syn$^+$ genes specifying VP16 was reported (Dalrymple *et al.,* 1985; Pellet *et al.,* 1985). Its designation is subject to considerable confusion. Initially, this protein was identified in purified virions and designated as VP16. Because HSV-1 proteins present in infected cells could not always be related to their virion counterparts, the protein present in infected cell lysates was designated as infected cell protein No. 25 (ICP25). Dalrymple et al. designated the protein as Vmw65, presumably after its apparent molecular weight in denaturing polyacrylamide gels (Dalrymple *et al.,* 1985). *α*TIF, the designation given it by Pellet et al. reflects its function as the transactivator of *α* genes (Pellet *et al.,* 1985).

Reports from several laboratories soon established that VP16 forms complexes with DNA, but only in the presence of Oct-1, or under conditions in which Oct-1 can participate in the reaction (Gerster and Roeder, 1988; McKnight *et al.,* 1987; O'Hare and Goding, 1988; O'Hare *et al.,* 1988; Preston *et al.,* 1988; Spector *et al.,* 1990; Treizenberg *et al.,* 1988). Specifically, (1) in-vitro synthesized, labeled VP16 was demonstrated to form a complex with unlabeled DNA in the presence of partially purified Oct-1 (Gerster and Roeder, 1988). and (2) the complex was retarded by antibody to VP16, but not by preimmune serum (Gerster and Roeder, 1988; McKnight *et al.,* 1987; Preston *et al.,* 1988;).

II. Structure of the VP16 Gene and the Protein it Encodes

The VP16 protein is encoded by the open reading frame designated as UL48, mapping between 0.688 and 0.679 and is predicted to contain a maximum of 490 amino acids with a translated molecular weight of 54,000 and a minimum of 479 amino acids if the first in-frame methionine codon is not used or is used intermittently (Dalrymple *et al.,* 1985; Pellet *et al.,* 1985). The sequence predicts an extremely acidic carboxyl half of the protein. A strong *α*4 protein binding site is contained in the leader sequence (Michael *et al.,* 1988). The gene is expressed as a γ_1 protein.

Several laboratories have attempted to delineate the functional domains of VP16 by deletion / insertion mutagenesis (Ace *et al.,* 1988; Greaves and O'Hare, 1989, 1990; Stern and Herr, 1991; Treizenberg *et al.,* 1988). The results of these studies may be summarized as follows:

(1). The domains of VP16 responsible for the formation of multiprotein complexes with DNA including Oct-1, map nearer the amino terminus of the protein than the domain responsible for the trans-activation. Specifically, the amino-terminal boundary for complex assembly mapped between residues 49 and 75 of the VP16 protein. Two other domains identified as important for VP16-induced complex formation include the region located between amino acids 173-241 (region I) and the region located between residues 378-389, near the carboxyl-terminal acidic activation domain (region II). Kristie and Sharp demonstrated that under certain conditions involving high VP16 concentrations, VP16-DNA complexes could be observed in mobility shift assays in the absence of additional host factors (Kristie and Sharp, 1990). Studies by Stern and Herr revealed that mutations within region I of VP16 strongly diminished the ability of the protein to interact with DNA in the absence or presence of the Oct-1 POU domain, suggesting that this region is required for DNA binding and could be involved directly in this activity (Stern and Herr, 1991). In fact, region I contains the most basic region of the VP16 polypeptide, and a synthetic peptide encompassing this region interacted avidly but nonspecifically with DNA. A 14-amino acid peptide derived from region II specifically inhibited assembly of the multiprotein complex but had no observed effect on the DNA binding of Oct-1 (Haigh *et al.,* 1990). Specific point mutations within this region also completely destroyed VP16-induced complex assembly (Greaves and O'Hare 1989). Stern and Herr recently identified region II as the domain of VP16 responsible for specific interaction with the Oct-1 homeo domain and postulated that it may contain a 12 amino-acid alpha helix (Stern and Herr, 1991). A 33 amino acid peptide derived from this domain formed specific complexes with the POU domain of either Oct-1 or Oct-2 in mobility-shift experiments. It is noteworthy that VP16 interacts specifically with the Oct-1 POU domain, and is incapable of interacting with a POU domain in which helix 2 of the Oct-2 protein has been substituted for the corresponding helix of Oct-1 (Stern *et al.,* 1989). Mutational analysis also suggests that region II is primarily responsible for VP16 interaction with the additional host factor, although mutations in region I can also affect this interaction (Stern and Herr, 1991). Since deletion of region II also strongly diminishes VP16-DNA interactions, it may be that this deletion generally inactivates VP16 by unfolding its native structure. Interestingly, region II of VP16 is absent from the VZV homologue of the protein encoded by the ORF10 gene, as is the entire carboxyl terminal acidic activating region of the protein. Consequently, the ORF10 protein has been demonstrated incapable of forming multiprotein complexes with Oct-1 (McKee *et al.,* 1990). The correlated presence and absence of the two domains in these proteins suggests that they have been acquired together by VP16 in evolution.

(ii). Proteins which lack the domain responsible for the trans-activation of α genes may retain their capacity to form complexes with Oct-1 and DNA. Conversely, deletion of the domain responsible for complex formation obliterates the capacity of VP16 to induce the reporter genes.

(iii). The domain responsible for the trans-activation is contained within the carboxyl-terminal 80 amino acids. This region is particularly acidic. Although deletion analyses suggested that induction of reporter genes required this domain, definitive evidence emerged when a chimeric protein consisting of the DNA binding domain of the yeast transcriptional activator GAL4 fused to the C-terminal domain of the VP16 gene induced a reporter gene to a very high level (Sadowski *et al.,* 1988). Detailed mutational analysis of this region has revealed that while net negative charge contributed to transcriptional activation, it was not sufficient for it. The observation that a phenylalanine residue at position 442 was markedly sensitive to mutation, and that transcriptional activators of several classes contained hydrophobic amino acids in patterns resembling that of VP16, suggests that the mechanism of transcriptional activation by VP16 could involve both ionic and specific hydrophobic interactions with target molecules (Cress and Treizenberg, 1991). The identity of this target molecule has received much recent attention. In one study, the VP16 activation domain was used as a column ligand for affinity chromatography, and was demonstrated to bind strongly and highly selectively to the human and yeast TATA-box binding factors, TFIID (Stringer *et al.,* 1990). In a more recent study, in which affinity-purified transcriptional complexes at various stages of assembly were analyzed, it was argued that VP16 stimulates transcription in the absence of ATP by increasing the number of functional preinitiation complexes through recruitment of the general transcription factor TFIIB to the promoter (Lin and Green, 1991). Protein affinity chromatography also demonstrated a specific interaction. In both of the above studies, mutants of the VP16 activation domain which demonstrated decreased abilities to transactivate correlated well with decreased affinities for the respective transcription factors (Ingles *et al.*, 1991; Lin and Green, 1991; Stringer *et al.*, 1990). Still other laboratories argue that the direct target of VP16 is not a general transcription factor but rather another factor referred to as an "adapter" or a "coactivator," which forms a bridge to one of the general transcription factors (Kelleher III *et al.*, 1990). Recently the gene for such a putative adapter molecule ADA2, has been isolated from yeast on the basis of its interaction with the acidic activation of VP16, and was shown to respond to some but not all acidic activator domains. (Berger *et al.*, 1992).

(iv). VP16 is a component of the tegument and in this function is an essential structural protein. Ace et al. analyzed a series of mutants derived by insertion of 12 bp into selected domains of the HSV-1(17)syn VP16 gene (Ace *et al.*, 1988). The mutants were tested for their capacity to induce, enter into protein:DNA complexes, and rescue an HSV-2 mutant with a *ts* mutation in the VP16 gene. With few exceptions, mutants which failed to enter DNA complexes and trans-induce also failed to rescue the *ts* mutant. These studies suggest that VP16 has two domains whose structure is critical for entry into protein:DNA complexes, one centering at approximately amino acid 150 and the other at amino acid 350. Definitive evidence emerged for the essential structural role of VP16 when the open reading frame was successfully deleted on a VP16 producing cell line (Weinheimer *et al.*, 1992). The resulting deletion virus induced nearly normal levels of viral DNA synthesis and capsid production, but induced slightly lower levels of viral DNA encapsidation and appeared by transmission electron microscopy to be defective in further steps of viral maturation.

(v). Ace et al. addressed the role of VP16 in the reproductive cycle by showing that a recombinant virus whose VP16 possessed an insertion, which abolished the ability of the protein to induce recorder genes, did not render the virus noninfectious, although its ability to multiply and spread at low multiplicity of infection was impaired (Ace *et al.*, 1989). The virus stocks were characterized by high particle/pfu ratios. In infected cells, the expression of $\alpha 0$ and $\alpha 27$ were reduced four- to five-fold, whereas that of $\alpha 4$ was not. Interpretation of these results presents two problems. Foremost, the promoter domains of α genes are laden with cis-acting sites for a variety of transcriptional factors and each gene is readily expressed in cultured cells, albeit at a reduced basal level, in the absence of VP16. High multiplicity of infection would compensate for weak promoter activity. A second, and perhaps more worrisome problem is the possibility that a very much reduced level of trans-activation, which might not be detectable in transfected cells, would be sufficient to enable the virus to multiply. Significantly, the virus appeared to be avirulent in mice, was able to establish latency in the trigeminal ganglion, and reactivated upon explantation. (Steiner *et al.*, 1990).

(iv). The role of VP16 in latency was also addressed by Sears et al, who constructed a recombinant virus containing a VP16 gene under the control of a metallothionein promoter (Sears *et al.*, 1991). Mice infected with this virus were administered cadmium in order to force expression of the VP16 gene. Cadmium injections did not preclude the establishment of latency nor did they induce reactivation of latent virus in the infected mice. In another series of experiments, transgenic mice expressing the metallothionein-driven VP16 did not differ from nontransgenic siblings with respect to the incidence of latent virus in trigeminal ganglia (Sears *et al.*, 1991). It was thus concluded that the absence of VP16 alone cannot account for the establishment of latency in sensory neurons.

III. Structure of the Oct-1 Gene and the Protein it Encodes

The Oct-1 protein was independently discovered in several laboratories studying a variety of biological systems. Consequently, the protein bears several names in the literature (NF-A1, OBP100, OTF-1, NFIII, Origin Recognition Protein C, αH1, and Oct-1) (Fletcher *et al.*, 1987; Kristie and Roizman, 1987, O'Neill and Kelly, 1988; Staudt *et al.*, 1986;

Sturm *et al.*, 1987, 1988). Its widespread recognition is no surprise; as a transcription factor, it interacts with the octamer motif of the cis-regulatory element in the promoters of ubiquitously expressed sequences such as those encoding the small nuclear RNAs and the histone H2b genes (LaBella *et al.*, 1988; Mattaj *et al.*, 1985). The identical octamer motifs present in cell-specific immunoglobulin promoters are thought to interact with another related transcription factor, Oct-2 (Scheidereit *et al.*, 1987). The octamer motif also exists in a number of DNA virus promoters. In addition to herpesviruses, it is also found in SV40 and adenovirus. Interestingly, Oct-1 has been demonstrated as a stimulatory replication factor in the adenovirus system (O'Neill and Kelly, 1988; Pruijn *et al.*, 1986).

With the isolation of the gene for Oct-1 by Sturm et al. came the recognition that Oct-1 is a homeobox protein (Sturm *et al.*, 1988). Homeobox proteins had long been suspected of functioning as transcription factors regulating the morphology, polarity, and differentiation of the organism. Along with two other mammalian transcription factors, Pit-1 and Oct-2, and the product of the nematode *C. elegans* Unc-86 gene, Oct-1 helped define a new class of homeobox proteins called POU proteins (for Pit, Oct, Unc) (Herr *et al.*, 1988). These four proteins share a highly related 150 to 160 amino acid-long region of sequence designated for the same reasons as the POU domain. PCR amplification using degenerate primers to the POU domain of human and rat brain cDNA and rat testes cDNA has allowed the discovery of four more POU proteins, (i.e., Brn-1, Brn-2, Brn-3, and Tst-1) (He *et al.*, 1989). All of these known mammalian POU proteins have been implicated by in situ hybridization as playing a role in the development of the nervous system. A number of other proteins have recently been discovered with homology to the known POU proteins and many more are certain to follow.

The POU domain can be further subdivided into the homeobox-related POU-homeo and the unique POU-specific domains. These regions are separated by a short 15 to 20 amino acid region of nonconserved linker. It is convenient to deal with both of these regions separately.

The specific DNA binding proteins of the homeodomain of many homeobox proteins, including both Oct-1 and Oct-2, has recently become the focus of much attention. It has been pointed out by several laboratories that the homeodomain is highly homologous to the helix-turn-helix motif present in the DNA-binding domains of many prokaryotic repressors (i.e., lambda, 434, and trp repressors) (Laughton and Scott, 1984; Garcia-Blanco *et al.*, 1989). These domains all include three α-helices, the latter two being the helix-turn-helix structure which makes contact with the DNA (Pabo and Sauer, 1984; Wharton and Ptashne, 1985). The similarities among homeodomains, however, extend beyond this prokaryotic motif and it is unlikely to be sufficient for all specific DNA binding (Garcia-Blanco *et al.*, 1989). In keeping with these predictions, the structure of the *Antennapedia* homeodomain was recently demonstrated by NMR spectroscopy in solution to possess the predicted helix-turn-helix structure of the prokaryotic repressors, but the molecular architecture differed markedly from there (Otting *et al.*, 1988; Quian *et al.*, 1989). In addition, a single point mutation in the proposed recognition helix (helix 3) of another drosophila homeoprotein *Prd*, was sufficient to confer upon it the DNA-binding specificity of a different homeodomain protein, *Ftz* (Treisman *et al.*, 1989).

The POU homeodomain subregion is more highly conserved among the POU proteins than with the homeodomains of other homeobox proteins. The homology between Oct-1 and Oct-2 in this region approaches 88%. Sturm et al. have demonstrated the importance of both the POU-homeo and the POU-specific domains of Oct-1 for DNA binding by creating triple alanine substitutions in each of these subregions (Sturm and Herr, 1988). Interestingly, the additional insertion of six alanines into the nonconserved linker joining these two domains had no effect on DNA binding. They thus suggested that the POU domain is a bipartite DNA-binding structure. Such flexibility in the DNA-binding domain might explain the many apparently dissimilar DNA sequences which OTF-1 appears to bind. Sequences which are identical in only 4 of 14 residues have been shown to bind, yet simple point mutations within some binding sites can abolish binding (Baumruker *et al.*, 1988).

The POU-homeo domain is involved in both DNA binding and in directing the formation of the multiprotein-DNA complex with VP16. Helix swap experiments between

Oct-1 and Oct-2 performed by Stern et al. demonstrate that exchanging helix 2 of Oct-1 for that of Oct-2 imparts upon Oct-2 the ability to form a complex with VP16 (Stern *et al.*, 1989). Conversely, replacing helix 2 of Oct-1 with that of Oct-2 destroys the ability of Oct-1 to form the complex. The ability of either Oct protein to bind DNA is not affected by the switch. It is now known that helix 2 of Oct-1 makes direct contact with the VP16 protein. The role the POU-specific domain plays in the formation of the VP16 complex is less clear as this region is 99% conserved between Oct-1 and Oct-2. If the POU-specific domain does play a role in complex formation, it must be at amino acids conserved between Pit-1 and Oct-1 because replacing the POU-specific domain of Oct-1 with that of Pit-1 reduced, but did not eliminate VP16 complex formation (Stern *et al.*, 1989).

The POU-specific region is 75 to 83 amino acids long and located in the amino terminal half of the overall POU domain. Its function is largely unknown, although as mentioned previously it contributes to DNA binding. At least in Oct-2, it cannot be essential for specific DNA binding, as clones which are missing over half of this region have been isolated on the basis of binding to the octamer site (LeBowitz *et al.*, 1988). Deletions and mutations in this region also lower the affinity of Oct-2 for the octamer site (LeBowitz *et al.*, 1989). Mutational analyses of Pit-1 proteins has revealed that while the POU-homeo domain is required and sufficient for low affinity binding, the POU-specific domain is necessary for high affinity binding and accurate sequence-specific recognition of natural Pit-1 response elements (Ingraham *et al.*, 1990). In addition, they have identified the POU-specific domain as being essential in mediating DNA-dependent protein-protein interactions between Pit-1 molecules. Recent work with both Oct-1 and Oct-2 indicate that these proteins may interact cooperatively with either themselves and/or with other proteins. Oct-1 and Oct-2 have been demonstrated to interact cooperatively at the octamer/heptamer sites present in the immunoglobulin heavy chain promoter (LeBowitz et al.,; Poellinger and Roeder, 1989; Poellinger *et al.*, 1989). The interactions of Oct-1 with VP16 involves similar cooperative interactions (Kristie *et al.*, 1989). Because no other known homeodomain proteins have exhibited such cooperativity, it is tempting to speculate on the role of the POU-specific domain in promoting such protein-protein interactions.

IV. The cis-Acting Site and its Properties

Mackem and Roizman noted that the promoters of all α genes shared sequences with the consensus 5' **GyATGnTAATGArATT** 3' (noncoding) and postulated that this sequence is the cis-site for induction by VP16 (Mackem and Roizman, 1982c). These studies also demonstrated that, at least in the case of some α genes, the promoter functions enabling the expression of α the genes were 3' and separate from the sequences required for the regulation of the genes as α genes (Mackem and Roizman, 1982b). Studies on chimeric proteins consisting of the promoter domain of the α4 gene preceded by small fragments containing the homologs of the consensus sequence present in the α0 (29bp) and α27 (48bp) genes confirmed the hypothesis that the sequence is the cis-acting site for the trans-induction of the α genes by the virion component (Kristie and Roizman, 1984). The cis-site acts in both native and inverted orientation. Gaffney et al. designated the cis-acting site as the sequence **TAATGARAT** (Gaffney *et al.*, 1985). For reasons discussed below, the cis-acting site is more extensive and extends upstream.

Gel retardation studies using the 29-bp fragment from the α0 gene (29Rα0), and the 48-bp fragment from the α27 gene (48Rα27) revealed that the cis-acting site binds host proteins (Kristie and Roizman, 1987). Kristie and Roizman determined by footprinting analysis and methylation interference studies that those purines whose methylation interfered with the binding of the largest of these host proteins, αH1, was part of the sequence **ATGCTAAT**, which resembled the consensus binding site **ATGCAAAT** of several proteins designated variously as nuclear factor III (NF-III), the octamer binding factor, OTF-1, or Oct-1 (Kristie and Roizman, 1988). NF-III and OTF-1 were shown to be identical, and soon after affinity-purified OTF-1 was demonstrated to mediate the DNA-protein complex with VP16, in conjunction with other unidentified cellular proteins present in mock extracts depleted of OTF-1 by specific antibody (Gerster and Roeder,1988, O'Neill *et al.*, 1988).

Extensive mutational analysis of the VP16 consensus site in the context of the viral genome was performed by Spector, et al. by constructing recombinant viruses containing a chimeric α27-tk gene consisting of a 325-bp fragment encoding the promoter-regulatory region of the α27 gene linked to a tk reporter gene (Spector et al., 1990, 1991). The mutated sequences were tested for their ability to form complexes with Oct-1 in nuclear extracts of mock-infected cells and with both Oct1 and VP16 in nuclear extracts of infected cells. The amount of α27-**tk** mRNA expressed by each virus was analyzed by RNAse protection assay under conditions of inhibition of protein synthesis by cycloheximide. A key question addressed in these studies centered on the observation that in the cis-acting site upstream of the α27 gene, the octamer binding site of the host protein, Oct-1, (**ATGCTAAT**) overlapped with the more highly conserved core consensus sequence, **TAATGARAT**; whereas in some cis-acting sites, especially those contained in the promoter of the α4 gene, the octamer sequence had degenerated so significantly that the question arose as to whether Oct-1 was binding to the degenerate sequence or the **TAATGARAT** itself during induction of the α genes. Transversion of the purines in the α27 sequence **ATATGCTAAT** located at the 5' end of the VP16 cis-acting site abolished the capacity of the sequence to form complexes containing Oct-1 or Oct-1 and VP16, and reduced the level of expression of the reporter gene to 6% that of wild-type virus. Transversion of the sequence **GATATG**, which was designed to destroy the consensus octamer binding site but preserve the **TAATGARAT** sequence also abolished high affinity binding of Oct-1 and VP16, but at vastly increased amount of probe, trace amounts of Oct-1 could be detected, presumably reflecting direct binding of the protein to the **TAATGARAT** sequence. Consequently, in this virus, α27-tk expression was reduced to only 20-30% of that of wild-type virus. Further studies in which recombinant viruses were constructed in which various elements of the VP16 cis-acting site of the α4 and α27 genes were employed in specific combinations revealed that expression of the α27-tk mRNA by the mutant promoters correlated well with their respective Oct-1 binding levels. The relative levels of Oct-1 binding and α27-tk gene induction decreased in the order **ATGCTAAT** > **TAATGAGATgc** > **TAATTAAATac** > no Oct-1 binding site, leading to the conclusion that the relative level of binding of Oct-1 is more critical in determining the level of induction than whether the protein is binding to the overlapping octamer element or the **TAATGARAT** sequence itself.

These results are congruent with the findings of other laboratories investigating the DNA binding properties of the Oct-1 protein, which has been footprinted to both the consensus octamer site and the **TAATGARAT** sequence in the context of various α gene promoters (Baumruker *et al.*, 1988; Kristie and Roizman, 1988; Preston *et al.*, 1988; Triezenberg *et al.*, 1988). ApRhys et al. addressed the question of binding, but not trans-activation, by using mutant synthetic oligonucleotides and concluded that the overlapping octamer and **TAATGARAT** sequences of the α0 cis-site represent distinct Oct-1 binding sites, the octamer sequence possessing the much higher affinity (ApRhys *et al.*, 1989). They reported that mutations within the **GARAT** sequences, of either α0 cis-sites with mutations in the octamer sequences or within an α4 cis-site, failed to bind Oct-1. Verrijzer et al. reported that whereas the entire POU domain has a 600-fold-higher affinity than the POU-homeo subdomain for the octamer motif, the two regions have nearly equal affinity for the α4 **TAATGARAT** sequence (Verrijer *et al.*, 1990). More recent studies have suggested that the POU-homeo domain of Oct-1 contacts the **TAAT** sequence of the VP16 cis-acting site, the POU-specific domain contacts the **ATGC** sequence immediately upstream of the **TAAT** sequence, and VP16 may even contact the **GARAT** sequences (Kristie and Sharp, 1990; Stern and Herr, 1991).

The function of the **GARAT** portion of the conserved VP16 cis-acting site was studied by site-specific mutagenesis in several laboratories, and it was observed that while mutations contained within this region had no effect on the ability of the sequence to bind Oct-1, they abolished the formation of VP16-containing complexes. Surprisingly, in the studies of Spector et al., the α27-tk mRNA levels induced in recombinant viruses containing such mutations were only reduced to 20-30% that of wild-type, significantly higher than the levels of α27-tk mRNA expressed when the Oct-1 consensus sequence was destroyed (6%) (Spector *et al.*, 1990). Importantly, the stringency of the requirement for a **GARAT** sequence varied significantly with cell type, with α gene expression from such mutants often as high as 80% that of wild-type in certain cell lines (Spector *et al.*, 1991). On the basis of these findings, it was postulated that an unknown cellular factor is capable of interacting with Oct-1 to induce gene expression, albeit to a lesser degree than when Oct-1 is complexed with VP16, that this factor has no dependence on the **GARAT** sequence as does VP16, and that certain cell lines possess more of this factor than others. Indeed, such cellular factors have recently been discovered (Murphy *et al.*, 1992).

In addition to the VP16-responsive cis-acting site, α gene promoters contain numerous other cis-acting sites, including well known binding sites for such factors as SP1 and CAAT-box binding proteins, some of which have been demonstrated to markedly affect gene expression (Spector *et al.*, 1990). The purine-rich element **GCGGAA** often present a variable number of times upstream of the transcribed domains of α genes has been reported to bind proteins by footprint analysis (Triezenberg *et al.*, 1988). The role of the repeats and their environment may be deduced from the observation that the proteins purified by affinity to this sequence do not bind to a single isolated hexanucleotide (LaMarco and McKnight, 1989). More recently, cDNA clones that encode two distinct subunits of this transcription factor have been isolated; the amino acid sequence of one protein exhibits similarity to the Ets DNA binding domain (LaMarco *et al.*, 1991). The function of the purine-rich GA region in the context of α gene promoters remains a mystery, however, inasmuch as mutations within this region of the α27 promoter, in the context of the viral genome, reduced the accumulation of α27-tk mRNA only slightly (Spector *et al.*, 1990).

V. Additional Components of the Multiprotein Oct-1/VP16 Complex

Gerster and Roeder were the first to note that the stable formation of a complex between purified Oct-1 and VP16 required the addition of mock extract depleted of Oct-1 by specific antibody, inferring a requirement for additional cellular factors within the complex (Gerster and Roeder, 1988). Kristie and Sharp demonstrated that a succession of protein-DNA complexes could be assembled in a stepwise manner on an α gene promoter (Kristie *et al.*, 1989; Kristie and Sharp, 1990). Specifically, Oct-1 binding to the octamer homolog in the 5' domain of the VP16 cis-acting site, was assembled into a slower migrating C1 complex by the addition of a crude S100 extract from insect cells infected with a recombinant baculovirus that overexpressed the VP16 protein. Partial purification of VP16 from these extracts suggested that the S100 extract supplied an additional factor, the C1 factor, which was required to assimilate VP16 into the C1 complex. Comparison of the metal ion cleavage patterns of the Oct-1 and C1 complexes showed that the former protected the octamer homolog while the latter protected the entire cis-acting site. An additional complex, the C2 complex, was formed when a nuclear extract of HeLa cells was added to the reaction containing Oct-1 and VP16. Therefore, a fourth component was hypothesized to bind to the C1 complex, resulting in the C2 complex. This complex did not alter the footprint of the C1 complex, suggesting that the C2 factor associated with the C1 complex primarily via protein-protein interactions. The C1 factor was present in both SF9 insect cell and HeLa cell extracts, whereas the C2 factor was only present in the HeLa cell nuclear extracts.

The C1 factor was partially purified from S100 extracts of SF9 cells and was shown to lack sequence-specific DNA binding ability. Using affinity chromatography, they determined that the C1 factor binds VP16 with high affinity in the absence of DNA and the Oct-1 protein. Conversely, this component had limited or no apparent affinity for the Oct-1 POU domain. Similarly, the C1 factor but not the C2 factor, present in HeLa cell nuclear extracts, also bound the VP16 matrix with high affinity and specificity. Other laboratories have also identified cellular factors with properties similar to those of C1 (Katan *et al.*, 1990; Xiao and Capone, 1990). Katan et al. have achieved 100-fold purification of this factor, which they refer to as complex forming factor (CFF), and have shown that the highly purified polypeptide profile consists of two major species with molecular masses of 80K and 70K (Katan *et al.*, 1990).

In summary, the multiprotein complex responsible for induction of the α genes is initiated through Oct-1 binding to the octamer element within the VP16 cis-acting site. The VP16/C1 factor complex binds to the Oct-1/DNA complex and is stabilized by specific interactions between the VP16 protein and the Oct-1 POU-homeo domain and between VP16 and the GARAT sequences of the cis-acting site. Although VP16 interacts directly with the POU-homeo domain, no interaction is detected in the absence of DNA, suggesting that this interaction is of low affinity. Likewise, the DNA binding affinity of VP16 is extremely low and may require an interaction with Oct-1 in vivo. The C1 factor may act to stabilize VP16 in a conformation with the highest affinity for either DNA binding or interaction with Oct-1. The interactions of the C2 factor remain unknown, but presumably involve protein-protein interactions as the binding of this component does not affect the DNA footprint of the complex.

HSV-1 Late Gene Regulation

I. Background

The γ or late genes encode proteins important for virion assembly, DNA packaging, virus egress, attachment, and trans-activation of α genes. The γ genes are defined by their dependence on the replication of the virus genome for their expression (Holland et al., 1980; Jones and Roizman, 1979). The genes are further divided into two subclasses, γ_1 and γ_2, The γ1 genes, also known as ßγ genes are expressed at low levels in the presence of DNA synthesis inhibitors, whereas the γ_2 genes also known as true late genes are expressed only in the absence of the inhibitor.

The exact mechanism for γ gene regulation has yet to be elucidated but probably involves expression of α genes, inhibition of expression early in infection. response elements in both 5' transcribed non coding and coding domains of γ genes, and activation of gene expression after the onset of viral DNA synthesis.

II. Environmental Constraints on γ Genes for Proper Expression

The initial studies on HSV-1 γ genes were done in transient expression systems transfected with chimeric genes consisting of a reporter gene under the control of a γ gene promoter. In two early studies the reporter gene was the HSV-1 tk fused to the γ_1VP5 (Dennis and Smiley, 1984), or the γ_2UL49.5 promoters (Silver and Roizman, 1985). In both cases the expression of the γ-tk gene transfected into thymidine kinase minus (TK$^-$) cells was low compared to that of the tk under its natural promoter. The expression was increased by infection of the cells with a tk$^-$ mutant. Silver and Roizman found that when cells were treated with the inhibitor of viral DNA synthesis, phosphonoacetate (PAA), and infected with a tk$^-$ mutant, the cells still produced high levels of TK activity. Thus the tk gene was regulated as a ß gene rather than as a γ gene. Similar results were obtained with cell lines that contain a stably integrated gC (a γ_2) gene (Arsenakis et al., 1986). In the environment of the cellular genome, the gene was not expressed unless the cells were infected with HSV. gC was produced in cells infected with a ts DNA$^-$ virus at the nonpermissive temperature or treated with PAA indicating that it was regulated as a β gene. These observations led to the conclusion that γ genes must be studied within the context of a viral genome in productive infection.

III. The Dependence of γ Gene Expression on Viral DNA synthesis

Among the hypotheses that could explain the dependence of γ gene expression on DNA synthesis, two were especially attractive. The first hypothesis is that γ genes are self-regulated during the early phase of viral infection by either the binding of trans-acting negative factors, or by constraints placed on the late genes by a particular DNA secondary structure formed in the vicinity of the γ genes. In either case, the process of viral DNA replication could relieve the block and allow normal expression of the γ genes. The second hypothesis involves the production or modification of a trans-acting factor during viral DNA replication that allows the activation of γ gene expression.

To differentiate between these two hypotheses, Mavromara-Nazos and Roizman (Mavromara-Nazos and Roizman, 1987) infected cells with a tk$^-$ mutant. Six hours later, after the onset of DNA synthesis, the cells were incubated in medium containing an inhibitor of viral DNA synthesis and superinfected with a virus containing the HSV tk gene under the control of the $\gamma_2$49.5 gene. Assays done on the doubly infected cells failed to detect the expression of the tk gene of the superinfecting virus and indicated that (i) the expression of γ genes is tightly linked to viral DNA synthesis and (ii) the effect of DNA synthesis is mediated by a cis-acting function.

It has been reported that the infected cell protein No. 8 (ICP8), the HSV-1 single stranded DNA binding protein essential for viral DNA synthesis, may be a negative regulator of γ gene expression. In addition, ICP8 is an essential protein for viral DNA synthesis inasmuch as the mutant HSV-1 KOS1.1 ts18, at the non permissive temperature, blocked viral DNA synthesis but allowed low level expression of the γ_2 gC gene even in the presence of inhibitors of DNA synthesis (Godowskie and Knipe, 1985).

IV. The Role of α Gene Products in γ Gene Regulation

Mutational analysis of the five α genes of HSV-1 have shown that α4 and α27 are essential for virus replication in cell culture (DeLuca *et al.*, 1985; Preston, 1979; Sacks and Schaffer, 1987), whereas α0, α22, and α47 are dispensable (Longnecker and Roizman, 1986; Sears et al., 1985; Stow and Stow, 1986). α0, α4, and α27 transactivate HSV-1 promoters in transient expression assays (Blair and Wagner, 1986; Costas *et al.*, 1985; Everett, 1986). In these experiments cells were transfected with plasmids containing a reporter gene under the control of an HSV-1 promoter along with plasmids containing one or more α genes. The significant observations were that whereas
α0 and α4 transactivated both γ_1 and γ_2 promoters, α27 transactivated γ_1 and repressed γ_2 promoters (Sekulovich *et al.*, 1988). Thus the chimeric γ_1VP5-chloramphenicol acetyl transferase (CAT) gene was expressed at low levels in the presence of α0 or α4 gene, slightly higher if both α27 and either α0 or α4 were present, but not in the presence of the α27 gene alone. In the case of the γ_2gC-CAT reporter gene, the α27 downregulated the enhancement seen with α0 and α4 genes. Deletion mutants mapped the repressor and activator domains to the carboxyl terminal 78 and 63 amino acids of the α27 protein, respectively (Hardwick *et al.*, 1989). The 63 amino acid sequence shared by the two domains contains a sequence compatible with a zinc finger. Although the results of the transient transfection studies are full of interesting revelations, they must be confirmed in studies with genetically altered viruses.

In order to determine the true effects of the α gene products on γ gene regulation, studies were done on viruses with mutations in the α genes. ts α4 mutants grown at the nonpermissive temperature, overexpressed α genes, and failed to express ß and γ genes (Dixon and Schaffer, 1980; Preston, 1979). Mutants lacking both copies of the α4 gene yielded similar results (DeLuca *et al.*, 1985). Viruses with mutations in the activation domain of the α4 protein were able to bind DNA and down regulate α gene expression but failed to express ß and γ genes. Heterodimers of α4 protein, where one subunit lacks the activation domain and the other the DNA binding domain, induced the expression of ß and γ_1 but not γ_2 genes and yielded reduced levels of progeny DNA (Shepard and DeLuca, 1989) These results were the basis of the conclusion that (i) functional α4 is essential for the expression of γ genes and (ii) viral DNA synthesis is necessary but not sufficient for the proper expression of γ genes.

Studies with ts and deletion mutants of α27 have shown that α27 is involved with the down-regulation of α and some ß genes, the activation of γ genes and is not essential for the production of new viral DNA (McCarthy *et al.*, 1989; Rice and Knipe, 1990; Sacks *et al.*, 1985). It has also been reported that α27 protein acts post transcriptionally to switch from β to γ gene expression (Smith *et al.*, 1992).

α0 has been postulated to play a role in γ gene regulation based mainly on its ability to transactivate γ promoters in transient transfection assays (Cai and Schaffer, 1989; Everett, 1988; O'Hare and Hayward, 1985a,b). $\alpha 0^-$ mutants of HSV-1 do replicate, but not as efficiently as wild type (Sacks and Schaffer, 1987; Stow and Stow, 1986). Recent studies with α0 mutants have shown that ICP0 up-regulated ß and γ gene expression but has a differential effect on the α genes (Cai and Schaffer, 1992; Chen and Silverstein, 1992). Cai and Schaffer measured the expression of a select group of α, ß and γ genes for several null and nonsense $\alpha0$ mutants. They also reported that mutants expressed normal levels of ICP4 but reduced levels of ß and γ gene products. In cells expressing the $\alpha0$ gene, the mutations were complemented but not to wild type levels. An $\alpha0^{+}$Lat^{-} repair mutant expressed the early and late proteins normally. Chen and Silverstein performed similar experiments with a different set of $\alpha0$ mutants and obtained similar results with a few exceptions. They too found that the $\alpha0$ mutants did not express ß and γ gene products as efficiently as wild type HSV-1, while the $\alpha4$ gene was overexpressed and the expression of the $\alpha27$ gene was delayed by two hours. Moreover, the expression of a selected set of α, ß and γ genes was shown to be both reduced and delayed. The hypothesis that $\alpha0$ regulates γ gene expression by regulating $\alpha27$ which in turn affects the switch from ß to γ gene expression is interesting but much more data will be needed to support it.

V. The Involvement of Cis-Acting Sites in γ Gene Regulation

There is a general agreement that the response elements required for the expression of β genes consists of binding sites for cellular transcriptional factors, a TATA box and cap site (McKnight and Kingsbury, 1982; McKnight and Tijan, 1986), although removal of all known response elements does not appear to negate β gene regulation (Imbalazo *et al.,* 1991). There is less agreement on the structure of γ gene promoters and particularly of γ_2 promoters. One series of studies suggested that they consist simply of a specific TATAA box (Flanagan *et al.,* 1991; Homa *et al.,* 1988; Johnson and Everett, 1986). Others, notably Mavromara-Nazos and Roizman (Mavromara-Nazos and Roizman, 1989) demonstrated that a reporter gene driven by the 5' nontranscribed sequences of the tk (a β) gene fused to the 5' transcribed non coding domain of the $U_L49.5$ (a γ_2) gene was regulated as a γ_1 gene, leading to the hypothesis that response elements of γ_2 genes are located in the 5' transcribed non coding domains and that fusion of the β 5' untranscribed domains to the γ_2 response elements yields promoters with characteristics of a γ_1 gene. Homa, et al. obtained similar results with a chimeric promoter consisting of the tk 5' untranscribed region fused to the gC (a γ_2) 5' transcribed non coding domain (Homa *et al.,* 1986; Weir and Narayanan, 1988). The situation may be far more complex. Thus the U_L24 gene was expressed when its TATAA box was replaced by that of the gC (a γ_2) gene but not when it was replaced by the TATAA box of the tk (a β) gene or that of U_S11 (a γ_2) gene (Kibler et al., 1991). These studies, (Mavromara-Nazos and Roizman, 1989) and those of Homa, et al. (Homa *et al.,* 1986) had concluded that the γ_2 response elements resided in the 5' transcribed non coding domains. Linker-scanning mutations in the gC and gH promoters led to the conclusion that three sites, i.e. the TATAA box, the cap site, and nucleotides +30 to +40 relative to the cap site at +1 played an important function for the expression of these genes (Steffy and Weir, 1991). The mutations in the TATAA box and cap site affected the transcription of these genes whereas the mutation in the leader sequence had little affect on transcription but had a dramatic effect on the amount of reporter protein produced. Although all three mutations reduced the expression of the reporter gene, none altered its temporal regulation.

It has been suggested that the ICP4 binding sites play an important role in the expression of HSV-1 genes (Michael *et al.,* 1988; Muller, 1987). The gD promoter has been reported to contain three ICP4 binding sites, two in the 5' untranscribed and one in the 5' transcribed noncoding domains. Smiley, et al. constructed a mutant virus lacking these three sites (Smiley *et al.,* 1992). Comparisons of the amount of gD RNA produced by the mutant and wild type viruses demonstrated that the lack of the three ICP4 binding sites did not affect the expression of gD RNA. Slightly different results were obtained with the $U_L49.5$ gene (Romanelli *et al.,* 1992). The $U_L49.5$ (a γ_2) promoter contains four know ICP4 binding sites, two in the 5' untranscribed and two in the 5' transcribed non coding domains. Destruction of one of the 5' transcribed non coding domain binding sites within the context of a chimeric promoter regulated as a γ_1 gene resulted in a significant decrease in the amount of accumulated mRNA (Romanelli *et al.,* 1992).

A more striking role of the γ_2 5' transcribed non coding domains emerged from studies of a series of chimeric genes consisting of the gC (γ_2) 5' untranscribed and transcribed non coding domains fused to the chick ovalbumin coding sequences and inserted into the viral genome. The key observations is that some genes with deletions within the 5' transcribed non coding domain were expressed in the presence of cycloheximide although the indicator gene was expressed to a higher level late in infection (R. King, M. Arsenakis, A. Poon and B. Roizman, manuscript in preparation).

At least two hypotheses could explain the observation that deletion mutants in the 5' transcribed non coding sequences of a γ_2 gene could be expressed as an α gene. First, the 5' non coding domain may contain a site that allows the binding of specific trans-acting proteins necessary for the regulation of γ_2 gene expression. A survey of several γ_2 genes (McGeoch *et al.*, 1988) failed to reveal a sequence conserved in all or most γ_2 genes. Second, the 5' transcribed non coding domain may form a secondary structure which affects expression of the γ genes. We have used the RNAFOLD program (Genetics Computer Group, Madison, WI) to analyze the ability of the γ_2 5' transcribed non coding domains to form stable secondary structures. The γ_2 genes studied in detail could form long and apparently stable stem-loop structures although evidence that they actually exist is lacking. Whereas the wild type sequence is potentially capable of forming a stem 25 bp long, the mutation which confers upon the chimeric gene the capacity to be expressed as an α gene lost the ability to form a stable stem-loop structure.

Conclusions

Although much remains to be done, studies on the regulation of gene expression in cells infected with HSV has led to numerous significant observations of relevance to studies on gene regulation in eukaryotic cells. There is still incomplete knowledge on the function of numerous genes whose products have a regulatory function and on the role of the conserved and non conserved sequences present both 5' and 3' of coding domains of viral genes.

Acknowledgements

These studies were aided by grants from the National Cancer Institute (CA47451), the United States Public Health Service. D.S. is a predoctoral trainee of the United States Public Health grant GM7281.

References

Ace, C.I., Dalrymple, M.A., Ramsay, F.H., Preston, V.G., and Preston, C.M. (1988). Mutational analysis of the herpes simplex virus 1 *trans*-inducing factor Vmw65. *J. Gen. Virol.* **69**, 2595-2605.

Ace, C.I., McKee, T.A., Ryan, M., Cameron, J.M., and Preston, C.M. (1989). Construction and characterization of a herpes simplex virus type 1 mutant unable to transinduce immediate-early gene expression. J. Virol. 63, 2260-2269.

ApRhys, C.M., Ciufo, D.M., O'Neill, E.A., Kelly, T.J., and Hayward, G.S. (1989). Overlapping octamer and TAATGARAT motifs in the vmw65-responsive elements in herpes simplex virus immediate-early promoters represent independent binding sites for cellular nuclear factor III. *J. Virol.* 63, 2798-2812.

Arsenakis, M., Tomasi, L.F., Speziali, V., Roizman, B., and Campadelli-Fiume, G.M. (1986). Expression and regulation of glycoprotein C genes of herpes simplex virus 1 resident in a clonal L-cell line. *J. Virol.* 66, 562-566.

Batterson, W., Furlong, D., and Roizman, B. (1983). Molecular genetics of herpes simplex virus. VII. Further characterization of a *ts* mutant defective in release of viral DNA and in other stages of viral reproductive cycle. *J. Virol.* 45, 397-407.

Batterson, W., and Roizman, B. (1983). Characterization of the herpes simplex virion-ass ociated factor responsible for the induction of α genes. *J. Virol.* 46, 371-377.

Baumruker, T., Sturm, R., and Herr, W. (1988). OBP100 binds remarkably degenerate octamer motifs through specific interactions with flanking sequences. *Genes Dev.* 2, 1400-1413.

Berger, S.L., Pina, B., Silverman, N., Marcus, G.A., Agapite, J., Regier, J.L., Triezenberg, S.J., and Guarente, L. (1992). Genetic isolation of ADA2: A potential transcriptional adapter required for the function of certain acidic activation domains. Cell 70, 251-265.

Blair, E.D., and Wagner, E.K. (1986). A single regulatory region modulates both cis activation and transactivation of the herpes simplex VP5 promoter in transient expression assays in vitro. *J. Virol.* 60, 460-469.

Cai, W., and Schaffer, P.A. (1989). Herpes simplex virus type 1 ICP0 plays a critical role in the de novo synthesis of infectious virus following transfection of viral DNA. *J. Virol.* 63, 4578-4590.

Cai, W., and Schaffer, P.A. (1992). Herpes simplex virus type 1 ICP0 regulates expression of immediate-early, early, and late genes in productively infected cells. *J. Virol.* 66, 2904-2915.

Campbell, M.E.M., Palfreyman, J.W., and Preston, C.M. (1984). Identification of herpes simplex virus DNA sequences which encode a *trans*-acting polypeptide responsible for stimulation of immediate early transcription. *J. Mol. Biol.* 180, 1-19.

Chen, J., and Silverstein, S. (1992). Herpes simplex viruses with mutations in the gene encoding ICP0 are defective in gene expression. *J. Virol.* 66, 2916-2927.

Costas, R.H., Draper, K.G., Devi-Rao, G., Thompson, R.L., and Wagner, E.K. (1985). Virus-induced modification of the host cell is required for expression of the bacterial chloramphenicol acetyltransferase gene controlled by a late herpes simplex virus promoter (VP5). *J. Virol.* 56, 19-30.

Cress, W.D., and Triezenberg, S.J. (1991). Critical Structural Elements of the VP16 Transcriptional Activation Domain. *Science* 251, 87-90.

Dalrymple, M.A., McGeoch D.J., Davison, A.J., and Preston, C.M. (1985). DNA sequence of the herpes simplex virus type 1 gene whose product is responsible for transcriptional activation of immediate early promoters. *Nucleic Acids Res.* 13, 7865-7879.

DeLuca, N.A., McCarthy, A.M., and Schaffer, P.A. (1985). Isolation and characterization of deletion mutants of herpes simplex virus type 1 in the gene encoding the immediate-early regulatory protein ICP4. *J. Virol.* 56, 558-570.

Dennis, D., and Smiley, J.R. (1984). Transactivation of a late herpes simplex virus promoter. *Mol. Cell. Biol.* 4, 544-551.

Dixon, R.A.F., and Schaffer, P.A. (1980). Fine structure mapping and functional analysis of temperature-sensitive mutant in the gene encoding the herpes simplex virus type 1 immediate early protein VP175. *J. Virol.* 36, 189-203.

Everett, R.D. (1986). The products of herpes simplex virus type 1 (HSV-1) immediate-early genes 1, 2, and 3 can activate HSV-1 gene expression in trans. *J. Gen. Virol.* 67, 2507-2513.

Everett, R.D. (1988). Analysis of functional domains of herpes simplex virus type 1 immediate-early polypeptide Vmw110. *J. Virol.* 202, 87-96.

Flanagan, W.M., Papavassiliou, A.G., Rice, M., Hecht, L.B., Silverstein, S., and Wagner, E.K. (1991). Analysis of the herpes simplex virus type 1 promoter controlling the expression of U_L38, a true late gene involved in capsid assembly. *J. Virol.* 65, 769-786.

Fletcher, C., Heintz, N., and Roeder, R.G. (1987). Purification and characterization of OTF-1, a transcription factor regulating cell cycle expression of a human histone H2b gene. Cell 51, 773-781.

Gaffney, D.F., McLauchlan, J., Whitton, J.L., and Clements, J.B. (1985). A modular system for the assay of transcription regulatory signals: the sequence TAATGARAT is required for herpes simplex virus immediate early gene activation. *Nucleic Acids Res.* 13, 7847-7863.

Garcia-Blanco, M.A., Clerc, R.G., and Sharp, P.A. (1989). The DNA-binding homeo domain of the Oct-2 protein. *Genes Dev.* 3, 739-745.

Gerster, T., and Roeder, R.G. (1988). A herpesvirus trans-activating protein interacts with transcription factor OTF-1 and other cellular proteins. *Proc. Natl. Acad. Sci. USA.* 85, 6347-6351.

Godowski, P.J., and Knipe, D.M. (1985). Identification of a herpes simplex virus function that represses late gene expression from parental viral genomes. *J. Virol.* 55, 357-365.

Godowski, P.J., and Knipe, D.M. (1983). Mutation in the major DNA-binding protein of herpes simplex virus type 1 result in increased levels of viral gene expression. *J. Virol.* 47, 478-486.

Greaves, R., and O'Hare, P. (1989). Separation of requirements for protein-DNA complex assembly from those for functional activity in the herpes simplex virus regulatory protein Vmw65. *J. Virol.* 63, 1641-1650.

Greaves, R.F., and O'Hare, P. (1990). Structural requirements in the herpes simplex virus type 1 transactivator Vmw65 for interaction with the cellular octamer-binding protein and target TAATGARAT sequences. *J. Virol.* 64, 2716-2724.

Haigh, A., Greaves, R., and O'Hare, P. (1990). Interference with the assembly of a virus-host transcription complex by peptide competition. *Nature* 344, 257-259.

Hardwicke, M.A., Vaughn, P.J., Sekulovich, R.E., O'Conner, R., and Sandri-Goldin, R.M. (1989). The regions important for the activator and repressor functions of herpes simplex virus type 1 a protein ICP27 map to the C-terminal half of the molecule. *J. Virol.* 63, 4590-4602.

He, X., Treacy, M.N., Simmons, D.M., Ingraham, H.A., Swanson, L.W., and Rosenfeld, M.J. (1989). Expression of a large family of POU-domain regulatory genes in mammalian brain development. *Nature* 340, 35-42.

Heine, J.W., Honess, R.W., Cassai, E., and Roizman, B. (1974). Proteins specified by herpes simplex virus. XII. The virion polypeptides of type 1 strains. *J. Virol.* 14, 640-651.

Herr, W., Sturm, R.A., Clerc, R.G., Corcoran, L.M., Baltimore, D., Sharp, P.A., Ingraham, H.A., Rosenfeld, M.G., Finney, M., Ruvkun, G., and Horvitz, H.R. (1988). The POU domain: a large conserved region in the mammaian *pit*-1, *oct*-1, *oct*-2, and *Caenorhabditis elegans unc*-86 gene products. *Genes Dev.* 2, 1513-1516.

Holland, L.E., Anderson, K.P., Shipman, C., and Wagner, E.K. (1980). Viral DNA synthesis is required for the efficient expression of specific herpes virus type 1 mRNA species. *Virology* 101, 40-53.

Homa, F.L., Glorioso, J.C., and Levine, M. (1988). A specific 15-bp TATA box promoter element is required for expression of a herpes simplex virus type 1 late gene. *Genes Dev.* 2, 40-53.

Homa, F.L., Otal, T.M., Glorioso, J.C. and Levine, M. (1986). Transcriptional control signals of a herpes simplex virus type 1 late (γ_2) gene lie within bases -34 to +124 relative to the 5' terminus of the mRNA. *Mol. Cell. Biol.* 6, 3652-3666.

Honess, R.W., and Roizman, B. (1974). Regulation of herpesvirus macromolecular synthesis. I. Cascade regulation of the synthesis of three groups of viral proteins. *J. Virol.* 14, 8-19.

Honess, R.W., and Roizman, B. (1975). Regulation of herpesvirus macromolecular synthesis: Sequential transition of polypeptide synthesis requires functional viral polypeptides. *Proc. Natl. Acad. Sci. USA* 72, 1276-1280.

Imbalzano, A.N., Coen, D.M., and DeLuca, N.A. (1991). Herpes simplex virus transactivator ICP4 operationally substitutes for the cellular transcription factor Sp1 for efficient expression of the viral thymidine kinase gene. *J. Virol.* 65, 565-574.

Ingles, C.J., Shales, M., Cress, W.D., Triezenberg, S.J., and Greenblatt, J. (1991). Reduced binding of TFIID to transcriptionally compromised mutants of VP16. *Nature* 351, 588-590.

Ingraham, H.A., Flynn, S.E., Voss, J.W., Albert, V.R., Kapiloff, M.S., Wilson, L., and Rosenfeld, M.G. (1990). The POU-specific domain of Pit-1 is essential for sequence-specific high affinity DNA binding and DNA-dependent Pit-1-Pit-1 interactions. *Cell* 61, 1021-1033.

Johnson, P.A., and Everett, R.D. (1986). The control of herpes simplex virus type 1 late gene expression: a TATA box/cap-site region is sufficient for fully efficient regulated activity. *Nucleic Acids Res.* 14, 8247-8264.

Jones, P.C., and Roizman, B. (1979). Regulation of herpesvirus macromolecular synthesis: VIII. The transcription program consists of three phases during which both extent of transcription and accumulation of RNA in the cytoplasm are regulated. *J. Virol.* 31, 299-413.

Katan, M., Haigh, A., Verrijzer, C.P., van der Vliet, P.C., and O'Hare, P. (1990). Characterization of a cellular factor which interacts functionally with Oct-1 in the assembly of a multicomponent transcription complex. *Nucleic Acids Res.* 18, 6871-6880.

Kelleher III, R.J., Flanagan, P.M., and Kornberg, R.D. (1990). A novel mediator between activator proteins and the RNA polymerase II transcription apparatus. *Cell* 61, 1209-1215.

Kibler, P.K., Duncan, J., Deith, B.D., Hupel, T., and Smiley, J.R. (1991). Regulation of herpes simplex virus true late gene expression: sequences downstream from the U_S11 TATA box inhibit expression from an unreplicated template. *J. Virol.* 65, 6749-6760.

Kristie, T.M., and Roizman, B. (1984). Separation of sequences defining basal expression from those conferring α gene recognition within the regulatory domains of herpes simplex virus 1 α genes, *Proc. Natl. Acad. Sci. USA* 81, 40654069.

Kristie, T.M., and Roizman, B. (1987). Host cell proteins bind to the cis-acting site required for virion-mediated induction of herpes simplex virus 1 α genes. *Proc. Natl. Acad. Sci. USA* 84, 71-75.

Kristie, T.M., and Roizman, B. (1988). Differentiation and DNA contact points of host proteins binding at the *cis* site for virion-mediated induction of α genes of herpes simplex virus 1. *J. Virol.* 62, 1145-1157.

Kristie, T.M., LeBowitz, J.H., and Sharp, P.A. (1989). The octamer-binding proteins form multi-protein-DNA complexes with the HSV aTIF regulatory protein. *EMBO* J. 8, 4229-4238.

Kristie, T.M., and Sharp, P.A. (1990). Interactions of the Oct-1 POU subdomains with specific DNA sequences and with the HSV α-trans activator protein. *Genes Dev.* 4, 2383-2396.

LaBella, F., Sive, H.L., Roeder, R.G., and Heintz, N. (1988). Cell-cycle regulation of a human histone H2b gene is mediated by the H2b subtype-specific consensus element. *Genes Dev.* 2, 32-39.

LaMarco, K.L., and McKnight, S.L. (1989). Purification of a set of cellular polypeptides that bind to the purine-rich *cis*-regulatory element of herpes simplex virus immediate early genes. *Genes Dev.* 3, 1372-1383. (1989).

LaMarco, K., Thompson, C.C., Byers, B.P., Walton, E.M., and McKnight, S.L. (1991). Identification of ets- and notch-related subunits in GA binding protein. *Science* 253, 789-791.

Laughon, S., and Scott, M.P. (1984). Sequence of a drosophila segmentation gene: protein structure homology with DNA binding proteins. *Nature* 310, 25-31.

LeBowitz, J.H., Kobayashi,T., Staudt, L., Baltimore, D., and Sharp, P.A. (1988) Octamer-binding proteins from B or HeLa cells stimulate transcription of the immunoglobulin heavy-chain promoter in vitro. *Genes Dev.* 2, 1227-1237.

LeBowitz, J.H., Clerc, R.G., Brenowitz, M., and Sharp, P.A. (1989). The Oct-2 protein binds cooperatively to adjacent octamer sites. *Genes Dev.* 3, 1625-1638.

Lin, Y.-S., and Green, M.R. (1991). Mechanism of action of an acidic transcriptional activator in vitro. *Cell* 64, 971-981.

Longnecker, R., and Roizman, B. (1986). Generation of an inverting herpes simplex virus 1 mutant lacking the L-S junction a sequences, an origin of DNA synthesis, and several genes including those specifying glycoproteins E and the α47 gene. *J. Virol.* 58, 583-591.

Mackem, S., and Roizman, B. (1982a). Regulation of α genes of herpes simplex virus: the α27 gene promoter-thymidine kinase chimera is positively regulated in converted L cells. *J. Virol.* 43, 1015-1023.

Mackem, S., and Roizman, B. (1982b). Differentiation between α promoter and regulator regions of herpes simplex virus1: the functional domains and sequence of a moveable α regulator, *Proc. Natl. Acad. Sci. USA* 79, 4917-4921.

Mackem, S., and Roizman, B. (1982c). Structural features of the α gene 4, 0, and 27 promoter-regulatory sequence which confer α regulation on chimeric thymidine kinase genes. *J. Virol.* 44, 939-949.

Mavromara-Nazos, P., and Roizman, B. (1987). Activation of herpes simplex virus 1 γ_2 genes by viral DNA replication. *Virology* 161, 593-598. (1987).

Mavromara-Nazos, P., and Roizman, B. (1989). Delineation of regulatory domains of early (ß) and late (γ_2) genes by construction of chimeric genes expressed in herpes simplex virus 1 genomes. *Proc. Natl. Acad. Sci. USA* 86, 4071-4079.

Mattaj, I.W., Lienhard, S., Jiricny, J., and DeRobertis, E.M. (1985). An enhancer-like sequence within the Xenopus U2 gene promoter facilitates the formation of stable transcription complexes. *Nature* 316, 163-167.

McCarthy, A.M., McMahan, L., and Schaffer, P.A. (1989). Herpes simplex virus type 1 ICP27 deletion mutants exhibit altered patterns of transcription and are DNA deficient. *J. Virol.* 63, 18-27.

McGeoch, D. J., Dalrymple, M.A., Davison, A.J., Dolan, A. Frame, M.C., McNab, D., Perry, L.J., Scott, J.E., and Taylor, P. (1988). The complete DNA sequence of the long unique region in the genome of herpes simplex virus type 1. *J. Gen. Virol.* 69, 1531-1574.

McKee, T.A., Disney, G.H., Everett, R.D., and Preston, C.M. (1990). Control of expression of the varicella-zoster virus major immediate early gene. *J. Gen. Virol.* 71, 897-906.

McKnight, J.L.C., Kristie, T.M., and Roizman, B. (1987). Binding of the virion protein mediating α gene induction in herpes simplex virus-1 infected cells to its cis-site requires cellular proteins. *Proc. Natl. Acad. Sci. USA* 84, 7061-7065.

McKnight, S.L., and Kingsbury, R. (1982).Transcriptional control signals of a eukaryotic protein-coding gene. *Science* 217, 316-324.

McKnight, S.L., and Tijan, R. (1986). Trancriptional selectivity of viral genes in mammalian cells. *Cell* 46, 795-805.

Michael, N., Spector, D., Mavromara-Nazos, P., Kristie, T.M., and Roizman, B. (1988). The DNA-binding properties of the major regulatory protein α4 of herpes simplex viruses. *Science* 239, 1531-1534.

Muller, M.T. (1987). Binding of the herpes simplex virus immediate-early gene product ICP4 to its own transcription start site. *J. Virol.* 61, 858-865.

Murphy, S., Yoon, J.-B., Gerster, T., and Roeder, R.G. (1992). Oct-1 and Oct-2 potentiate functional interactions of a transcription factor with the proximal sequence element of small nuclear RNA genes. *Mol. Cell. Biol.* 12, 3247-3261.

O'Hare, P., and Goding, C.R. (1988). Herpes simplex virus regulatory elements and the immunoglobulin octamer domain bind a common factor and are both targets for virion transactivation. Cell 52, 435-445.

O'Hare, P., Goding, C.R., and Haigh, A. (1988) Direct combinatorial interaction between a herpes simplex virus regulatory protein and a cellular octamer-binding factor mediates specific induction of virus immediate-early gene expression. *EMBO J.* 7, 4231-4238.

O'Hare, P., and Hayward, G.S. (1985a). Evidence for a direct role for both the 175,000 and 110,000-molecular weight immediate-early proteins of herpes simplex virus in the transactivation of delayed-early promoters. *J. Virol.* 53, 751-760.

O'Hare, P. and Hayward, G.S. (1985b). Three trans-acting regulatory proteins of herpes simplex virus modulate immediate-early gene expression in a pathway involving positive and negative feedback regulation. *J. Virol.* 56, 723-733.

O'Neill, E.A., Fletcher, C., Burrow, C.R., Heintz, N., Roeder, R.G., and Kelly, T.J. (1988). Transcription factor OTF-1 is functionally identical to the DNA replication factor NF-III. *Science* 241, 1210-1213.

O'Neill, E.A. and Kelly, T.J. (1988). Purification and characterization of nuclear factor III (origin recognition protein C), a sequence-specific DNA binding protein required for efficient initiation of adenovirus DNA replication. *J. Biol. Chem.* 263, 931-937.

Otting, G., Qian, Y., Muller, M., Affolter, M., Gehring, W., and Wuthrich, K. (1988). Secondary structure determination for the *Antennapedia* homeodomain by nuclear magnetic resonance and evidence for a helix-turn-heliX motif. *EMBO J.* 7, 4305- .

Pabo, C.O., and Sauer, R.T. (1984). Protein-DNA recognition. *Annu. Rev. Biochem.* 53, 293-321.

Pellet, P.E., McKnight, J.L.C., Jenkins, F.J., and Roizman, B. (1985). Nucleotide sequence and predicted amino acid sequence of a protein encoded in a small herpes simplex virus DNA fragment capable of *trans*-inducing α genes. *Proc. Natl. Acad. Sci. USA* 82, 5870-5874.

Poellinger, L., and Roeder, R.G. (1989). Octamer transcription factors 1 and 2 each bind to two different functional elements in the immunoglobulin heavy-chain promoter. *Mol. Cell. Biol.* 9, 747-756.

Poellinger, L., Yoza, B.K., and Roeder, R.G. (1989). Functional cooperativity between protein molecules bound at two distinct sequence elements of the immunoglobulin heavy chain promoter. *Nature* 337, 573-576.

Post, L.E., Mackem, S., and Roizman, B. (1981). Regulation of α genes of herpes simplex virus: expression of chimeric genes produced by fusion of thymidine kinase with α gene promoters. *Cell* 24, 555-565.

Preston, C.M. (1979). Control of herpes simplex virus type 1 mRNA synthesis in cells infected with wild type virus or the temperature sensitive mutant tsK. *J. Virol.* 29, 275-284.

Preston, C.M., Frame, M.C., and Campbell, M.E.M. (1988). A complex formed between cell components and an HSV structural polypeptide binds to a viral immediate early gene regulatory DNA sequence. *Cell* 52, 425-434.

Pruijn, G.J., van Driel, W., and van der Vliet, P.C. (1986). Nuclear factor III, a novel sequence-specific DNA-binding protein from HeLa cells stimulating adenovirus DNA replication. *Nature* 322, 656-659.

Quian, Y.Q., Billeter, M., Otting, G., Muller, M., Gehring, W., and Wuthrich, K. (1989). The structure of the *Antennapedia* homeodomain determined by NMR spectroscopy in solution: comparison with prokaryotic repressors. *Cell* 59, 573-582.

Rice, S.A., and Knipe, D.M. (1990). Genetic evidence for two distinct transactivation functions of the herpes simplex virus a protein ICP27. *J. Virol.* 64, 1704-1715.

Romanelli, M.G., Mavromara-Nazos, P., Spector, D., and Roizman, B. (1992). Mutational analysis of the ICP4 binding sites in the 5'transcribed non-coding domains of herpes simplex virus type 1 U_L49.5 γ_2 gene. *J. Virol.* 66, 4855-4863.

Sacks, W.R., Greene, C.C., Aschman, D.P., and Schaffer, P.A. (1985). Herpes simplex virus type 1 ICP27 is an essential regulatory protein. *J. Virol.* 55, 796-805.

Sacks, W.R., and Schaffer, P.A. (1987). Deletion mutants in the gene encoding the herpes simplex virus type 1 immediate-early protein ICP0 exhibit impaired growth in cell culture. *J. Virol.* 61, 829-839.

Sadowski, I., Ma, J., Triezenberg, S., and Ptashne, M. (1988). GAL4-VP16 is an unusually potent transcriptional activator. *Nature* 335, 563-564.

Scheidereit, C., Heguy, A., and Roeder, R.G. (1987). Identification and purification of a human lymphoid-specific octamer binding protein (OTF-2) that activates transcription of an immunoglobulin promoter *in vivo*. *Cell* 51, 783-793.

Sears, A.E., Halliburton, I.W., Meignier, B., Silver, S., and Roizman, B. (1985). Herpes simplex virus 1 mutant deleted in the α22 gene: growth and gene expression in permissive and restrictive cells and establishment of latency in mice. *J. Virol.* 55, 338-346.

Sears, A.E., Hukkanen, V., Labow, M.A., Levine, A.J., and Roizman, B. (1991). Expression of the herpes simplex virus 1 α transinducing factor (VP16) does not induce reactivation of latent virus or prevent the establishment of latency in mice. *J. Virol.* 65, 2929-2935.

Sekulovich, R.E. Leary, K., and Sandri-Goldin, R.M. (1988). The herpes simplex virus type 1 alpha protein ICP27 can act as a trans-repressor or a trans-activator in combination with ICP4 and ICP0. *J. Virol.* 62, 4510-4522.

Shepard, A.A., and DeLuca, N.A. (1989). Intragenic complementation among partial peptides of herpes simplex virus regulatory protein ICP4. *J. Virol.* 63, 1203-1211.

Silver, S., and Roizman, B. (1985). γ_2-thymidine kinase chimeras are identically transcribed but regulated as γ_2 genes in herpes simplex virus genomes and as ß genes in cell genomes. *Mol. Cell. Biol.* 5, 518-528.

Smiley, J.R., Johnson, D.C., Pizer, L.I., and Everett, R.D. (1992). The ICP4 binding sites in the herpes simplex virus type 1 glycoprotein D (gD) promoter are not essential for efficient gD transcription during virus infection. *J. Virol.* 66, 623-6

Smith, I.L., Hardwicke, M.A., and Sandri-Goldin, R.M. (1992). Evidence that the herpes simplex virus immediate early protein ICP27 acts post-transcriptionally during infection to regulate gene expression. *Virology* 186, 74-85.

Spear, P.G., and Roizman, B. (1972). Proteins specified by herpes simplex virus. V. Purification and structural proteins of the herpes virion. *J. Virol.* 9, 431-439.

Spector, D., Purves, F., and Roizman, B. (1990). Mutational analysis of the promoter region of the α27 gene of herpes simplex virus 1 within the context of the viral genome. *Proc. Natl. Acad. Sci. USA* 87, 5268-5272.

Spector, D., Purves, F., and Roizman, B. (1991). Role of α-transinducing factor (VP16) in the induction of α genes within the context of viral genomes. *J. Virol.* 65, 3504-3513.

Staudt, L.M., Singh, H., Sen, R., Wirth, T., Sharp, P.A., and Baltimore, D. (1986). A lymphoid specific protein binding to the octamer motif of immunoglobulin genes. *Nature* 323, 640-643.

Steffy, K.R., and Weir, J.P. (1991). Mutational analysis of two herpes simplex virus type 1 late promoters. *J. Virol.* 65, 6454-6460.

Steiner, I., Spivack, J.G., Deshmane, S.L., Ace, C.I., Preston, C.M., and Fraser, N.W. (1990). A herpes simplex virus type 1 mutant containing a nontransinducing Vmw65 protein establishes latent infection in vivo in the absence of viral replication and reactivates efficiently from explanted trigeminal ganglia. *J. Virol.* 64, 1630-1638.

Stern, S., and Herr, W. (1991). The herpes simplex virus trans-activator VP16 recognizes the Oct-1 homeo domain: evidence for a homeo domain recognition subdomain. *Genes Dev.* 5, 2555-2566.

Stern, S., Tanaka, M., and Herr, W. (1989). The Oct-1 homeodomain directs formation of a multiprotein-DNA complex with the HSV transactivator VP16. *Nature* 341, 624-630.

Stow, N.D., and Stow, E.C. (1986). Isolation and characterization of a herpes simplex virus type 1 mutant containing a deletion within the gene encoding the immediate early polypeptide Vmw110. *J. Gen. Virol.* 67, 2571-2585.

Stringer, K.F., Ingles, C.J., and Greenblatt, J. (1990). Direct and selective binding of an acidic transcriptional activation domain to the TATA-box factor TFIID. *Nature* 345, 783-786.

Sturm, R., Baumruker, T., Franza Jr, B.R., and Herr, W. (1987). A 100-kD HeLa cell octamer binding protein (OBP100) interacts differently with two separate octamer-related sequences within the SV40 enhancer. *Genes Dev.* 1, 1147-1160.

Sturm, R., Das, G., and Herr, W. (1988). The ubiquitous octamer binding protein Oct-1 contains a POU domain with a homeobox subdomain. *Genes Dev.* 2, 1582-1599.

Sturm, R.A., and Herr, W. (1988). The POU domain is a bipartite DNA-binding structure. *Nature* 336, 601-604.

Treisman, J., Gonczy, P., Vashishtha, M., Harris, E. and Desplan, C. (1989). A single amino acid can determine the DNA binding specificity of homeobinding proteins. *Cell* 59, 553-562.

Triezenberg, S.J., Kingsbury, R.C., and McKnight, S.L. (1988). Functional dissection of VP16, the *trans*-activator of herpes simplex virus immediate early gene expression. *Genes Dev.* 2, 718-729.

Triezenberg, S.J., LaMarco, K.L., and McKnight, S.L. (1988) Evidence of DNA:protein interactions that mediate HSV-1 immediate early gene activation by VP16. *Genes Dev.* 2, 730-742.

Verrijzer, C.P., Kal, A.J., and van der Vliet, P.C. (1990). The oct-1 homeo domain contacts only part of the octamer sequence and full oct-1 DNA- binding activity requires the POU-specific domain. *Genes Dev.* 4, 1964-1974.

Weinheimer, S.P., Boyd, B.A., Durham, S.K., Resnick, J.L., and O'Boyle II, D.R. (1992). Deletion of the VP16 open reading frame of herpes simplex virus type 1. *J. Virol.* 66, 258-269.

Weir, J.P., and Narayanan, P.R. (1988). The use of B-galactosidase as a marker gene to define the regulatory sequences of the herpes simplex virus type 1 glycoprotein C in recombinant herpesvirus. *Nucleic Acids Res.* 16, 10267-10282.

Wharton, R.P., and Ptashne, M. (1985). Changing the binding specificity of a repressor by redesigning an α-helix. *Nature* 316, 601-605.

Xiao, P., and Capone, J.P. (1990). A cellular factor binds to the herpes simplex virus type 1 transactivator Vmw65 and is required for Vmw65- dependent protein-DNA complex assembly with Oct-1. *Mol. Cell. Biol.* 10, 4974-4977.

TRANSCRIPTIONAL ACTIVATION BY THE ADENOVIRUS E1A PROTEINS

Brian A. Lewis and Thomas Shenk

Howard Hughes Medical Institute
Department of Molecular Biology
Princeton University
Princeton, NJ 08544-1014

INTRODUCTION

The adenovirus E1A gene products can transactivate viral and many cellular promoters while repressing others, drive quiescent cells to enter the cell cycle and synthesize DNA, and contribute to the oncogenic transformation of cells. These diverse effects of the E1A proteins appear to result from their ability to bind to a variety of cellular proteins, altering the activity of several different transcription factors.

There are two predominant E1A polypeptides. They contain the same amino and carboxyl termini, but differ at an internal site due to differential splicing of E1A mRNAs. The larger, 289 amino acid E1A polypeptide is termed the 13S E1A protein (due to the sedimentation coefficient of its mRNA), and the smaller 243 residue polypeptide is the 12S E1A protein. The 12S protein lacks 46 amino acids present in the larger protein. Sequence comparison of the E1A proteins encoded by different adenovirus serotypes indicates three blocks of similar amino acid sequence. Conserved region 1 contains residues 41-80, conserved region 2 contains residues 121-139 and conserved region 3 is the 13S-specific 46 residues, from 140-188 (Kimmelman *et al*, 1985). A zinc finger motif is found in CR3. The conserved regions are important functional domains within the E1A proteins, mediating binding interactions with a variety of cellular proteins.

Regulation of Gene Expression in Animal Viruses, Edited by
L. Carrasco *et al.*, Plenum Press, New York, 1993

A large body of experimentation has demonstrated E1A's ability to transactivate viral and cellular promoters (see Flint and Shenk, 1989; 1991 and Nevins, 1989 for reviews). This work has led to the identification of several promoter sequences that respond to E1A transactivation, and, more recently, the transcription factors that bind to these promoter elements have been identified and shown to be directly influenced by E1A proteins. Here we will discuss the mechanism of E1A transcriptional activation through three cellular transcription factors: E2F, TFIID (TBP) and YY1.

E2F

E1A binding proteins were first identified by co-immunoprecipitation. Harlow *et al.* (1986) described a series of co-immunoprecipitating polypeptides, including 105kDa and 107kDa species. The 105kDa protein has since been identified as the product of the retinoblastoma (RB) gene (Whyte *et al.*, 1988). The 107kDa protein has been cloned and its primary sequence is similar to that of the RB protein, including a "pocket" domain that is essential for binding E1A (Ewen *et al.*, 1991; Kaelin *et al.*, 1991). Several groups have demonstrated that the cellular E2F transcription factor binds to the RB protein (Chellappan *et al.*, 1991; Bagchi *et al.*, 1992; Chittenden *et al.*, 1991; Cao *et al.*, 1992; Bandara and LaThangue, 1991). E2F was originally described as an activity necessary for E1A transactivation of the adenovirus E2 promoter. Since then, other genes, including genes involved in growth regulation (e.g. c-myc, DHFR), have been found to have E2F sites.

E1A can disrupt an RB/E2F complex as a result of its ability to bind the RB protein (Bagchi *et al.*, 1990; Chellappan *et al.*, 1991). Transfection of an RB expression vector reduces expression from the adenovirus E2 promoter, fused to a CAT reporter gene, while cotransfection of RB and an E1A 12S expression vector alleviates this downregulation (Hiebert *et al.*, 1992). Further, the domains necessary for E1A's ability to bind RB, CR1 and CR2, are also responsible for alleviating this repression (Raychauduri *et al.*, 1991). These data suggest that E2F is a transcriptional activator and that RB represses transcription by binding E2F.

In addition to the RB protein, p107 is able to bind E2F (Cao *et al.*, 1992; Shirodkar *et al..*, 1992), and cyclin A, which binds to p107, is present in this complex (Faha *et al.*, 1992; DeVoto *et al.*, 1992; Ewen *et al.*, 1992). E1A can also disrupt the p107/E2F complex.

RB protein and p107 appear to bind E2F in a cell cycle-specific fashion (Shirodkar *et al.*, 1992; Mudryj *et al.*, 1991). The E2F/p107/cyclin A complex was predominantly restricted to S-phase cells while the E2F/RB binding activity was found in cells in G1 and S but disappeared during G2/M. These data raise the possibility that E2F regulates different genes at different points in the cell cycle and perhaps RB and p107 are involved in this cell cycle specificity. By interfering with this regulation, E1A protein might free E2F to act constitutively. This could, in turn, at least partly account for the ability of the E1A protein to drive quiescent cells into active growth and to oncogenically transform cells.

TFIID (TBP)

Several groups have described the ability of E1A to transactivate through the TATA box (see Flint and Shenk, 1989; 1991 and Nevins 1989 for reviews), the sequence motif to which the TFIID basal transcription factor binds. However, not all TATA boxes are receptive to transactivation (Simon et al., 1988; 1989), implying that different TFIID complexes exist, some of which are refractory to E1A transactivation.

Two groups have provided data to explain E1A transactivation through the TATA box (Horikoshi *et al.*, 1991; Lee *et al.*, 1991). Both have shown that E1A physically interacts with the TATA-binding protein, TBP (TBP is the DNA-binding subunit of TFIID, which includes a variety of polypeptides, termed TAFs, whose functions are not yet understood). Lee *et al.* (1991) further show that the 13S E1A has a greater affinity for TBP than does 12S, correlating with the 13S protein's greater transactivation ability, and that point mutations within CR3, that interfere with E1A activation through TATA motifs, abolish or greatly diminish TBP binding.

How might the binding of E1A to TBP activate transcription? There are three possible mechanisms. First, the E1A protein might encode a function enabling it to stimulate transcriptional initiation when present in the TFIID complex. Second, E1A binding might displace a negatively acting polypeptide from the TFIID complex. Third, E1A could serve as an adaptor, allowing a positively acting polypeptide to bridge to TBP and enter the TFIID complex.

YY1

Chang *et al.* (1989) demonstrated that E1A transactivation of the AAV P5 promoter is mediated by a *cis* element at -60 from the initiation site at +1. Shi *et al.* (1991) cloned the protein that bound to the -60 site, a 414 amino acid zinc finger protein called YY1, and demonstrated that it can repress transcription. E1A transactivates through YY1 binding sites by relieving the repression and further activating transcription through the protein. Recent work (B. Lewis, E. Seto, Y. Shi, N. Horikoshi, R. Weinmann, and T. Shenk unpublished) indicates that E1A and YY1 can physically interact, and that the 13S E1A protein has a greater affinity for YY1 than does 12S E1A. This correlates well with data showing that the 13S protein is better able to transactivate through YY1 binding sites than is the 12S protein (Shi et al., 1991). Deletion analysis of YY1 has identified two functional domains. The first domain, comprised of two and one-half zinc fingers at the carboxyl terminus, is sufficient to repress a heterologous promoter. The second domain, the site of interaction with E1A proteins that is required for E1A transactivation, maps to the amino terminal side of the zinc fingers (B. Lewis *et al.*, unpublished).

Thus, E1A relieves YY1-mediated repression by interacting at a site distinct from the repression domain. How does the E1A-YY1 interaction relieve repression? E1A binding could physically cover a repression function encoded in the repression domain of YY1, blocking it from acting; or it could induce a confirmational change in YY1, burying the repression domain. Alternatively, the region in YY1 that we have mapped as a repression domain could be a site of interaction with another cellular factor that encodes the repression function.

E1A not only relieves YY1-mediated repression, but it can further activate transcription through YY1. In fact, it can activate through a deleted YY1 variant lacking its carboxyl terminal repression domain (B. Lewis *et al.*, unpublished), suggesting that the activation event is distinct from that which relieves repression. Conceivably, E1A serves to bridge between YY1 and another transcriptional activator, and in this fashion E1A induces transcription through YY1 in addition to relieving repression.

CONCLUSIONS

Two main conclusions can be drawn at this point regarding the mechanism of E1A action. First, the E1A proteins exert their activities by binding to a variety of cellular proteins. It is possible that the only functions encoded in E1A proteins are protein-protein binding functions. E1A proteins might modulate the activity of their target cellular proteins simply by altering the constellation of polypeptides with which the cellular proteins can interact. Second, the E1A proteins appear to target transcriptional regulatory proteins. It seems likely that E1A proteins mediate their diverse effects on cell growth and regulation by modulating the activity of a series of key transcription factors.

Acknowledgments

B. A. Lewis is a predoctoral fellow of the New Jersey Commission on Cancer Research. T. Shenk is an American Cancer Society Professor and an Investigator of the Howard Hughes Medical Institute.

References

Bagchi, S., Raychaudhuri S., and Nevins, J. (1990). Adenovirus E1A proteins can dissociate heteromeric complexes involving the E2F transcription factor: a novel mechanism for E1A transactivation. *Cell* **62,** 659-669.

Bagchi, S., Weinmann, R., and Raychaudhuri, P. (1991). The retinoblastoma protein copurifies with E2F-I, an E1A-regulated inhibitor of the transcription factor E2F. *Cell* **65**, 1068-1072.

Bandara, L. R., and La Thangue, N. B. (1991). Adenovirus E1A prevents the retinoblastoma gene product from complexing with a cellular transcription factor. *Nature* **351**, 494-497.

Cao, L., Faha, B., Dembski, M., Tsai, L., Haarlow, E. and Dyson, N. (1992). Independent binding of the retinoblasoma protein and p107 to the transcription factor E2F. *Nature* **355**, 176-179.

Chang, L., Shi, Y., Shenk, T. (1989). Adeno-associated virus P5 promoter contains an adenovirus E1A-inducible element and a binding site for the major late transcription factor. *J. Virol.* **63**, 3479-3488.

Chellappan, S. P., Hiebert, S., Mudryj, M., Horowitz, J. M., and Nevins, J. R. (1991). The E2F transcription factor is a cellular target for the RB protein. *Cell* **65**, 1053-1061.

Chittenden, T., Livingston, D. M., and Kaelin, W. G. Jr. (1991). The T/E1A-binding domain of the retinoblastoma product can interact selectively with a sequence -specific DNA-binding protein. *Cell* **65**, 1073-1082.

Devoto, S.H., Mudryj, M., Pines, J., Hunter, T., and Nevins, J. R. (1992). A cyclin A-protein kinase complex possesses sequence-specific DNA binding activity: p33cdk2 is a component of the E2F-cyclin A complex. *Cell* **68**, 167-176.

Ewen, M. E., Xing, Y., Lawrence, J. B., and Livingston, D. M. (1991). Molecular cloning, chromosomal mapping, and expression of the cDNA for p107, a retinoblastoma gene product-related protein. *Cell* **66**, 1155-1184.

Ewen, M., Faha, B., Harlow, E., and Livingston, D. (1992). Interaction of p107 with cyclin A independent of complex formation with viral oncoproteins. *Science* **255**, 85-87.

Faha, B., Ewen, M. E., Tsai, L., Livingston, D. M., and Harlow, E. (1992) Interaction between human cyclin A and adenovirus E1A-associated p107 protein. *Science* **255**, 87-90.

Flint, J., and Shenk, T. (1989). Adenovirus E1A protein paradigm viral transactivator. *Annu. Rev. Genet..* **23**, 141-161.

Harlow, E., Whyte, P., Franza, Jr., B. R., and Schley, C. (1986). Association of adenovirus early-region 1A proteins with cellular polypeptides. *Mol. Cell. Biol.* **6**, 1579-1589.

Hiebert, S. W., Chellappan, S. P., Horowitz, J. M., and Nevins, J. R. (1992). The interaction of RB with E2F coincides with an inhibition of the transcriptional activity of E2F. *Genes and Dev.* **6**, 177-185.

Horikoshi, N., Maguire, K., Kralli, A., Maldonado, E., Reinberg, D. and Weinmann, R. (1991). Direct interaction between adenovirus E1A protein and the TATA box binding transcription factor IID. *Proc. Natl. Acad. Sci. USA* **88**, 5124-5128.

Kaelin Jr., W. G., Pallas, D. C., DeCaprio, J. A., Kaye, F. J., and Livingston, D. M. (1991). Identification of cellular proteins that can interact specifically with the T/E1A-binding region of the retinoblastoma gene product. *Cell* **64**, 521-532.

Kimelman, D., Miller, J. S., Porter D., Roberts, B. E. (1985). E1A regions of the human adenovirus and the highly oncogenic simian adenovirus 7 are closely related. *J Virol.* **53**, 399-409.

Lee, W. S., Kao, C. C., Bryant, G. O., Liu, X., and Berk, A. J. (1991). Adenovirus E1A activiation domain binds the basic repeat in the TATA box transcription factor. *Cell* **67**, 365-376.

Mudryj, M., Devoto, S. H., Hiebert, S. W., Hunter, T., Pines, J., and Nevins J. R. (1991). Cell cycle regulation of the E2F transcription factor involves an interaction with cyclin A. *Cell* **65**, 1243-1253.

Nevins, J. R. (1989). Mechanisms of viral-mediated trans-activation of transcription. *Adv. Virus Res.* **37**, 35-83.

Raychaudhuri, P., Bagchi, S., Devoto, S. H., Kraus, V. B., Moran, E., and Nevins, J. R. (1991). Domains of the adenovirus E1A protein required for oncogenic activity are also required for dissociation of E2F transcription factor complexes. *Genes and Dev.* **5**, 1200-1211.

MECHANISMS REGULATING NUCLEOCAPSID FORMATION OF THE HEPATITIS B VIRUSES

Ralf Bartenschlager,[1,2] and Heinz Schaller[2]

[1]F. Hoffmann-La Roche Ltd., Pharmaceutical Research-New Technologies,
4002-Basel, Switzerland
[2]ZMBH, University of Heidelberg,
Im Neuenheimer Feld 282, 6900 Heidelberg, Germany

INTRODUCTION

The hepatitis B viruses are a group of small enveloped DNA viruses which are characterized by a distinct liver tropism and a narrow host range. The most important member of the hepadnavirus group is the human pathogenic hepatitis B virus (HBV; Dane et al., 1970). In addition, four animal hepatitis B viruses have been described. The woodchuck hepatitis B virus (WHV; Summers et al., 1978), the ground squirrel hepatitis B virus (GSHV; Marion et al., 1980), the duck hepatitis B virus (DHBV; Mason et al., 1980) and the heron hepatitis B virus (HHBV; Sprengel et al., 1988). Characteristic for this virus group is their mode of replication which includes the reverse transcription of an RNA genome. For this reason, hepadnaviruses are classified, along with retroviruses, as viral retroelements. However, while in retroviral replication the onset of DNA synthesis occurs after infection of the host cell, DNA synthesis in hepadnaviruses commences before release of virus particles. Therefore, retrovirus particles contain an RNA genome, hepadnavirus particles a DNA genome, leading to their classification as pararetroviruses.

Formation of replication competent nucleocapsids requires selective and ordered interactions between its individual components: the core protein, the P (*pol*) protein and the RNA pregenome. Although of central importance, only recently considerable insight into the mechanisms governing hepadnaviral nucleocapsid assembly was achieved. In this article we will review recent progress in this field of hepadnavirus research and discuss the implications for regulation of viral gene expression. We will focuss primarily on the HBV system and include data from other hepadnaviruses only when necessary.

VIRION STRUCTURE AND GENOME ORGANIZATION

The morphology of hepatitis B viruses resembles that of other enveloped DNA viruses (Fig. 1A). In case of HBV the infectious virion, also called Dane particle (Dane et al., 1970) has a diameter of 42 nm and consists of an outer envelope composed of host derived lipids and 3 variant viral surface proteins encoded by the preS/S gene. Of these the "small S-protein" (HBsAg) is the predominant species whereas the "middle S-protein" (preS2/S) and the "large S-protein" (preS1/S2/S) are less abundant. The preS1 domain of the large S-protein most likely represents the host cell receptor binding site on the virion (Neurath et al., 1986; Klingmüller et al., 1992). Internal to the envelope is the nucleocapsid or core particle composed of 180 subunits of the core protein. Within the capsid is the circular DNA genome of about 3200 nucleotides in length. Its minus strand, which is slightly longer than one genome length, has a fixed 3' end and a terminal protein covalently attached to its 5' end. The plus strand, which spans the interruption in the minus strand, has a heterogenous 3' end and carries a short oligoribonucleotide at its 5' end. This peculiar DNA structure reflects the unusual mode of replication described below.

Together with the viral genome, a DNA-polymerase activity was found to be associated with the nucleocapsid (Kaplan et al., 1973). In agreement with earlier homology comparisons between the retroviral and the hepadnaviral *pol* gene products (Toh et al., 1983) it could recently be shown by genetic analysis that the core-associated polymerase activity is encoded by the P gene (Schlicht et al., 1989[a]; Chang et al., 1989; Radziwill et al., 1990).

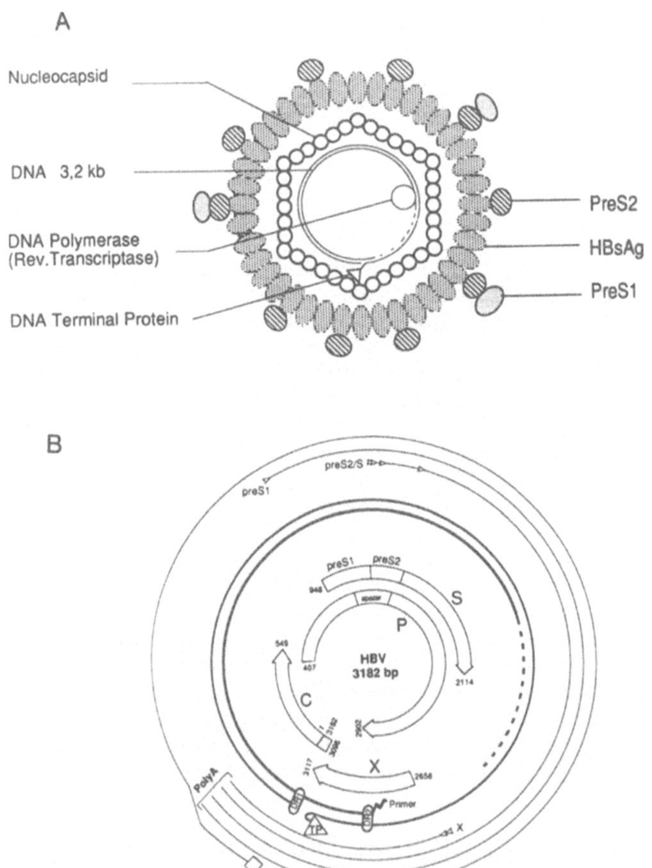

Figure 1. Hepatitis B virus structure, genome organization and transcriptional map. **A.** Schematic representation of the HBV particle. For details see text. **B.** Genetic, physical and transcriptional map of the human hepatitis B virus. The heavy double circle represents the DNA genome with the terminal protein (dotted triangle) at the 5' end of the complete (minus) strand and the RNA primer (wavy line) at the 5' end of the incomplete (plus) strand. The signals important for replication (DR1 and DR2) are shown as ovals. The open arrows depict the different open reading frames. The outer circles represent the viral transcripts. They initiate at different positions (arrowheads) and all terminate at a common poly adenylation signal. The box labeled "ε" indicates the functional copy of the RNA encapsidation signal. Numbers are given according to the nomenclature of Pasek et al. (1979) starting with the A residue of the core gene initiation codon.

In addition to these structural proteins, mammalian hepadnaviruses encode in the X gene a transcriptional transactivator (for a recent review see Schek et al., 1991). The precise role of this protein for the viral life cycle remains to be elucidated. X protein appears not to be essential for the formation of replication competent virus-like particles, since X null mutants are not severely affected in particle formation in a transient expression system (Yaginuma et al., 1987; Blum et al., 1992).

Apart from virus particles, infected cells secret S-particles which are envelopes devoid of the nucleocapsid and furthermore a non-particulate variant of the core protein termed HBeAg. The biosynthesis of the latter protein will be described below in more detail.

With the smallest DNA genome for animal viruses known, hepatitis B viruses have developed highly specialized mechanisms for optimal use of the restricted genome space (Fig. 1B). These include extensively overlapping reading frames, transcriptional control regions which are themselves actively transcribed during viral replication and which are also used for translation of viral gene products and differential transcription initiation, giving rise to four classes of mRNAs: the 3.5 kb genomic RNA which is used both for translation of core and P proteins and as template for reverse transcription (therefore called C-mRNA/pregenome), the 2.1 kb preS2/S transcript used for translation of the small and middle surface proteins, the 2.4 kb preS1 transcript utilized for synthesis of the large surface protein, and an X transcript of about 0.9 kb length which so far could only be seen in certain experimental systems (for a review see Schaller and Fischer, 1991). In addition, 3.5 kb genomic transcripts were identified containing 28 - 35 extra nucleotides at their 5' ends (Yaginuma et al., 1987; Junker-Niepmann et al., 1990). These slightly longer transcripts are designated as precore mRNAs since they are used for the translation of the precore protein which is the precursor of the secreted HBeAg.

Apart from these transcripts, several spliced RNAs have been identified in transfected cells and in the livers of chronically infected patients (Chen et al., 1989; Suzuki et al., 1989; Su et al., 1989; Terré et al., 1991). In addition, low levels of a spliced RNA species have also been found in naturally infected Pekin ducks (Deindl, 1991). However, mutations of splice donor or acceptor sites or of the putative open reading frame did not affect viability of the virus neither in tissue culture nor in animal models (Wu et al., 1991; Deindl, 1991). Thus, the biological relevance of these spliced transcripts remains unclear.

HEPADNAVIRUS REPLICATION: AN OVERVIEW

Following uptake of virions, viral nucleocapsids are transported to the nucleus where their genomes are converted to covalently closed circular DNA (ccc DNA, Fig. 2). These molecules are used as template for synthesis of the viral transcripts by host RNA polymerase II. While subgenomic RNAs and the preC-mRNA are exclusively used for viral protein synthesis, genomic RNAs are bifunctional, serving for synthesis of the core and P protein as well as template for reverse transcription. Upon encapsidation of the RNA along with P protein(s) into the nucleocapsid, reverse transcription is initiated, most likely by a protein-priming mechanism, at the 3' copy of the direct repeat 1 sequence (DR1*; Fig. 3) allowing continuous synthesis of DNA minus strand (Seeger and Maragos, 1990; Seeger and Maragos, 1991). An RNaseH activity in the P protein degrades the RNA template from the resulting DNA:RNA hybrid probably concomitant with DNA synthesis leaving an RNA oligomer of about 18 nucleotides from the very 5' end of the pregenome including the cap structure (Lien et al., 1986; Will et al., 1986; Loeb et al., 1991). This oligoribonuleotide is then transferred by an unknown mechanism to the complementary DR2 sequence located close to the 3' end of the molecule where it is used to prime DNA plus strand synthesis. As the DNA minus strand has a terminal redundancy of about 8 nucleotides, this allows circularization and thereby continuation of plus strand synthesis on the circular DNA genome. For unknown reasons, synthesis of DNA plus strand ceases before completion and therefore a gap of variable length remains.

In a certain fraction of the molecules the RNA primer is not translocated but rather used to prime plus strand synthesis from the 3' end of minus strand DNA (Fig. 3, left panel; Staprans et al., 1991; Loeb et al., 1991). The resulting linear double strands are not infectious as the genomic RNA promotor can only function in the circular molecule.

Maturation of the nucleocapsids occurs by budding into the endoplasmatic reticulum (ER), where viral surface proteins are cotranslationally inserted (Fig. 2). The mechanism initiating this budding process is not known. One attractive possibility would be that synthesis of the viral DNA genome transduces a signal for core particle envelopement to the nucleocapsid surface (Summers and Mason, 1982). In favour of this idea, Yu and Summers (1991) could show that DHBV core gene mutants defective in synthesis of mature DNA genomes were also defective in assembly of enveloped virions. However, results from an analogous approach in HBV demonstrate that synthesis of complete DNA genomes and envelopement of nucleocapsids are not mutually dependent (M. Nassal, 1992). This result is supported by the finding that enveloped core particles are formed in transfection experiments with constructs lacking essential replication functions (genomic RNA or P protein; Birnbaum, 1992). Whether this reflects the different properties of the DHBV and HBV core proteins or whether this indicates different signals for envelopement of nucleocapsids in DHBV and HBV remains to be established.

CORE PARTICLE FORMATION: CAPSID COMPONENTS

Formation of replication competent nucleocapsids requires the correct assembly of at least three viral components: the core protein, the P protein and the genomic RNA template. In this section we will discuss the structure and function of each constituent and how they participate in the assembly process.

Core Antigens

The major core protein forming the viral nucleocapsid is encoded by the C open reading frame (ORF, Fig. 1B). This protein, serologically defined in case of HBV as HBcAg, has several properties characteristic for capsid proteins of other enveloped viruses. Most importantly, it can *self* assemble into nucleocapsids i.e. in the absence of other viral proteins. This is clearly shown by the detection of particulate core protein in *Escherichia coli* recombinants bearing the HBV core gene (Pasek et al., 1979; Stahl et al., 1982). However, core particle assembly appears to be very dependent on core protein concentration. Using microinjection of synthetic core mRNA into oocytes of *Xenopus laevis*, Zhou et al. (1992) could show that core particle formation only occurs when high concentrations of core protein

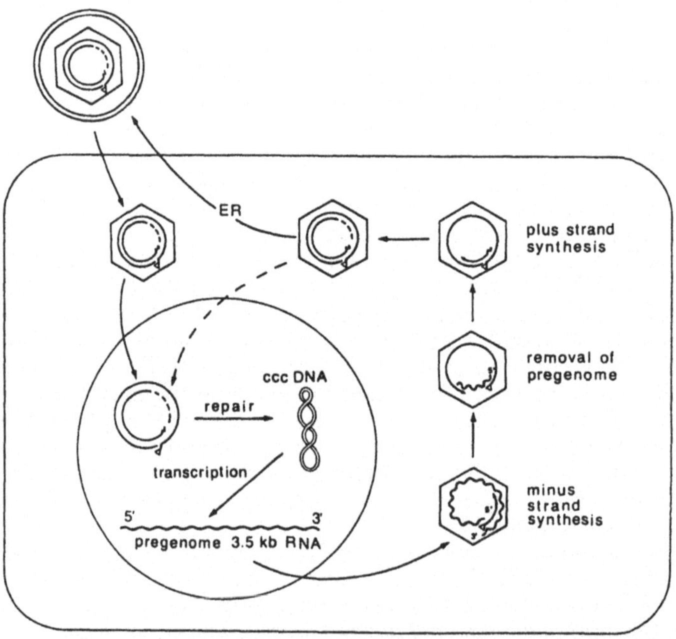

Figure 2. Schematic presentation of the hepadnaviral life cycle. ER, endoplasmatic reticulum. For details see text.

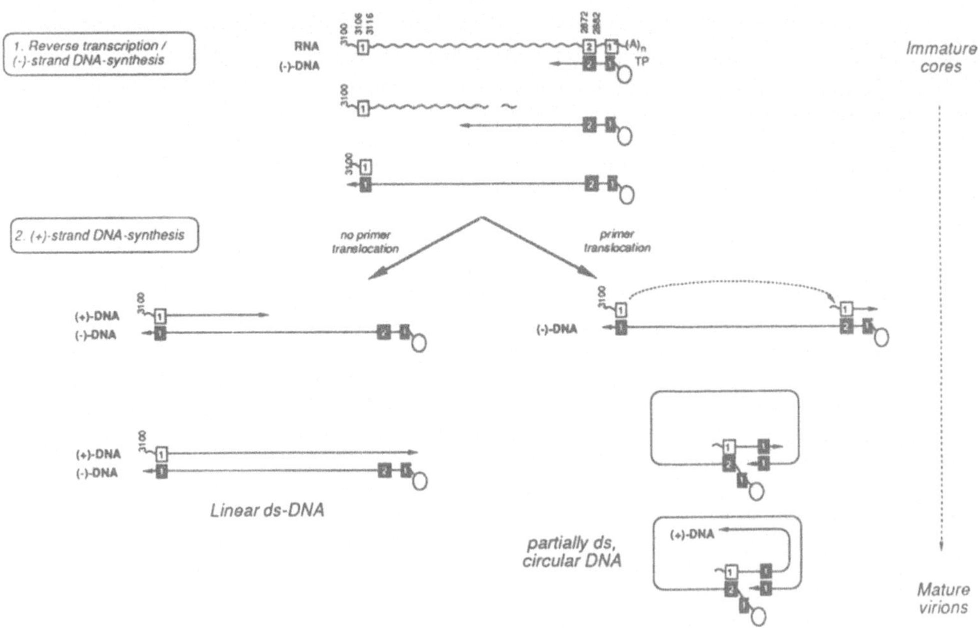

Figure 3. Replication of the hepadnaviral pregenome. Reverse transcription is initiated at the 3' copy of direct repeat 1 sequence (DR1*), most likely by the terminal protein (TP). Minus strand synthesis proceeds towards the 5' end of the pregenome concomitant with the degradation of the RNA by the RNaseH activity in the P protein. Productive replication requires transfer of the RNA plus strand primer to the DR2 sequence (right panel) and synthesis of DNA on the minus strand up to its 5' end. Circularization is enabled via the complementary sequence between the 3' ends of plus and minus strand DNA. In a certain fraction of minus strand DNA molecules the RNA primer is not translocated but instead extended from the 3' end (left panel).

are present; lowering the concentration of microinjected mRNA drastically reduced the conversion of free, non-aggregated core polypeptides into nucleocapsids. The implication of this property on core particle assembly will be discussed below.

Self assembly requires only the first 144 amino acids (aa) of the 183 aa core protein (Birnbaum and Nassal, 1990). Of the residual 39 aa, 16 arginines are occurring primarily in clusters of three or four. This so-called protamine-like region, which forms a structurally and functionally distinct domain (Birnbaum and Nassal, 1990), has the ability to interact with nucleic acids. Removal of this region from the core protein drastrically reduces its nucleic acid binding capacity *in vitro* (Petit and Pillot, 1985; Gallina et al., 1989) although a report to the contrary has appeared (Matsuda et al., 1988). Binding appears to be non-selective as heterologous nucleic acids are bound as well *in vitro*, although *Escherichia coli* produced core particles are enriched for HBV specific transcripts (Birnbaum and Nassal, 1990). However, as these RNAs lack a functional encapsidation signal (see below) this result may reflect a preferential encapsidation of the RNA which is spatially closest to the nascent polypeptide, i.e. the core gene mRNA.

The importance of the basic carboxyterminus of the core protein for encapsidation of the RNA pregenome could recently be demonstrated by a mutational analysis (M. Nassal, 1992). Transient expression of a series of HBV genomes directing the synthesis of carboxyterminally truncated core proteins revealed that a variant lacking the basic C-terminus was defective for genomic RNA encapsidation. Most interestingly however, a variant containing half of the protamine-like region (aa 1 - 164) was fully competent for RNA packaging but displayed an aberrant pattern of replicative DNA intermediates with drastically reduced amounts of plus strand DNA and no relaxed circular molecules (M. Nassal, 1992). These results strongly suggest that the basic carboxyterminus of the core protein has a dual function: to provide interactions between the core protein and the RNA pregenome for packaging and an auxilliary function during DNA synthesis (possibly like a histone or a single-stranded DNA binding protein).

Like many other nucleocapsid proteins, the HBcAg is a phosphoprotein. Incubation of partially purified core particles with [^{32}P]-γATP leads to further phosphorylation at serine residues (Albin and Robinson, 1980; Gerlich et al., 1982) located at the carboxyterminus (Yeh and Ou, 1981; Schlicht et al., 1989b). The relevance of phosphorylation for virus replication is unclear but the finding that serum-derived DHB virions were underphosphorylated relative to cytoplasmic cores isolated from infected cells suggests that core dephosphorylation is involved in triggering core particle envelopment and export (Pugh et al., 1989).

The responsible kinase is not intrinsic to the C gene product itself since procaryotically expressed nucleocapsids are not phosphorylated. It is also unlikely that the responsible kinase activity resides in the X gene product since avian hepadnaviruses (DHBV, HHBV) lack the X gene although DHBcAg is highly phosphorylated (Schlicht et al., 1989[b]).

Upstream of the 5′ end of ORF C is a short in phase open reading frame (Fig. 1B) which is used for the synthesis of a "pre-core" (preC) protein (Ou et al., 1986). Candidate RNAs (preC-mRNAs) encoding this protein are produced by the transcriptional machinery initiating transcription at different positions upstream of the preC AUG codon (the majority of transcripts initiates within the preC-ORF and therefore directs core gene translation). Expression from this initiation codon generates a signal sequence for protein translocation through the ER (Ou et al., 1986). Subsequently the protein enters the secretory pathway undergoing cleavage at the aminoterminus and at the carboxyterminus. The secreted protein lacks the protamine-like region and has 10 extra aa at the aminoterminus. It appears in the bloodstream of infected patients with an altered antigenicity termed HBeAg (Standring et al., 1988; Salfeld et al. 1989).

The biological role of this protein so far remained elusive. Based on the finding that the preC-region is highly conserved in all hepadnaviruses and that the presence of HBeAg in the blood correlates with elevated virus titres, it was assumed that the preC-region is essential for the formation of infectious virus. However, this hypothesis was ruled out by the results of a mutational analysis in the DHBV system showing that the preC-region is not essential for virus infectivity (Schlicht et al., 1987; Chang et al., 1987). In addition, preC-deficient mutants have subsequently been detected in the circulation of some HBV carriers (Brunetto et al., 1989; Okamoto et al., 1990). Although a recent report suggests that a function of the HBeAg could be to induce immunologic tolerance *in utero* (Milich et al., 1990), the relevance of these results for the *in vivo* situation remains to be examined.

In addition to circulating HBeAg, core gene products with HBeAg reactivity have been detected on the surface of cells infected with HBeAg-expressing vaccinia viruses (Schlicht et al., 1989[c]). This cell surface expressed "protein" is recognized by anti-HBe antibodies in patient sera (Schlicht et al., 1991) and appearance of such antibodies usually correlates with virus elimination. Thus, why preC-expression has been retained in viral evolution remains enigmatic.

The P protein

During the past few years many efforts have been undertaken to characterize the P protein, which is of central importance for the hepadnaviral replication cycle. However, biochemical studies have so far been hindered by its low abundance in infected cells and virus particles and by the failure to solubilize this enzyme from virions (Radziwill et al., 1988). Thus, analysis of P protein structure and function relied exclusively on indirect measurements, in particular the endogenous polymerase reaction which utilizes the ability of the nucleocapsid-associated P protein to incorporate radioactively labelled deoxyribonucleotides into the gapped genome (Fig. 1). Using this assay in combination with transient expression of mutated DHBV or HBV genomes it could be shown that the *pol* gene is translated from an internal initiation codon at the 5' end of the P ORF (Schlicht et al., 1989[a]; Chang et al. 1989). Thus, despite the retroviral genetic organization of overlapping core (*gag*) and P (*pol*) open reading frames (Fig. 1) the hepadnaviral P protein is not expressed as a *gag*/*pol* fusionprotein, created by ribosomal frameshifting or terminator suppression as in case of retroviruses (for review: Hatfield and Oroszlan, 1990), but instead as a separate entity.

This result raised the question how the freely diffusible P protein is targeted specifically to its template. An initial answer came from analyses designed to identify the origin of the terminal protein (TP) assumed to prime reverse transcription. Using a panel of antisera raised against different segments of the DHBV P gene product, specific precipitation of DNA minus strand from a preparation of denatured core particles could be demonstrated (Bosch et al., 1988; Bartenschlager et al., 1988). In addition, limited digestion of the replication complex with protease V8 prior to immunoprecipitation did not interfere with its reactivity with antisera raised against aminoterminal P protein fragments (Bartenschlager et al., 1988). This result suggested that the terminal protein is a P gene product encoded by the 5' region of the P gene and it implied a (high affinity) binding of the P protein to the 3' copy of DR1 on the RNA pregenome where it initiates reverse transcription (Fig. 3).

However, this idea soon had to be given up because of the findings that (1) the RNA encapsidation signal, which earmarks the pregenome for packaging into the core particle (see below) is sufficient to mediate RNA encapsidation (Junker-Niepmann et al., 1990) and (2) the P gene product is essential of the encapsidation for genomic RNA (Hirsch et al., 1990; Bartenschlager et al., 1990). This result suggested a direct or indirect interaction of the P protein with the encapsidation signal ahead of initiation of reverse transcription. Since the known enzymatic activities of the P protein are not essential for the packaging function the P protein appears to participate as a structural component in the encapsidation reaction (Hirsch et al., 1990; Bartenschlager et al., 1990).

A surprising property of the P protein is its *cis* preference, i.e., genomic RNAs from which P protein is translated are encapsidated preferentially (Hirsch et al., 1990; Bartenschlager et al., 1990). The mechanism underlying this finding is not understood. It could simply reflect very low P protein quantities favouring conditions under which P protein would interact with the target sequence in the inmediate vicinity. However, *cis* preference is not obligatory as it can be overcome by P-transcomplementing constructs lacking a functional ε sequence (Bartenschlager et al., 1990).

In principal each reverse transcriptase (RT) must contain three enzymatic activities: an RNA-dependent DNA polymerase, a DNA-dependent DNA polymerase and an RNaseH activity. The corresponding regions of the P protein, initially predicted on the basis of sequence comparisons between retroviral and hepadnaviral *pol* gene products, were recently identified by mutational analysis (Radziwill et al., 1990; Fig. 4). Missense mutations introduced into highly conserved sequence motifs known to be crucial for RT function and RNaseH function abolished the corresponding activities. A first clue about the structural organization of the P protein came from the observation that the terminal protein, located in the aminoterminus of the P protein and the RT/RNaseH activities, located more to the carboxyterminus are separated by a spacer region of about 100 aa which can be deleted without affecting the enzymatic activities (Radziwill et al., 1990; Bartenschlager et al., 1988). Interestingly, this region encodes in the +1 frame the aminoterminal region of the large surface protein which harbours the viral putative attachment site for the host cell (Neurath et al., 1986). Thus, by introducing a non-essential spacer element into this position of the P protein the overlapping preS1 region can undergo mutational changes apparently selected for by the host immune system without affecting P protein function.

Despite this progress in understanding P protein expression and function its physical nature could not be analyzed due to the lack of a sufficiently sensitive detection method. We have recently overcome this problem by introducing a target sequence for protein kinase A (PKA) into the P protein (Bartenschlager and Schaller, 1992[a]). By introducing the sequence RRXSX (Edelman et al., 1987) into the carboxyterminus of the P protein and expressing this protein with recombinant vaccinia viruses we could show that this method increases sensitivity over standard techniques by two orders of magnitude, and allows to detect amounts of P protein as low as 5 picogramm. The biological activity of the such modified P protein was subsequently demonstrated in transient expression of an HBV genome carrying the appropriate modification in the P gene in that core particles produced from this construct displayed somewhat reduced endogenous polymerase activity, but were otherwise indistinguishable from nucleocapsids produced from a wild type construct. After *in vitro* phosphorylation the nucleocapsid-associated P protein was then characterized to be a 90 kDa full length protein with a certain fraction to be linked at least initially to the viral DNA. Although several other reports described P proteins of about 65 - 75 kDa on the basis of their reactivity with P-specific antisera (Mack et al., 1988) or their RT/DNA polymerase reactivity (Bavand and Laub, 1989; Guilhot et al., 1992) or by transcomplementation of P null rmutants with constructs expressing the functional domains separately (Wu et al., 1991), we were not

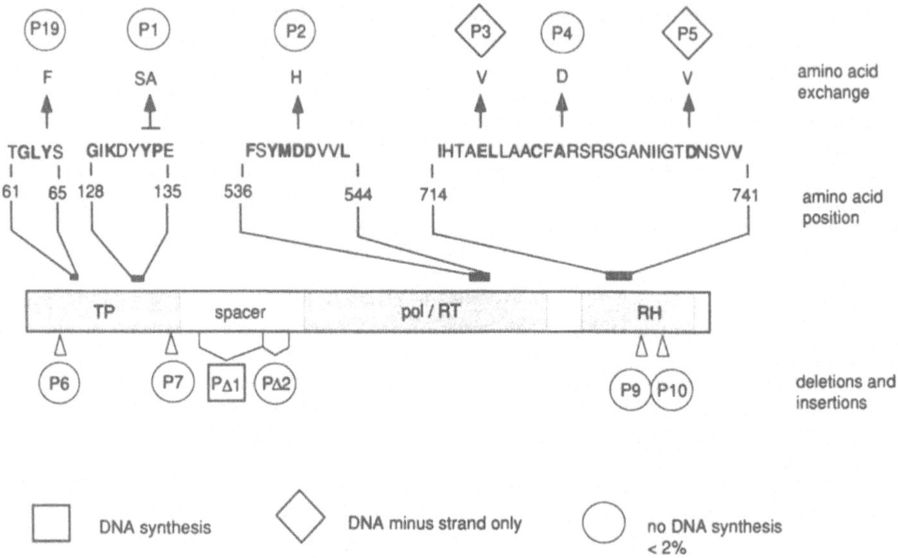

Figure 4. Domain structure of the hepadnaviral P protein. Mutations affecting amino acid motifs that were shown to be essential for retroviral *pol* protein function or highly conserved between different hepadnaviruses were analyzed for their effects on the endogenous polymerase reaction (Radziwill et al., 1990; Bartenschlager and Schaller, 1992[b]). The observed phenotypes are indicated by the different symbols. Note that deletion mutant PΔ1 is replication competent defining the deleted region as dispensable spacer. TP, terminal protein; pol/RT, DNA polymerase/reverse transcriptase; RH, RNaseH.

able to detect phosphorylatable P proteins smaller than 90 kDa. However, the identification of full length P gene products, linked as such to the viral DNA is in agreement with the results that antisera directed against aminoterminal as well as carboxyterminal P protein fragments efficiently precipitate DNA minus strand (see above) and it suggests that the hepadnaviral P protein, in contrast to the retroviral *pol* protein, can perform its functions as a multidomain polypeptide without requiring proteolytic processing. On the other hand as processed or modified P gene products might be labelled only weakly and are therefore not detected, further studies are required.

The RNA pregenome

Viral RNA isolated from the livers of infected animals contains three major transcripts about 3.5 kb, 2.4 kb and 2.1 kb in length (Fig. 1B). Of these, only the first has the physical structure required for reverse transcription. However, as this RNA has heterogenous 5' ends, the nature of the pregenome could not simply be inferred from 5' end mappings of cytoplasmic RNA. Initially, an answer came from the analysis of the mechanism of plus strand DNA synthesis. As described above, plus strand synthesis is primed by a short oligoribonucleotide derived from the 5' end of the RNA pregenome. Direct sequence determination of this RNA revealed that it corresponds to the 5' end of the major transcript initiated within the preC-region (Lien et al., 1986). More recently, these results could be confirmed with RNase protection analyses of transiently expressed HBV genomes which allowed the direct detection of the fraction of genome length transcripts packaged into nucleocapsids (Junker-Niepmann et al., 1990; Nassal et al., 1991). While preC-mRNAs as well as C-mRNAs were clearly present in the cytoplasm of transfected cells, exclusively C-mRNAs were detected within the nucleocapsids. The mechanism underlying this stringent discrimination will be discussed below.

In addition to its use as template for reverse transcription, the pregenome also serves as mRNA for translation of the core protein. Furthermore, since subgenomic or spliced RNAs that would allow the expression of a full length P protein are not detectable it was assumed that the pregenome also serves as mRNA for P gene translation. Substantial support for this idea has recently been provided by Huang and Summers (1991). Transfection of a chicken hepatoma cell line with DHBV RNA pregenomes synthesized *in vitro* allowed the production of complete infectious virus. Thus, the pregenome can express all the proteins required for DNA synthesis.

First evidence that the 5'end of genomic RNA contributes to the signal required for its selective encapsidation were obtained from the GSHV system (Enders et al., 1987) demonstrating that only the shortest of the ca. 3.5 kb RNAs was packaged into the nucleocapsid. To define more precisely the *cis* acting element that earmarks the pregenome for encapsidation, Junker-Niepmann et al. (1991) analyzed a series of deletion mutants for their effects on pregenome packaging. They thus identified a 137 nucleotides long region near the 5' end of the RNA pregenome as being necessary for packaging of the molecule into the core particle. Furthermore, by fusing the segment to the 5' end of a heterologous RNA they could show that the encapsidation signal (termed ε) is sufficient to mediate packaging of a non-viral RNA in *cis*. As expected, the ε-sequence contains a number of inverted repeat sequences with the potential to form stem-loop structures. In Fig. 5 one such stem-loop structure is depicted. It is conserved in all hepadnaviruses and despite significant sequence divergence between mammalian and avian hepadnaviruses some features are unchanged: the number and lenghts of the stem-loop structures, a CUGU-motif in the central loop and a pyrimidine rich bulge loop. This structure prediction is supported by data from RNAse digestion of *in vitro* transcribed ε-RNAs (T. Knaus and M. Nassal, unpublished observations).

While the position of the ε-sequence within the pregenome explains the discrimination against packaging of cellular RNAs, it does not so for subgenomic RNAs. As the encapsidation signal is located within the terminal redundancy it occurs twice in the RNA (cf. Fig. 1B). This implies that regulatory mechanisms exist to inactivate the function of the 3' copy of ε.

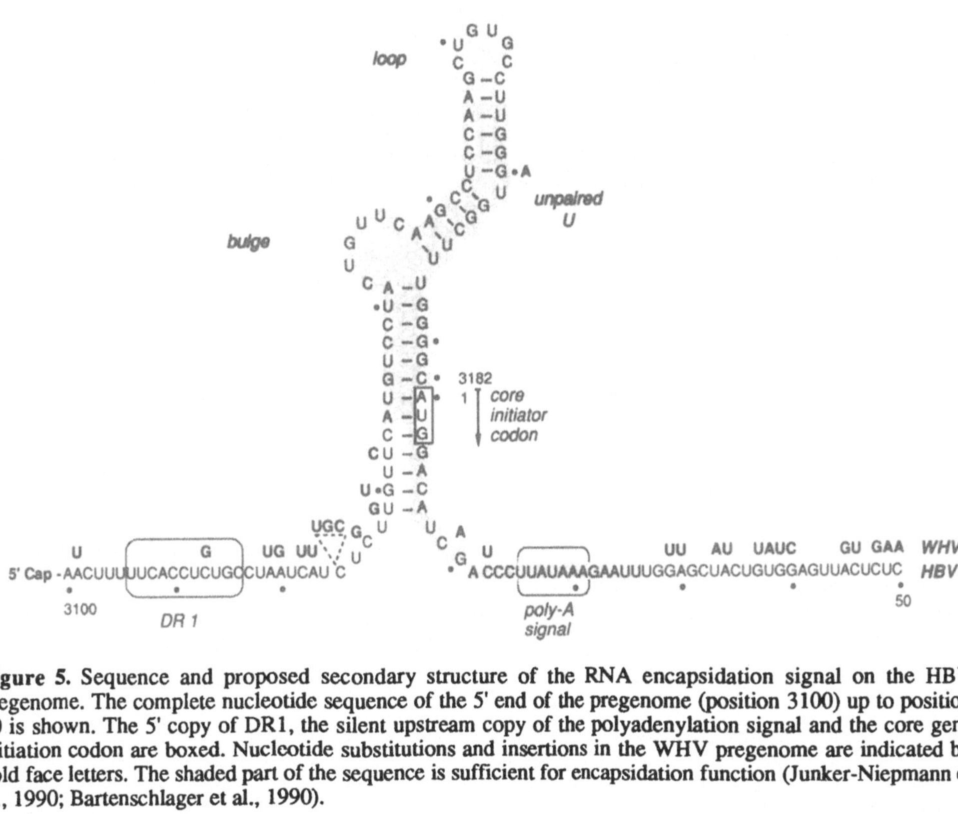

Figure 5. Sequence and proposed secondary structure of the RNA encapsidation signal on the HBV pregenome. The complete nucleotide sequence of the 5' end of the pregenome (position 3100) up to position 50 is shown. The 5' copy of DR1, the silent upstream copy of the polyadenylation signal and the core gene initiation codon are boxed. Nucleotide substitutions and insertions in the WHV pregenome are indicated by bold face letters. The shaded part of the sequence is sufficient for encapsidation function (Junker-Niepmann et al., 1990; Bartenschlager et al., 1990).

Another intriguing question that remained was why preC RNAs, which differ from the RNA pregenome only by the presence of about 30 extra nucleotides at the 5' end, are not encapsidated. A first hint to answer this question came from the observation that deletion of the precore start codon releases the block for preC-mRNA packaging (Junker-Niepmann et al., 1990). This result suggested that translation through the preC region interferes with pregenome packaging. A final prove of this assumption was recently provided by Nassal et al. (1991). Mutational ablation of preC-translation restored preC-mRNA packaging irrespective of the translated reading frame (Fig. 6). However, a stop codon close to the beginning of the preC region (mutant TAA-4 in Fig. 6) did not impair preC-mRNA exclusion, demonstrating that preC-mRNAs are encapsidated in the absence of *extended* preC translation.

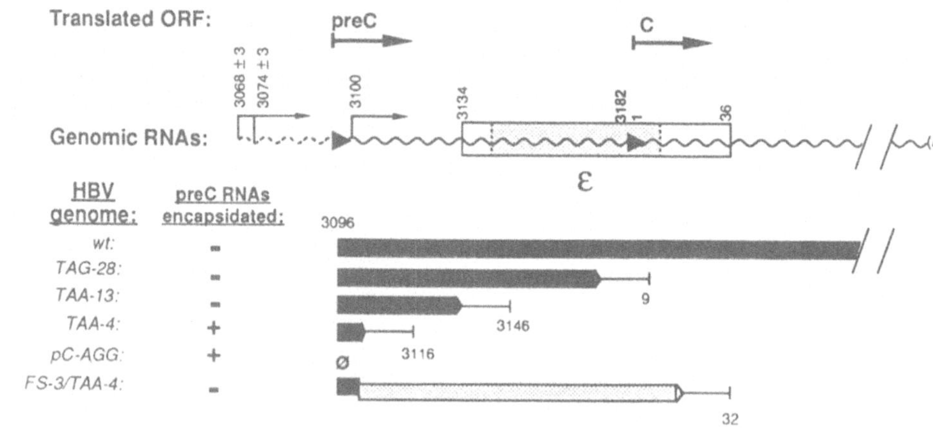

Figure 6. Summary presentation of the mutational analysis showing that translation through the preC region inactivates the encapsidation signal (Nassal et al., 1990). The preC and C-initiation codons are shown as black arrowheads, the initiation sites of the preC RNAs (dashed wavy line) and the RNA pregenome (position 3100) by thin arrows. The functional encapsidation signal is indicated by the box; its precise 5' and 3' ends are located between the outer limits and the shadded area. On the RNA pregenome translating 80S ribosomes assemble only on the C gene initiation codon within the ε-sequence whereas in case of the preC RNAs 29 aa of the preC ORF precede the C gene start codon. Mutations affecting translation of the preC ORF (TAG-28, TAA-13, TAA-4), exchanging the preC AUG codon to AGG or introducing a frame-shift after amino acid position 2 were analyzed for their effects on preC RNA packaging into nucleocapsids. PreC RNAs are encapsidated when translation of the preC ORF is aborted at all (pC-AGG) or very early (TAA-4); they are excluded when translation proceeds into or through the ε-sequence irrespective of the translated reading frame (FS-3/TAA-4). The thin bars indicate the physical extension of the ribosome that halts on the introduced stop codons.

To understand the model derived from these results we should point out that in the case of the mammalian hepadnaviruses the core gene AUG is an integral part of the ε-sequence located within the basal stem structure (Fig. 5). Obviously 80S ribosomes engaged in preC translation on the preC-mRNA compete much more effectively with RNA packaging than 40S ribosomal subunits scanning the pregenome for the C gene AUG. However, as core protein is efficiently translated from the pregenome it is not the RNA structure of the ε-sequence as such that blocks the scanning 40S subunits. Instead, proteins binding to the ε-sequence [P protein(s), perhaps in conjunction with core protein(s)] appear to interfere with scanning 40S subunits. In contrast, translating 80S ribosomes would prevent binding of the viral proteins to the ε-sequence or displace bound proteins and in this way inhibit preC-mRNA encapsidation.

Why did hepadnaviruses develop a mechanism to exclude these RNAs from packaging? As described above, synthesis of DNA plus strand requires transfer of the RNA primer from the DR1 sequence at the 3'end of DNA minus strand to the DR2 sequence (Fig. 3). Furthermore, this oligoribonucleotide is generated sequence independent by the RNaseH activity degrading the RNA up to the very last 18, or so, nucleotides at the 5' end of the pregenome (Loeb et al., 1991). Thus, plus strand primers derived from pregenomic RNAs which contain extra sequences at their 5' end could no longer basepair with DR2 due to the lack of complementary sequences. Instead, the non-translocated primer would initiate plus strand synthesis from the 3' end of DNA minus strand and thereby create linear, non-infectious genomes (Staprans et al., 1991). Thus, hepadnaviruses utilize translational inactivation of preC-mRNA packaging function to discriminate against encapsidation of a subset of RNAs that would lead to non-infectious genomes.

NUCLEOCAPSID ASSEMBLY

Segregation of reverse transcription into a subviral particle is a general feature shared by all retroelements. Formation of this structure requires the P protein to be packaged along with the RNA pregenome into the nucleocapsid shell. Recent genetic and biochemical studies have shown that RNA encapsidation in retroviruses is principally mediated by the *gag* protein (Gorelick et al., 1988; Meric and Goff, 1989; Luban and Goff, 1991) and that products of the *pol* gene are not required for this process (Shields et al., 1978; Oertle and Spahr, 1990). In addition, as the retroviral *pol* protein is expressed as a *gag/pol* fusion protein its packaging is mediated by coassembly of this fusion protein along with *gag* protein molecules via the *gag* moiety and is independent from genomic RNA packaging (Levin et al., 1974). In contrast, the hepadnaviral P protein is expressed unlinked to the core protein raising the question for its encapsidation mechanism. As described above, the finding that the P gene product is essential for packaging of RNAs carrying a functional ε-sequence suggested a direct or indirect binding of the P protein to this sequence to ensure encapsidation of the RNA and possibly also of the P protein. In an alternative scenario P protein(s) could interact with core protein subunit(s) prior to or during nuleocapsid assembly forming a protein complex that is encapsidated in a way reminiscent to the retroviral situation i.e. independent of the viral RNA. A first clue came from transfection experiments utilizing constructs that express phosphorylatable P proteins. Density gradient purification of core particles produced from these constructs revealed that only a certain fraction of the capsids contained viral DNA and that the P protein was detectable only in these nucleocapsids whereas core particles lacking HBV-specific DNA were devoid of P protein (Bartenschlager and Schaller, 1992[b]). As the RNA pregenome is the precursor of viral DNA this result suggested that genomic RNA is essential for P protein packaging and argued against a mechanism depending exclusively on P protein/core protein interactions. Further support came from the observation that a missense mutation in the P gene inactivating RNA packaging function of the P protein also impaired P protein encapsidation. Most importantly however, mutational inactivation of the ε-sequence abolished P protein packaging but could be restored by transcomplementation with RNA molecules carrying a functional encapsidation signal (Bartenschlager and Schaller, 1992[b]). These results strongly suggest that formation of replication competent nucleocapsids is initiated by the direct binding of P protein(s) (in association with core protein molecule(s) ?) to the encapsidation signal on the RNA pregenome (Fig. 7). Additional support for this assumption was recently obtained by Köchel et al. (1991) who identified the ε-sequence in an *in vitro* transcribed HBV-RNA as a specific binding site for a procaryotically expressed P-fusionprotein.

Formation of such an assembly complex would have a dual function. First, binding of core protein is likely to stabilize the RNA/P protein complex; rapid addition of core polypeptides to the assembly complex would explain that the P protein acts primarily in *cis*, i.e. packaging competent RNAs expressing the P protein are encapsidated preferentially, even in case of a low affinity binding of the P protein to the ε-sequence. Second, it could trigger the addition of further core protein molecules. As described above conversion of free core protein to core particles critically depends on a high concentration of core polypeptides (Zhou et al., 1992). The trigger function of the assembly complex would be to lower this concentration and thereby favour formation of complete replication-competent nucleocapsids and prevent assembly of "empty" core particles. Addition of further core protein molecules to the assembly complex then would be driven by the contacts arising between individual core protein subunits and by their nucleic acid binding capacity until the complete capsid structure is formed.

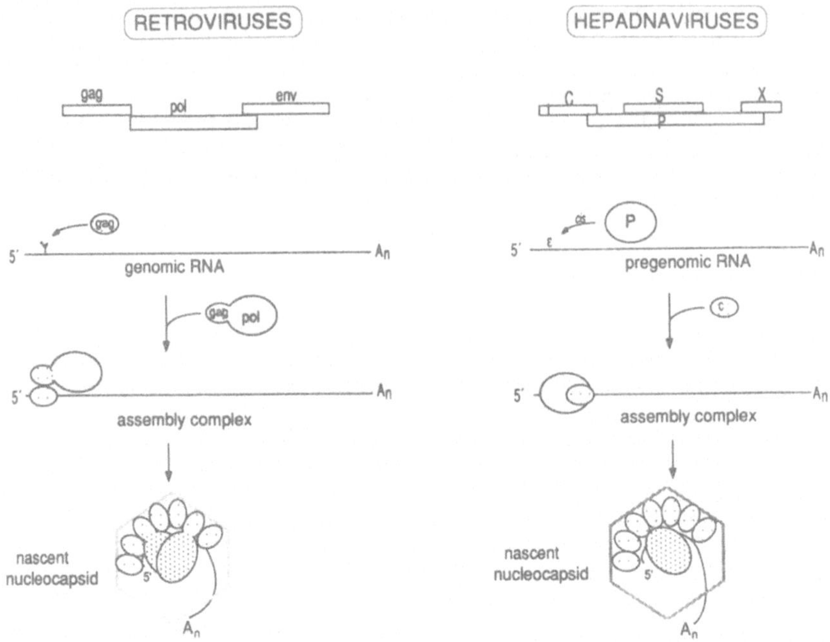

Figure 7. Schematic comparison of the early steps of retroviral and hepadnaviral nucleocapsid assembly. For details see text.

As known so far, this mechanism of core particle formation is without precedence among retroviruses. Here, packaging of the *pol* protein is accomplished by synthesis of *gag/pol* fusionproteins and their coassembly via the *gag* moiety (Fig. 7, left panel) explaining why packaging of the *pol* protein can occur in the absence of genomic RNA (Levin et al., 1974). Thus, retroviruses contain about 20 - 70 *pol* protein molecules per particle (Panet et al., 1975; Krakower et al., 1977) whereas in the case of hepadnaviruses this number appears to be much lower. Using the newly developed *in vitro* phosphorylation technique of the HBV-P protein we calculated that there are probably only one or two P protein molecules present per particle (Bartenschlager and Schaller, 1992b). Although this calculation bears some uncertainties, it is in agreement with the assumed low copy number of this molecule and with the assembly model in Fig. 7, implicating that direct binding of the P protein, as monomer or dimer, to the ε-sequence regulates the number of encapsidated molecules.

Another interesting feature of this assembly scheme concerns the utilization of the RNA pregenome during the replication cycle. Early in replication, this molecule is used for translation of core proteins and, to a much lesser extend, of the P protein. However, as the core gene initiation codon is located within the ε-sequence (Fig. 5), P proteins binding at this site should repress core gene translation. Indeed, preliminary data based on cotransfection experiments of a P null mutant with increasing amounts of a plasmid directing the expression of an altered P protein suggest that the P protein acts as a translational repressor of the core gene (R. Bartenschlager, unpublished) and withdraws the pregenome from the pool of translated RNAs. In addition, as the P gene is translated from the same RNA, initiation of packaging by P protein binding to the ε-sequence autoregulates P gene translation. Thus, the critical parameter that switches the function of genomic RNA from CP-mRNA to pregenome is the concentration of P protein(s) and a critical concentration of core protein which is not sufficient for assembly without the triggering function of the assembly complex.

PERSPECTIVES

During the past few years hepatitis B virus research has greatly profited from the development of tissue culture systems supporting the production of virus particles from transfected genomes. Several aspects of replication, viral gene expression and virion morphogenesis have been clarified. However, many open questions remain, most of which can only be answered in the appropriate cell free system. For instance, which regions of the P protein interact with which sites in the pregenome during the encapsidation reaction? Is it P protein alone or a core/P complex? What is the mechanism by which reverse transcription is initiated? Is it really primed by the terminal protein or is this protein linked to nascent minus strand chains early after initiation as speculated to be the case for poliovirus VPg (Tobin et al., 1989) ? How is initiation of reverse transcription at the DR1 sequence near the 3' end of the pregenome accomplished when initial binding of the P protein occurs at the ε-sequence near the 5' end of the RNA? Equally important are questions concerning envelopement of core particles. Among these are questions dealing with the mechanism by which cores trigger budding or the nature of the interactions between the core protein and the surface proteins.

A more detailed understanding of these and other problems might define targets for an antiviral therapy which so far is missing. Although the general replication scheme of the hepatitis B viruses follows that of retroelemts, the last few years have shown that several unique properties exist which could be used to develop drugs that specifically interfere with virus replication. With about 300 millions of chronically infected patients the need for a targeted therapy is obvious.

ACKNOWLEGEMENTS

We are grateful to Dr. Christa Kuhn for the preparation of the figures and Karin Coutinho for preparation of the manuscript. Work in the authors' laboratory was supported by grants from the Bundesministerium für Forschung und Technologie (BCT 318-5) and the Deutsche Forschungsgemeinschaft (SFB 229).

REFERENCES

Albin, C., and Robinson, W.S., 1980, Protein kinase activity in hepatitis B virus. *J. Virol.* 34:297.

Bartenschlager,R. and Schaller,H., 1988, The amino terminal domain of the hepadnaviral P-gene encodes the terminal protein (genome-linked protein) believed to prime reverse transcription. *EMBO J.,* 7:4185.

Bartenschlager,R., Junker-Niepmann,M. and Schaller,H., 1990, The P gene product of hepatitis B virus is required as a structural component for genomic RNA encapsidation. *J. Virol.,* 64:5324.

Bartenschlager,R., Kuhn,C. and Schaller,H., 1992[a], Expression of the P-protein of the human heoatitis B virus in a vaccinia virus system and detection of the nucleocapsid-associated P-gene product by radiolabelling at newly introduced phosphorylation sites. *Nucl.Acids Res.*, 20:195 .

Bartenschlager R. and Schaller, H., 1992[b], Hepadnaviral assembly is initiated by polymerase binding to the encapsidation signal in the viral RNA genome. *EMBO J.* (in press)

Bavand, M., Feitelson, M., and Laub, O., 1989, The hepatitis B virus recerse transcriptase is encoded by the viral pol gene. *J. Virol.* 63:1019.

Birnbaum, F., 1992, Hepatitis B virus-Reifung: Untersuchungen zum Umhüllungsprozeß des viralen Kapsids. Thesis, University of Heidelberg.

Birnbaum, F. and Nassal, M., 1990, Hepatitis B viurus nucleocapsid assembly: primary structure requirements in the core protein.. *J. Virol.* 64:3319.

Blum, H., Zhen-Sheng, Z., Galun, E., von Weizsäcker, F., Garner, B., Liang, T.J., & Wands, J.R., 1992, Hepatitis B virus X protein is not central to the viral life cycle *in vitro*. *J. Virol.* 66:1223.

Bosch,V., Bartenschlager,R., Radziwill,G. and Schaller,H., 1988, The duck hepatitis B virus P gene codes for protein strongly associated with the 5'-end of the viral minus DNA strand.*Virology,* 166:475.

Brunetto, M.R., Stemler, M., Schödel, F., Will, H., Ottobrelli, M., Rizzetto, M., Verme, G., and Bonino, F., 1989, Identification of HBV variants which cannot produce precore derived HBeAg and may be responsible for severe hepatitis. *Ital. J. Gastroenterol.* 21:151.

Chang, C., K. Jeng, C. Hu, S. J. Lo, T. Su, L.-P. Ting, C.-K. Chou, S. Han, E. Pfaff, J. Salfeld, and Schaller, H., 1987, Production of hepatitis B virus *in vitro* by transient expression of cloned HBV DNA in a hepatoma cell line. *EMBO J.* 6:675.

Chang, L.-J., Pryciak, P., Ganem, D., & Varmus, H., 1989, Biosynthesis of reverse transcriptase of hepatitis B viruses involves *de novo* translational initiation not ribosomal frameshifting. *Nature* (London) 337:364.

Chen, P.J., Chen, C.R., Sung, J.L., and Chen, D.S., 1989, Identification of a doubly spliced viral transcript joining the separated domains for putative protease and reverse transcriptase of hepatitis B virus. *J. Virol.* 64:4165.

Dane, D.S., Cameron, D.H., and Briggs, M., 1970, Virus-like particles in serum of patients with Australia antigen-associated hepatitis. *Lancet* 1: 695.

Deindl, E., 1991, Nachweis und Charakterisierung eines gespleißten Transkripts bei Enten Hepatitis B Viren. Diplomarbeit, Unviersity of Heidelberg.

Edelman, A. M., Blumenthal, D., & Krebs, E., 1987, Protein serine/threonine kinases. *Annu. Rev. Biochem.* 56:567.

Enders, G.H., Ganem, D., and Varmus, H.E., 1987, 5'-Terminal sequences influence the segregation of ground squirrel heaptitis virus RNAs into polyribosomes and viral core particles. *J. Virol.* 61:35.

Gallina, A., Bonelli, F., Zentilin, L., Rindi, G., Muttini, M., and Milanesi, G., 1989, A recombinant hepatitis B core antigen polypeptide with the protamine-like domain deleted self-assembles into capsid particles but fails to bind nucleic acid. *J. Virol.* 63: 4645.

Gerlich, W., Goldmann, U., Müller, R., Stibbe, W., and Wolff, W., 1982, Specificity and localization of the hepatitis B virus associated protein kinase. *J. Virol.* 42:761.

Gorelick,R., Henderson,L., Hanser,J., and Reins,A., 1988, Point mutant of Moloney murine leukemia virus that fail to package viral RNA: evidence for specific RNA recognition by "zinc finger like" protein sequence. *Proc. Natl. Acad. Sci.* USA, 85:8420.

Guilhot, S., Fowler, P., Portillo, G., Margolskee, R.F., Ferrari, C., Bertoletti, A., and Chsiari, F.F., 1992, Hepatitis B virus (HBV)-specific cytotoxic T-Cell response in humans: production of target cells by stable expression of HBV-encoded proteins in immortalized human B-cell lines. *J. Virol.* 66: 2670.

Hatfield, D. and Oroszlan, S., 1990, The where, what and how of ribosomal frameshifting in retroviral protein synthesis. *TIBS* 15:186.

Hirsch,R., Lavine,J., Chang,L.-J., Varmus,H.E., and Ganem,D., 1990, Polymerase-gene products of hepatitis B virus are required for genomic RNA packaging as well as for reverse transcription. *Nature (London)*,344:522.

Huang, M.J., and Summers, J., 1991, Infection initiated by the RNA pregenome of a DNA virus. *J. Virol.* 65:5435.

Junker-Niepmann, M., Bartenschlager, R., and Schaller, H., 1990, A short cis-acting sequence is required for hepatitis B virus pregenome encapsidation and sufficient for packaging of foreign RNA. *EMBO J.* 9:3389.

Kaplan, P., Greenman, R., Gerin, J., Purcell, R., & Robinson, W., 1973, DNA polymerase associated with human hepatitisd B virus antigen. *J. Virol.* 12:995.

Klingmüller, U. and Schaller, H., **1992,** The pre-S domain of avian hepadna virus is the major attachment site for binding and infection. *J. Virol.* submitted.

Krakower, J., Barbacid, M., & Aaronson, S., 1977, Radioimmunassay for mammalian type C viral reverse transcriptase. *J.Virol.* 22:331.

Levin, J., Grimley, P., Ramseur, J., & Berezesky, I., 1974, Deficiency of 60 to 70S RNA in murine leukemia virus particles assembled in cells treated with actinomycin D. *J. Virol.* 14:152.

Lien, J., Aldrich, C.E., and Mason, W.S., 1986, Evidence that a capped oligoribonucleotide is the primer for duck hepatitis B virus plus strand DNA synthesis. *J. Virol.* 57:229.

Loeb, D.D., Hirsch, R.C., and Ganem, D., 1991, Sequence-independent RNA cleavages generate the primers for plus-strand DNA synthesis in hepatitis B viruses: implications for other reverse transcribing elements. *EMBO J.* 10:3533.

Luban, J., and Goff, S.P., 1991, Binding of human immunodeficiency virus type 1 (HIV-1) RNA to recombinant HIV-1 gag polyprotein. *J. Virol.* 65:3203.

Mack,D.H., Bloch,W., Nath,N. and Sninsky,J., 1988, Hepatitis B virus particles contain a polypeptide encoded by the largest open reading frame: a putative reverse transcriptase. *J. Virol.*, 62: 4786.

Marion, P. L., Oshiro, L. S., Regnery, D.C., Scullard, G.H., and Robinson, W.S., 1980, A virus in Beechey ground squirrels that is related to hepatitis B virus of humans, *Proc. Natl. Acad. Sci. U.S,A.*, 77: 2941.

Mason W. S., Seal, G., and Summers, J., 1980, Virus of Pekin ducks with structural and biological relatedness to human hepatitis B virus, *J. Virol.* 36: 829.

Matsuda, K., Satoh, S., and Ohori, H., 1988, DNA-binding activity of hepatitis B e antigen polypeptide lacking the protaminelike sequence of nucleocapsid protein of human hepatitis B virus. *J. Virol.* 62: 3517.

Meric, C. and Goff,S., 1989, Characterization of Moloney Murine Leukemia virus mutants with single-amino-acid substitutions in the cis-his box of the nucleocapsid protein. *J.Virol.*, 63:1558.

Milich, D.R., Jones, J.E., Hughes, J.L., Price, J., Raney, A.K., and McLachlan, A., 1990, Is a function of the secreted hepatitis B virus e antigen to induce immunologic tolerance in utero? *Proc. Natl. Acad. Sci. USA* 87:6599.

Nassal, M., 1992, Conserved cysteines of the Hepatitis B virus core protein are not required for assembly of replication-competing core particles nor for their envelopment. *Virology* (in press).

Neurath, A. R., Kent, S. B. H., Strick, N., and Parker, K., 1986, Identification and chemical syntesis of a host cell receptor binding site on hepatitis B virus. *Cell* 46:429.

Oertle, S. and Spahr, P.-F., 1990, Role of gag polyprotein precursor in packaging and maturation of Rous sarcoma virus genomic RNA. *J.Virol.*, 64:5757.

Okamoto, H., Yotsumoto, S., Akahane, Y., Yamanaka, T., Miyazaki, Y., Sugai, Y., Tsuda, F., Tanaka, T., Miyakawa, Y., and Myaumi, M., 1990, Hepatitis B viruses with precore region defects prevail in persistently infected hosts along with seroconversion to the antibody against e antigen. *J. Virol.* 64:1298.

Ou, J.-H., Laub, O., and Rutter, W., 1986, Hepatitis B virus gene function: the region targets the core antigen to cellular membranes and causes the secretion of the e-antigen. *Proc. Natl. Acad. Scie. U.S.A.*, 83:1578.

Panet, A., Baltimore, D., & Hanafusa, H., 1975, Quantitation of avian RNA tumor virus reverse transcriptase by radioimmunoassay. *J.Virol.* 16:146.

Pasek,M., Goto,T., Gilbert,W., Zink,B., Schaller,H., MacKay,P., Leadbetter,G. and Murray,K., 1979, Hepatitis B virus genes and their expression in *E. coli. Nature (London)* 282:575.

Petit, M.-A., and Pillot, J., 1985, HBc and HBe antigenicity and DNA-binding activity of major core protein P22 in hepatitis B virus core particles isolated from the cytoplasm of human liver cells. *J. Virol.*, 53:543.

Pugh, J. C. and Summers J.W., 1989, Infection and uptake of duck hepatitis B virus by duck hepatocytes maintained in the presence of dimethyl sulfoxide. *Virology* , 172:564.

Radziwill, G., Zentgraf, H., Schaller, H. and Bosch, V., 1988, The duck hepatitis B virus DNA polymerase is tightly associated with the viral core structure and unable to switch to an exogenous template. *Virology,* 163:123.

Radziwill, G., Tucker, W. , and Schaller, H., 1990, Mutational analysis of the hepatitis B virus P gene product: domain structure and RNase H activity. *J. Virol* 64:613.

Salfeld, J., Pfaff, E., Noah, M., and Schaller, H., Antigenic determinants and functional domains in core and e-antigen from hepatitis B virus. *J. Virol.* 63: 798.

Schaller H., and Fischer, M., 1991, Transcription control of hepadnaviral gene expression. in: *Current Topics in Microbiology and Immunology,* Vol. 168. Springer Verlag, Berlin - Heidelberg, pp. 21.

Schek, N., Fischer, M., and Schaller, JH. 1991. The hepadnaviral X protein. in: *Molecular Biology of the Hepatitis B virus* (McLachlan, A. ed.) CRC Press, Inc., Boca Raton, Fl., USA, pp.181.

Schlicht, H.J., Salfeld, J., and Schaller, H., 1987, The duck hepatitis B virus pre-C region encodes a signal sequence which is for synthesis and secretion of processed core proteins but not for virus formation. *J. Virol.* , 61:3701.

Schlicht,H.-J., Radziwill, G., and Schaller,H., 1989[a], Synthesis and encapsidation of duck hepatitis B virus reverse transcriptase do not require formation of core-polymerase fusion proteins.*Cell*, 56:85.

Schlicht,H.-J., Bartenschlager,R., and Schaller,H., 1989[b], The duck hepatitis B virus core protein contains a highly phosphorylated C-terminus essential for replication but not for RNA packaging. *J. Virol.,* 63:2995.

Schlicht,H.-J. and Schaller,H., 1989c, The secretory core protein of human hepatitis B virus is expressed on the cell surface. *J. Virol.,* 63:5399.

Schlicht,H.-J., von Brunn, A., and Theilmann, L. 1991, Antibodies in anti-HBe positive patient sera bind to a HBe protein expressed on the cell surface of human hepatoma cells: implications for virus clearance. *Hepatology,* 13: 57

Seeger, C. and Maragos, J., 1990, Identification and characterization of the woodchuck hepatitis virus origin of DNA replication. *J. Virol.* 64:16.

Seeger, C. and Maragos, J., 1991, Identification of the signal necessary for initiation of reverse transcription of the hepdnavirus genome. *J. Virol.* 65:5190

Shields,A., Witte,O.,Rothenberg,E. and Baltimore,D., 1978, High frequency of aberrant expression of Moloney murine leukemia virus in clonal infections. *Cell,* 14:601.

Standring, D.N., Ou, J.-H., Masiarz, F.R., and Rutter W.J., 1988, A signal peptide encoded within the precore region of hepatitis B virus directs the secretion of a heterogeneous population of e antigens in Xenopus oocytes. *Proc. Natl. Acad. Sci. USA* 85:8405.

Sprengel, R., Kaleta, E.F., and Will, H., 1988, Isolation and characterization of a hepatitis B virus endemic in he4rons. *J. Virol.* 62: 3832.

Sprengel, R., Kuhn, C., Manso, C., and Will, H., 1984, Cloned duck hepatitis B virus DNA is infectious in Pekin ducks. *J. Virol.* 52:932.

Stahl, S., Mackay, P., Magazin, M., Bruce, S.A., and Murray, K., 1982, Hepatitis B virus core antigen: synthesis in *E. coli* and application in diagnosis, *Proc. Natl. Acad. Sci. U.S.A.,* 79:1606.

Staprans, S., Loeb, D.D., and Ganem. D., 1991, Mutations affecting hepadnavirus plus-strand DNA synthesis dissociate primer cleavage from translocation and reveal the origin of linear viral DNA. *J. Virol.* 65:1255.

Su, T.-S., Lai, C.J., Huang, J.-L., Lin, L.-H., Yauk, Y.-K., Chang, C., Lo, S.J., and Han, S.H., 1989, Hepatitis joining the separated domains for putative protease and reverse transcriptase of hepatitis B virus. *J. Virol.* 63:4011.

Summers, J. and Mason,W., 1982, Replication of the genome of a hepatitis B-like virus by reverse transcription of an RNA intermediate. *Cell,* 29:403.

Summers, J., Smith, P. M. , Huang, M. , and Yu, M., 1991, Morphogenetic and regulatory effects of mutations in the envelope proteins of an avian hepadnavirus. *J. Virol.* 65:1310.

Suzuki, T., Masui, N., Kajino, K., Saito, I., and Miyamura, T., 1989, Detection and mapping of spliced RNA from transfected hepatitis B virus genome in a human hepatoma cell line, *Proc. Natl. Acas. Sci. U.S.A.,* 86:8422.

Terré, S., Petit, M.-A., and Brechot, C., 1991, Defective hepatitis B virus particles are generated by packaging and reverse transcription of spliced viral RNAs *in vivo. J. Virol.* 65:5539.

Tobin, G. J., Young, D.C., and Flanegan, J.B., 1989, Self-catalyzed linkage of polivirus terminal protein VPg to poliovirus RNA. *Cell* 59:511.

Toh, H., Hayashida, H., and Miyata, T., 1983, Sequence homology between retroviral reverse transcriptase and putative polymerases of hepatitis-B virus and cauliflower mosaic virus. *Nature (London)* 305:827.

Will,H., Salfeld,J., Pfaff,E., Manso,C., Theilmann,L. and Schaller,H., 1986, Putative reverse transcriptase intermediates of human hepatitis B virus in primary liver carcinomas. *Science*, 231:594 .

Wu, H.-L., Chen, P.-J., Tu, S.-J., Lin, M.-H., Lai, M.-Y., and Chen, D.-S., 1991, Characterization and genetic analysis of alternatively spliced transcripts of hepatitis B virus in infected human liver tissues and transfected HepG2 cells. *J. Virol.* 65:1680 .

Wu T.-T., Condreay, L.D., Coates, L., Aldrich, C., and Mason, W., 1991, Evidence that less-than-full-length *pol* gene products are functional in hepadnavirus DNA synthesis, *J. Virol.* 65:2155.

Yaginuma, K., Shirakata, Y., Kobayashi, M., and Koike, K., 1987, Hepatitis B virus (HBV) particles are produced in a cell culture system by transient expression of transfected HBV DNA. *Proc. Natl. Acad. Sci. USA* 84:2678.

Yeh, C.-t., and Ou, J.-h., 1991, Phosphorylation of hepatitis B virus precore and core proteins. *J. Virol.* 65:2327.

Yu, M. and Summers, J., 1991, A domain of the hepdanvirus capsid protein is specifically required for DNA maturation and virus assembly. *J. Virol.* 65:2511.

Zhou, S., Yang, S.Q., and Standring, D., 1992, Characterization of hepatitis B virus capsid particle assembly in Xenopus oocytes. *J. Virol.* 66:3086.

TRANSCRIPTIONAL ACTIVATION BY THE HEPATITIS B VIRUS X PROTEIN

Robert Lucito and Robert J. Schneider

Department of Biochemistry and Kaplan Cancer Center
NYU Medical Center
New York, NY 10016

BIOLOGY OF HBV INFECTION

General Remarks

Human HBV is a member of a family of hepatotropic viruses with double-stranded DNA genomes (Robinson et al., 1982), which include woodchuck hepatitis virus (WHV, Summers et al., 1978), ground squirrel hepatitis virus (GSHV, Lee et al., 1988), and duck hepatitis virus (DHBV, Mason et al., 1980). All members maintain persistent infections, but the human and woodchuck viruses appear to cause the most significant chronic active infection and hepatocellular carcinoma (HCC, reviewed in Ganem and Varmus, 1987; Tiollais et al., 1985). The correlation between chronic HBV infection and development of HCC has been well established, primarily from epidemiological and clinical evidence (Beasley et al., 1981; Szumness, 1978; reviewed in Ganem and Varmus, 1987; Popper et al., 1987).

There is now a general understanding of viral replication, the genetic organization of the genome, and the structure of the virus particle. Elements involved in gene expression that have been described include two enhancer elements, a polyadenylation signal, and promoters for the Core, Surface and X genes (reviewed in Ganem and Varmus, 1987; Tiollais et al., 1985).

Disease Related To HBV Infection

Infection by HBV represents a health problem of enormous proportions, with at least 200 million chronic HBV carriers worldwide (Szumness, 1978), of whom a significant percentage will die of liver cirrhosis and HCC. Although poorly understood, the underlying cause of disease and carcinoma has been linked to chronic liver inflammation, cell proliferation and necrosis, probably due in part to immune attack of infected tissue

(Chisari et al., 1989; Bohman, 1990; Moriyama et al., 1990; Popper et al., 1987). In fact, experimental animal models have shown that replication of virus in and of itself is not directly cytopathic to hepatocytes (Chisari et al., 1989; Bohman, 1990). The majority of HCCs that arise from HBV infection contain integrated viral DNA (Ganem and Varmus, 1987). Nevertheless, the mechanisms involved in malignant transformation by HBV are not understood. Integration of HBV genomes into cellular DNA is thought to be comparable to the initiation step of carcinogenesis, since it is required but not sufficient for hepatocyte transformation (Brechot et al., 1981; Dejean et al., 1986; Hsu et al., 1988). This notion is also supported, at least superficially, by the recent demonstration that NIH-3T3 cells immortalized with SV40 T-antigen can be malignantly transformed by transfection with HBV DNA (Hohne et al., 1990). HBV does not appear to encode a transforming gene, and only infrequently has activation of cellular proto-oncogenes been found to occur, either by integration of HBV DNA or extensive rearrangement of cellular sequences (e.g. Dejean et al., 1986; Hsu et al., 1988).

HBV X GENE ENCODES A TRANSACTIVATOR OF TRANSCRIPTION

Because it is clear that there is little evidence supporting cis-activation of cellular proto-oncogenes by HBV integration, several groups investigated whether HBV encodes transacting functions. Based upon a genomic comparison between HBV and HTLVs, Miller and Robinson (1986) initially suggested that the X protein may be a regulator of gene expression. It was then demonstrated by several groups that the HBV X ORF encodes a transcriptional activator (transactivator) for several enhancer-promoter combinations (Spandau and Lee, 1988; Twu and Schloemer, 1987; Zahm et al., 1988), which have been extended to include those of HBV and a number of cellular and viral genes (Colgrove et al., 1989; Levrero et al., 1990; Seto et al., 1988; Siddiqui et al., 1989; Twu and Robinson, 1989). The sequence of the X ORF is very highly conserved in HBV serotypes (Lo et al., 1989) and can encode a 17kd polypeptide of 154 amino acids. This ORF is also conserved in WHV and GSHV (Ganem and Varmus, 1987). Although the polypeptide itself has not been detected in tumor derived samples, there is evidence that the X gene is expressed during at least some stage of infection, since many infected individuals produce antibodies to the X protein (Elfassi et al., 1986; Landshulz et al., 1988; Meyers et al., 1986). In fact, it is unlikely that full-length X polypeptides are produced in the majority of HCC tumors since integration and rearrangement of HBV sequences usually occurs through the cohesive overlap region located within the X gene (reviewed in Ganem and Varmus, 1987), thereby deleting either the 3' or 5' regions. It has been reported that expression of the X gene in transgenic mice does not promote serious liver disease (Lee et al., 1990), casting some doubt on its role in development of HCC. In contrast, another recent report has found a strong correlation between constitutive expression of the X gene and liver cancer in transgenic mice (Kim et al., 1991). The authors suggested that their results are related to higher and more prolonged expression of the X gene throughout the lifetime of the animals. The biological significance of either transgenic mouse model with respect to the generation of HCC now remains to be determined.

A second distinct HBV transactivating function has recently been described, composed of a specifically truncated portion of the preS/S region of genomic viral DNA (Castleman et al., 1990; Kekule et al., 1990). At this time, however, it is not known whether this novel transactivator is expressed in cells infected or transformed by HBV. It has been suggested that although formation of the preS/S transactivator probably occurs infrequently, it may have implications in tumor development (Castleman et al., 1990).

Mechanism Of X Protein Activity Is Unknown

Some transcription regulators have been shown to function by interacting directly with specific DNA sequences present in promoter and enhancer elements (reviewed in Ptashne, 1988). Some regulators do not bind DNA directly, but presumably act by modifying transcription factors or by associating with DNA binding transcription factors (i.e., "adaptor molecules", Martin et al., 1990; reviewed in Flint and Shenk, 1989). A range of structural features in some of these regulators has been revealed by cloning and mutagenic analysis. Domains involved in DNA binding include cysteine-rich "zinc-finger" regions (reviewed in Evans and Hollenberg, 1988), helix-loop-helix motifs (Pabo and Sauer, 1984), acidic α-helical regions (Gill and Ptashne, 1987; Hope and Struhl, 1986), and periodic leucine-rich "zipper" motifs which function by facilitating dimerization (Landshulz et al., 1988). In addition, activity of some binding proteins has been localized to regions rich in glutamine and proline residues (e.g. Bohman et al., 1987). Some transcription regulators, such as the adenovirus E1a protein, modulate transcription of a variety of genes and act on both pol II and III promoters (reviewed in Flint and Shenk, 1989). The mechanism of E1a action is not known, but it likely functions by interacting with cellular proteins involved in transcriptional regulation. It has been suggested that E1a may contain two important regulatory regions, one involved in transcriptional activation and another in promoter association that directs E1a to the target promoter through interaction with cellular DNA binding proteins (Liu and Green, 1990; Martin et al., 1990; Flint and Shenk, 1989).

The X protein does not appear to bind DNA (Siddiqui et al., 1987; Wu et al., 1990), and examination of its amino acid sequence does not reveal any predictable structural motifs which would classify it as a DNA binding protein. X protein has no recognizable sequence homology to any of the known factors involved in transcription (unpublished observations). However, our computer modeling analysis predicts that a C-terminal acidic region may form a strongly conserved α-helix (Kwee et al., 1992). This region may be vital to X protein activity because mutants which lack it are fully defective in transactivation function (Kwee et al., 1992; Facktor and Shaul, 1990). Although the mechanism of X protein activity is unknown, it is reasonable to suspect that X protein may function through activation of a signal transduction pathway. Recent evidence for this (Lucito and Schneider, 1992) will be discussed in a following section, but is based on the fact that transcription factor NF-kB is activated by both X protein (Siddiqui et al., 1989; Twu and Robinson, 1989) and phorbol esters (Baeurle et al., 1988). Alternatively, X protein may also function at the promoter itself, as shown recently in an in vitro system by its ability to interact with the cellular transcription factor ATF/CREB (Maguire et al., 1991).

Generally speaking, X and Ad E1a proteins are members of a group of transactivators that act promiscuously on diverse promoter elements. Several mechanisms, which are not mututally exclusive, can therefore be envisioned for transactivation by X protein. (a) *Indirect transactivation.* Cellular transcription factors can be enhanced in activity or concentration (reviewed in Berk and Schmidt, 1990). The activity or availability of transcription factors can be controlled by phosphorylation events, as suspected for activation of TFIIIC (Hoeffler et al., 1988; White et al., 1990), and NF-kB (Lenardo and Baltimore, 1989). As discussed next, both factors are also stimulated by X protein (Aufiero and Schneider, 1990; Siddiqui et al., 1989; Twu and Robinson, 1989; Lucito and Schneider, 1992). Activation of NF-kB occurs through a signal transduction pathway probably mediated by the protein kinase C (PKC) family and acted on by tumor promoters (Kikkawa et al., 1989). (b) *Direct transactivation.* Transcription stimulation may occur through protein interactions at the promoter, as suggested for the activation of ATF/CREB by Ad E1a (Flint and Shenk, 1989; Lillie and Green, 1989) and now for X protein

(Maguire et al., 1991). Recent evidence demonstrates that the X gene is expressed and functions as an efficient transactivator during productive viral infection (Colgrove et al., 1989), and it has now been implicated in the enhancement of viral transcription and replication. Thus, it seems likely that the X gene plays a role during the life-cycle of HBV because it is expressed during at least some stage of virus infection (Kay et al., 1985; Meyers et al., 1986; Liang et al., 1988). Nevertheless, such a role does not preclude other functions for X protein during HBV infection or liver disease.

X Protein Activates Transcription By RNA Polymerases II And III

Since the X gene can enhance transcription of a number of different viral and cellular transcription elements, it was asked whether it could also stimulate transcription of class III promoters which are transcribed by RNA polymerase III. The ability to regulate such a diverse group of transcriptional elements could indicate whether the X protein acts directly on promoter elements or indirectly by activating other proteins which modify transcription factors.

It was shown by Aufiero and Schneider (1990) that the X gene transactivates a variety of RNA pol III transcribed promoters when expressed in transfected cells, or when stably integrated in a hepatocyte cell line. Transactivation was shown to take place at the level of transcription initiation, and that the extent of stimulation and the kinetics of activation were almost indistuinguishable from that of E1a. It was also shown that X protein caused an increase in the assembly of stable preinitiation complexes on pol III promoters, probably by acting to enhance the level of active transcription factor TFIIIC. Factor TFIIIC has been shown to be activated by phosphorylation (Hoeffler et al., 1988; White et al., 1990), implicating X protein in the activation of protein kinases.

The biological significance of class III promoter transactivation by X protein is unknown. Ad E1a protein, SV40 t antigen and HBV X protein all stimulate RNA pol III activity (Aufiero and Schneider, 1990; Hoeffler and Roeder, 1985; Loeken et al., 1988; White et al., 1990; Yoshinaga et al., 1986). At a minimum, transactivation of RNA pol III transcription may be the fortuitous result of X protein stimulation of a number of cellular protein kinase activities. In this respect, activation of class III promoters by SV40 t antigen appears to be a general feature of transformation by the virus (White et al., 1990). Stimulation of RNA pol III transcription by X protein may therefore be indicative of its enhancement of other cellular kinase activities involved in regulation of transcription or transformation.

Given the variety of transcriptional elements stimulated by X protein it would make good sense if it possessed protein kinase activity or regulated such activities. However, this is currently an area of considerable controversy. Several groups showed that X protein is not phosphorylated and does not display detectable kinase activities *in vivo* (Levrero et al., 1990; Siddiqui et al., 1989). Our unpublished computer analysis failed to disclose elements that resemble an ATP or GTP binding site, or any of the consensus sequences found in known protein kinases. In contrast, Wu et al. (1990) recently described a strong autophosphorylating serine/threonine kinase activity for X protein expressed in bacteria or reticulocyte lysates. There is currently no good explaination for the difference in results, although it is possible that X protein studied by Wu et al. (1990) may have strongly associated and copurified with any of several resident protein kinases.

X PROTEIN ACTIVATES TRANSCRIPTION FACTOR NF-*K*B IN A PKC-INDEPENDENT PATHWAY

As noted earlier, the X protein is a promiscuous transactivator of class II transcriptional elements including HBV enhancer I, the SV40 enhancer, the HIV LTR and major histocompatability complex I genes (Seto et al., 1988; Spandau and Lee, 1988; Twu and Robinson, 1989; Twu and Schloemer, 1987; Zahm et al., 1988). The stimulation of some of these promoters has also been shown to correlate with the activation of transcription factors NF-*k*B and AP1, and accordingly these sites were also found to be required for transactivation by X protein (Siddiqui et al., 1989; Twu and Robinson, 1990; Seto et al., 1990). Both factors AP1 and NF-*k*B are regulated through second messenger pathways which are induced by tumor promoters, growth factors and cytokines. In particular, NF-*k*B is stimulated by activators of PKC (Sen and Baltimore, 1986), protein kinase A (Shirakawa and Mizel, 1989) and the double stranded RNA activated kinase DAI (Visvanathan and Goodburn, 1989). NF-*k*B is regulated by sequestration in the cytoplasm in an inactive complex with its inhibitor (I*k*B), but is released and transported to the nucleus after inducer action. Initial studies conducted *in vitro* suggested that the direct phosphorylation of I*k*B by PKC might mediate the release and therefore activation of NF-*k*B (Ghosh and Baltimore, 1990; Shirakawa and Mizel, 1989). However, several studies now indicate that activation of NF-*k*B may not involve the direct phosphorylation of I*k*B by PKC. For instance, activation of PKC and NF-*k*B by TNF alpha are not coupled in HL-60 cells (Hohman et al., 1991; Meichle et al., 1990). In addition, evidence has recently been presented that reactive oxygen intermediates, which are generated by the action of a variety of biological agents including phorbol esters (reviewed in Cerutti, 1985), may serve in some manner to mediate the release of I*k*B (Schreck et al., 1991). Therefore, there are likely to be several pathways for activation of NF-*k*B, some of which are independent of PKC activity.

Lucito and Schneider (1992) examined the activation of NF-*k*B responsive promoters by the X protein, specifically to determine whether X protein acts on cell signalling pathways. They showed that X protein activates inducible NF-*k*B through a pathway which is stimulated by tumor promoters but does so independently of PKC. Several lines of evidence were provided to support this conclusion. These include the inability to prevent X protein function with inhibitors of PKC, and the failure of active X protein to cause the translocation of PKC from the cytoplasm to the plasma membrane, which is correlated with PKC activation (Kraft and Anderson, 1983; reviewed in Kikkawa et al., 1989). Some evidence in support of these conclusions is outlined below.

Inducible NF-*k*B Is Activated By X Protein. Lucito and Schneider (1992) investigated whether X protein acts through PKC to activate NF-*k*B, as described for the HIV*tat* protein (Jakobovits et al., 1990), or in a PKC-independent manner as described for TNF alpha (Meichle et al., 1990). Figure 1 shows that the target of X protein is the inducible form of NF-*k*B. X protein was found to activate transcription of plasmid p*k*B-CAT (which is solely dependent on NF-*k*B) almost as well as did stimulation by TPA. Treatment of cells with a high concentration of TPA depletes the level of PKC (Rodriguez-Pena and Rozengurt, 1984), and results in re-sequestration of NF-*k*B in an inactive cytoplasmic complex that is resistent to further stimulation *in vivo* (Beauerle et al., 1988; Sen and Baltimore, 1986). X protein, like phorbol esters, was shown to be unable to reverse the block to further stimulation of NF-*k*B created by prolonged treatment with TPA. It was concluded that X protein and phorbol esters stimulate transcription by activating the pool of NF-*k*B, and the activity of both are blocked by prolonged exposure of cells to TPA.

X Protein Activation Of NF-*k*B Is Not Blocked By PKC Inhibitors

Studies were conducted using H7 (Hidaka et al., 1989) and calphostin C (Kobayashi et al., 1989) which are inhibitors of PKC that have been widely used to assess PKC involvement in mediating signal transduction pathways. H7 was shown to prevent transactivation by HIV*tat* protein (Jakobovits et al., 1990), demonstrating that an active PKC system is

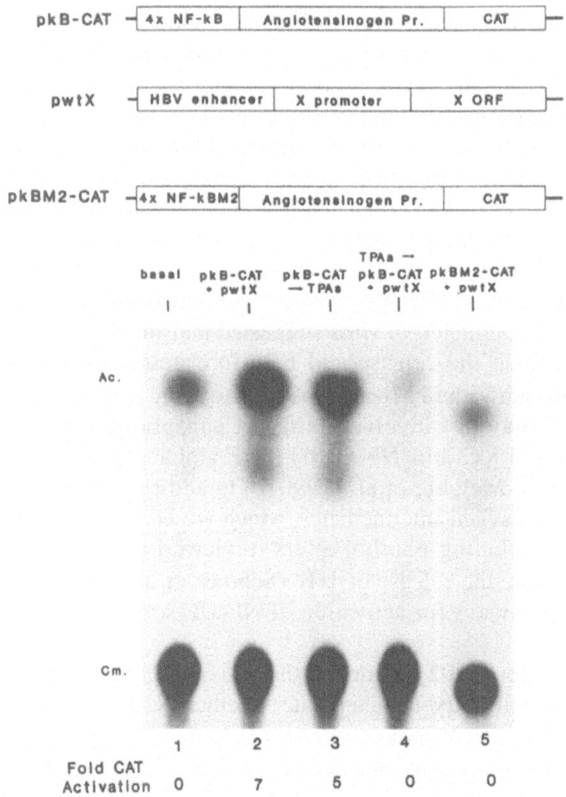

Figure 1. X protein transactivation of latent pool of NF-*k*B. Plasmid p*k*B-CAT contains 4 NF-*k*B binding sites. Plasmid p*k*BM2-CAT contains a mutated binding site for NF-*k*B. HepG2 cells were transfected with plasmids and CAT enzyme activity levels determined. Stimulation of PKC by TPA (TPAs) and ablation (TPAa) were performed in the indicated order. Fold activation of CAT is presented relative to the basal level (p*k*B-CAT alone).

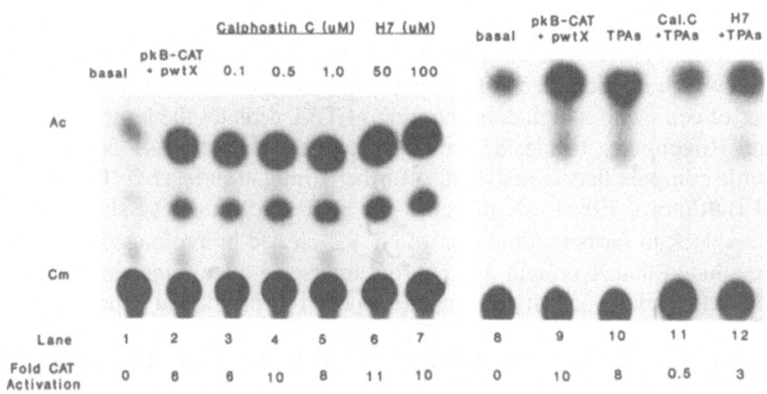

Figure 2. X protein activity is not blocked by PKC inhibitors. HepG2 cells were transfected with the indicated plasmids and treated with PKC inhibitors at concentrations shown. PKC inhibitors were added to medium 6hr after transfection.

required for *tat* function. Neither inhibitor was found to block X protein activation of NF-*k*B (Figure 2). In contrast, calphostin C and H7 impaired the ability of TPA to induce PKC activation of p*k*B-CAT. Thus, X protein was shown to activate NF-*k*B without an apparent requirement for functional PKC.

X Protein Activates NF-*k*B Without Translocation Of PKC. Activation of PKC correlates with its translocation from the cytoplasm to the plasma membrane (Kraft et al., 1983), and provides a reliable marker for the detection of PKC activation. The subcellular distribution of PKC in cells expressing biologically active X protein was therefore examined (Lucito and Schneider, 1992). The X gene was constructed into a replication-defective adenovirus vector which was then used to infect cells in culture. X protein expressed from the Ad-X recombinant was first shown to be biologically active and to strongly activate NF-*k*B (Lucito and Schneider, 1992). The subcellular distribution of PKC was examined in cells expressing X protein by infecting with either Ad-X or control *dl*312 viruses by subjecting cytoplasmic and membrane fractions to immunoblot analysis with an antiserum directed against the three major isoforms of PKC. The compartmentalization of PKC in non-stimulated and TPA stimulated cells, and the correlation between PKC activation and translocation to the plasma membrane was shown to occur as expected in these cells (Figure 3). In cells infected with *dl*312, PKC is quantitatively retained in the cytoplasmic fraction, as was observed in cells infected with the Ad-X virus. X protein activation of NF-*k*B was therefore shown to occur without concomitant activation or translocation of PKC. Thus, X protein was shown to regulate the second messenger system which controls NF-*k*B, but without activating PKC.

X mRNA SYNTHESIZES MULTIPLE POLYPEPTIDES

Some transactivators such as the E2 protein from bovine papilloma virus (BPV), are comprised of full-length and amino-truncated polypeptides which display different transcriptional regulatory activities through formation of hetero- and homodimers (Spalholz

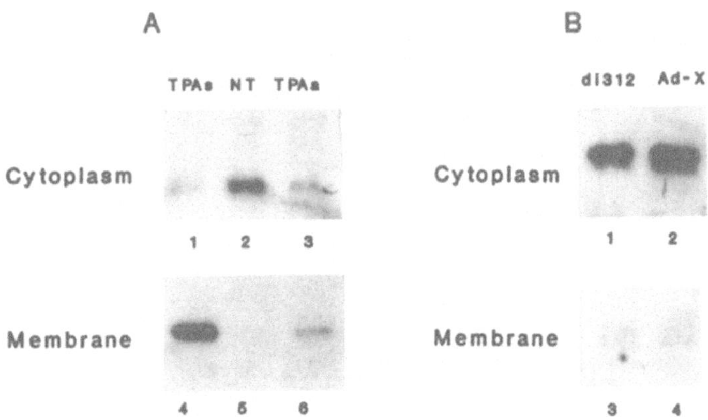

Figure 3. PKC is not translocated by X protein. (A) Cells were treated with TPA to stimulate or ablate PKC, then cytoplasmic and membrane fractions probed by immunoblot analysis for PKC distribution. (B) Cells were infected with Ad-X or control dl312 viruses and analyzed as in panel A.

et al., 1987; Lambert et al., 1987). Accordingly, transactivation of class II and III promoters by the HBV X gene product(s) might reside in different activation domains or encoded proteins. The X gene mRNA, for example, could potentially encode a full-length and two related amino-terminally truncated proteins by translation initiating at either of two internal AUG codons. This possibility is particularly attractive since the virus generates multiple proteins from the envelope, core and polymerase open reading frames in such a manner (reviewed in Ganem and Varmus, 1987).

Kwee et al. (1992) demonstrated by genetic and mutational studies that the X gene is similar to other HBV genes in encoding multiple products by alternate translation initiation on a single mRNA species. All three X polypeptides, when expressed individually, were found to be sufficient for transactivation of class III promoters, but transactivation of several different class II elements displayed varying requirements for the different X proteins. It was suggested that transcriptional regulation of class II promoters by the HBV X gene involves several distinct activities contributed by the different X proteins. Some evidence in support of this conclusion will be presented next.

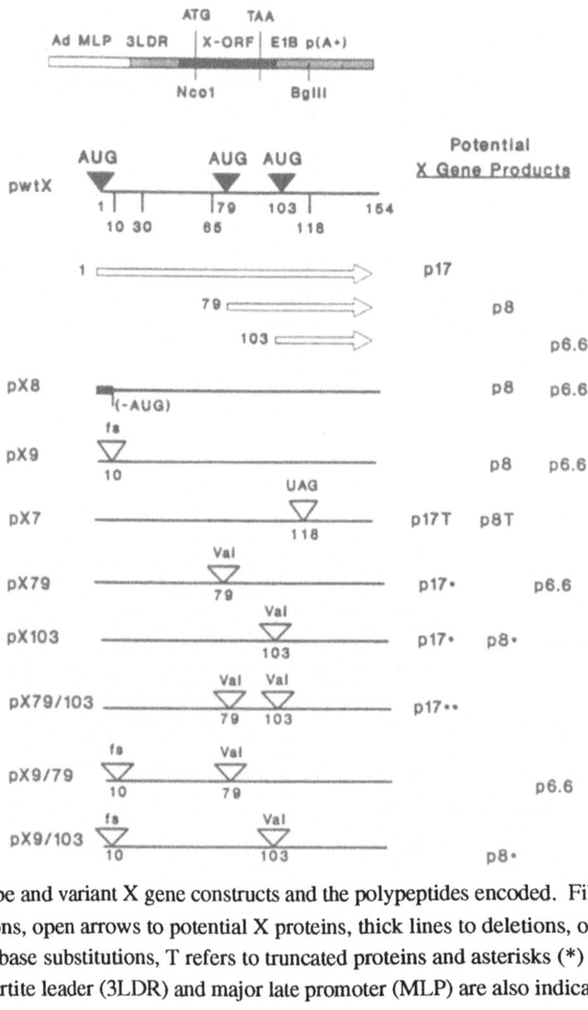

Figure 4. Wild-type and variant X gene constructs and the polypeptides encoded. Filled triangles refer to in-frame AUG codons, open arrows to potential X proteins, thick lines to deletions, open triangles to insertions or single base substitutions, T refers to truncated proteins and asterisks (*) to the met to val mutations. The tripartite leader (3LDR) and major late promoter (MLP) are also indicated.

Genetic And Transactivational Analysis Of Additional X Proteins

Mutations were employed to test for the ability to transactivate promoters transcribed by RNA pol II and III (Figure 4). All constructs contained the full-length X open reading frame (ORF) controlled by the adenovirus major late promoter, which is not responsive to the X protein, thereby eliminating complications due to autoregulation. The 5' proximal AUG codon was deleted in variant pX8, whereas variant pX9 contains a frame-shift mutation introduced after amino acid 10, resulting in termination of translation 37 amino acids downstream The remaining variants contained single nucleotide mutations which convert the in-frame AUG (Met) codons at positions 79 or 103 to GUG (Val), to test for the synthesis of additional X polypeptides.

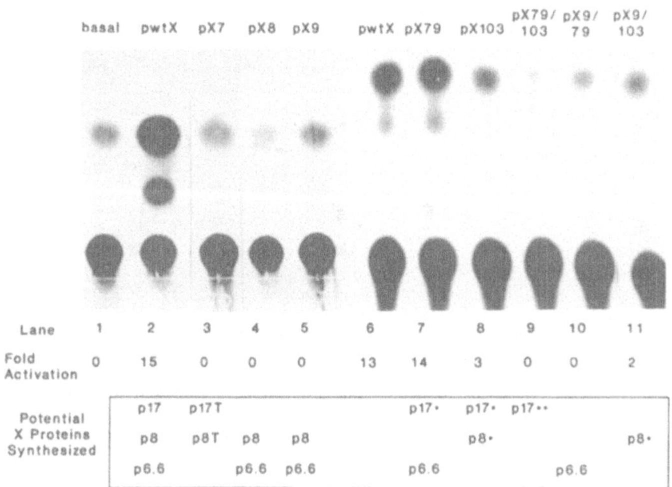

Figure 5. Effect of X gene mutations on transcription of SV40 enhancer. Cells were transfected with the indicated X gene mutations and CAT assays performed on extracts. Data are presented as the fold activation relative to the basal control (pSV2-CAT alone).

X gene variants were tested for the ability to transactivate class II transcriptional elements by cotransfection with the plasmid pSV2-CAT, which expresses the chloramphenicol acetyl transferase (CAT) gene linked to the SV40 enhancer and early promoter (Figure 5). The X protein was previously shown to act predominantly on the SV40 enhancer rather than promoter element (Spandau and Lee, 1988; Aufiero and Schneider, 1990). Both variants pX8 and pX9 were shown to be unable to stimulate transcription of pSV2-CAT (Figure 5), in agreement with results published previously (Spandau and Lee, 1988; Colgrove et al., 1989), and demonstrating that transactivation of the SV40 enhancer/early promoter apparently requires synthesis of full-length 17kd X protein. In contrast, both variants pX8 and pX9 were found to efficiently stimulate transcription from the VAI RNA class III promoter (Figure 6), suggesting that a second translation product must be produced *in vivo.*

By comparing the human HBV sequences it was suggested that two fully conserved internal AUG codons in all isolates of the four viral serotypes could serve to initiate additional X polypeptides (Kwee et al., 1992). Furthermore, AUG 1 was shown to be conserved in a rather poor codon context (Kozak, 1986) (UccAUGG) and therefore should not fully impede the initiation of translation downstream at the internal AUG at codon 79 or

103. AUG codon 79 is conserved in an equally poor codon context as AUG 1, whereas the AUG at 103 is conserved in a slightly better recognition context (GcaAUGU), providing the potential for synthesis of up to two additional in-frame, amino-truncated polypeptides. It was noted that there are no potential AUG codons in other reading frames of the X gene which might also serve to initiate translation.

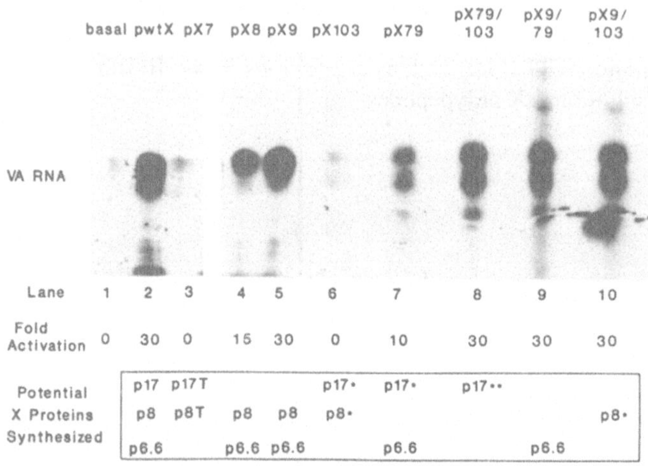

Figure 6. Effect of X gene mutations on transactivation of RNA pol III transcribed promoters. Cells were transfected with the indicated X gene mutations and plasmid pVA1 which transcribes VA1 RNA. RNA levels were measured by RNAse protection assay, and autoradiograms quantitated by densitometry.

Variant pX79 (Figure 5) stimulated transcription of the SV40 enhancer/early promoter (RNA pol II) and VA1 RNA gene (Figure 6; RNA pol III). These results indicated that the methionine at position 79 is not required for activity of the full-length 17kd protein, and that if the 8kd X protein is synthesized *in vivo*, it is not essential for transactivation of either of the class II or III promoters that were tested. However, variant pX103 failed to stimulate transcription by either RNA pol II or III. It was then shown by cotransfection analysis using X mRNA variants that can synthesize only a single X polypeptide that both the 17kd and 6.6kd X polypeptides are required for transactivation of the SV40 enhancer/early promoter (Figure 7). It was shown that the VAI RNA class III promoter could be activated by each of the individual X proteins. The failure of mutant pX103 to transactivate VA1 RNA transcription was related to the co-synthesis of variant 17kd and 8kd polypeptides (Kwee et al., 1992). It was concluded that the 17kd, 8kd and 6.6kd X proteins each independently stimulated transcription by RNA pol III, and that at least one additional X polypeptide, either the 8kd or 6.6kd species, is normally synthesized from the X mRNA *in vivo*.

Transcriptional activation studies using plasmid p*k*B-CAT, which contains four tandem copies of the NF-*k*B binding sequence inserted upstream of the rat minimal angiotensinogen promoter and CAT gene, demonstrated that it was activated by pX79/103 which synthesizes only the 17kd X protein (Kwee et al., 1992). However, mutant pX9, which does not synthesize the 17kd form of X protein, failed to stimulate NF-*k*B dependent transcription. These results were also supported by a study in which it was shown that full-length 17kd X protein produced in bacteria could stimulate *in vitro* transcription by the HIV LTR (Jameel et al., 1990). Thus, activation of NF-*k*B dependent transcription was shown to be carried out only by the full-length 17kd X protein.

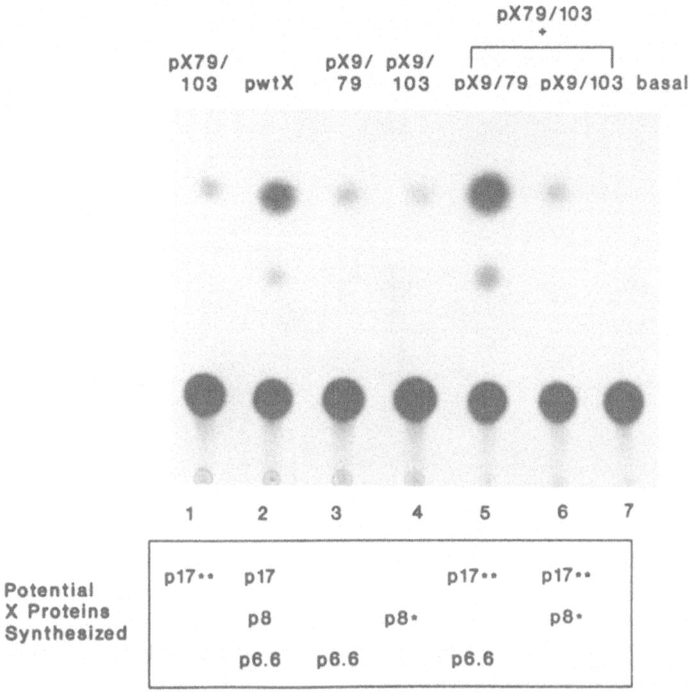

Figure 7. X protein requirment for activation of SV40 enhancer. Cells were transfected with equal amounts of the indicated plasmids and reporter plasmid pSV2-CAT, then CAT assays performed.

In Vitro Translation Of X mRNA

The X mRNA was translated *in vitro* to corroborate the genetic analysis by determining whether multiple polypeptides are synthesized initiating at the internal AUG codons. X mRNAs produced *in vitro* from the T7 RNA polymerase promoter were used to program rabbit reticulocyte lysates. ^{35}S-methionine labeled X proteins were then immunoprecipitated with antisera generated against bacterially produced X protein and subjected to SDS-polyacrylamide gel electrophoresis and autoradiography. Two polypeptides were routinely observed migrating at 17kd and about 6.6-7.0 kd when lysates were programmed with wild-type X mRNA. The two polypeptides were then mapped to their respective initiation codons using mutants X7, X8 and X9. Mutant X8, which lacks a 5' AUG codon, failed to synthesize the full-length 17kd protein but retained the ability to synthesize the smaller protein. Mutant X9, which contains a frame-shift mutation shortly downstream of AUG1, also synthesized the smaller polypeptide. This result correlates with the genetic and functional analysis of X protein mutants and demonstrates that ribosomes bypass the weaker AUG 1 and initiate at a downstream AUG codon. Mutant X7 truncates the C-terminus of the full-length X protein as shown in lane 3, and eliminates synthesis of the smaller polypeptide as expected if initation occurs at AUG 2 or 3. Based on the apparent molecular weight, it seems most likely that alternate translation initation at AUG3 (position 103) is utilized to produce the smaller polypeptide rather than at AUG 2 (position 79). However, it cannot be excluded that AUG 2 (position 79) produces a ~8kd polypeptide as well, which may comigrate with the smaller species.

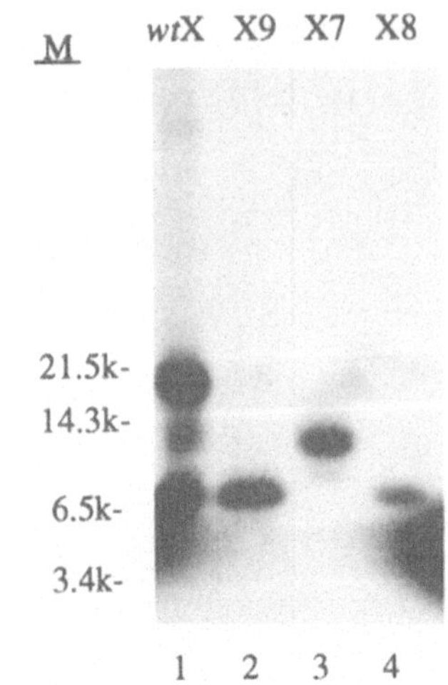

Figure 8. In vitro transcription/translation analysis of X mRNA. A single X mRNA species was syntheiszed from a vector directed by T7 RNA polymerase, purified and used to program rabbit reticulocyte lysates. Translation was performed in the presence of ^{35}S-methionine and products purified by immunoprecipiattion with X-specific antisera. Immunoprecipitates were resolved by SDS-polyacrylamide gel electrophoresis and autoradiography. Molecular weights were determined relative to marker proteins (not shown).

In summary, a genetic analysis showed that the HBV X gene encodes several transcriptional transactivating proteins; a full-length 17kd protein, an amino-truncated 6.6kd polypeptide and possibly an 8kd amino-truncated species. Different class II transcription elements were found to display varying requirements for the multiple X proteins. Synthesis of the full-length 17kd X protein was required to stimulate several class II transcriptional elements that were tested, including the SV40 enhancer/early promoter and an NF-*k*B dependent promoter. However, the 6.6kd X protein was additionally required to activate the SV40 enhancer. All three X proteins were found to be capable of independently activating class III promoters.

ACKNOWLEDGMENTS

The work presented in this report was supported by a grant from the National Cancer Institute (CA54525) and from the American Cancer Society (MV394).

REFERENCES

Aufiero, B. and Schneider, R.J. (1990) *EMBO J.* **9**, 497-504.

Baeurle, P.A., Lenardo, M., Pierce, J.W. and Baltimore, D. (1988). *Cold Spring Harbor Symp. Quant. Biol.* **53**, 789-798.

Beasley, R.P., Hwang, L.Y., Lin, C.C. and Chien, C.S. (1981) *Lancet* **2**, 1129-1133.

Berk, A.J. and Schmidt, M.C. (1990) *Genes and Devel.* **4**, 151-155.

Bohmann, D., Bos, T.J., Admon, A., Nishimura, T., Vogt, P.K. and Tjian, P. (1987) *Science* **238**, 1386-1392.

Bohman, D. (1990). *Cancer Cells* **2**, 337-344.

Brechot, C., Scotto, J., Charnay, P., Hadchovel, M., Degos, J., Trepo, C. and Tiollais, P. (1981) *Lancet* **2**, 765-768.

Caselmann, W.H., Meyer, M., Kekule, A.s., Lauer, U., Hofschneider, P.H. and Koshy, R. (1990) *Proc. Natl. Acad. Sci. USA* **87**, 2970-2974.

Cerutti, P.A. (1985). *Science* **227**, 375-381.

Chisari, F.V., Klopchin, K., Moriyana, T., Pasquinelli, C., Dunsford, H.A., Sell, S., Pinkert, C.A., Brinster, R.L. and Palmiter, R.D. (1989) *Cell* **59**, 1145-1156.

Colgrove, R., Simon, G. and Ganem, D. (1989) *J. Virol.* **63**, 4019-4026.

Dejean, A., Bougueleret, L., Grzeschik, K.H. and Tiollais, P. (1986) *Nature* **322**,70-72.

Elfassi, E., Haseltine, W.A. and Dienstag, J.L. (1986) *Proc. Natl. Acad. Sci. USA* **83**, 2219-2222.

Evans, R.M. and Hollenberg, S.M. (1988) *Cell* **52**, 1-3.

Flint, J. and Shenk, T. (1989). *Ann. Rev. Genet.* **23**, 141-161.

Ganem, D. and Varmus, H.E. (1987) *Ann. Rev. Biochem.* **56**, 651-693.

Gill, G. and Ptashne, M. (1987) *Cell* **51**, 121-126.

Ghosh, S. and Baltimore, D. (1990). *Nature* **344**, 678-682.

Hidaka, H., Inagaki, M., Kawamoto, S. and Sasari, Y. (1989). *Biochem.* **23**, 5036-5041.

Hoeffler, W.K. and Roeder, R.G. (1985). *Cell* **41**, 955-963.

Hoeffler, W.K., Kovelman, R. and Roeder, R.G. (1988). *Cell* **53**, 907-920.

Hohman, H.P., Kolneck, R., Remy, R. and Van Loon, A.P.G.M. (1991). *Mol. Cell. Biol.* **11**, 2315-2318.

Hohne, M., Schaefer, S., Seifer, M., Feitelson, M.A., Paul, D. and Gerlich, W.H. (1990) *EMBO J.* **9**, 1137-1145.

Hope, I.A. and Struhl, K. (1986) *Cell* **46**, 885-894.

Hsu, T.Y., Moroy, T., Etiemble, J., Louise, A., Trepo, C., Tiollais, P. and Buendia, M.A. (1988) *Cell* **55**, 627-635.

Jakobovits, A., Rosenthal, A. and Gapon, D.J. (1990). *EMBO J.* **9**, 1165-1170.

Jameel, S., Siddiqui, A., Maguire, H.F. and Rao, K.U.S. (1990) *J. Virol.* **64**, 3963-3966.

Kikkawa, V., Kishimoto, A. and Nishizuka, Y. (1989). *Ann. Rev.Biochem.* **58**, 31-44.

Kay, A., Mandart, E., Trepo, C., and Galibert, F. (1985) *EMBO J.* **4**, 1287-12.

Kekule, A.S., Laver, U., Meyer, M., Caselmann, W., Hofschneider, P.H. and Koshy, R. (1990) *Nature* **343**, 457-461.

Kobayashi, E.H., Nakano, H., Morimoto, M. and Tamaoki, T. (1990). *Biochem. Biophys. Res. Commun.* **159**, 548-533.

Kozak, M. (1986) *Cell* **44**, 283-292.

Kraft, A.S. and Anderson, W.B. (1983). *Nature* **301**, 621-623.

Lambert, P.F., Spalholz, B.A.,and Howley, P.M. (1987) *Cell* **50**, 69-78.

Landschulz, W.H., Johnson, P.F. and McKnight, S.L. (1988) *Science* **240**, 1759-1764.

Lee, T.H., Finegold, M.J., Shen, R.F., DeMayo, J.L., Woo, S.L.C. and Butel, J.S. (1990). *J.Virol.* **64**, 5939-5947.

Lee, H-S., Rajagopalan, M.S. and Vyas, G.N. (1988) *Hepatol.* **8**, 1116-1120.

Lenardo, M.J. and Baltimore, D. (1989). *Cell* **58**, 227-229.

Levrero, M., Jean-Jean, O., Balsano, C., Will, YH. and Perricudet, M. (1990) *Virol.* **174**, 299-304.

Liang, X.H., Stemler, M., Will, H., Braun, R., Tang, Z-Y., and Schroeder, C.H. (1988). *J.Med. Virol.* **25**, 329-337.

Lillie, J.W. and Green, M.R. (1989). *Nature* **341**, 279-280.

Liu, F. and Green, M.R. (1990) *Cell* **61**, 1217-1224.

Lo, S.J., Chien, M.L., Lee, Y.H.W. (1988) *Virol.* **167**, 289-292.

Loeken, M., Bikel, I., Livingston, D.M. and Brady, J. (1988). *Cell* **55**, 1171-1177.

Martin, K.J., Lillie, J.W. and Green, M.R. (1990). *Nature* **346**, 147-152.

Mason, W.S., Seal, G. and Summers, J. (1980) *J. Virol.* **35**, 829-836.

Meichle, A., Schutze, S., Hensel, G., Brunsing, D. and Kronke, M. (1990). *J. Biol. Chem.* **265**, 8339-8343.

Meyers, M.L., Trepo, L.V., Nath, N. and Sninsky, J.J. (1986) *J. Virol.* **57**, 101-109.

Miller, R.H. and Robinson, W.S. (1986) *Proc. Natl. Acad. Sci. USA* **83**, 2531-2535.

Moriyama, T., Guilhot, S., Klopchin,K., Moss, B., Pinkert, C., Palmiter, R.D., Brinster, R.L., Kanagawa, O. and Chisari, F.V. (1990) *Science* **248**, 361-364.

Pabo, C.O. and Sauer, R.T. (1984) *Ann. Rev. Biochem.* **53**, 293-321.

Popper, H., Shafritz, D.A. and Hoofnagle, J.H. (1987) *Hepatol.* **7**, 764-772.

Ptashne, M. (1988) *Nature* **335**, 683-689.

Robinson, W.S., Marion, P., Feitelson, M. and Siddiqui, A. (1982) in *Viral Hepatitis 1981 Int. Symp.* eds Szmuness, W., Alter, H.J. and Maynard, J.E., pp.57-68. Franklin Inst. Press, Philadelphia, Pa.

Rodriguez-Pena, A. and Rozengurt, E. (1984). *Biochem. Biophys. Res. Commun.* **120**, 1053-1059.

Schreck, R., Rieber, P. and Baeurle, P.A. (1991). *EMBO J.* **10**, 2247-2258.

Sen, R. and Baltimore, D. (1986). *Cell* **47**, 921-928.

Seto, E., Yen, T.S.B., Peterlin, B.M. and Ou, J.H. (1988) *Proc. Natl. Acad. Sci. USA* **85**, 8286-8290.

Seto, E., Mitchell, P.J. and Yen, T.S.B. (1990). *Nature* **344**, 72-74.

Shirakawa, F., Chedid, M., Suttles, J., Pollol, B.A. and Mizel, S.B. (1989). *Mol. Cell. Biol.* **9**, 959-964.

Shirakawa, F and Mizel, S.B. (1989). *Mol. Cell. Biol.* **99**, 2424-2430.

Siddiqui, A., Jameel, S and Mapoles, J. (1987) *Proc. Natl. Acad. Sci. USA* **84**, 2513-2517.

Siddiqui, A., Gaynor, R., Srinivasan, A., Mapoles, J. and Farr, R.W. (1989) *Virol.* **169**, 479-484.

Spandau, D.F. and Lee, C.H. (1988) *J. Virol.* **62**, 427-434.

Summers, J., Smolec, J. and Snyder, R. (1978) *Proc. Natl. Acad. Sci. USA* **75**, 4533-4537.

Szmuness, W. (1978) *Prog. Med. Virol.* **24**, 40-69.

Tiollais, P., Pourcel, C. and Dejean, A. (1985) *Nature* **317**, 489-495.

Twu J.S. and Schloemer, R.H. (1987) *J. Virol.* **61**, 3448-3453.

Twu, J-S. and Robinson, W.S. (1989) *Proc. Natl. Acad. Sci. USA* **86**, 2046-2050.

Visvanathan, K.V. and Goodburn, S. (1989). *EMBO J.* **8**, 1129-1138.

White, R.J., Stott, D. and Rigby, P.W.J. (1990). *EMBO J.* **9**, 3713-3721.

Wu, J.Y., Zhou, Z-Y., Judd, A., Cartwright, C.A. and Robinson, W.S. (1990). *Cell* **63**, 687-695.

Yoshinaga, S., Dean, N., Han, M. and Beck, A.J. (1986). *EMBO J.* **5**, 343-354.

Zahm, P., Hofschneider, P.H. and Koshy, R. (1988) *Oncogene* **3**, 169-177.

Maguire, H.F., Hoeffler, J.P. and Siddiqui, A. (1991). *Science* **252**, 842-844.

Kim, C-M., Koike, K., Saito, I., Miyamura, T. and Jay, G. (1991). *Nature* **351**, 317-320.

REGULATION OF HUMAN IMMUNODEFICIENCY VIRUS STRUCTURAL PROTEIN EXPRESSION AND VIRION FORMATION

Hans-Georg Kräusslich

Angewandte Tumorvirologie, Deutsches
Krebsforschungszentrum, D-6900 Heidelberg, Germany

INTRODUCTION

The replication cycle of retroviruses, and in particular of complex retroviruses like human immunodeficiency virus (HIV), encompasses several unusual or unique features. After adsorption and penetration of the infecting virus, the viral plus-stranded RNA genome is reverse transcribed into a linear double-stranded cDNA which is integrated into the host cell genome as a provirus. While these early replication steps are mainly achieved by the action of viral enzymes associated with the incoming virion, transcription of the integrated provirus is performed by cellular polymerase II complexes. In the case of the simple animal retroviruses, cytoplasmic concentration of viral mRNAs is mainly regulated by host cell transcription control and by incomplete splicing of the primary transcript which also serves as the genomic RNA. Complex retroviruses, on the other hand, provide a multitude of *cis*-acting elements and *trans*-acting factors which regulate gene expression at a transcriptional and/or post-transcriptional level. In contrast to these transcriptional and post-transcriptional control mechanisms which have been scrutinized extensively in recent years, the late stages of assembly and maturation of infectious progeny virus are poorly understood. The structural proteins of the virion are synthesized as polyproteins which are transported to the plasma membrane via different cellular pathways. The various protein components and the genomic RNA congregate at the site of assembly leading to the formation of the viral bud and the extrusion of an immature virion. Proteolytic cleavage of structural polyproteins to yield the functional components is achieved by the viral proteinase and occurs predominantly after all virion components are confined in the assembled particle. Given the complexity of this assembly mechanism, it appears likely that, besides the regulation of mRNA synthesis, regulation of virion assembly provides an additional control mechanism in viral replication which may also serve as an important target for the development of novel antiviral compounds.

STRUCTURE OF THE VIRAL GENOME

Analysis of the genetic structure of the HIV-1 genome reveals that, in addition to the characteristic features of all replication competent retroviruses - i.e. the *gag*, *pol*, and *env* genes coding for the structural proteins of the viral core, the viral enzymes, and the glycoproteins of the viral envelope, respectively, and the long terminal repeats (LTRs) - there are at least six additional proteins which are encoded in the 3' part of the genome and are believed to function predominantly as regulatory proteins (Fig. 1). The presence of these auxiliary gene products, all of which are expressed from multiply spliced RNAs (Fig. 1), has led to the designation of HIV as a complex retrovirus (Cullen, 1991). In general, these auxiliary gene products have been grouped into those that are essential for viral replication

Regulation of Gene Expression in Animal Viruses, Edited by
L. Carrasco *et al.*, Plenum Press, New York, 1993

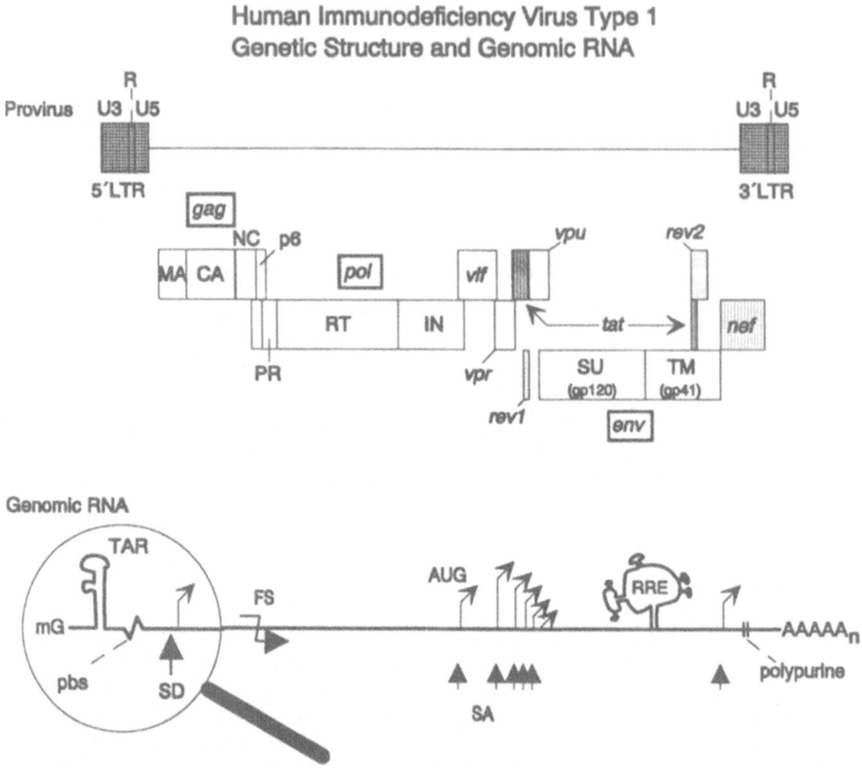

Figure 1. Schematic representation of the genetic structure of human immunodeficiency virus 1 and its genomic RNA. The upper part depicts the proviral DNA; the open reading frames are indicated as open (late genes) or shaded (early genes) boxes. Viral genes are identified in italic letters, different protein domains encoded as polyprotein segments from the *gag*, *pol*, and *env* genes are represented in capital letters. The viral LTRs are shown as hatched boxes. In the lower part, the genomic RNA is depicted as solid line. Important *cis* regulatory elements on the RNA include initiation codons for viral proteins (bent arrows), splice donor and splice acceptor sites (straight arrows), the signal for translational frameshifting (FS), the primer binding site (pbs), and the trans-activation response (TAR) and rev responsive elements (RRE). The 5' cap (mG) and 3' polyadenylation are also indicated. For clarity, the 5' terminal segment is enlarged. MA, matrix protein; CA, capsid protein; NC, nucleocapsid protein; PR, proteinase; RT, reverse transcriptase; IN, integrase; SU, surface glycoprotein; TM, transmembrane glycoprotein (nomenclature according to Leis et al., 1988).

in tissue culture (i.e. Tat and Rev) and those that are not (i.e. Nef, Vif, Vpu, and Vpr). It is likely, however, that this classification does not necessarily reflect their true biological function in the infected host (*in vivo*) but may also be caused by different requirements for viral replication to occur depending on the tissue culture cells used. Accordingly, recent evidence indicates a strong selective pressure for maintenance of the *nef* open reading frame *in vivo* and a dramatically altered phenotype manifested by an animal infected with a *nef* deletion mutant of simian immunodeficiency virus (SIV; Kestler et al., 1991). Besides the observed differences between replication *in vivo* and in tissue culture, characterization of the functional contributions of different HIV gene products may also be complicated by host range effects (i.e. use of various T-lymphoma cell lines or peripheral blood mononuclear cells [PBMC]) and by the selection of tissue culture adapted variants that may not always reflect the phenotype of the primary isolate.

Analysis of the functional roles of the different HIV regulatory genes has been quite successful regarding *tat* and *rev* gene functions (see below), but has been rather controversial for those HIV genes which are dispensable for replication in tissue culture. The conservation of the different genes in various lentiviral genomes and the importance of a functional *nef* gene product for SIV replication and pathogenesis *in vivo* (Kestler et al., 1991) indicate that at least Nef, but most likely also the other regulatory proteins, serve important functions in the infected host. However, the functional contributions of the gene products to viral replication and their precise mechanism of action have not been established unequivocally for any one of these proteins. Although several potentially relevant functions have been determined using *in vitro* assay systems, definitive assignment of their contribution to viral replication *in vivo* requires additional experimental evidence.

Transcription of the integrated proviral DNA from the viral LTR promoter yields the full length genomic RNA which is capped at the 5' end and polyadenylated at the 3' end (Fig. 1). The unspliced RNA can serve either as the viral genome or as mRNA for the structural proteins encoded by the *gag* and *pol* genes, or can be spliced to yield more than twenty subgenomic mRNAs encoding all other HIV gene products. All HIV-specific RNAs are identical in their 5' terminal sequence until the major splice donor, located in the 5' untranslated region upstream of the initiation codon for the Gag polyprotein (Fig. 1). Spliced mRNAs are derived from utilization of a multitude of splice acceptor and additional splice donor sequences in the 3' half of the genome (straight arrows in Fig. 1) and almost all viral proteins can be encoded by several alternatively spliced RNAs with the respective splice acceptor generally immediately preceding the translation initiation codon (bent arrows in Fig. 1). Several HIV gene products, including the Tat and Rev proteins, are translated from multiply spliced RNAs. Besides its protein coding capacity, the genomic RNA also contains several *cis*-acting signals for viral replication and gene expression. The primer binding site (pbs; Fig. 1) serves as the template for annealing of the $tRNA^{lys3}$ primer required for the initiation of minus-strand cDNA synthesis whereas plus-strand cDNA synthesis starts from a poly-purine tract in the 3' segment of the genome (Fig. 1). Also indicated in Fig. 1 are the target sequences for the viral Tat and Rev proteins (TAR and RRE, respectively) which are involved in the regulation of viral gene expression (see below), and the signal regulating translational frameshifting from the *gag* to the *pol* reading frame (FS).

THE REPLICATION CYCLE OF HIV

Analysis of the HIV replication cycle has revealed many similarities to that of other retroviruses and enveloped viruses in general. In the context of this review, however, I shall only discuss a few selected aspects which are specifically relevant for HIV replication (Fig. 2; for a detailed review: Coffin, 1990).

The tropism of HIV is primarily determined by the viral surface glycoprotein gp120 (SU) which interacts with the CD4 receptor expressed on T-helper cells, monocytes/macrophages, and other antigen presenting cells (reviewed in Sattentau and Weiss, 1988). In addition, HIV can also infect certain CD4-negative cells and candidate molecules which may serve as alternate receptors have been identified recently (Bhat et al., 1991; Harouse et al., 1991). Analogous to other enveloped viruses, the uptake of HIV may involve either direct fusion of the viral envelope with the plasma membrane of the target cell or endocytosis of the complete virion and subsequent fusion in the endosomal compartment (Fig. 2). The observations that HIV entry can be pH-independent (Stein et al., 1987) and does not require uptake of the CD4 receptor (Bedinger et al., 1988; Maddon et al., 1988) certainly argue for fusion at the cell surface. However, both pathways of uptake have been detected by electron microscopy (Stein et al., 1987; Goto et al., 1988; reviewed in Gelderblom, 1991) and, while it is clear that HIV-infection can occur via fusion at the plasma membrane, receptor-mediated endocytosis may represent a second, potentially equally important, mechanism of entry. Whatever mechanism finally leads to productive infection, the process of fusion itself is mediated by a fusogenic hydrophobic domain on the transmembrane glycoprotein gp41 (Gallagher, 1987) and results in the uptake of the viral core into the cytoplasm of the target cell. Subsequent steps, including reverse transcription of the viral RNA genome and integration of the proviral cDNA into the host cell genome (Fig. 2), have been studied extensively in *in vitro* systems using purified viral enzymes obtained from recombinant sources, but are poorly understood in the infected cell.

A particular feature of all retroviral infections is the potential for latent infection which can be reactivated at later times. Latency may occur at two different levels, namely before and after integration of the provirus into the host cell genome. Most retroviruses can

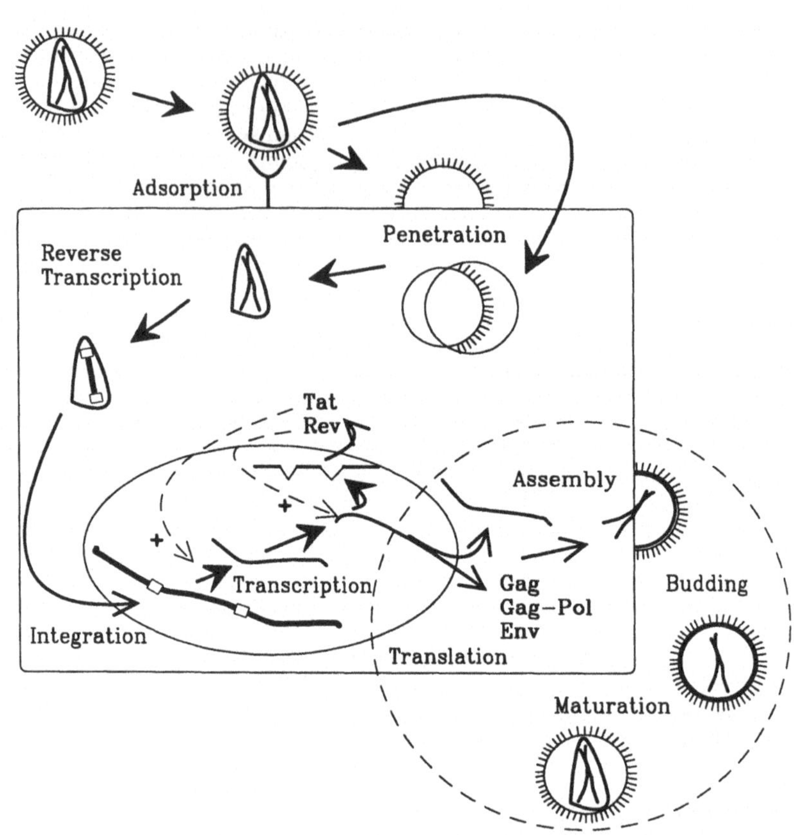

Figure 2. Schematic representation of the human immunodeficiency virus replication cycle. A complete infectious cycle of HIV is depicted. Solid arrows indicate the normal pathway, double arrows represent alternative options, and broken arrows represent the action of viral trans-acting proteins. The late phase of the replication cycle (see Fig. 3) is separated by a broken circle.

only productively infect dividing cells and replication in non-dividing cells is blocked after reverse transcription but prior to integration of the viral cDNA into the host cell genome. Consistent with this observation, the HIV replication cycle is arrested in quiescent T-cells with the proviral DNA still unintegrated (Stevenson et al., 1990; Zack et al., 1990) but may be induced to a complete infectious cycle or may result in an abortive infection (reviewed in Levy, 1990). In contrast to animal onco-retroviruses, HIV can also productively infect terminally differentiated cells of the human macrophage lineage and other human cells experimentally arrested in the cell cycle (Bukrinsky et al., 1992; Lewis et al., 1992). It may thus be that HIV integration requires metabolically active rather than actively dividing cells (Bukrinsky et al., 1992). Integration of the viral cDNA into the host cell genome has long been thought to occur at random. Recent evidence, however, indicates a marked preference for nucleosomal regions over nucleosome-free regions with favored integration at positions where the major groove is on the exposed face of the nucleosomal DNA, yielding a 10 base pair periodicity of preferred sites (Pryciak and Varmus, 1992). In addition, integration efficiency appears to be modulated by the actual sequence at the integration sites (Pryciak and Varmus, 1992). Following integration into the host cell genome, the provirus may remain transcriptionally silent for long periods of time -leading to post-integration latency- or may be actively transcribed. Transcription is mediated by cellular RNA Pol II complexes and is, at least initially, controlled by cellular *trans*-acting factors binding to the LTR promoter/enhancer element. In the case of complex retroviruses like HIV, there is an additional very potent system of virus-specified regulatory elements leading to the controlled and consecutive expression of defined genetic elements needed for production of infectious virus (Fig. 2).

Applying a terminology that had already been coined for the replication of DNA viruses, gene expression of HIV has been divided into an early phase, giving rise to the production of viral regulators, and a late phase, yielding large scale production of structural proteins of the virion and viral replication enzymes. In contrast to DNA viruses where early and late reflects expression before and after replication of the viral genome, respectively, in HIV these terms refer to a biphasic mode of gene expression that occurs only after reverse transcription of the infecting virus's genome and after integration of the proviral DNA into the host cell genome. Moreover, whereas in the case of DNA viruses differential gene expression is mainly achieved by the use of multiple promoters, the HIV genome is expressed as a single transcriptional unit and regulated expression of all viral proteins requires a multitude of *cis*-acting control elements as well as viral and cellular *trans*-acting factors, most notably the viral Tat and Rev proteins.

The HIV Tat and Rev proteins are small nuclear proteins which are expressed from multiply spliced RNAs (Fig. 1) and bind with high affinity to specific RNA elements. The target sequence for Tat is the *trans*-activator response element (TAR), an RNA helix-loop structure located at the 5' end of all viral RNAs (Fig. 1). Tat binds directly to TAR at a U-rich three nucleotide bulge and binding requires local distortion of the major groove of the RNA helix (Weeks et al., 1991) as well as a stretch of basic residues with a single essential arginine residue on the Tat protein (Calnan et al., 1991; reviewed in Kräusslich, 1992). The HIV Rev protein also binds to a highly structured RNA element, the Rev response element (RRE) located in the 3' part of the RNA genome (Fig. 1). Rev binding also appears to require local distortion of the sugar-phosphate backbone of a type A helix, similar to what was observed for the Tat-TAR interaction (reviewed in Kräusslich, 1992). While the precise mechanism of action has not been elucidated for either Tat or Rev, it is clear that Tat serves as a potent *trans*-activator of transcription from the viral LTR promoter, most likely by promoting elongation of nascent RNA transcripts (Feinberg et al., 1991; reviewed in Kräusslich, 1992). Rev-RRE interaction, on the other hand, does not increase the rate of transcription but has the unusual and novel property of selectively promoting the synthesis of late viral proteins from incompletely spliced HIV mRNAs. Several investigators have reported that Rev function requires multiple Rev molecules bound to the RRE (reviewed in Kräusslich, 1992) and the requirement for cooperative binding to the RNA may explain the non-linear relationship between Rev concentration and Rev function that is observed in transfected cells (Pomerantz et al.,1990). These observations suggest that Rev may accomplish its function only after achieving a certain threshold concentration. Taken together with recent experiments on *tat* minus mutants (Feinberg et al., 1991), these results can be explained by a two step molecular switch model for the productive phase of the viral replication cycle (Fig. 2). Initial synthesis of Tat leads to a rapid and significant increase in full length primary transcripts that are completely spliced and an almost concomitant burst of early viral protein synthesis, with the subsequent induction of Rev function and the delayed accumulation of late viral protens (after 12 - 24 h). This regulated expression allows the synchronized accumulation of viral structural proteins and genomic RNA culminating in morphogenesis and maturation of infectious progeny virus.

POLYPROTEIN SYNTHESIS AND VIRAL ASSEMBLY

In the late phase of the HIV replication cycle, the genomic RNA follows three predominant pathways. First, the unspliced RNA transcript can function as mRNA for the Gag and Gag-Pol polyproteins which are translated on cytoplasmic free polysomes (Fig. 3). Second, the primary transcript can be transported to the site of assembly and two molecules of genomic RNA are subsequently packaged into the viral particle yielding the diploid plus strand RNA genome. Third, the primary transcript can be spliced to give the singly spliced mRNA for the glycoproteins of the viral envelope. The *env* gene products are translated on ER-associated ribosomes and are processed through the secretory pathway, eventually leading to a multimeric complex of surface and transmembrane glycoproteins inserted into the plasma membrane (Fig. 3).

All products of the *gag*, *pol* and *env* genes are translated as polyprotein precursors and are proteolytically processed to yield the mature proteins found in the infectious virion. The gp160 precursor of the viral glycoproteins is cleaved by a cellular proteinase in the Golgi apparatus. By contrast, the intact Gag (pr55) and Gag-Pol (pr160; derived from translational frameshifting in *gag*; Fig. 1; Jacks et al., 1987) polyproteins are transported to

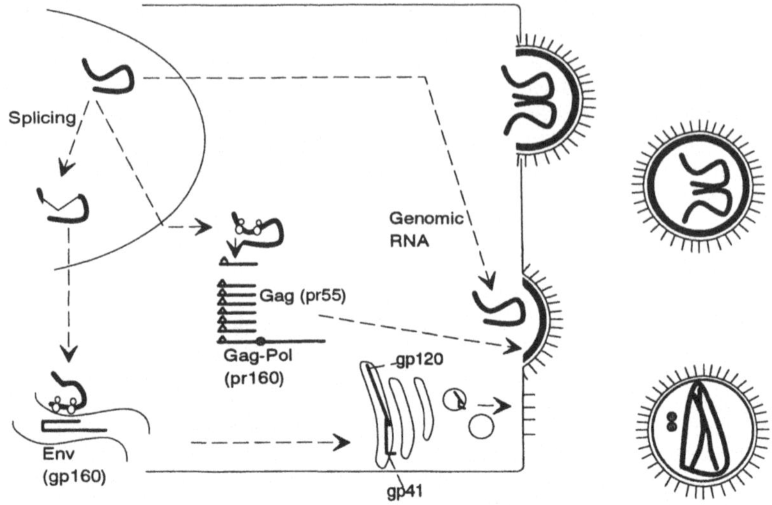

Figure 3. Late phase of human immunodeficiency virus type 1 replication. The primary transcript of the HIV genome is shown in the cell nucleus as a solid line; also indicated are its three functions as genomic RNA and as mRNA for the Gag-Pol and Env proteins, respectively, leading to viral assembly, budding, and release. Putative N-terminal targeting signals on the Gag domain (including myristoylation) are shown as triangles, the PR domain is depicted as a circle and the PR dimer in the mature virion as a double circle.

the plasma membrane where they interact with each other, with the viral glycoproteins, and with the genomic RNA to assemble the immature viral particle (Fig. 4a and b). Bud formation in the case of HIV (and other lentiviruses and type C oncoviruses) occurs concomitant with the assembly of the viral core (Fig. 4a and b) whereas in type B and D oncoviruses and spumaviruses budding occurs from preformed cores (reviewed in Gelderblom, 1991). Early immature retroviral cores are spherical, about 80 nM in diameter, and are enveloped by a host derived lipid bilayer which contains the viral glycoproteins (Fig. 4b). Condensation of the immature core shell to the typical lentiviral cone-shaped core is necessary for the virion to become infectious and depends on proteolytic processing of the precursor proteins by the viral proteinase (Fig. 5; reviewed in Kräusslich and Wimmer, 1988). Morphogenesis of the infectious virus therefore requires the morphopoetic function of the structural components of the viral core (encoded by the *gag* gene) as well as the incorporation of functional enzymes (proteinase, PR; reverse transcriptase, RT; integrase, IN; derived from the *pol* gene) and the viral glycoproteins, gp120 (SU) and gp41 (TM; Fig. 5). In contrast to the requirement for *pol*- and *env*-gene products for viral infectivity, formation of non-infectious, morphologically immature virus-like particles depends solely on the expression of the Gag polyprotein and does not require proteolytic processing or the interaction with glycoproteins. Accordingly, several groups have recently reported HIV polyprotein synthesis and production of recombinant subviral particles following transient or stable expression of *gag* gene products from viral or plasmid vectors (Mergener et al., 1992 and references therein). Although retroviral glycoproteins are not essential for particle formation, the presence and subcellular localization of *env* gene products may influence the site of virion assembly. Owens et al. (1991) have shown that in polarized cells virus-like particles were released randomly in the absence of *env* gene products, whereas in their presence, budding occured almost exclusively at the basolateral membrane where the HIV-1 glycoproteins were localized.

The $pr55^{gag}$ precursor can be divided into the four separate functional domains matrix (MA), capsid (CA), nucleocapsid (NC), and p6 (Fig. 1), which need to become proteolytically separated before maturation of the virion. Cotranslational myristoylation of the N-terminal MA domain which is closely apposed to the inner surface of the lipid envelope of the virus (Fig. 5; Gelderblom et al., 1987) has been shown to be essential for production of extracellular particles (Bryant and Ratner, 1990; Göttlinger et al., 1989). It is likely, however, that besides N-terminal acylation, additional signals governing membrane targeting reside on MA (see below). Furthermore, although the requirements for specific incorporation of viral glycoproteins into virion particles are not understood in any viral system, it is likely that an interaction between the MA and TM proteins plays a role in this process. This hypothesis is derived from the close proximity of these proteins in the mature virion (Gelderblom et al., 1987; Fig. 5) and is supported by experiments showing that the Rous sarcoma virus TM protein can be chemically crosslinked to the MA protein (Gebhardt et al., 1984), and that in frame deletions within the MA domain of HIV-1 abolish incorporation of glycoproteins (Yu et al., 1992).

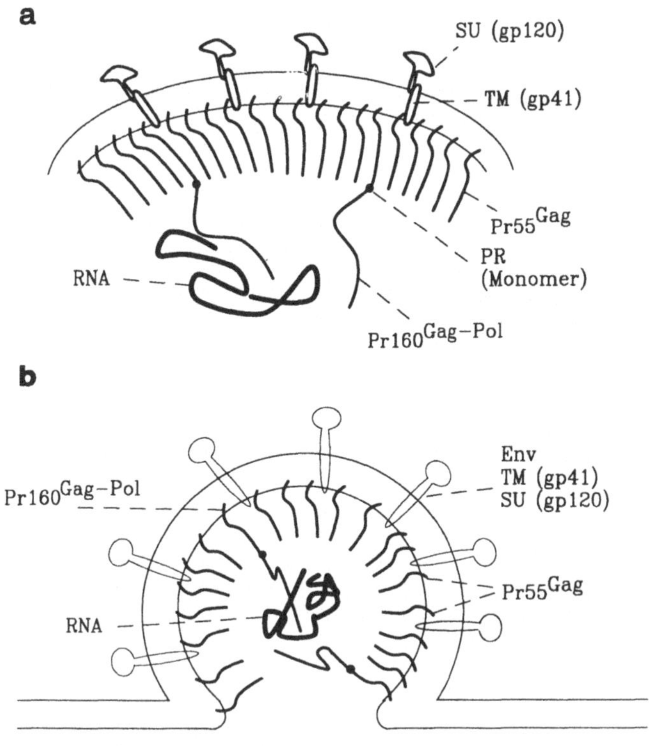

Figure 4. Schematic representation of human immunodeficiency virus morphogenesis. The viral Gag- and Gag-Pol polyproteins are depicted as solid lines, the genomic RNA is shown as a thick line and the PR domain is indicated as a circle. **(a)** Assembly at the plasma membrane; **(b)** formation of the viral bud.

The CA protein (p24) forms the eccentric, cone-shaped core shell of the mature virion which encases a ribonucleoprotein complex consisting of the dimeric viral RNA genome associated with the NC protein (Fig. 5) and probably also contains the viral enzymes. Mutations that prevent either N-terminal or C-terminal cleavage of CA have been shown to interfere with virion maturation and to abolish viral infectivity (Göttlinger et al., 1989; M. Fäcke and H.-G.K.; unpublished results). Furthermore, C-terminal processing of CA involves cleavage at two different sites which are only fourteen amino acids apart and mutational analysis suggests that these two sites are neither functionally equivalent nor

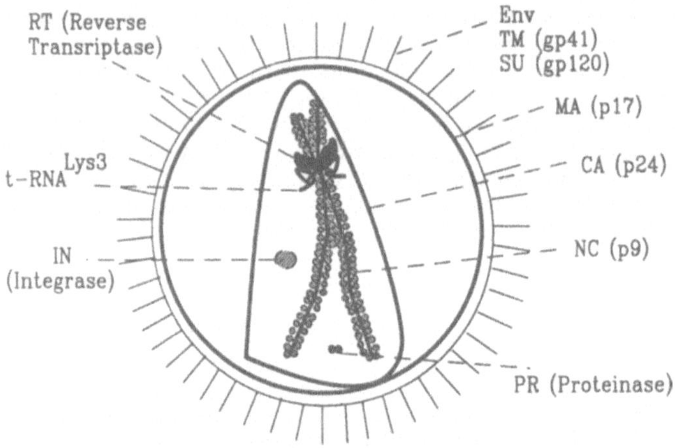

Figure 5. Schematic representation of mature infectious human immunodeficiency virus. The viral proteins as well as the lipid envelope and the RNA components are depicted.

redundant (M. Fäcke and H.-G.K., unpublished results). Sequential processing of these sites may therefore serve an important, as yet undefined function in HIV morphogenesis (see below).

While it is well established that the MA and CA domains or segments thereof are required for assembly and budding at the plasma membrane, the contributions of the C-terminal NC and p6 domains are less well understood. NC is a small basic nucleic acid binding protein that, in the case of HIV, contains two zinc-finger motives and is believed to be involved in packaging of the viral genome. However, mature NC exhibits very little if any binding specificity for the putative packaging signal on the viral RNA (located in the 5' untranslated and *gag* region; Lever et al., 1989; Luban and Goff, 1991) and RNA binding appears to be mediated by basic amino acids in NC and not by the Cys-His boxes (De Rocquigny et al., 1992). Besides its proposed function in RNA packaging, NC is also believed to be important in the initiation of reverse transcription in the infected cell. Accordingly, purified NC protein and NC-derived peptides have been shown to stimulate cDNA synthesis in an *in vitro* assay employing a HIV-specific RNA template, t-RNA^{lys3} primer, and purified HIV RT (De Rocquigny et al., 1992; B. Deckert and H.-G.K., unpublished results). However, neither the mechanism of NC function in RNA packaging and in reverse transcription nor the potential contribution of the zinc-finger motives are presently understood. It is also not clear whether there is a structural contribition of NC in viral morphogenesis.

All lentiviruses and D-type oncoviruses, but not the other retroviruses, encode an additional domain of unknown function at the C-terminus of their Gag polyprotein (reviewed in Gelderblom, 1991). In the case of HIV, this p6 domain (Fig. 1) has been proposed to form a specific core-envelope-link (CEL; Höglund et al., 1992) in the mature virion but currently there is no direct evidence supporting this hypothesis. The p6 domain does not appear to be required for assembly but, using transient transfection of proviral constructs in COS cells, Göttlinger et al. (1991) have shown that a deletion of p6 had a marked effect in the late stage of HIV budding, effectively preventing release of assembled immature particles from the outside of the plasma membrane. By contrast, expression of a truncated Gag polyprotein lacking p6 and the twelve C-terminal amino acids of NC from a recombinant vaccinia virus vector did not appear to reduce particle release or to alter particle morphology (Hoshikawa et al., 1991). The same result was obtained following expression of a similarly truncated Gag polyprotein from a recombinant baculovirus vector (Royer et al., 1992). Currently, it is not clear whether the observed differences are due to a contribution of additional HIV sequences not present in the heterologous expression systems, to subtle differences in the various deletions made, or to the different host cells used. At least in the baculovirus expression sytem, additional deletion of the entire NC domain also did not appear to affect release of virus-like particles, although in this case no electron microscopic analysis was performed (Royer et al., 1992). Further deletions, however, removing the short sequence between the two C-terminal cleavage sites of CA yielded a dramatically altered phenotype leading to the complete loss of virus-like particle formation (Gheysen et al., 1989; Hoshikawa et al., 1991). Truncation of the Gag polyprotein after the C-terminal amino acid of mature CA (p24, as opposed to CA^+, p25) abolished regular lentiviral budding of spherical immature particles and led to the assembly of large tubular protrusions connected to the plasma membrane (Gheysen et al., 1989). These observations argue for an important function of the C-terminal sequence of CA^+ in spherical particle formation, probably as a component of the Gag polyprotein. Assembly of rod shaped structures was also observed following in vitro incubation of purified completely processed CA (p24) protein (Ehrlich et al., 1992) and may thus be an inherent property of HIV-1 CA. However, the relationship between the in vitro assembled rods and the tubular structures obtained in recombinant baculovirus infected cells is currently not understood.

Assembly of viral polyproteins may be a simple and efficient way of ensuring that different components of the virion are all localized to the assembly site using a single targeting signal. Evidence for such a signal has been reported for the matrix domain of a D-type retrovirus. In this case, specific mutations in MA were found to prevent targeting of preformed cores to the plasma membrane, stable interaction with the lipid bilayer, or extrusion of the viral bud (Rhee and Hunter, 1991). Moreover, a single point mutation within MA changed the budding phenotype of the virus, indicating that intracellular transport and interaction with cellular structures and proteins may be determined by the MA domain (Rhee and Hunter, 1990). Furthermore, a large deletion within the MA domain of HIV-1 reduced particle release by approximately 90% and led to intracellular budding of morphologically immature viruses into the ER (M. Fäcke and H.-G.K., unpublished observations). Thus, it appears likely that besides acylation of the N-terminus, "signal sequences" within MA are important in determining the intracellular transport and association of retroviral polyproteins. These conclusions can be summarized in the following model: Retroviral polyproteins interact through their MA domains with cellular receptor proteins which may be associated with the ER or may be free in the cytoplasm. Subsequently, the protein complex migrates to the plasma membrane. This can either be achieved by membrane-associated transport (which may also explain the role of glycoproteins in determining the budding site) or by bulk flow of cytoplasmic proteins. Besides its function in targeting, formation of a protein complex may also prevent premature activation of viral enzymes which would be deleterious for viral replication (see below). At the budding site, intermolecular association of viral polyproteins may destabilize complex interactions, thereby eliminating cellular proteins from the nascent particle. Additional important factors in this process may be local changes in ionic environment or pH. This model predicts the existence of specific cellular Gag-receptors and, while there is currently no direct evidence for their existence, future experiments will address the identification and characterization of such molecules.

Besides a specific signal, targeting of stable polyproteins, which are proteolytically cleaved only when confined to the budding particle, also requires that a specific proteinase be targeted to the assembly site which becomes activated only in the budding particle. Controlled limited proteolysis can be achieved by synthesizing an inactive form of the viral proteinase as part of the polyprotein which is activated upon assembly of the viral particle. At the site of assembly, cellular or viral factors may be encountered that lead to activation of an inactive zymogen possibly by post-translational modification or by alteration of the pH or ionic environment. Alternatively, there might not be activation of a zymogen but rather relief of inactivation by dissociating a cellular or viral inhibitor bound to the polyprotein. Conceivably, this inhibition may be achieved by association of viral and cellular proteins in the process of protein targeting as discussed above. Such polyprotein complexes may prevent proteolytic activity by local or general unfolding or by binding and obliterating the active site or substrate binding pocket. A third non-exclusive possibility would be that initiation of the proteolytic cascade requires cooperative action of multiple molecules of the viral proteinase and, thus, a critical concentration can only be achieved in the budding particle. HIV PR is translated as a component of the Gag-Pol polyprotein but is made as an enzymatically inactive monomer which requires dimerization to achieve catalytic competence. Expression of a covalently linked, fully competent proteinase dimer as part of the viral polyprotein has been shown to induce premature processing, thereby dissociating the components of the virion prior to reaching the assembly site (Kräusslich, 1991; Kräusslich, 1992). A similar phenotype has been observed following overexpression of PR as component of an in frame Gag-Pol fusion protein (Mergener et al., 1992). Concentration-dependent dimerization of PR therefore appears to be an important, albeit not the only, factor controlling the initiation of proteolysis.

Besides inducing premature cleavage of viral polyproteins, overexpression of PR is also exquisitely cytotoxic (Kräusslich, 1992a). Intracellular activation of PR which is observed in lytically infected cells (Kaplan and Swanstrom, 1991) may therefore contribute to HIV-induced cell killing and may play a role in HIV induced pathogenesis. Inactivation of PR (Kohl et al., 1988), on the other hand, or partial inhibition of PR activity (Kräusslich, 1992a) also abolish viral infectivity. Moreover, PR-defective HIV particles display a pronounced morphological alteration of the immature core shell and a marked increase in intracellular budding structures (Peng et al., 1989; H.-G.K., unpublished observation). We interpret these results as evidence for a requirement of HIV PR activity prior to formation of the immature spherical core. This activity may be directed against HIV polyproteins or against cellular proteins at the assembly site. In summary, both activation as well as inactivation of HIV PR and even subtle changes of PR activity can lead to a dramatic alteration of viral morphogenesis and to loss of virion infectivity suggesting that tight regulation of PR activity is an important control mechanism in the viral replication cycle.

CONCLUSIONS

Retroviruses have evolved a complex mode of regulation of gene expression and even the simpler animal oncoviruses utilize most of the currently known strategies for the expression of multiple proteins from a single primary transcript. A more complicated pattern emerges from the analysis of complex retroviruses with HIV-1 as their prototypic example. In this case, more than twenty different mRNAs, some of them bicistronic, are synthesized. In addition, the regulation of HIV gene expression involves novel *cis*-acting elements and *trans*-acting factors, permitting the controlled and consecutive expression of defined genetic elements after acute infection, a situation that may be accurately recapitulated upon reactivation of latent genomes.

While recent experiments have shed some light on the relative importance of certain gene products and have provided information regarding their biological consequences, the molecular principles underlying their function are poorly understood. Future studies are likely to identify a multitude of cellular factors involved in their mechanism of action. Besides the extensively studied transcriptional and post-transcriptional control mechanisms, recent experiments also indicate a complex mode of virion morphogenesis and maturation providing an additional important control mechanism in viral replication. Finally, detailed analysis of the molecular mechanisms involved in HIV protein expression and virion formation will almost certainly permit identification of novel virus-specific targets for the development of selective and effective antiviral compounds.

ACKNOWLEDGMENTS

I thank V. Vogt for valuable suggestions and V. Bosch, B. Deckert, M. Fäcke, H. Kottler, and A.-M. Traenckner for discussion. I am also grateful to H. zur Hausen for support and interest. Work in the author's laboratory is supported by a grant from the Ministery for Research and Technology (FG5-1075).

REFERENCES

Bedinger, P., Moriarty, A., Von Borstel, R.C. II, Donovan, N.J., Steimer, K.S., and Littman, D.R. 1988. Internalization of the human immunodeficiency virus does not require the cytoplasmic domain of CD4. *Nature* **334**, 162-16 .

Bhat, S., Spitalnik, S.L., Gonzalez-Scarano, F., and Silberberg, D.H. 1991. Galactosyl ceramide or a derivative is an essential component of the neural receptor for human immunodeficiency virus type 1 envelope glycoprotein gp120. *Proc. Natl. Acad. Sci.* USA **88**, 7131-7134.

Bryant, M.L. and Ratner, L. 1990. Myristoylation-dependent replication and assembly of human immunodeficiency virus 1. *Proc. Natl. Acad. Sci.* USA **87**, 523-527.

Bukrinsky, M.I., Sharova, N., Dempsey, M.P., Stanwick, T.L., Bukrinskaya, A.G., Haggerty, S., and

Stevenson, M. 1992. Active nuclear import of human immunodeficiency virus type 1 preintegration complexes. *Proc. Natl. Acad. Sci.* USA **89**, 6580-6584.

Coffin, J.M. 1990. Retroviridae and their replication. In: *Virology* (eds. B.N. Fields, D.M. Knipe), Raven Press, New York, pp. 1437-1500.

Calnan, B.J., Tidor, B., Biancalana, S., Hudson, D., and Frankel, A.D. 1991. Arginine-mediated RNA recognition: the arginine fork. *Science* **252**, 1167-1171.

Cullen, B.R. 1991 Human inmmunodeficiency virus as a prototypic complex retrovirus. *J. Virol.* **65**, 1053-1056.

De Rocquigny, H., Gabus, C., Vincent, A., Fournié-Zaluski, M.-C., Roques, B., and Darlix, J.-L. 1992. Viral RNA annealing activities of human immunodeficiency virus type 1 nucleocapsid protein require only peptide domains outside the zinc fingers. *Proc. Natl. Acad. Sci.* USA **89**, 6472-6476.

Ehrlich, L.S., Agresta, B.E., and Carter, C.A. 1992. Assembly of recombinant human immunodeficiency virus type 1 capsid protein in vitro. *J. Virol.* **66**, 4874-4883.

Feinberg, M.B., Baltimore, D., and Frankel, A.D. 1991. The role of tat in the human immunodeficiency virus life cycle indicates a primary effect on transcriptional elongation. *Proc. Natl. Acad. Sci.* USA **88**, 4045-4049.

Gallagher W.R. 1987. Detection of a fusion peptide sequence in the transmembrane protein of human immunodeficiency virus. *Cell* **50**, 327-328.

Gebhardt, A., Bosch, J.V., Ziemiecki, A., and Friis, R.R. 1984. Rous sarcoma virus p19 and gp35 can be chemically cross-linked to high molecular weight complexes: An insight into virus assembly. *J. Mol. Biol.* **174**, 297-317.

Gelderblom, H.R. 1991. Assembly and morphology of HIV: potential effect of structure on viral function. *AIDS* **5**, 617-638.

Gelderblom, H.R., Hausmann, E.H., Özel, M., Pauli, G., and Koch, M.A. 1987. Fine structure of human immunodeficiency virus (HIV) and immunolocalization of structural proteins. *Virology* **156**, 171-176.

Gheysen, D., Jacobs, E., de Foresta, F., Thiriart, C., Francotte, M., Thines, D., and deWilde, M. 1989. Assembly and release of HIV-1 precursor pr55gag virus-like particles from recombinant baculovirus-infected insect cells. *Cell* **59**, 103-112.

Göttlinger, H.G., Dorfman, T., Sodroski, J.G., and Haseltine, W.A. 1991. Effect of mutations affecting the p6 gag protein on human immunodeficiency virus particle release. *Proc. Natl. Acad. Sci.* USA **88**, 3195-3199.

Göttlinger, H.G., Sodroski, J.G., and Haseltine, W.A. 1989. Role of capsid precursor processing and myristoylation in morphogenesis and infectivity of human immunodeficiency virus type 1. *Proc. Natl. Acad. Sci.* USA **86**, 5781-5785.

Goto, T., Harada, S., Yamamoto, N., and Nakai, M. 1988. Entry of human immunodeficiency virus (HIV) into MT-2, human T-cell leukemia virus carrier cell line. *Arch. Virol.* **102**, 29-38.

Harouse, J.M., Bhat, S., Spitalnik, S.L., Laughlin, M., Stefano, K., Silberger, D.H., and Gonzalez-Scarano, F. 1991. Inhibition of entry of HIV-1 in neural cell lines by antibodies against galactosyl ceramide. *Science* **253**, 320-323.

Hoshikawa, N., Kojima, A., Yasuda, A., Takayashiki, E., Masuko, S., Chiba, L., Sata, T., and Kurata, T. 1991. Role of the gag and pol genes of human immunodeficiency virus in the morphogenesis and maturation of retrovirus-like particles expressed by recombinant vaccinia virus: an ultrastructural study. *J. Gen. Virol.* **72**, 2509-2517.

Höglund, S., Öfverstedt, L.G., Nilsson, A., Lundquist, P., Gelderblom, H., Özel, M., Winkel, T., and Skoglund, U. 1992. Spatial visualization of the maturing HIV-1 core and its linkage to the envelope. *AIDS Res. Human Retrovir.* **8**, 1-7.

Jacks, T., Power, M.D., Masiarz, F.R., Luciw, P.A., Barr, P.J., and Varmus, H.E. 1988. Characterization of ribosomal frameshifting in HIV-1 gag-pol expression. *Nature* (London) **331**, 280-283.

Kaplan, A.H. and Swanstrom, R. 1991. Human immunodeficiency virus type 1 gag proteins are processed in two cellular compartments. *Proc. Natl. Acad. Sci.* USA **88**, 4528-4532.

Kohl, N.E., Emini, E.A., Schleif, W.A., Davis, L.J., Heimbach, J.C., Dixon, R.A.F., Scolnick, E.M., and Sigal, I.S. 1988. Active human immunodeficiency virus protease is required for viral infectivity. *Proc. Natl. Acad. Sci.* USA **85**, 4686-4690.

Kräusslich, H.-G. 1992a. Genetic analysis and gene expressionn of human immunodeficiency virus. *Curr. Opin. Genetics Dev.* **2**, 82-89.

Kräusslich, H.-G. 1991. Human immunodeficiency virus proteinase dimer as component of the viral polyprotein prevents particle assembly and viral infectivity. *Proc. Natl. Acad. Sci.* USA **88**, 3213-3217.

Kräusslich, H.-G. 1992b. Specific inhibitor of human immunodeficiency virus proteinase prevents the cytotoxic effects of a single-chain proteinase dimer and restores particle formation. *J. Virol.* **66**, 567-572.

Kräusslich, H.-G. and Wimmer, E. 1988. Viral proteinases. *Annu. Rev. Biochem.* **57**, 701-754.

Kestler, H.W., Ringler, D.J., Mori, K., Panicall, D.L., Sehgal, P.K., Daniel, M.D., and Desrosiers, R.C. 1991. Importance of the nef gene for maintenace of high virus loads and for development of AIDS. *Cell* **65**, 651-662.

Leis, J., Baltimore, D., Bishop, J.M., Coffin, J., Fleissner, E., Goff, S.P., Oroszlan, S., Robinson, H., Skalka, A.M., Temin, H.M., and Vogt, V. 1988. Standardized and simplified nomenclature for proteins common to all retroviruses. *J. Virol.* **62**, 1808-1809.

Lever, A., Göttlinger, H., Haseltine, W., and Sodroski, J. 1989. Identification of a sequence required for efficient packaging of human immunodeficiency virus type 1 RNA into virions. *J. Virol.* **63**, 4085-4087.

Levy, J.A. 1990. Changing concepts in HIV infection: challenges for the 1990s. *AIDS* **4**, 1051-1058.

Lewis, P., Hensel, M., and Emerman, M. 1992. Human immunodeficiency virus infection of cells arrested in the cell cycle. *EMBO J.* **11**, 3053-3058.

Luban, J. and Goff, S.P. 1991. Binding of human immunodeficiency virus type 1 (HIV-1) RNA to recombinant HIV-1 gag polyprotein. *J. Virol.* **65**, 3203-3212.

Maddon, P.J., McDougal, J.S., Clapham, P.R., Dalgleish, A.G., Jamal, S., Weiss, R.A., and Axel, R. 1988. HIV infection does not require endocytosis of its receptor, CD4. *Cell* **54**, 865-874.

Mergener, K., Fäcke, M., Welker, R., Brinkmann, V., Gelderblom, H.R., and Kräusslich, H.-G. 1992. Analysis of HIV particle formation using transient expression of subviral constructs in mammalian cells. *Virology* **186**, 25-39.

Owens, R.J., Dubay, J.W., Hunter, E., and Compans, R.W. 1991 Human immunodeficieny virus envelope protein determines the site of virus release in polarized epithelial cells. *Proc. Natl. Acad. Sci.* USA **88**, 3987-3991.

Peng, C., Ho, B.K., Chang, T.W., and Chang, N.T. 1989. Role of human immunodeficiency virus type1-specific protease in core protein maturation and viral infectivity. *J. Virol.* **63**, 2550-2556.

Pomerantz, R.J., Trono, D., Feinberg, M.B., and Baltimore, D. 1990. Cells nonproductively infected with HIV-1 exhibit an aberrant pattern of viral RNA expression: a molecular model for latency. *Cell* **61**, 1271-1276.

Pryciak, P.M. and Varmus, H.E. 1992. Nucleosomes, DNA-binding proteins, and DNA sequence modulate retroviral integration target site selection. *Cell* **69**, 769-780.

Rhee, S.S. and Hunter, E. 1990. A single amino acid substitution within the matrix protein of a type D retrovirus converts its morphogenesis to that of a type C retrovirus. *Cell* **63**, 77-86.

Rhee, S.S. and Hunter, E. 1991. Amino acid substitutions within the matrix protein of type D retroviruses affect assembly, transport and membrane association of a capsid. *EMBO J.* **10**, 535-546.

Royer, M., Hong, S.-S., Gay, B., Cerutti, M., and Boulanger, P. 1992. Expression and extracellular release of human immunodeficiency virus type 1 Gag precursors by recombinant baculovirus-infected cells. *J. Virol.* **66**, 3230-3235.

Sattentau, Q.J. and Weiss, R.A. 1988. The CD4 antigen: physiological ligand and HIV receptor. *Cell* **52**, 631-633.

Stein, B.S., Gowda, S.D., Lifson, J.D., Penhallow, R.C., Bensch, K.G., and Engleman, E.G. 1987. pH-independent HIV entry into CD4-positive T cells via virus envelope fusion to the plasma membrane. *Cell* **49**, 659-668.

Stevenson, M., Stanwick, T., Dempsey, M., and Lamonica, C. 1990. *EMBO J.* **9**, 1551-1560.

Weeks, K.M. and Crothers, D.M. 1991. RNA recognition by Tat-derived peptides: interaction in the major groove. *Cell* **66**, 577-588.

Yu, X., Yuan, X., Matsuda, Z., Lee, T.-H., and Essex, M. 1992. The matrix protein of human immunodeficiency virus type 1 is required for incorporation of viral envelope protein into mature virus. *J. Virol.* **66**, 4966-4971.

Zack, J.A., Arrigo, S.J., Weitsman, S.R., Go, A.S., Haislip, A., and Chen, I.S.Y. 1990. *Cell* **61**, 213-222.

TRANSCRIPTION FACTORS OF THE ETS FAMILY : THE EXAMPLE OF C-*ETS1*

Kay Macleod, Bernard Vandenbunder and Dominique Stehelin

CNRS URA 1160
Laboratory of Molecular Oncology
Pasteur Institute
59019 Lille Cedex - France

INTRODUCTION

Over the past few years, the fast growing ETS gene family has attracted considerable interest. Ets proteins have a conserved DNA-binding domain and regulate transcriptional initiation from a variety of cellular and viral gene promoter and enhancer elements. Some members of the Ets family, Ets-1 and Ets-2, cooperate in transcription with the AP-1 transcription factor, the product of the proto-oncogene families, *fos* and *jun,* while others, Elk-1 and SAP-1, form ternary complexes with the serum response factor. Some *ets* gene family members possess transforming activity or are located close to chromosome breakpoints detected in specific leukaemias.

The v-*ets* oncogene, the first member of the *ets* family to be described, was discovered in the E26 ("*E-t*wenty-*s*ix") acutely transforming retrovirus of chicken, from which it derives its name (Leprince et al., 1983; Watson et al., 1990). Other family members cloned from a variety of organisms (Leprince et al., 1983; Watson et al., 1990; Seth and Papas, 1990; Leprince et al., 1988; Koizumi et al., 1990; Karim et al., 1990; Boulukos et al., 1990; Reddy and Rao, 1991; Rao and Reddy, 1992; Hipskind et al., 1991; Ben-David et al., 1991; Pongubala et al., 1992; Thompson et al., 1991; LaMarco et al., 1991; Pribyl et al., 1991; Thompson et al., 1992; Dalton and Treisman, 1992; Xin et al., 1992; Waterman et al., 1991), are listed in Table 1. In most family members, the conserved DNA binding domain (the ETS domain), is localized in the carboxyl terminus of the protein, with the exception of Elk-1 and SAP-1 where it is found in the amino terminus (Rao et al., 1992 ; Hipsking et al., 1991 ; Dalton and Treisman, 1992). The ETS domain, which covers approximately 85 amino acids (figure 1), does not bear any structural homology to other known DNA-binding motifs such as the zinc finger,

Regulation of Gene Expression in Animal Viruses, Edited by
L. Carrasco *et al.*, Plenum Press, New York, 1993

homeodomain, leucine zipper or helix-turn-helix motifs (Karim et al., 1990) but has limited homolgy to the DNA binding domain of the c-*myb* oncogene product. In c-Myb, 9 tryptophan repeats are separated by 18 or 19 amino acids spaced around a predicted a-helical structure (Karim et al., 1990) but the ETS domain contains only 2 or 3 conserved tryptophan repeats (figure 1) (Karim et al., 1990).

The sequence conservation of members of the *ets* family and their chromosome locations provide clues to gene duplication and divergence events that gave rise to this gene family. For example, *ets*-1 and *fli*-1 both map to the same region of human chromosome 11q23 while *ets*-2 and *erg* both map to 21q22 (Watson et al., 1990; Ben-David et al., 1991). However, the homology between *ets*-1 and *ets*-2 and between *erg* and *fli*-1 is greater than the homology between those genes mapping next to each other on the same chromosome (Watson et al., 1990; Ben-David et al., 1991). This suggests that duplication of an ancestral *ets*-related gene may have occurred giving rise to two sister genes which subsequently diverged over time. This was followed by duplication of the entire locus to another chromosome.

The ETS Domain

```
                 ** *** **          * *     *B      *       **  B  B  ***    B * ***      B        * *
Ets-1    GSGP-IQLWQFLLELLTDKS-CQSFISWT-GDGWEFKLSDPDEVA--RRWGKRK-NKPKMNYEKLSRGLRYYYDKNIIHKTAGKRYVYRFV
Ets-2    GSGP-IQLWQFLLELLSDKS-CQSFISWT-GDGWEFKLADPDEVA--RRWGKRK-NKPKMNYEKLSRGLRYYYDKNIIHKTSGKRYVYRFV
GABP-α   NNGQ-IQLWQFLLELLTDKD-ARDCISWVGDEEGEFKLWQPELVA--QKWGQRK-NKPTMNYEKLSRALRYYYDGDMICKVQGKRFVYKFV
Erg      GSGQ-IQLWQFLLELLSDSS-NSSCITWE-GTNGEFKMTDPDEVA--RRWGERK-SKPNMNYDKLSRALRYYYDKNIMTKVHGKRYAYKFD
Fli-1    GSGQ-IQLWQFLLELLSDSA-NASCITWE-GTNGEFKMTDPDEVA--RRWGERK-SKPNMNYDKLSRALRYYYDKNIMTKVHGKRYAYKFD
D-Elg    NNGQ-VQLWQFLLEILTDCE-HTDVIEWV-GTEGEFKLTDPDRVA--RLWGEKK-NKPAMNYEKLSRALRYYYDGDMISKVSGKRFAYKFD
E74A     GST--TYLWEFLLKLLQDREYCPRFIKWTNREKGVFKLVDSKAVS--RLWGMHK-NKPDMNYETMGRALRYYYQRGILAKVDGQRLVYQFV
Elf-1    GNT--IYLWEFLLALLQDKATCPKYIKWTQREKGIFKLVDSKAVS--RLWGKHK-NKPDMNYETMGRALRYYYQRGILAKVEGGRLVYQF-
SAP-1    -MDSAITLWQFLLQLLQKPQ-NKHMICWTS-MDGQFKLLQAEEVA--RLWGIRK-NKPNMNYDKLSRALRYYYVKNIIKKVNGQKFVYKFV
Elk-1    PS---VTLWQFLLQLLREQG-NGHIISWTSADGGEFKLVDAEEVA--ALWGLAK-NKTNMNYDKLSRALAYYYDKNIIIAKVSGQKFVYKV
PEA 3    QRRGALQLWQFLVALLDDPT-NAHFIAWTG-RGMEFKLIEPEEVA--RLWGIQK-NRPAMNYDKLSRSLRYYYEKGIMQKVAGERYVYKFV
PU.1     GSKKKIRLYQFLLDLLRSGD-MKDSIWWVDKDKGTFQFSSKHKEALAHRWGIQKGNRKKMTYQKMARALRNYG-KTGEVKKVKKKLTYQFS
```

Figure 1 : Alignment of the primary amino acid sequences of the ETS domains. The conserved tryptophan residues (W) are boxed. Other highly conserved amino acids are indicated by asterisk (*) while conserved basic amino acid residues are indicated by (B).

REGULATION OF ETS EXPRESSION AND ACTIVITY

Ets gene expression is controlled at the level of initiation of transcription (Mavrothalassitis and Papas, 1991), alternative splicing (Leprince et al., 1988 ; Koizumi et al., 1990), post-translational modification (Boulukos et al., 1990) and protein stability (Pognonec et al., 1989). Expression patterns of different members of the *ets* gene family vary between tissues as indicated in Table 1.

In T-cells, expression of *ets*-1 is highest in quiescent cells and decreases following T-cell activation (Bhat et al., 1990). Conversely, *ets*-2 gene expression increases from a low level following T-cell activation to reach maximal levels 1-2 hours after induction (Bhat et al., 1990). The mechanism

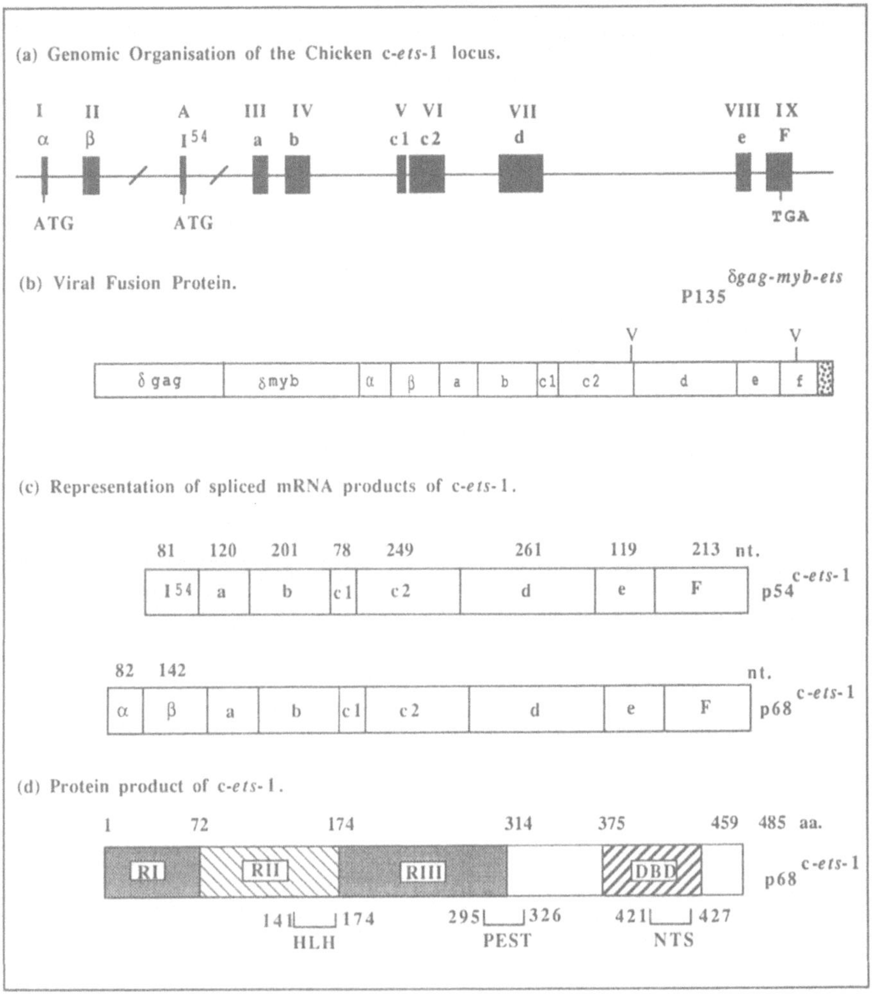

Figure 2 : (a) Genomic organization of the chicken c-*ets*-1 locus. Chicken exons (a, b, I^{54}, a, b, c1, c2, d, e and F) are shown as black boxes, along with an alternative nomenclature used for the chicken and human exons (I, II, A, III, IV, V, VI, VII, VIII, IX). Two translation start signals (ATG) are found in exons a and I^{54} while the stop signal (TGA) is in exon F.
(b) The 135 kD viral fusion protein of the E26 virus showing the organisation of *ets* sequences relative to those of *gag* and *myb* and the E26 virus has transduced the a and b but not the I^{54} exon. The positions of non-conservative mutations (V) in v-*ets* compared with c-*ets*-1 and a change in carboxy terminal sequences are also shown.
(c) The organisation and sizes of the different exons in the alternatively spliced c-*ets*-1 mRNAs which give rise to p54$^{c\text{-}ets\text{-}1}$ and p68$^{c\text{-}ets\text{-}1}$. The sizes in nucleotides of the exons are indicated.
(d) Functional domains of the p68 $^{c\text{-}ets\text{-}1}$ protein. The amino acid residues are numbered and aligned with the nucleotide sequences of the p68 mRNA in (c) above, thus indicating the exons that code for each functional domain of the protein. RI, region I (transactivation domain); RII, region II (regulatory domain); RIII, region III (transactivation domain); DBD, DNA binding domain ; HLH, helix-loop-helix domain; PEST, region rich in proline, glutamic acid, serine and threonine riesidues which may target protein cleavage to this region of the protein ; NTS, nuclear targeting signal.

Table 1. The Ets Gene Family.

Gene	Source	Protein(kD)	% amino acid homology to ETS domain of Ets-1	Human chromosomal location.	Expression/Features[Ref.]
Ets-1	chicken human mouse	54/68 kD 39-52 kD 63 kD	100%	11q23	elevated expression in thymus and endothelial cells, phosphorylated, alternatively spliced, positively autoregulates transcription[1,2,3,4,5,6].
Ets-2	chicken human	58/62 kD	90%	21q22	expression induced following macrophage differentiation and T cell activation, alternatively spliced, phosphorylated[2,6,7].
Erg	human	41/52 kD	70%	21q22	alternatively spliced, 98% homologous to Fli-1 [6,8]
Fli-1	mouse human	51 kD	68%	11q23	activated by proviral insertion of Friend MuLV, 98% homologous to Erg [11]
Elk-1	human	60 kD	76%	Xp11.2	ETS domain located in the amino terminus of protein, forms ternary complex with the serum response factor, is conserved in three domains to SAP-1 6,9,10
SAP-1a/b	human	58/52 kD	75%		SRF accessory protein 1, which like Elk-1 forms ternary complex with SRF over the c-*fos* SRE, contains three regions of homology to Elk-1 including the ETS domain which is located in amino terminus of the protein, two isoforms, SAP-1a and SAP-1b which differ in their carboxyl termini[17]
Spi-1/PU1	mouse human	30 kD	38%	11p11.22	activated in Friend erythroleukemia by proviral insertion of SFFV, normal expression of the PU-1 transcription factor is restricted to B cells and macrophages[6, 12]
E74A/B	Drosophila	110/120 kD	50%	Drosophila chr. 3L74EF	E74A is induced by ecdysone and regulates the expression of E74B, which is also *ets*-related [6]
Elf-1	human	68 kD	50%		the ETS domain of this protein is the human homologue of the E74A protein of Drosophila, binds to the NF-AT and NFIL-2B sites in the interleukin-2 promoter and the human immunodeficiency virus II LTR[16]
GABP-α	rat	51 kD	82%		high level expression in rat thymus, complexes with GABP-β which contains ankyrin repeats and is related to the *notch* protein[13,14]
D-Elg	Drosophila	15 kD	64%	Drosophila chr. 3R97D	contains only a DNA binding domain, maternally expressed message and also expressed throughout embryogenesis[15]
PEA 3	mouse	68 kD	63%		Polyomavirus Enhancer Activator 3, expressed in mouse brain and epididymis, in fibroblast and epithelial cell lines, down-regulated in embryonic cell lines in response to retinoic acid induced differentiation[18]
TCF1-α	human	55 kD			very limited homology to ETS domain exists within the HMG box of this factor, expression is restricted to the thymus and is induced following T cell activation, regulates activity of the TCRα enhancer[19]

of coordination of the expression of these two genes during T-cell differentiation is unclear.

Regulation of transcriptional initiation is best understood for the human *ets*-2 gene. In common with all other *ets* promoters described and many other oncogene promoters, the *ets*-2 promoter lacks the classical TATA and CCAAT box sequences found in many RNA polymerase II transcribed genes. A putative Ets binding site (EBS) has been identified downstream of the promoter (Table 2) and *ets*-2 may be negatively autoregulated. Adjacent to the EBS is a putative AP-1 site and factors bound at these sites have been shown to cooperate in transactivation (Mavrothalassitis and Papas, 1991). These EBS and AP-1 sites bind activator or repressor factors in a cell-specific fashion (Mavrothalassitis and Papas, 1991).

The promoter of the human *ets*-1 gene contains putative AP-1, EBS, AP-2 and SP1 sites. Expression of exogenous AP-2 and Ets-1 augments transcription from the *ets*-1 promoter implying an activating role for AP-2 in *ets*-1 expression and positive autoregulation of *ets*-1 transcription (Seth and Papas, 1990 ; Oka et al., 1991).

Chicken and human *ets*-1 gene RNAs are alternatively spliced (Leprince et al., 1988 ; Koizumi et al., 1990). In chicken, two forms of Ets-1 mRNA are generated by alternative splicing of 5' exons; a and b of p68 are replaced in p54 by exon I^{54} (figure 2). p68, unlike p54, is not expressed in the thymus or T cells but is expressed in endothelial cells (Leprince et al., 1988). The differential function of these proteins is not clear but may be associated with the different transactivation domains encoded by the 5' exons (Schneikert et al., 1992). Exons a and b have not been identified in any mammalian species (O. Albagli *pers comm*). In human *ets*-1, alternative splicing of exon IV and of exon VII has been observed (Koizumi et al., 1990). Exon VII encodes a domain whose negative regulatory effect on the DNA binding domain may be the result of phosphorylation of serine residues in this region (Koizumi et al., 1990).

Phosphorylation of both chicken and human Ets-1 and Ets-2 results in the loss of non-specific DNA binding activity (Pognonec et al., 1989). The cellular kinases responsible for phosphorylating Ets-1 and Ets-2 proteins have not yet been identified although putative sites of phosphorylation for protein kinase C, casein kinase II and calmodulin kinase have been identified. Following activation of T-cells, in addition to the reduced level of *ets*-1 transcription, there is an increase in phosphorylation of Ets-1 that results in reduced DNA binding activity (Pognonec et al., 1989). Phosphorylation of Ets-2 also increases following T-cell activation (Fujiwara et al., 1988). Phosphorylation appears to increase the half-life of the Ets-2 protein, from approximately 20 minutes to greater than 2 hours. Protein kinase C inhibitors block this effect suggesting that PKC directly/indirectly modifies the phosphorylation state of Ets-2 (Fugiwara et al., 1988).

PEST sequences which may play a role in protein turnover are found in both PU-1 (from amino acids 121-159) (Pongubala et al., 1992) and $p68^{c\text{-}ets\text{-}1}$(from amino acids 295-326) proteins. The PEST sequences are located between putative transactivation domains and DNA binding domains and cleavage may create a DNA-binding domain free from the regulatory restraints imposed by amino-terminal domains.

SPECIFICITY OF ETS BINDING TO DNA

The $p68^{c\text{-}ets\text{-}1}$ protein binds to the PEA 3 element of the polyoma virus enhancer (Wasylyk et al., 1990). The DNA sequence specifically recognized by $p68^{c\text{-}ets\text{-}1}$ in the polyoma enhancer, AGCAGGAAGT, binds $p68^{c\text{-}ets\text{-}1}$ with moderate affinity as determined by electrophoretic mobility shift assay (EMSA). Mutational analysis of the PEA 3 element has identified higher affinity Ets binding sites (EBS), such as AGCCGGAAGT (Wasylyk et al., 1991; Fisher et al, 1991), in which the second A has been replaced by C. Selection and amplified binding studies further identified mutations which increase the binding affinity of the EBS for p54$^{c\text{-}ets\text{-}1}$, such as a G to A transition in addition to the A to C transversion to generate AACCGGAAGT (Woods et al., 1992). p54$^{c\text{-}ets\text{-}1}$, p58$^{c\text{-}ets\text{-}2}$, Erg and Elk-1 show similar affinities for the same sequences as p68$^{c\text{-}ets\text{-}1}$ (Reddy and Rao, 1991; Rao and Reddy, 1992; Wasylyk et al., 1991) while p30$^{spi\text{-}1}$ and Elf-1 appear to have a different sequence specificities (Thompson et al., 1992 ; Wasylyk et al., 1991).

EBS have been identified in other cellular and viral gene regulatory sequences (Table 2) including the long terminal repeats (LTR) of the human T-cell leukaemia virus-I (HTLV-I) (Gitlin et al., 1991), the Moloney Sarcoma Virus (MSV) LTR (Gunther et al., 1990) and the rat stromelysin 1 promoter (Wasylyk et al., 1991). A combination of EMSA and methylation interference studies on the binding of Ets proteins to the MSV (Gunther et al., 1990) and HTLV-I (Gitlin et al., 1991) LTRs has shown that the GGAA purine core is essential for the specific binding of Ets related proteins to these sequences. However, sites within certain T-cell specific enhancers, such as the enhancers of the T-cell receptor-a (Ho et al., 1990) and b genes (Dalton and Treisman, 1992), diverge from this rule. In these cases, an A to T transversion in the purine core to generate GGAT is frequently observed (Table 2).

When viral point mutations and the carboxyl terminus of the v-*ets* protein were introduced into a 35 kD version of the p68$^{c\text{-}ets\text{-}1}$ protein, the altered protein lost its ability to bind to the EBS (Leprince et al., 1992). This suggested that the v-*ets* protein has a different DNA binding specificity to p68$^{c\text{-}ets\text{-}1}$ or that the p68$^{v\text{-}ets}$ protein does not bind DNA at all (Leprince et al., 1992). Conflicting evidence has emerged showing that the v-*ets* protein has a stronger binding affinity for the EBS than the full-length p68$^{c\text{-}ets\text{-}1}$ protein and that cleavage of c-Ets-1 generates a truncated protein that binds strongly to the EBS (Lim et al., 1992). It has been suggested that a negative regulatory domain in the amino-terminus of Ets-1 that interacts with the carboxyl-terminus of the protein to prevent DNA binding. Cleavage of the protein would derepress DNA-binding by p68$^{c\text{-}ets\text{-}1}$ (Lim et al., 1992). The v-*ets* protein has a different carboxyl-terminus and it has been proposed that the negative regulatory domain of the protein would not function in the viral protein since it would not be able to associate with carboxy-terminal sequences present only in the cellular protein (Lim et al., 1992). It will be interesting to determine whether the viral fusion protein active in erythroleukaemia, P135$^{gag\text{-}myb\text{-}ets}$ binds the EBS or not.

ETS TARGET GENES

The identification of EBS in different promoters/enhancers has permitted identification of genes which may be regulated by Ets-related proteins and expression of these genes often correlates with the expression of Ets-related proteins. A significant number of lymphoid-specific genes contain EBS in their regulatory sequences, including the enhancer of the T-cell receptor-a (Ho et al., 1990) and the enhancer of the T-cell receptor-b (M. Owen, *pers comm*). The EBS of the TCR-b enhancer mediates both basal and phorbol ester/T-cell activator induced expression of the TCR-b gene. Ets-1 expression represses both basal and inducible activity of the TCR-b enhancer. Expression of Ets-2 has no effect on enhancer activity. Derepression of TCR-b transcription may be mediated by inactivation of Ets-1 DNA binding activity, possibly by phoshorylation.

A new member of the *ets* gene family, *elf*-1, has recently been cloned from a human T-cell library and shown to bind to the EBS in both the interleukin-2 enhancer and the human immunodeficiency virus-II LTR (Thompson et al., 1992). The interleukin-2 enhancer contains two EBS, previously shown to be bound by the NFAT-1 and NFIL-2B regulatory transcription factor complexes (Thompson et al., 1992). These sites both bind Elf-1 but not Ets-1 or Ets-2 (Thompson et al., 1992). Other EBS have been identified in lymphoid-specific promoters/enhancers, such as the IgH enhancer, the MHC II 1-Ab enhancer, the HTLV-I LTR, the CD2 enhancer, the CD3 enhancer, the *mb-1* promoter and the terminal deoxynucleotide transferase promoter.

Table 2. Regulatory elements containing Ets binding sites. The EBS sequence is shown and the Ets protein known to bind to each site is indicated. The purine core of the EBS is highlighted.

Regulatory Element[ref.]	Sequence	Ets Protein which Binds
ets-2 promoter[22]	TGGA**GGAA**GT	Ets-1/Ets-2
interleukin-2 enhancer (NFAT-1)[13]	AAGGA**GGAA**A	Elf-1
polyomavirus enhancer PEA 3[27]	AGCA**GGAA**GT	Ets-1/Ets-2/Erg/Elk-1
Moloney Sarcoma Virus LTR[32]	GAGC**GGAA**GC	Ets-1/Ets-2
SV40 enhancer[12]	AAGA**GGAA**CT	PU-1
ICP4 promoter[13,14]	6X (C**GGAA**A/G)	GABP-α (binding depends on GABP-β)
HTLV I LTR[31]	GGA**GGAA**AT	Ets-1/Ets-2
HTLV I LTR[31]	CCG**GGAA**GC	Ets-1/Ets-2
stromelysin-1 promoter[28]	GCA**GGAA**GC	Ets-1/Ets-2
stromelysin-1 promoter[28]	CCA**GGAA**AT	Ets-1/Ets-2
c-*fos* promoter[10]	CA**GGAT**GT	SAP-1 (binding depends on SRF)
T-cell receptor α enhancer[33]	CAGA**GGAT**GT	Ets-1

Genes involved in the degradation of the extracellular matrix (ECM), such as stromelysin-1, collagenase and urokinase plasminogen activator (Wernert et al., 1992) contain EBS in their regulatory flanking sequences. Ets-1 activates expression of rat stromelysin-1 (Wasylyk et al., 1991) and may thus play a role in coordinating events that lead to degradation of the ECM, an important process in metastasis (Wernert et al., 1992).

Ets-1 can act either as an activator of stromelysin-I in HeLa cells or as a repressor of TCR-b in quiescent Jurkat cells depending on promoter/enhancer context. Analysis of transactivation of different target genes suggests a role for Ets-1 in both T-cell differentiation and in control of ECM degradation.

ETS AS A TRANSCRIPTION MODULATOR

Transactivation by p68$^{c\text{-}ets\text{-}1}$ from the PEA 3 site of the polyomavirus enhancer is cooperative with activation by the transcription factor, AP-1 bound to the adjacent PEA 1 site. Cooperativity is not mediated at the level of DNA binding since altering the DNA binding affinity of either the PEA 1 or PEA 3 sites did not alter the ability of these factors to cooperate in transcription (Wasylyk et al., 1990). Over-expression of p68$^{c\text{-}ets\text{-}1}$ resulted in inhibition of transactivation by AP-1 of the oncogene response unit, mutated in its PEA 3 site (Wasylyk et al., 1990). This suggested that the cooperativity between AP-1 and p68$^{c\text{-}ets\text{-}1}$ is mediated by protein-protein interactions (Wasylyk et al., 1990). p54$^{c\text{-}ets\text{-}1}$ and p58$^{c\text{-}ets\text{-}2}$ of chicken also bind to the PEA 3 site and effect cooperative transactivation with AP-1 (Fujiwara et al., 1988). Recent evidence has demonstrated cooperative transactivation by c-Myb and Ets-2 of the *mim*-1 (*myb*-induced myeloid-1) promoter (Dudek et al., 1992). Ets-1 failed to cooperate with c-Myb in transactivation (Dudek et al., 1992). Cooperativity in transactivation has been observed between two palindromic EBS in the rat stromelysin I gene promoter (Wasylyk et al., 1991). The ability of a single EBS to mediate transcriptional activation linked to a minimal promoter is negligible (Wasylyk et al., 1989) and as for p75$^{c\text{-}myb}$, multiple binding sites are required to observe transcriptional activation.

Domain swap experiments involving fusion of deletion mutants of p68$^{c\text{-}ets\text{-}1}$ to the DNA binding domain of the bacterial *lex* A protein have identified three transactivation domains, RI, RII and RIII, in the amino-terminal region of p68$^{c\text{-}ets\text{-}1}$ (Schneikert et al., 1992) (figure 2). When fused to the *lex* A DNA binding domain, RI and RIII can both activate transcription from promoters containing the *lex* A operator (Schneikert et al., 1992). RII cannot transactivate alone, but has a negative regulatory effect on the activity of RI and a positive effect on the activity of RIII. RII, which contains a helix-loop-helix motif (Seith and Papas, 1990), is conserved between Ets-1 and Ets-2 suggesting that RII may mediate common control interactions of Ets-1 and Ets-2. Ets-1 and Ets-2 compete with LexA-Ets-1/-2 fusion proteins in transactivation suggesting that Ets-1 and Ets-2 proteins interact with a common factor.

The transactivation domain of PU.1 has both a glutamine rich region (mapping from amino acids 74-93) and an acidic region (amino acids 1-165), which may act as transactivation domains (Pongubala et al., 1992). Recent work has identified a role for PU-1 in recruiting other transcription factors, such as NF-EM5, a B-cell specific factor, to the 3' enhancer of the mouse immunoglobulin kappa light chain gene where binding to the enhancer by NF-EM5 depends upon prior binding of PU-1 to the enhancer (Pongubala et al., 1992). The functional domains involved in these interactions have not been identified.

The serum response factor (SRF) binds to the serum response element (SRE) in the promoters/enhancers of serum-inducible genes, such as c-*fos*. SRF forms a ternary complex with p62^{TCF} which is essential for complex formation on the SRE. p62^{TCF} does not bind DNA alone. Elk-1 is functionally equivalent to p62^{TCF}, in that it forms a ternary complex with SRF (Hipskind et al., 1991) and similar results have been obtained with an Elk-related protein, SAP-1 (Dalton and Tresiman, 1992). SAP-1 forms a ternary complex with SRF over the SRE of the c-*fos* promoter and like Elk-1, its DNA binding activity depends upon interaction with the SRF. Two domains including the ETS domain of SAP-1 are necessary for ternary complex formation (Dalton and Treisman, 1992).

A new member of the *ets* gene family, GABP-a forms the a-subunit of the GA-Binding Protein (GABP) which binds to the purine rich sequence, CGGAAG/A of the ICP4 gene of *Herpes simplex* virus (Seth and Papas, 1990; LaMarco et al., 1991). GABP-a mediates DNA binding by the multi-subunit protein but GABP-b is also necessary for binding (Thompson et al., 1991 ; LaMarco et al., 1991). The amino-terminus of the b-subunit contains four 33-amino-acid ankyrin repeats found in the Notch, Lin-12, cdc 10, NF-kB and Ik-B proteins (Thompson et al., 1991 ; LaMarco et al., 1991). These 33 amino acid repeats are necessary for interaction of GABP-b with GABP-a. The ETS domain of the a-subunit is required for its interaction with the b-subunit .

The transcription factor E4TF1 regulates transcription of the adenovirus-5 E4 gene and binds to a sequence CGGAAGT, a high-affinity EBS, in the E4 promoter (Watanabe et al., 1990). E4TF1 is composed of two subunits, p53 and p60 and addition of purified p53 increases the binding affinity of p60 for this site. Neither p53 nor p60 are able to transactivate alone but together mediate a 10-fold increase in transcription mediated by the CGGAAGT site (Watanabe et al., 1990). p60 is a good candidate member of the Ets family.

ETS IN TRANSFORMATION

The v-*ets* oncogene is cotransduced in the E26 retrovirus with the v-*myb* oncogene and the E26 genome expresses a 135 kD fusion protein, $P135^{gag\text{-}myb\text{-}ets}$ (Leprince et al., 1983 ; Watson et al., 1990). The acute myeloblastosis virus (AMV), which has transduced the v-*myb* oncogene alone, induces myeloid leukaemias (Metz and Graf, 1991) whereas E26 induces erythroleukaemias. It was proposed that v-*ets* confers upon the E26 virus, the ability to transform immature erythroid progenitor cells. Both v-*ets* and v-*myb* retain a reduced capacity to transform erythroid cells individually but when coexpressed as separate proteins, there is an increased self-renewal rate of transformed erythroid cells (Metz and Graf, 1991). Clearly, the transforming ability of v-*ets* and v-*myb* is additive suggesting that v-*myb* and v-*ets* oncogene products may cooperate in transcription (Dudek et al., 1992 ; Metz and Graf, 1991). This transforming activity is superceded by the transforming potential of *myb-ets* fusion proteins which generate erythroleukemia with a much reduced latency and higher frequency (Metz and Graf, 1991). Fusion of the v-*ets* and v-*myb* oncogenes may generate a novel transactivator with an altered target gene specificity or altered interactions with other regulatory factors.

A common mechanism results in the activation of two different *ets* genes in virally induced murine erythroleukemias. Proviral integration by the Friend Murine Leukemia Virus and the Spleen Focus Forming Virus into flanking sequences activates expression of the *fli-1* and *spi-1* genes respectively.

The human *ets*-1 and *ets*-2 genes are translocated in certain types of acute leukaemia from chromosomes 11 to 4 and 21 to 8, respectively, although neither gene spans the chromosome breakpoint (Watson et al., 1990). Similarly, the *elk*-1 gene maps close to chromosome breakpoints in synovial sarcomas (Watson et al., 1990). The significance of these translocations for Ets function is unclear.

The role of *ets* gene family members as oncogenes has been tested in classical transformation experiments in which over-expression of the human c-*ets*-1 oncogene in NIH-3T3 mouse fibroblasts generates transformed foci in low serum conditions (Seth and Papas, 1990). These transformed cells grew in soft agar and formed tumours in nude mice. The transforming capacity of the c-*ets*-1 proto-oncogene may be associated with its ability to positively auto-regulate its own expression (Seth and Papas, 1990).

Increased binding to the PEA 3 element of the polyoma enhancer occurs when fibroblasts are stimulated with serum or phorbol esters or transformed by v-*src*, v-*raf*, polyoma middle T antigen, v-*mos* or c-Ha-*ras* (Dudek et al., 1992) This induction of binding activity is independent of protein synthesis inhibitors suggesting that it is mediated by post-translational modifications, such as phosphorylation or cofactors interactions.

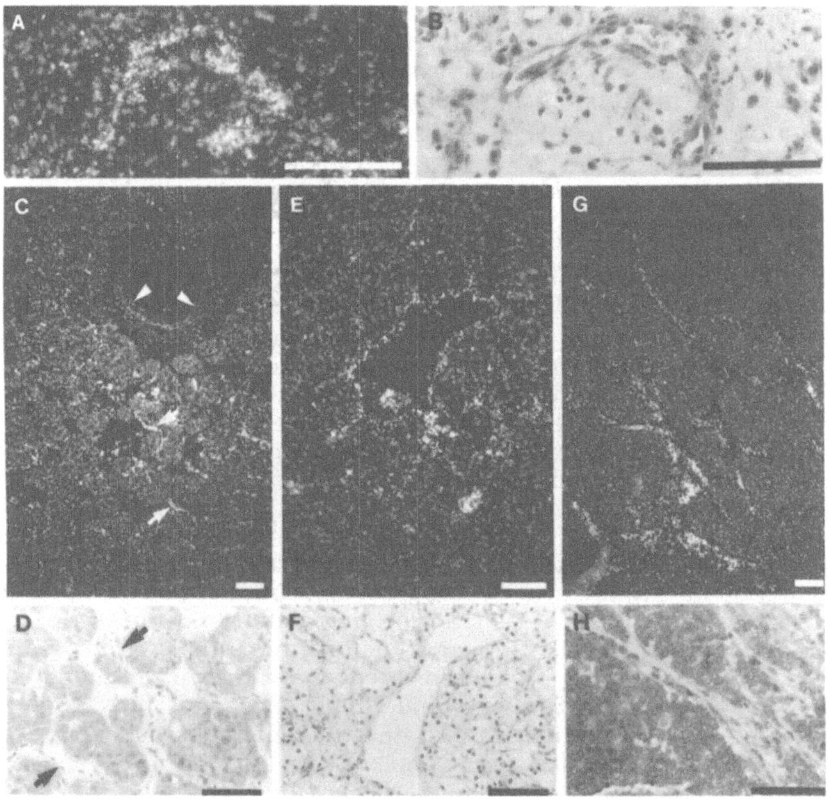

Figure 3 : c-*ets1* transcripts in granulation tissue and in tumors.
A,C,E,G : dark fields and epifluorescence views of tissue sections hybridized with the antisense c-*ets1* RNA probe ; B,D,F,H : bright fields of neighbouring sections stained with hematoxylin eosin. bar = 100 mm.
A,B : granulation tissue forming in the skin during reparation after severe burning. A capillary showing a strong signal within its endothelial cells.
C,D : benign renal oncocytoma. Around the tumor cells with an eosinophilic cytoplasm the supporting stroma is filled with capillaries (arrows) positive for the c-*ets1* mRNA. Note that endothelial cells of the large vessel (arrowheads) are negative.
E,F : well differentiated clear cell-carcinoma of the kidney. The tumor cells have a water-clear cytoplasm and the stroma is well vascularized with many capillaries (F). c-*ets1* mRNA is expressed in the endothelial cells (E).
G,H : renal metastasis of a pulmonary small cell carcinoma. The stroma forms septa with numerous fibrocytes (H). c-*ets1* mRNA is abundant in the stromal cells, whereas tumor cells are negative.

ETS1, ANGIOGENESIS AND TUMOUR INVASION

An interesting new aspect concerning the role of *ets* in growth control has emerged from studies of Ets-1 expression in human endothelial cells (Wernert et al., 1992). By Northern blot and immunoprecipitation, it has been shown that c-*ets1* mRNAs and proteins are abundant in both T and B cells (Wernert et al., 1992). However, *in situ* hybridization experiments have revealed a more widespread expression in a variety of cells of the mesodermal lineage at the beginning of organogenesis in chick embryos (Vandenbunder et al., 1989). c-*ets1* was most highly expressed within endothelial cells during blood vessel formation. Endothelial cell expression was detected in embryos irrespective of the process through which the vessels were forming, i.e. vasculogenesis (emergence of blood vessels occurring *de novo*) or angiogenesis (budding of new blood vessels from preexisting ones). The different mRNAs and proteins generated from the chicken c-*ets1* locus are differentially expressed in lymphoid cells and in endothelial cells.

In man, c-*ets1* transcripts have been detected not only in the endothelia of developing vessels of embryos (figure 3 A,B), but also in blood vessels of adult tissues when angiogenesis resumes, as in granulation tissue, and especially during the vascularisation of tumors (figure 3 C,D ; E,F). In contrast, mature vessels in adult tissues did not exhibit detectable levels of c-*ets1* mRNA.

It has been shown that the transcriptional stimulation of the u-PA gene in murine keratinocytes by phorbol ester or epidermal growth factor is mediated by a *ets* binding site (EBS) (Rorth et al., 1990). The *ets* binding site PEA3, acting cooperatively with AP1 is also involved in the induction of both human collagenase and stromelysin 1 promotors (Wasylyk et al., 1991). During angiogenesis the production of matrix degrading proteases by endothelial cells and the subsequent degradation of the basement membrane that surrounds preexisting capillaries or venules are essential to ensure the invasive properties of endothelial cells and the formation of a new blood vessel (Folkman, 1985). Therefore it is tempting to speculate that c-*ets1* might stimulate the transcription of matrix degrading proteases genes in endothelial cells during angiogenesis. *In vitro* results are consistent with this hypothesis. TNFa, which is an angiogenic factor, transiently increases steady-state levels of both c-*ets1* and u-PA mRNA after several hours of treatment in confluent human umbilical vein endothelial cells. The hypothesis that c-*ets1* proteins may regulate matrix degrading protease gene expression is also supported by the detection of c-*ets1* transcripts in various other processes that involve extensive extracellular matrix remodeling, such as the migration of neural crest cells in chick embryos (Vandenbunder et al., 1989), and new bone formation in human embryos (Wernert et al., unpublished results).

In this respect, it is interesting that c-*ets1* mRNAs are also detected in the fibrocytic stroma of invasive tumors where it might play the same role as in endothelial cells (figure 3 G,H). As reviewed by Liotta et al. (Liotta et al., 1991), extracellular matrix degradation is known to be essential for tumor invasion. Furthermore it has been reported that tumor cells are able to stimulate the expression of collagenase I in cultured fibroblasts (Biswas, 1984). More recently, a new metalloproteinase gene, stromelysin 3, has been identified that is specifically expressed in stromal cells surrounding invasive breast carcinoma (Basset et al., 1990), and u-PA has been localized by immunohistochemistry in stromal cells of colon adenocarcinomas (Grondahl-Hansen, 1991). Further studies will also have to analyze the expression of c-*ets1* mRNA in a greater number of tumors, including common carcinomas like breast and lung cancers.

PERSPECTIVES

The *ets* gene family is a rapidly growing gene family and the encoded proteins function as regulators of cellular and viral gene transcription. Some Ets proteins, such as Ets-1 and Ets-2, interact with the same DNA recognition sites and may function antagonistically, either activating or repressing transcription of the same target gene, depending on promoter context, cell-type or differentiation-stage. Other Ets proteins, such as Elf-1 interact with different, although similar sites, and probably regulate different target genes. How the expression and activity of different Ets proteins are coordinated with respect to cell-type, in differentiation, during the cell cycle and in development is unclear. The capacity of certain Ets proteins to interact with regulatory cofactors and to be regulated by phosphorylation suggests a role for Ets proteins in coordinating changes in gene expression in response to second messengers and extracellular signals. Ets proteins may be modulated differently by interactions with different cofactors and what role these interactions may have in the regulatory pathways of cell-cycle control and differentiation needs to be assessed.

ACKNOWLEDGEMENTS

The authors would like to thank Drs. Yvan de Launoit, Olivier Albagli, Patrick Martin and Simon Saule for critical reading of the manuscript and Dr. Mike Owen, ICRF for communication of unpublished data. K.M. is financially supported by the Commission of the European Community, D.L. and D.S. are funded by the Centre National de la Recherche Scientifique, D.S. is Professor of Molecular Oncology, Pasteur Institute, Lille.

REFERENCES

Only a limited list of references is indicated because of space availability but much of the relevant work will be found referenced in those papers.

Basset, P., Bellocq, J.P., Wolf, C., Stoll, I., Hutin, P., Limacher, J.M., Podhajcer, O.L., Chenard, M.P., Rio M.C., and Chambon, P., 1990, A novel metalloproteinase gene specifically expressed in stromal cells of breast carcinomas, *Nature* 348:699-704.

Ben-David, Y., Giddens, E.B., Letwin, K., and Bernstein, A., 1991, Erythroleukemia induction by Friend murine leukemia virus : insertional activation of a new member of the ets gene family, fli-1, closely linked to c-ets1, *Genes Dev* 5:908-918.

Bhat, N.K., Thompson, C.B., Lindsten, T., June, C.H., Fujiwara, S., Koizumi, S., Fisher, R.J., and Papas, T.S., 1990, Reciprocal expression of human ets1 and ets2 genes during T-cell activation : regulatory role for the proto-oncogenes ets1, *Proc Natl Acad Sci* USA 87:3723-3727.

Biswas, C.J., 1984, Collagenase stimulation in cocultures of human fibroblasts and human tumor cells, *Cancer Lett.* 24:201-207.

Boulukos, K.E., Pognonec, P., Sariban, E., Bailly, M., Lagrou C. and Ghysdael, J., 1990, Rapid and transient expression of ets2 in mature macrophages following stimulation with cMGF, LPS and PKC activators, *Genes Dev* 4:401-409.

Dalton S., and Treisman, R., 1992, Characterization of SAP-1, a protein recruited by serum response factor to the c-fos serum response element, *Cell* 68:597-612.

Dudek, H., Tantravahi, R.V., Rao, V.N., Reddy, E.S.P., and Reddy, E.P., 1992, Myb and Ets proteins cooperate in transcriptional activation of the mim-1 promoter, *Proc Natl Acad Sci USA* 89:1291-1295.

Fisher, R.J., Mavrothalassitis, G., Kondoh, A., and Papas, T.S., 1991, High-affinity DNA-protein interactions of the cellular ETS1 protein : the determination of the ETS binding motif, *Oncogene* 6:2249-2254.

Folkman, J., 1990, What is the evidence that tumors are angiogenesis dependent ? *Perspectives in Biology and Medicine* 29:10-36.

Fujiwara, S., Fisher, R., Initiales N.K.Bhat, S.Moreno Diaz de la Espina, 1988, and T.S.Papas, A short-lived nuclear phosphoprotein encoded by the human ets-2 proto-oncogene is stabilized by activation of protein kinase C, *Mol Cell Biol* 8:4700-4706.

Gitlin, S.D., Bosselut, R., Gégonne, A., Ghysdael, J., and Brady, J.N., 1991, Sequence-specific interaction of the Ets1 protein with the long terminal repeat of the human T-lymphotropic virus type 1, *Journal of Virology* 65:5513-5523.

Grondahl-Hansen, E.Ralfkiser, L.T.Kirkeby, L.R.Lund and K.Dand, 1991, Localization of urokinase-type plasminogen activator in stromal cells in adenocarcinomas of the colon in humans, *Am. J. Pathol.* 138 : 111-117.

Gunther, J., Nye, J.A., Bryner, R.S., and Graves, B.J., 1990, Sequence-specific DNA binding of the proto-oncoprotein ets-1 defines a transcriptional activator sequence within the long terminal repeat of the Moloney murine sarcoma virus, *Genes Dev* 4:667-679.

Hipskind, R.A., Rao, V.N., Mueller, C.G.F., Reddy, E.S.P., and Nordheim, A., 1991, Ets-related protein Elk-1 is homologous to the c-fos regulatory factor $p62^{TCF}$, *Nature* 354:531-534.

Ho, I-C., Bhat, N.K., Gottschalk, L.R., Lindsten, T., Thompson, C.B., Papas, T.S., and Leiden, J.M., 1990, Sequence-specific binding of human Ets-1 to the T cell receptor a gene enhancer, *Science* 250:814-818.

Karim, F.D., Urness, L.D., Thummel, C.S., Klemsz, M.J., McKercher, S.R., Celada, A., Van Beveren, C., Maki, R.A., Gunther, C.V., Nye, J.A., and Graves, B.J., 1990, The ETS-domain : a new DNA-binding motif that recognizes a purine-rich core DNA sequence, *Genes Dev* 4:1451-1453.

Koizumi, S., Fisher, R.J., Fujiwara, S., Jorcyk, C., Bhat, N.K., Seth A., and Papas, T.S., 1990, Isoforms of the human ets-1 protein : generation by alternative splicing and differential phosphorylation, *Oncogene* 5:675-681.

LaMarco, K., Thompson, C.C., Byers, B.P., Walton, E.M., and McKnight, S.L., 1991, Identification of Ets- and notch-related subunits in GA binding protein, *Science* 253:789-791.

Leprince, D., Gégonne, A., Coll, J., de Taisne, C., Scheenberger, A., Lagrou, C., and Stéhelin, D., 1983, A putative second cell-derived oncogene of the avian leukaemia retrovirus E26, *Nature* 306:395-397.

Leprince, D., Duterque-Coquillaud, M., Li, R-P., Henry, C., Flourens, A., Debuire, B., and Stéhelin, D., 1988, Alternative splicing within the chicken c-ets-1 locus : implications for transduction within the E26 retrovirus of the c-ets proto-oncogene, *Journal of Virology* 62:3233-3241.

Leprince, D, Crépieux, P., and Stéhelin, D., 1992, c-ets-1 DNA-binding to the PEA3 motif is differentially inhibited by all the mutations found in v-ets, *Oncogene* 7:9-17.

Lim, F., Kraut, N., Frampton, J., and Graf, T., 1992, DNA binding by c-Ets-1, but not v-Ets, is repressed by an intramolecular mechanism, *EMBO J* 11:643-652 .

Liotta, L.A., Steeg, P.S., and Stetler-Stevenson, W.G., 1991, Cancer metastasis and angiogenesis : an imbalance of positive and negative regulation, *Cell* 64:327-336.

Mavrothalassitis, G., and Papas, T.S., 1991, Positive and negative factors regulate the transcription of the ETS2 gene via an oncogene-responsive-like unit within the ETS2 promoter region, *Cell Growth & Differentiation* 2:215-224.

Metz T., and Graf, T., 1991, Fusion of the nuclear oncoproteins v-myb and v-ets is required for the leukemogenicity of E6 virus, *Cell* 66:95-105.

Oka, T., Rairkar , A., and Chen, J.H., 1991, Structural and functional analysis of the regulatory sequences of the ets-1 gene, *Oncogene* 6:2077-2083.

Pognonec, P., Boulukos K.E.., and Ghysdael, J., 1989, The c-ets1 protein is chromatin associated and binds to DNA in vitro, *Oncogene* 4:691-697.

Pongubala, J.M.R., Nagulpalli, S., Klemz, M.J., McKercher, S.R., Maki, R.A., and Atchison, M.L., 1992, PU.1 recruits a second nuclear factor to a site important for immunoglobulin k 3' enhancer activity, *Mol Cell Biol* 12:368-378.

Pribyl, L.J., Watson, D.K., Schulz, R.A., and Papas, T.S., 1991, D-elg, a member of the Drosophila ets gene family : sequence, expression and evolutionary comparison, *Oncogene* 6:1175-1183.

Rao, V.N., and Reddy, E.S.P., 1992, A divergent ets-related protein, Elk-1, recognizes similar c-ets-1 proto-oncogene target sequences and acts as a transcriptional activator, *Oncogene* 7:65-70.

Reddy, E.S.P., and Rao, V.N., 1991, erg, an ets-related gene, codes for sequence-specific transcriptional activators,*Oncogene* 6:2285-2289.

Rorth, P., Nerlov, C., Blasi, F., and Johnsen, M., 1990, Transcription factor PEA3 participates in the induction of urokinase plasminogen activator transcription in murine keratinocytes stimulated with epidermal growth factor or phorbol-ester, *Nucl. Acids. Res.* 18:5009-5017.

Schneikert, J., Lutz, Y., and Wasylyk, B., 1992, Two independent activation domains in c-ets1 and c-ets2 located in non-conserved sequences of the ets gene family, *Oncogene* 7:249-256.

Seth, A., and Papas, A., 1990, The c-ets-1 proto-oncogene has oncogenic activity and is positively autoregulated, *Oncogene* 5:1761-1767.

Thompson, C.B., Wang, C., Ho, I., Bohjanen, P.R., Petryniak, B., June, C.H., Miesfeldt, S., Zhang, L., Nabel, G.J., Karpinski, B., and Leiden, J.M., 1992, cis-acting sequences required for inducible interleukin-2 enhancer function bind a novel Ets-related protein, Elf-1, *Mol Cell Biol* 12:1043-1053.

Thompson, C.C., Brown, T.A., and McKnight, S.L., 1991, Convergence of Ets- and notch-related structural motifs in a heteromeric DNA binding complex, *Science* 253:762-768.

Vandenbunder, B., Pardanaud, L., Jaffredo, T., Mirabel, M.A., and Stéhelin, D., 1989, Complementary patterns of expression of c-ets-1, c-myb and c-myc in the blood-forming system of the chick embryo, *Development* 107:265-274.

Wasylyk, B., Wasylyk, C., Flores, P., Bègue, A., Leprince, D., and Stéhelin, D., 1990, The c-ets proto-oncogenes encode transcription factors that cooperate with c-fos and c-jun for transcriptional activation, *Nature* 346:191-193.

Wasylyk, C., Gutman, A., Nicholson, R., and Wasylyk, B., 1991, The c-ets oncoprotein activates the stromelysin promoter through the same elements as several non-nuclear oncoproteins, *EMBO J* 10:1127-1134.

Wasylyk, C., Flores, P., Gutman, A., and Wasylyk, B., 1989, PEA3 is a nuclear target for transcription activation by non-nuclear oncogenes, *EMBO J* 8:3371-3378.

Watanabe, H., Wada, T., and Handa, H., 1990, Transcription factor E4TF1 contains two subunits with different functions, *EMBO J* 9:841-847.

Waterman, M.L., Fischer, W.H., and Jones, K.A., 1991, A thymus-specific member of the HMG protein family regulates the human T cell receptor Ca enhancer, *Genes Dev* 5:656-669.

Watson, D.K., Ascione, R., and Papas, T.S., 1990, Molecular Analysis of the Ets Genes and their products, *Critical Reviews in Oncogenesis* 1:409-436.

Wernert, N., Raes, M.B., Lassalle, P., Dehouck, M.P., Gosselin, B., Vandenbunder, B., and Stéhelin, D., 1992, The c-ets-1 proto-oncogene is a transcription factor expressed in endothelial cells during tumor vascularisation and other forms of angiogenesis in man, *American Journal of Pathology* 140:119-127.

Woods, D.B., Ghysdael, J., and Owen, M.J., 1992, Identification of nucleotide preferences in DNA sequences recognised specifically by c-Ets-1 protein, *Nucl Acids Res* 20:699-704.

Xin, J-H., Cowie, A., Lachance, P., and Hassell, J.A., 1992, Molecular cloning and characterization of PEA3, a new member of the Ets oncogene family that is differentially expressed in mouse embryonic cells, *Genes Dev* 6:481-496.

INFECTIOUS INFLUENZA VIRUSES FROM cDNA-DERIVED RNA: REVERSE GENETICS

Adolfo García-Sastre and Peter Palese

Department of Microbiology, Mount Sinai School of Medicine
One Gustave L. Levy Place, New York, NY 10029

INTRODUCTION

Influenza viruses are enveloped animal viruses that contain a segmented RNA genome of negative polarity (Palese, 1977). Recombinant DNA technology has been widely used to study the biological and molecular properties of the genes and proteins of most DNA and positive-strand RNA viruses (Ahlquist et al., 1984; Peden et al., 1980; Post and Roizman, 1981; Racaniello and Baltimore, 1981; Weir et al., 1982; Xiong et al., 1989). However, negative-strand RNA viruses have been refractory to genetic manipulation because of the non-infectious nature of their naked RNAs. In contrast to the RNA of positive-strand RNA viruses, the viral RNA (vRNA) of negative-strand RNA viruses is not able to initiate translation by itself, and it must first be transcribed by its own RNA polymerase. Moreover, it is thought that the template recognized by the viral RNA polymerase *in vivo* is not naked RNA, but a complex between the vRNA and the viral nucleoprotein (ribonucleoprotein complex, RNP) (Honda et al., 1988). Until the development of an appropriate system of introducing cDNA-derived RNA into viral particles, it has not been posible to use reverse genetics for the study of influenza virus and other negative-strand RNA viruses. This was first achieved in 1989 by Luytjes et al. (1989). An RNP-transfection method was developed which allowed the amplification, expression, and packaging of a synthetic gene by influenza virus. The novel approach of this method involved the *in vitro* incubation of purified influenza virus polymerase proteins (NP, PB1, PB2 and PA) with cDNA-derived RNA. This synthetic RNA contained the reporter gene chloramphenicolacetyltransferase (CAT) in negative polarity flanked by the non-coding sequences of one of the influenza virus genes. The RNP complexes were biologically active, and when transfected into influenza virus infected cells CAT activity was detected. In addition, the recombinant RNA was also found to be packaged into the viral progeny, and was maintained for several passages of the virus in tissue culture. These experiments not only demonstrated that the signals for RNA replication, transcription , and packaging in influenza vRNAs were contained within their short non-coding regions, but

Regulation of Gene Expression in Animal Viruses, Edited by
L. Carrasco *et al.*, Plenum Press, New York, 1993

also established a powerful system for the study of *cis*-acting signals and opened for the first time the possibility to genetically manipulate the influenza virus genome.

STUDIES OF THE CIS SIGNALS FOR REPLICATION AND TRANSCRIPTION OF THE INFLUENZA GENOME

Replication and transcription of the influenza genome takes place in the nucleus of infected cells. The negative sense vRNAs coated with the NP protein are used as templates by the viral polymerase, which is a complex of three proteins, PB1, PB2 and PA (Krug et al., 1989; Lamb, 1989; Luo and Palese, 1992). Interestingly, two different RNA products can be synthesized by the viral polymerase from the vRNA: i) full-length complementary RNAs (cRNA), which can also be used as templates for the synthesis of new vRNA molecules, and ii) viral mRNAs, which are capped, polyadenylated and lack the 3' 17-22 nucleotides of the cRNA (Figure 1). It is not clear what signals the switch between these two different template activities, but it is known that the PB1 protein is required for the initiation and elongation of the synthesized RNA; the PB2 protein is responsible for the cleavage of cellular mRNA cap structures, which then are used as primers for viral mRNA synthesis; and the PA protein has kinase activity and seems, like the NP protein, to be involved in viral RNA replication (for a review, see Krug et al., 1989; Lamb, 1989; Luo and Palese, 1992; Palese and García-Sastre, 1992). These four proteins are able to support viral replication in the absence of other viral proteins (Huang et al., 1990; Kimura et al., 1992).

The reconstitution of fully active RNP complexes allowed the study of *cis* sequences responsible for RNA transcription, both *in vivo* and *in vitro*. It was postulated that the first

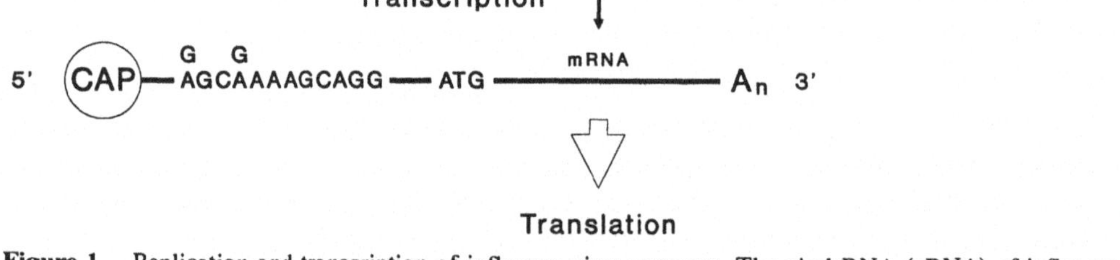

Figure 1. Replication and transcription of influenza virus genomes. The viral RNA (vRNA) of influenza virus is transcribed by the viral RNA polymerase yielding mRNA. The 5' end of the viral mRNA is derived from 5' cap structures of cellular mRNAs. Transcription of the vRNA stops at a stretch of 5-7 uridines located in the 5' noncoding region and is followed by addition of the polyA tail. mRNAs are synthesized in the nucleus of the infected cell, posttranscriptionally processed in some cases, transported to the cytoplasm and translated by the cellular ribosomes to produce viral proteins. Viral replication also takes place in the nucleus of the infected cell. In this case, the viral RNA polymerase uses again the vRNA as template and synthesizes full-length complementary RNA (cRNA) molecules. cRNAs are also used as templates to direct the synthesis of new vRNA molecules. 3' and 5' nucleotide positions which are conserved among all the genes and strains of influenza A virus in the noncoding regions of the different viral RNA species are indicated. The first ATG initiation codon and the uridine stretch of the vRNA are also represented.

12 or 13 terminal nucleotides of the 3' noncoding region of vRNA and cRNA, which are highly conserved in all RNA segments of influenza A viruses, function as the promoter for recognition by the viral RNA polymerase (for a review, see Luo and Palese, 1992). This hypothesis has been tested by several groups using different approaches (Li and Palese, 1992a; Parvin et al., 1989; Piccone et al., 1992; Seong and Brownlee, 1992; Yamanaka et al., 1991), and several interesting features have emerged: 1) A free 3' end is required for efficient promoter activity for both the cRNA and the vRNA promoters. 2) Nucleotide substitutions and deletions within the first 11 nucleotide positions for the cRNA promoter and the first 14 for the vRNA promoter have a significant effect in promoter activity. Interestingly, some substitutions in the cRNA promoter enhanced its activity *in vitro*, while such an observation was not made following mutational analysis of the vRNA promoter.

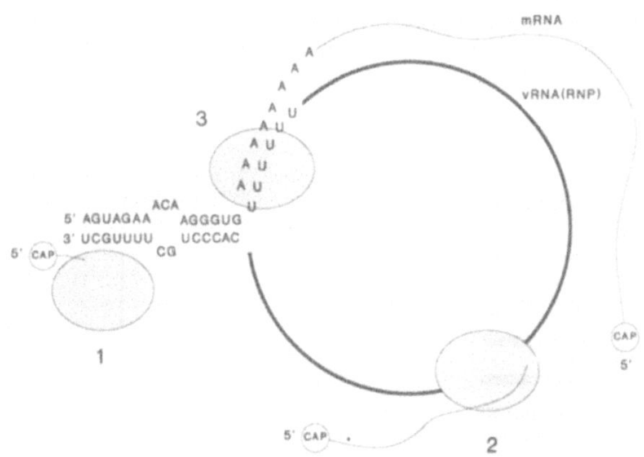

Figure 2. Proposed model of mRNA synthesis in influenza virus infected cells. 3' and 5' noncoding regions of the vRNA molecules form a panhandle structure which confers a circular conformation to the vRNA. In the figure, the panhandle structure of the NS gene of influenza A virus is represented as obtained by computer RNA modelling. vRNAs are found in the form of a ribonucleoprotein (RNP) complex which includes the viral NP and polymerase proteins. The viral RNA polymerase is a complex of three viral proteins, PB1, PB2 and PA, and is represented as grey ellipse in the figure. **1. Initiation.** The viral RNA polymerase interacts with the promoter at the 3' end of the vRNA. This protein complex is responsible of the cleavage of the 5' cap ends of cellular mRNAs. The host-derived cap structures are then used for initiation of viral mRNA synthesis. **2. Elongation.** The viral polymerase moves along the RNP complex whose vRNA is used as template for mRNA elongation. **3. Polyadenylation.** A polyA tail is added to the 3' end of the mRNA by stuttering mechanism involving the stretch of uridines which is close to the panhandle structure of the vRNA.

3) Although the 3' and 5' ends of the vRNA form a panhandle structure due to their partial complementarity in virions and infected cells (Hsu et al., 1987), the vRNA and cRNA promoters for viral replication do not require this double-stranded structure for *in vitro* activity. 4) Some nucleotide substitutions which do not significantly change *in vitro* promoter activity, abolish *in vivo* expression of a reporter gene in influenza virus infected cells. This is probably due to a decrease of mRNA transcription or translation levels *in vivo*, suggesting that the nucleotides which constitute the viral RNA polymerase promoter are also involved in other *cis*-acting signals required for mRNA synthesis or for translation.

mRNA polyadenylation is another interesting aspect of influenza virus transcription which can now be studied using reverse genetics (Luo et al., 1991). The use of mutagenized vRNA shows that polyadenylation of mRNAs synthesized from vRNAs is totally dependent on the presence of 5 to 7 uridine residues in the 5' noncoding region of the vRNA. This stretch of uridine residues must be located close to the panhandle structure formed by the 3' and 5' ends of the vRNA, in order to constitute a functional polyadenylation signal. This finding supports the notion that the vRNA panhandle acts as a physical barrier for the viral RNA polymerase which stutters at the stretch of uridine residues and synthesizes the polyA tail of the mRNA (Figure 2). The formation of the panhandle structure by itself could be the *cis* signal for the switch between cRNA and mRNA synthesis, in accordance with the finding that the double stranded panhandle structure is not required for cRNA synthesis *in vitro*.

INFLUENZA A VIRUS TRANSFECTANTS

The RNP transfection methodology has also successfully been used for the exchange of an influenza virus RNA segment by a cDNA-derived synthetic gene (Figure 3) (Enami

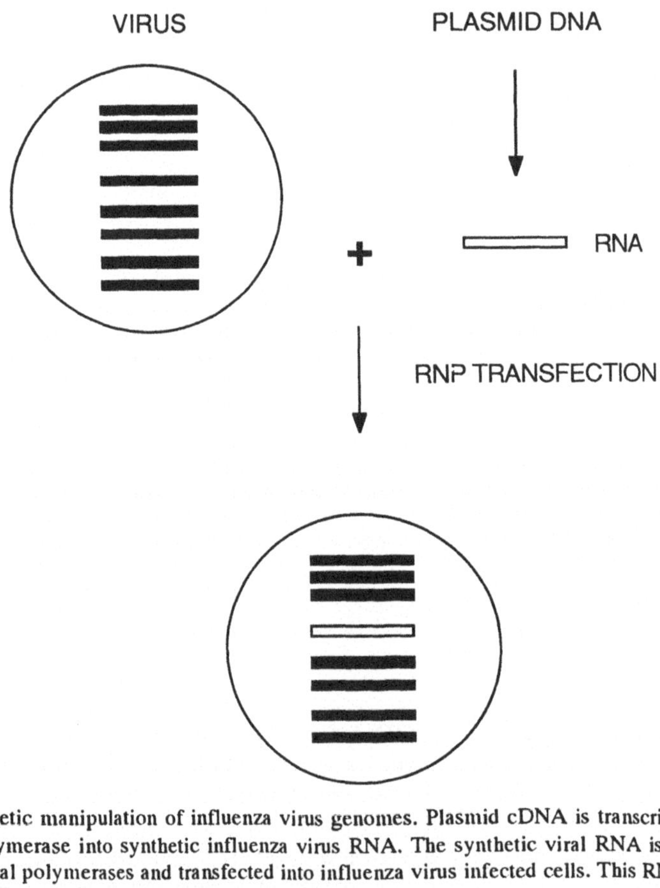

Figure 3. Genetic manipulation of influenza virus genomes. Plasmid cDNA is transcribed *in vitro* by T3 or T7 RNA polymerase into synthetic influenza virus RNA. The synthetic viral RNA is incubated *in vitro* with purified viral polymerases and transfected into influenza virus infected cells. This RNA segment can be rescued into the virus progeny (for details, see text).

et al., 1990). This approach allowed us for the first time to genetically manipulate the influenza virus genome. The newly generated viruses containing the RNP-transfected RNA were called transfectants. Exchanging an RNA segment in influenza virus by the described RNP transfection method is dependent on the availability of a good selection system (Enami and Palese, 1991). For example, the neuraminidase (NA) gene of an influenza virus can be exchanged by RNP transfection of an NA gene which confers to the virus the ability to grow in MDBK cells. Another approach of rescuing synthetic RNA into viral genomes involves antibody selection. It is possible to introduce a recombinant HA gene into virions by RNP transfection if the helper virus containing a different HA subtype is suppressed in the presence of neutralizing antibodies directed against the helper HA. Other influenza virus genes, like the NS gene, can be exchanged using temperature sensitive (*ts*) helper viruses which contain the *ts* lesion in the corresponding gene; in this case the selection is performed at the non-permissive temperature.

The RNP transfection methodology can also be used for the introduction of changes into the noncoding regions of viral genes. For example, a transfectant influenza A virus (NA/B-NS virus) was obtained after RNP transfection of a recombinant NA gene (Muster et al., 1991). The noncoding sequences of this NA gene were derived from the NS gene of influenza B virus. The noncoding sequences of influenza B virus genes are similar but not identical to those of influenza A viruses, and they differ in several nucleotide positions which are conserved among all influenza A virus. It was not known if the lack of reassortants between influenza A and B viruses (i.e., viruses containing different parental genes), was due to an incompatibility on the protein level or on the RNA level. The rescue of the NA/B-NS transfectant virus revealed that the noncoding sequences of influenza B genes contain functional signals for replication, transcription and packaging which are recognized by influenza A virus. This finding thus suggests that the restriction of reassortment between influenza A and B viruses is due to constraints in the interaction of viral proteins.

The biological properties of the NA/B-NS virus were, however, not like those of the wild-type virus. The chimeric strain was compromised in its ability to grow in tissue culture and in embryonated eggs, and was attenuated in mice. We also investigated the reasons for the attenuated phenotype of the NA/B-NS virus (Luo et al., 1992). Our studies revealed that the noncoding RNA region of the influenza B virus gene is a weak promoter for the RNA polymerase of influenza A virus. Therefore, the levels of the chimeric NA gene of the NA/B-NS virus were drastically reduced in infected cells in comparison with those of the other normal genes. Interestingly, this reduction was also reflected in the progeny virions, indicating that most of the virus particles budding from the host cell did not contain the NA gene, and behaved as defective particles. The low NA gene levels were not correlated with a similar reduction in the amount of the NA protein synthesized, suggesting that the attenuated phenotype is mostly due to the generation of defective particles lacking the NA gene. This mechanism of attenuation may show a novel way to generate attenuated virus strains and may also be used to decrease the pathogenicity of other segmented RNA viruses.

One of the intriguing aspects of influenza virus replication is how the virus manages to package 8 different genes (7 in the case of influenza C virus) into the same particle. Although the packaging signals are not well delineated, it is known that they are located in the noncoding regions of the vRNAs (Luytjes et al., 1989). Two models have been proposed for influenza virus vRNA packaging, which can also be applied to other viruses containing a segmented genome (Compans and Choppin, 1975). The selective model presumes a different packaging signal for each vRNA. The random packaging hypothesis assumes that the RNA segments are randomly chosen among a pool of 8 different vRNAs which carry the same packaging signal. The latter model requires that each virus particle contains on average 11-12 RNA segments in order to allow in each replicative cycle the generation of a high percentage (5-10%) of infectious virus particles; only viruses with the correct complement of 8 segments are infectious. The previously mentioned correlation of the levels of the NA vRNA in the attenuated transfectant NA/B-NS virus and in infected cells supports the mechanism for a random packaging of RNAs (Luo et al., 1992). Moreover, an influenza A virus transfectant was constructed whose genome consists of 9 different RNA segments (Enami et al., 1991). This transfectant virus contains two NS genes, which encode the NS1 and NS2 viral proteins, respectively. These two proteins are translated from the same gene in wild-type influenza A viruses. The random packaging hypothesis would predict that a 9-segmented influenza virus would show a lower percentage of infectious particles than the wild-type virus. Obviously, viruses with the correct complement of 9 different RNA segments would be generated at a lower frequency than those containing only 8 different RNA segments. This prediction turned out to be the case for the 9-segmented transfectant virus, which was shown to have a lower ratio of infectious particles to physical virus particles than the 8-segmented wild-type virus.

The genetic manipulation of influenza virus genomes might also be used for the generation of transfectant viruses which express foreign epitopes and genes. These new transfectant viruses could be employed to induce a protective immune response against

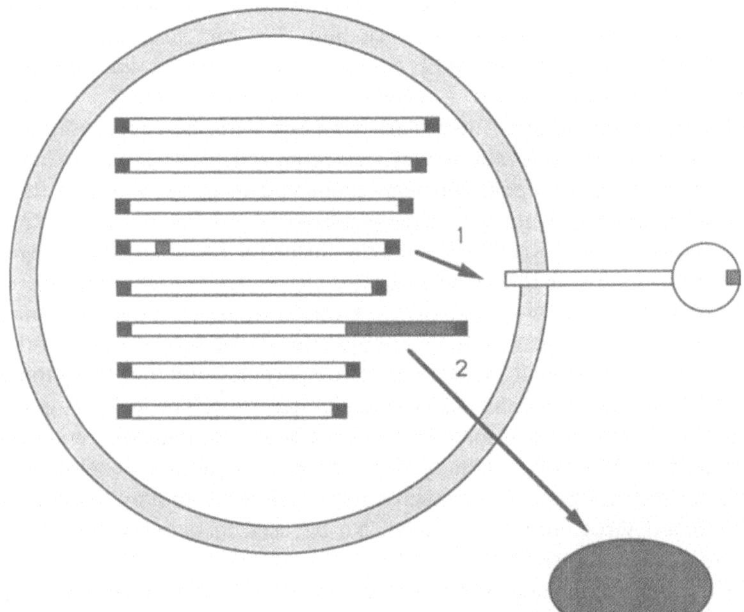

Figure 4. Models of influenza A virus expression vectors. **1. Epitope grafting.** A foreign sequence (represented by the grey square) coding for a B cell or T cell epitope of a known protein is introduced in frame into the coding region of an influenza A virus gene. The foreign epitope is now presented by the chimeric influenza virus protein. In the case of B cell epitope expresion, a surface-exposed region of an influenza virus envelope protein is chosen. **2. Bicistronic genes.** An influenza virus gene is engineered to code for two different proteins. The first open reading frame (ORF) directs the synthesis of an essential viral protein. The second ORF (represented by the grey rectangle) is used for the synthesis of a foreign protein (grey ellipsi) in influenza virus infected cells. Black squares represent noncoding regions of influenza virus genes.

different pathogenic agents. Since the expression of a foreign gene flanked by the noncoding sequences of influenza virus vRNAs in virus infected cells is only temporary (Luytjes et al., 1989), we are investigating two kinds of approaches in order to use influenza virus as an expression vector. The first involves the construction of transfectant influenza viruses with chimeric proteins in which a foreign epitope has been introduced. The second effort aims at generating transfectant viruses which contain a bicistronic gene with the coding capacity for a foreign protein in addition to that of an essential viral protein (Figure 4). Recently, we obtained an influenza A virus in which a portion of the antigenic site B of its subtype 1 HA (H1) was replaced by the corresponding domain of a subtype 3 HA (H3) (Li et al., 1992). The novel transfectant virus was recognized by monoclonal antibodies directed against subtype H1 and subtype H3 viruses. Moreover, this virus was able to induce in mice neutralizing antibodies against influenza A virus of both H1 and H3 subtypes. It was also possible to exchange the antigenic site E of different HA subtypes. In addition to these homologous replacements, in which a domain of the HA was replaced by the corresponding domain of another HA belonging to a different subtype, antigenic sites B and E of the HA can also be used for the presentation of different B-cell and T-cell epitopes whose amino acid sequences are unrelated to those of HA proteins (Guinea and Palese, 1992; Li and Palese, 1992b). Furthermore, the construction of bicistronic genes in influenza virus seems to be possible, since we were able to increase the length of the NA gene of a transfectant virus by almost 1,000 nucleotides (García-Sastre et al., 1992).

SUMMARY AND FUTURE PERSPECTIVES

The RNP transfection methodology has shown to be a powerful tool for the study of influenza virus replication. We now have a better understanding of the *cis*-signals responsible for replication, transcription, polyadenylation and packaging of influenza virus RNAs, and we believe that this knowledge will be expanded in the near future by the characterization of the proteins and/or protein domains involved in these mechanisms. Moreover, genetically altered influenza viruses have been obtained which exhibit unique features. Viruses have been constructed which posses novel attenuation characteristics and other transfectants have been isolated which express immunologically active foreign epitopes in their surface proteins. These approaches could result in a new generation of influenza virus vaccines, as well as in the use of influenza virus as an expression vector with clinical and industrial applications. Finally, the studies described here provide a useful paradigm to explore other negative-strand RNA viruses, which should result in a better understanding of tneir molecular and biological properties.

BIBLIOGRAPHY

Ahlquist, P., French, R., Janda, M., and Loesch-Fries, L.S., 1984, Multicomponent RNA plant virus infection derived from cloned viral cDNA, *Proc. Natl. Acad. Sci. USA* 81:7066.

Compans, R.W., and Choppin, P.W., 1975, Reproduction of Myxoviruses, *in*: "Comprehensive Virology", H. Fraenkel-Conrat and R.R. Wagner, eds., Plenum Press, New York.

Enami, M., and Palese, P., 1991, High-efficiency formation of influenza virus transfectants, *J. Virol.* 65:2711.

Enami, M., Luytjes, W., Krystal, M., and Palese, P., 1990, Introduction of site-specific mutations into the genome of influenza virus, *Proc. Natl. Acad. Sci. USA* 87:3802.

Enami, M., Sharma, G., Benham, C., and Palese, P., 1991, An influenza virus containing nine different RNA segments, *Virology* 185:291.

García-Sastre, A., Barclay, W., Percy, N., and Palese, P., 1992, unpublished results.

Guinea, R., and Palese, P., 1992, unpublished results.

Honda, A., Uéda, K., Nagata, K., and Ishihama, A., 1988, RNA polymerase of influenza virus: role of NP in RNA chain elongation, *J. Biochem.* 104:1021.

Hsu, M.-T., Parvin, J.D., Gupta, S., Krystal, M., and Palese, P., 1987, Genomic RNAs of influenza viruses are held in a circular conformation in virions and in infected cells by a terminal panhandle, *Proc. Natl. Acad. Sci. USA* 84:8140.

Huang, T.-S., Palese, P., and Krystal, M., 1990, Determination of influenza virus proteins required for genome replication, *J. Virol.* 64:5669.

Kimura, N., Nishida, M., Nagata, K., Ishihama, A., Oda, K., and Nakada, S., 1992, Transcription of a recombinant influenza virus RNA in cells that can express the influenza virus RNA polymerase and nucleoprotein genes, *J. Gen. Virol.* 73:1321.

Krug, R.M., Alonso-Caplen, F.V., Julkunen, I., and Katze, M.G., 1989, Expression and replication of the influenza virus genome, *in*: "The Influenza Viruses", R.M. Krug, ed., Plenum Press, New York.

Lamb, R.A., 1989, Genes and proteins of the influenza viruses, *in*: "The Influenza Viruses", R.M. Krug, ed., Plenum Press, New York.

Li, X., and Palese, P., 1992a, Mutational analysis of the promoter required for influenza virus virion RNA synthesis, *J. Virol.* 66:4331.

Li, S., and Palese, P., 1992b, unpublished results.

Li, S., Schulman, J.L., Moran, T., Bona, C., and Palese, P., 1992, Influenza A virus transfectants with chimeric hemagglutinins containing epitopes from different subtypes, *J. Virol.* 66:399.

Luo, G., and Palese, P., 1992, Genetic analysis of influenza virus, *Curr. Opinion Genet. Dev.* 2:77.

Luo, G., Luytjes, W., Enami, M., and Palese, P., 1991, The polyadenylation signal of influenza virus RNA involves a stretch of uridines followed by the RNA duplex of the panhandle structure, *J. Virol.* 65:2861.

Luo, G., Bergmann, M., García-Sastre, A., and Palese, P., 1992, Mechanism of attenuation of a chimeric influenza A/B transfectant virus, *J. Virol.* 66:4679.

Luytjes, W., Krystal, M., Enami, M., Parvin, J.D., and Palese, P., 1989, Amplification, expression, and packaging of a foreign gene by influenza virus, *Cell* 59:1107.

Muster, T., Subbarao, E.K., Enami, M., Murphy, B.R., and Palese, P., 1991, An influenza A virus containing influenza B virus 5' and 3' noncoding regions on the neuraminidase gene is attenuated in mice, *Proc. Natl. Acad. Sci. USA* 88:5177.

Palese, P., 1977, The genes of influenza virus, *Cell* 10:1.

Palese, P., and García-Sastre, A., 1992, Influenza virus and its replication, *in*: "Encyclopedia of Virology", R.G. Webster and A. Granoff, eds., Academic Press, London.

Parvin, J.D., Palese, P., Honda, A., Ishihama, A., and Krystal, M., 1989, Promoter analysis of influenza virus RNA polymerase, *J. Virol.* 63:5142.

Peden, K.W.C., Pipas, J.M., Pearson-White, S., and Nathans, D., 1980, Isolation of mutants of an animal virus in bacteria, *Science* 209:1392.

Piccone, M.E., Fernández-Sesma, A., and Palese, P., 1992, unpublished results.

Post, L.E., and Roizman, B., 1981, A generalized technique for deletion of specific genes in large genomes: α gene 22 of herpes simplex virus 1 is not essential for growth, *Cell* 25:227.

Racaniello, V.R., and Baltimore, D., 1981, Cloned poliovirus complementary DNA is infectious in mammalian cells, *Science* 214:916.

Seong, B.L., and Brownlee, G.G., 1992, A new method for reconstituting influenza polymerase and RNA *in vitro*: a study of the promoter elements for cRNA and vRNA synthesis *in vitro* and viral rescue *in vivo*, *Virology* 186:247.

Weir, J.P., Bajszár, G., and Moss, B., 1982, Mapping of the vaccinia virus thymidine kinase gene by marker rescue and by cell-free translation of selected mRNA, *Proc. Natl. Acad. Sci. USA* 79:1210.

Xiong, C., Levis, R., Shen, P., Schlesinger, S., Rice, C.M., and Huang, H.V., 1989, Sindbis virus: an efficient, broad host range vector for gene expression in animal cells, *Science* 243:1188.

Yamanaka, K., Ogasawara, N., Yoshikawa, H., Ishihama, A., and Nagata, K., 1991, *In vivo* analysis of the promoter structure of the influenza virus RNA genome using a transfection system with an engineered RNA, *Proc. Natl. Acad. Sci. USA* 88:5369.

STRUCTURE AND FUNCTION OF THE VESICULAR STOMATITIS VIRUS RNA-DEPENDENT RNA POLYMERASE

Sailen Barik, Adrienne Takacs, Tapas Das, David Sleat, and Amiya K. Banerjee

Department of Molecular Biology, Research Institute
The Cleveland Clinic Foundation, 9500 Euclid Ave.
Cleveland, OH 44195, USA

INTRODUCTION

The RNA-dependent RNA polymerase of vesicular stomatitis virus consists of two proteins: the large protein L (242 kD) and the phosphoprotein P (29 kD). Together, these two proteins transcribe the negative strand genome RNA wrapped with the nucleocapsid protein N to produce the viral messages (see Banerjee and Barik, 1992, for a recent review). Considerable knowledge on the structure and function of these two proteins has been gathered primarily due to the availability of an in vitro transcription system and the advent of recombinant DNA technology. This article will summarize some of the recent findings from our laboratory.

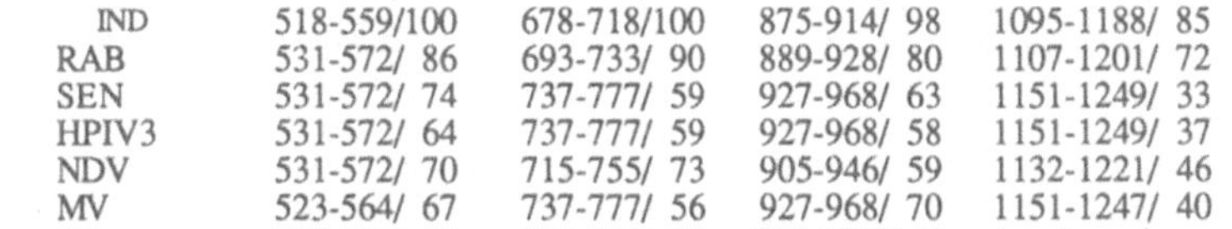

Virus	BoxA	BoxB	BoxC	BoxD
VSV				
OG	518-559/100	678-718/100	875-914/100	1095-1188/100
HZL	518-559/100	678-718/100	875-914/ 95	1095-1188/ 97
IND	518-559/100	678-718/100	875-914/ 98	1095-1188/ 85
RAB	531-572/ 86	693-733/ 90	889-928/ 80	1107-1201/ 72
SEN	531-572/ 74	737-777/ 59	927-968/ 63	1151-1249/ 33
HPIV3	531-572/ 64	737-777/ 59	927-968/ 58	1151-1249/ 37
NDV	531-572/ 70	715-755/ 73	905-946/ 59	1132-1221/ 46
MV	523-564/ 67	737-777/ 56	927-968/ 70	1151-1247/ 40
RS	607-637/ 55	775-815/ 56	976-1010/ 49	low homology

Figure 1. Regions of similarity in L proteins of nonsegmented negative-strand RNA viruses. Amino acid number at the two ends of each box and the per cent similarity with VSV(NJ) (Ogden) is shown. Virus names are abbreviated as follows: OG and HZL = Ogden and Hazelhurst subtypes, respectively, of NJ serotype VSV; IND = Indiana serotype VSV; RAB = Rabies; SEN = Sendai; HPIV-3 = human parainfluenza virus-3; NDV = Newcastle disease virus; MV = measles virus; RS = respiratory syncytial (Stec et al., 1991). Details in Barik et al. (1990).

L PROTEIN

Previous genetic and biochemical studies had suggested that the L protein encodes the RNA polymerase and its accessory activities, such as capping, methylation and polyadenylation of the nascent mRNA (reviewed in Banerjee and Barik, 1992), and a putative protein kinase activity that phosphorylates P protein (Sanchez et al , 1985; Masters and Banerjee, 1986; Barik and Banerjee, 1992a; Hammond et al., 1992). Thus, the P protein is generally regarded as a transcription factor essential for L function. However, the fact that these activities are all coupled to transcription which in turn requires both L and P proteins has made assignment of a particular function to any protein difficult. In one approach, we have compared the available primary sequences of L proteins of nonsegmented negative strand RNA viruses (rhabdo- and paramyxoviruses); results showed that the L proteins are highly similar in size, overall amino acid composition, and net charge. In addition, pairwise and multiple alignments resulted in the identification of conserved domains (Boxes A through D; Fig. 1) which also encoded common motifs postulated to be essential for functions such as binding to ATP and the RNA template (Barik et al., 1990; Poch et al., 1990). The functional importance of these boxes is underscored by their conservation during evolution despite the high mutability of RNA genomes in general.

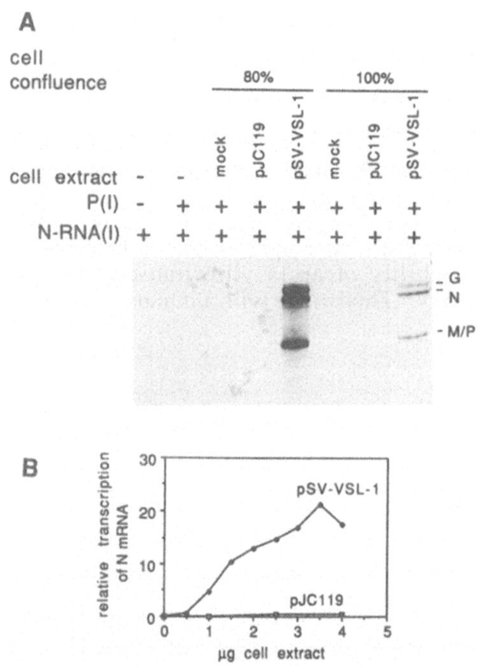

Figure 2. Expression of transcriptionally active recombinant L protein. COS cells were grown to indicated confluence and transfected with either pSV-VSL-1 (Schubert et al., 1985) or the parental plasmid pJC119 (as control). Following expression of the proteins (24-48 hr), cells were harvested by scraping into a hypo-osmotic buffer and incubated in ice for 5 min. The cytoplasmic extract, prepared by disrupting the cells in a Dounce homogenizer, was stored frozen at -80^{0}C in small portions until use. VSV transcription reactions (25 µl) were reconstituted essentially as described (Barik and Banerjee, 1991) and labeled transcripts were analyzed on 6% polyacrylamide gels containing urea. **A:** Reconstitution of transcription with extracts from confluent cells transfected with the indicated plasmid. (I) stands for Indiana serotype. Where indicated, 2 µg of cell extract was used. An autoradiograph of the labeled transcripts is shown. The various gene transcripts are indicated (G, N, M, P). **B:** Transcription is proportional to the amount of extract used (µg protein per reaction). Transcription was quantitated by densitometric scanning of the N message in an autoradiogram.

It became obvious that mapping of the various domains of the multifunctional L protein would require the use of recombinant L protein expressed from a cDNA clone. To this end, we have used the L gene of VSV_{IND} cloned in a SV40 vector (Schubert et al., 1985); in earlier studies this clone was shown to restore the growth of a temperature-sensitive L mutant of VSV at non-permissive temperature. As shown in Fig. 2, cytoplasmic extracts made from COS cells transfected with this plasmid were found to contain active L protein as judged by its ability to restore VSV transcription reaction reconstituted in vitro with N-RNA template and P protein. Using this in vitro system to assay recombinant L protein, we are currently studying the roles of the various domains of L (Fig. 1) by site-directed mutagenesis.

P PROTEIN

In comparison to L, our knowledge of the various domains of P protein and their functions is much more complete, due primarily to the small size of the protein, its relative stability, and its recent expression in *E. coli* in large quantities. The P polypeptide consists of 274 amino acids with a calculated molecular weight of ~32 kD (Gill and Banerjee, 1985; Hudson et al., 1986). Overall, the protein is highly acidic, with Asp and Glu residues constituting 32 out of the first 100 residues in the amino-terminal half of the polypeptide. It is also one of the most highly phosphorylated proteins known, Ser and Thr residues constituting a sixth of the total number of amino acids. Phosphoamino acid analyses have established Ser as the major phosphate acceptor (Clinton and Huang, 1981), indicating that serine kinase(s) is involved in the phosphorylation process.

Comparison of deduced amino acid sequences of various serotypes of P protein has revealed a surprising dissimilarity (30% overall similarity), concentrated mostly in three regions, viz., domain I (residues 1-137), domain II (residues 214-247), and a very poorly conserved region between these domains (Gill and Banerjee, 1985; Rae and Elliott, 1986). In contrast, domain III (residues 250-274), consisting of about 25 amino acids at the C-terminal end, is highly basic and conserved (90% similarity). Essentially identical domains have been identified recently on the basis of deduced sequence comparison of P proteins of 18 different isolates the NJ serotype of VSV (Bilsel et al., 1990). Domain I is highly acidic and is phosphorylated by a cellular protein kinase (see below). Deletion studies have identified this domain to be important in binding L protein (Emerson and Schubert, 1987). Domain II, on the other hand, is phosphorylated by a L-associated kinase at specific serine residues (236 and 242 in the NJ serotype P protein), as determined by mutational analysis of recombinant P proteins (Chattopadhyay and Banerjee, 1987). Domain III appears to be the strongest template-binding domain since its deletion in both serotypes of P abolishes binding to homologous N-RNA templates (Paul et al., 1988; Gill et al., 1986). In recent years, major advances have been made in the elucidation of the role of phosphorylation in P protein function, some of which are presented below.

Essential Role of CKII: Cascade Phosphorylation

We showed that bacterially expressed P protein is phosphate-free and thus provided an ideal substrate in studies of phosphorylation in vitro, the results of which (Barik and Banerjee, 1992a) are summarized as follows. Using anion-exchange (DEAE-cellulose) chromatography, three forms of P protein (NJ serotype), phosphorylated to various degrees, could be separated from one another: the phosphate-free, bacterially expressed form (P_0); a semi-phosphorylated intermediate (P_1); and a fully phosphorylated form (P_2). In vitro, uninfected BHK cell extracts phosphorylated P_0 and converted it into P_1, but not to P_2, demonstrating that P_1 is the end product of cell kinase-mediated phosphorylation. A highly purified L protein preparation failed to phosphorylate P_0, suggesting that it is devoid of the cellular kinase. This L protein, however, phosphorylated P_1 to produce P_2. Therefore, the cellular and L(-associated) kinases are highly specific for their respective substrates, viz., P_0 and P_1. Heparin completely inhibited cell kinase but had little effect on L kinase. Protamine, on the other hand, strongly inhibited L kinase but stimulated cell kinase about three-fold. The inevitable corollary of the substrate specificity is that phosphorylation mediated by cell kinase must precede that by L kinase, i.e., the two kinases must act sequentially, catalyzing a cascade phosphorylation pathway: $P_0 \dashrightarrow P_1 \dashrightarrow P_2$. Recently, a stepwise phosphorylation of the P protein have also been demonstrated by using viral RNP as the source of the substrate P protein as well as the protein kinases (Beckes and Perrault, 1992).

By far the most important outcome of these studies was the finding that transcription reaction reconstituted with host kinase-free L protein, N-RNA template and phosphate-free P protein (P_0) was nonfunctional. However, addition of cell kinase or substitution of P_0 with P_1 fully restored transcription, demonstrating for the first time the essential role of a cell kinase-mediated phosphorylation in the activation of a RNA viral transactivator (see below). In subsequent studies (Barik and Banerjee, 1992b), we have characterized this cellular kinase as casein kinase II (CKII), a cyclic AMP-independent kinase identified in a variety of animal tissues (Tuazon and Traugh, 1991). The kinase was purified from BHK cell extracts by monitoring its ability to phosphorylate P_0 in vitro. Biochemical properties of the purified kinase were found to be indistinguishable from those of authentic CKII, e.g., similar elution profiles in various ion-exchange matrices, ability to utilize both ATP and GTP as phosphate donors and both casein and P_0 as phosphate acceptors, strong inhibition by heparin, and stimulation (three- to four-fold) by basic polyamines, such as protamine and spermine.

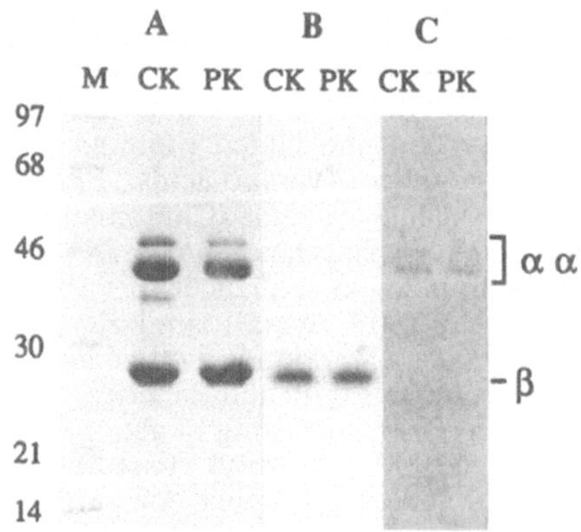

Figure 3. Protein profile, autophosphorylatin and western blot of purified protein kinase (PK) and bovine casein kinase II (CK). Purification of P protein kinase from BHK cell cytoplasmic extract is described in detail elsewhere (Barik and Banerjee, 1992b). **A:** Silver-stained protein profile following SDS-PAGE. **B:** Self-phosphorylation of the kinases in the presence of [γ-^{32}P]ATP; an autoradiograph of the labeled protein bands is shown. **C:** Western Blot using an antibody raised against a synthetic peptide corresponding to the catalytic domain (residue 70-91) of human CKII α peptide. See text for details.

The close identity of the two enzymes was further confirmed by a comparison of their polypeptide profiles. SDS-PAGE followed by silver staining revealed that both enzymes contain two major classes of polypeptides: larger subunit(s) ($\alpha\alpha'$) in the Mr range ~45 kD, and a smaller subunit (β) of Mr ~25 kD (Fig. 3, A). When the enzymes were allowed to autophosphorylate in the absence of added substrate, the smaller polypeptides (β) were phosphorylated in both cases (Fig. 3, B). An antibody raised against a 23-mer synthetic peptide representing the catalytic domain of the α subunit cross-reacted with the major 45 kD doublet of both preparations but did not react with the β subunit of either (Fig. 3, C). Taken together, these results demonstrate the immunological and catalytic similarity of the P_0 kinase with CKII. Some of these properties, e.g., heparin sensitivity and ability to use both ATP and GTP are in fact diagnostic of CKII.

Finally, VSV transcription, defective when reconstituted with phosphate-free P protein (P_0), host kinase-free N-RNA template, and L (Fig. 4, lane 1), could be restored by the addition of purified CKII (lanes 2, 3). In another set of reactions, P_0 was first phosphorylated by PK or CKII and the phosphorylated product (P_1), purified through DEAE-cellulose chromatography, was used to reconstitute transcription. Transcription was again functional (lanes 4, 5), demonstrating that a cell kinase-free VSV transcription can be reconstituted if the P protein has already been phosphorylated by the cell kinase (Barik and Banerjee, 1992a; Massey et al., 1990). Heparin, a specific inhibitor of PK and CKII, also inhibited transcription reactions programmed with phosphate-free P protein and either of these kinases (Fig. 4, lanes 6, 7). At the same concentration, heparin had little effect on L kinase, measured by the ability of the latter to phosphorylate P_1. As expected, heparin had no effect on transcription reactions that were independent of CKII, i.e., reconstituted with phosphorylated P proteins, P_1 or P_2 (lanes 8, 9). Protamine, a specific inhibitor of L kinase-mediated phosphorylation of P_1 to P_2 (Barik and Banerjee, 1992b) inhibited transcription

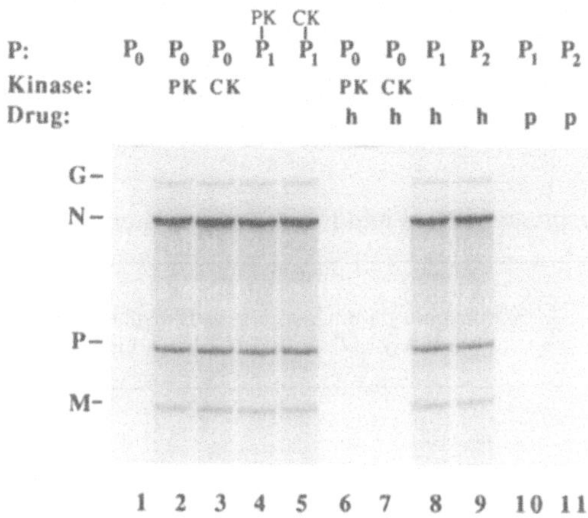

Figure 4. Essential role of casein kinase II in VSV transcription. Transcription reactions were reconstituted in vitro using the indicated species of P protein and host-kinase-free L and N-RNA template as described (Barik and Banerjee, 1992a); an autoradiograph of the labeled transcripts is shown. Where indicated, 10 ng (in a 20 μl reaction) of either purified P_0 kinase (PK) or CKII (CK) was also included. Reactions marked 'h' and 'p' received 20 ng of heparin and protamine, respectively. The various gene transcripts are indicated (G, N, P, M). See text for details.

reactions programmed with either P_0 or P_1 (lane 10), consistent with an essential role of L kinase in P function. Interestingly, protamine also inhibited transcription reactions programmed with P_2, the fully phosphorylated form of P protein (lane 11). This latter finding suggests that the P_2-specific phosphates, which were shown to undergo a turnover during VSV transcription (Barik and Banerjee, 1992a), must be continually replaced by the L kinase in order that transcription may progress. Taken together, these results and the earlier ones (Chattopadhyay and Banerjee, 1987) demonstrate that phosphorylation of the P protein by casein kinase II and L-associated kinase is essential for the transcriptional activation of the P protein.

Sites of Phosphorylation

Early proteolytic mapping of phosphorylated peptides in viral P protein of IND serotype localized the majority of phosphates between residues 35 and 70 in domain I, although the nature of the kinase(s) and their relative contribution were not determined (Marnell and Summers, 1984; Bell and Prevec, 1985; Hsu and Kingsbury, 1985). Comparison of the primary sequences of IND and NJ serotype P proteins revealed that five potential phosphate acceptor sites (Ser and Thr) are conserved in this region (corresponding to Thr^{49}, Ser^{51}, Ser^{59}, Ser^{61}, Thr^{63} in the NJ serotype P). In order to map the exact sites of phosphorylation of the P protein by the cellular kinase we altered specific serine and threonine residues within this area to alanine by site-directed mutagenesis. The mutated P proteins were then tested to determine what effect these mutations had on the phosphorylated state of the protein in vivo as well as its transcriptional activity in vitro (A. T., S. B., T. D., and A. K. B., submitted). To confirm that phosphorylation in vivo was catalyzed by casein kinase II, the relevant mutant P proteins were also expressed in bacteria in phosphate-free form. The bacterially produced purified proteins were tested as phosphorylation substrates in an in vitro kinase reaction using purified casein kinase II. Results of these studies are summarized in Table 1 which shows that mutants in which *both* Ser59 and Ser61 are mutated (P59/61,

P59/61/63, P4+63, P5) are highly deficient in phosphate. However, mutation of any one of these serines (P59 or P61) did not reduce the level of phosphorylation. Although we do not know the exact reason behind it, we speculate that this may be due to a compensatory increase in phosphorylation at one site in absence of the other. Alternately, other (secondary) sites may be phosphorylated in the absence of one of these primary sites. Nonetheless, it is clear that regardless of their phosphorylation status, mutant P proteins lacking either serine 59 or 61 are transcriptionally defective. Thus, both serines 59 and 61 are required for CKII to correctly phosphorylate the P protein in domain I. The strong correlation of the in vivo data with results obtained in vitro using bacterially expressed proteins and purified CKII (Table 1) strongly suggest that CKII is the only protein kinase phosphorylating P protein in vivo.

Table 1. Phosphate content and transcriptional activity of mutant P(NJ) proteins

P Protein	% Phosphorylation[a] in vivo	% Phosphorylation[b] in vitro	% Transcription[c] in vitro
wt	100	100	100
P13	103	ND	96
P49	86	ND	93
P59	88	90	<1
P61	109	ND	<1
P63	98	ND	104
P49/51	84	95	73
P49/51/59	70	ND	<1
P59/61	7	2	<1
P61/63	80	ND	<1
P59/61/63	6	1	<1
P4+59	106	ND	<1
P4+61	138	95	<1
P4+63	8	ND	<1
P5	7	<1	<1

[a] Details of the procedure will be published elsewhere. In brief, COS cells were transfected with P plasmids (P genes cloned in pECE) and labeled with ^{32}P-orthophosphate. Parallel cultures were labeled with ^{35}S-methionine to quantitate P protein synthesis. Expressed P proteins were immunoprecipitated with anti-VSV antibody and the precipitate analyzed by SDS-PAGE. ^{32}P content of the P protein bands were quantitated by densitometric scanning and normalized against unit amount of protein. [b] Phosphorylation in vitro was carried out with purified casein kinase II and bacterially expressed P proteins. [c] VSV transcription was reconstituted in vitro with viral L protein, N-RNA template, and the various mutant P proteins translated in wheat germ lysate (which contains CKII activity; data not shown). Transcription was quantitated as described for Fig. 2. ND=not done.

An examination of the amino acid sequence around the two critical serines (EEESS^{59}DS^{61}DTDY) revealed an excellent agreement with consensus CKII site (SXXD/E) whereby the phosphate-acceptor Ser residue (at position 'n') should have acidic residues (Asp/Glu) around it, especially at position 'n+3' (Roach, 1991). It is interesting to note that the properties of these phosphates almost perfectly fit the general nature of CKII-mediated phosphorylation events known to date, e.g., the relative stability of these phosphates, their regulatory role in nucleic acid metabolism (Tuazon and Traugh, 1991), and their occurrence as the first step of a multiple and hierarchal (cascade) phosphorylation pathway (Roach, 1991). The exact nature of host kinase(s) packaged in the mature virion (Clinton et al., 1982; Harmon et al., 1983; Massey et al.. 1990; Beckes and Perrault, 1991) still remains unknown.

Phosphorylation-Dependent Dimerization of P Protein

Since the activities of a number of transcription factors have been shown to depend on their ability to form homodimers (Mitchell and Tjian, 1989), we tested whether P protein can oligomerize in vitro. Bacterially expressed, purified, unphosphorylated P protein (P_0) at various concentrations was chromatographed through a Sephadex G-100 gel filtration column along with standard proteins of known molecular size. Fractions were analyzed by SDS-PAGE and the P protein band was recognized by its characteristic mobility (Mr ~46,000) as well as by western blotting with specific antibody. Densitometric scan of a representative fractionation shows (Fig. 5) that at a low concentration (5 µg in 100 µl), essentially all of the P protein eluted as a 29 kD monomoeric polypeptide (top panel, right peak). However, at a four-fold higher concentration (20 µg in 100 µl), P_0 fractionated mostly as a ~62 kD species, i.e., a dimer (top panel, left peak). We then phosphorylated the P_0 protein to P_1 by purified casein kinase II and $[\gamma\text{-}^{32}P]$ATP in vitro; the oligomerization of the ^{32}P-labeled protein was then tested. Interestingly, P_1 formed a dimer even at a low concentration of (~8 µg in 100 µl) (middle panel) In the same fractionation, small amounts of left-over unphosphrylated P_0 behaved as a monomer as expected. P 59/61 mutant protein, defective in phosphorylation (Table 1), remained a monomer even at the high concentration used (bottom panel).

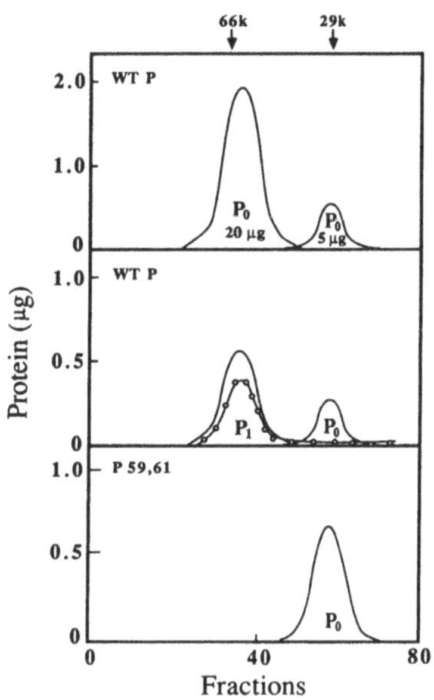

Fig. 5. Efficient dimerization of P protein induced by CKII-mediated phosphorylation. P_{NJ} protein of the indicated kind was fractionated through a Sephadex G-100 column (18 ml) equilibrated with VSV transcription buffer (Barik and Banerjee, 1991); alternate fractions were monitored by SDS-PAGE, followed by silver staining (and autoradiography, where applicable). Positions of the two standard size markers (66k and 29k) are shown on top. Top panel: Fractionation of either 5 or 20 µg wild type P protein. Middle: Fractionation of a mixture of unphosphorylated (P_0) and phosphorylated (P_1) wild type P protein. Bottom: Fractionation of mutant P 59, 61 protein. P_1 peak (middle panel) was monitored by silver staining (—) as well as ^{32}P counting (--o--o--, arbitrary scale not shown).

Based on these results, we propose that the wild type P protein, even in its unphosphorylated form, has a natural propensity to form dimer according to the reversible reaction: $2(P_0) \rightleftharpoons (P_0)_2$, such that a high concentration of the monomer drives the reaction to the right by the law of mass action. The inability of P 59/61 mutant to dimerize suggest that Ser^{59} and Ser^{61} play important roles in the phosphorylation-independent dimerization. Phosphorylation of these residues further facilitates the dimerization, possibly by improving the association constant of the monomers so that the reaction $2(P_1) \rightleftharpoons (P_1)_2$ takes place at a relatively low protein concentration. We are currently testing whether the various phosphorylation-defective mutants of P protein (Table 1) also dimerize poorly and whether the efficiency of dimerization correlates with the transcriptional activity of the P protein. It would also be interesting to find out whether P proteins in which the acidic domain I has been replaced by apparently unrelated acidic polypeptides such as tubulin, are also able to dimerize. Earlier, such chimeric P proteins were shown to possess VSV transcriptional activity in vitro (Chattopadhyay and Banerjee, 1988).

CONCLUSION

Expression of transcriptionally active L and P proteins from cloned genes has opened up new directions of research on the relationship between structure and function of the different domains of these multipotential proteins. Domains contributing to capping, methylation, and protein kinase activities of the L protein, for example, can now be characterized using the in vitro complemention assay of recombinant L described here (Fig. 2). Our results strongly suggest that phosphorylation by cellular casein kinase II induces efficient dimerization of the P protein (Fig. 5). Since this phosphorylation is also an essential prerequisite to the transcriptional activation of the P protein (Fig. 4), we speculate that the active form of the P protein is its phosphorylated dimer. This would constitute a unique mechanism of activation of a RNA-dependent RNA transcription factor by CKII. Further studies are needed to determine which of the many functions of P protein is facilitated by dimerization, such as binding to L or N-RNA template, phosphorylation by the L-associated kinase, etc. In this connection, it is interesting to note that genetic complementation data and the presence of excess P protein (over L protein) in the viral transcription complex have led to the suggestion that the functional VSV polymerase may consist of one polypeptide of L and two of P, i.e., $L(P)_2$ (reviewed by Emerson, 1987). The dimerization property of P protein makes an important prediction: One should be able to generate mutant P proteins which might form transcriptionally inactive heterodimers with the wild-type polypeptide and hence act as transdominant negative mutants. Our preliminary results have in fact demonstrated the existence of P mutants that inhibit VSV transcription reconstituted with wild type P protein in vitro (S. B. and T. D., unpublished). Finally, it remains to be seen whether phosphoproteins of other nonsegmented negative strand RNA viruses are also regulated in a similar manner.

ACKNOWLEDGMENTS

We thank Drs. Edwin G. Krebs and David Litchfield (Howard Hughes Institute, Seattle, Washington, USA) for the purified bovine casein kinase II. Work in the authors' laboratory was supported by a USPHS grant AI-26585 (A. K. B.).

REFERENCES

BANERJEE, A. K., AND BARIK, S. (1992). Gene expression of vesicular stomatitis virus genome RNA (Minireview). Virology **188:** 417-428.

BARIK, S., and BANERJEE, A. K. (1991). Cloning and expression of the vesicular stomatitis virus phosphoprotein gene in *Escherichia coli:* Analysis of phosphorylation status versus transcriptional activity. J. Virol. **65:** 1719-1726.

BARIK, S., and BANERJEE, A. K. (1992a). Sequential phosphorylation of the phosphoprotein of vesicular stomatitis virus by cellular and viral protein kinases is essential for transcription activation. J. Virol. **66:** 1109-1118.

BARIK, S., and BANERJEE, A. K. (1992b). Phosphorylation by cellular casein kinase II is essential for the transcriptional activity of vesicular stomatitis virus phosphoprotein P. Proc. Natl. Acad. Sci. USA, in press.

BARIK, S., RUD, E. W., LUK, D., BANERJEE, A. K., and YONG KANG, C. (1990). Nucleotide sequence analysis of the L gene of vesicular stomatitis virus (New Jersey serotype): Identification of conserved domains in L proteins of nonsegmented negative-strand RNA viruses. Virology **175:** 332-337.

BECKES, J. D., and PERRAULT, J. (1991). Two distinct protein kinase activities in vesicular stomatitis virions phosphorylate the NS transcription factor. Virology **184:** 383-386.

BECKES, J. D., and PERRAULT, J. (1992). Stepwise phosphorylation of vesicular stomatitis virus P protein by virion-associated kinases and uncoupling of second step from *in vitro* transcription. Virology **188:** 606-617.

BELL, J.C., and PREVEC, L. (1985). Phosphorylation sites of phosphoprotein NS of vesicular stomatitis virus. J. Virol. **54:** 697-702.

BILSEL, P. A., ROWE, J. E., FITCH, W. M., and NICHOL, S. T. (1990). Phosphoprotein and nucleocapsid protein evolution of vesicular stomatitis virus New Jersey. J. Virol. **64:** 2498-2504.

CHATTOPADHYAY, D., and BANERJEE, A. K. (1987). Phosphorylation within a specific domain of the phosphoprotein of vesicular stomatitis virus regulates transcription in vitro. Cell **49:** 407-414.

CHATTOPADHYAY, D., and BANERJEE, A. K. (1988). NH_2-terminal acidic region of the phosphoprotein of vesicular stomatitis virus can be functionally replaced by tubulin. Proc.Natl. Acad. Sci. USA. **85:** 7977-7981.

CLINTON, G. M., GUERINA, N. G., GUO, H., and HUANG, A. S. (1982). Host dependent phosphorylation and kinase activity associated with vesicular stomatitis virus. J. Biol. Chem. **257:** 3313-3319.

CLINTON, G. M. and HUANG, A. S. (1981). Distribution of phosphoserine, phosphothreonine, and phosphotyrosine in proteins of vesicular stomatitis virus. Virology **108:** 510-514.

EMERSON, S. U. (1987). Transcription of vesicular stomatitis virus. In "The Rhabdoviruses" (R. R. Wagner, Ed.), pp. 167-243. Plenum Press, New York.

EMERSON, S. U., and SCHUBERT, M. (1987). Location of the binding domains for the RNA polymerase L and the ribonucleocapsid template within different halves of the NS phosphoprotein of vesicular stomatitis virus. Proc. Natl. Acad. Sci. USA. **84:** 5655-5659.

GILL, D. S., and BANERJEE, A. K. (1985). Vesicular stomatitis virus NS proteins: Structural similarity without extensive sequence homology. J. Virol. **55:** 60-66.

GILL, D. S., CHATTOPADHYAY, D., and BANERJEE, A. K. (1986). Identification of a domain within the phosphoprotein of vesicular stomatitis virus that is essential for transcription in vitro. Proc. Natl. Acad. Sci. USA. **83:** 8873-8877.

HAMMOND, D. C., HALEY, B. E., and LESNOW, J. A. (1992). Identification and characterization of serine/threonine protein kinase activity intrinsic to the L protein of vesicular stomatitis virus New Jersey. J. Gen. Virol. **73:** 67-75.

HARMON, S. A., MARNELL, L. L., and SUMMERS, D. F. (1983). The major ribonucleoprotein associated protein kinase of vesicular stomatitis virus is a host protein. J. Biol. Chem. **258:** 15283-15290.

HSU, C.-H., and KINGSBURY, D. W. (1985). Constitutively phosphorylated residues in the NS protein of vesicular stomatitis virus. J. Biol. Chem. **260:** 8990-8995.

HUDSON, L. D., CONDRA, C., and LAZZARINI, R. A. (1986). Cloning and expression of a viral phosphoprotein : Structure suggests vesicular stomatitis virus NS may function by mimicking an RNA template. J. Gen. Virol. **67:** 1571-1579.

MARNELL, L. L., and SUMMERS, D. F. (1984). Characterization of the phosphorylated small enzyme subunit, NS, of the vesicular stomatitis virus RNA polymerase. J. Biol. Chem. **259:** 13518-13524.

MASTERS, P. S., and BANERJEE, A. K. (1986). Phosphoprotein NS of vesicular stomatitis virus: Phosphorylated states and transcriptional activities of intracellular and viral forms. Virology **154:** 259-270.

MASSEY, D. M., DEANS, N., and LENARD, J. (1990). Phosphorylation of NS protein by vesicular stomatitis virus nelcocapsids: lack of effect during RNA synthesis and separation of kinase from L protein. J. Virol. **64:** 3259-3264.

MITCHELL, P. J., AND TJIAN, R. (1989). Transcriptional regulation in mammalian cells by sequence-specific DNA binding proteins. Science **245:** 371-378.

PAUL, P. R., CHATTOPADHYAY, D., and BANERJEE, A. K. (1988). The functional domains of phosphoprotein (NS) of vesicular stomatitis virus (Indiana serotype). Virology **166:** 350-357.

POCH, O., BLUMBERG, B. M., BOUGUELERET, L., and TORDO, N. (1990). Sequence comparison of five polymerases (L proteins) of unsegmented negative-strand RNA viruses: Theoretical assignment of functional domains. J. Gen. Virol. **71:** 1153-1162.

RAE, B. P., AND ELLIOTT, R. M. (1986). Conservation of potential phosphorylation sites in the NS proteins of the New Jersey and Indiana serotypes of vesicular stomatitis virus. J. Gen. Virol. **67,** 1351-1360.

ROACH, P. J. (1991). Multisite and hierarchal protein phosphorylation. J. Biol. Chem. **266:** 14139-14142.

SANCHEZ, A., DE, B. P., and BANERJEE, A. K. (1985). In vitro phosphorylation of the NS protein by the L protein of vesicular stomatitis virus. J. Gen. Virol. **66:** 1025-1036.

SCHUBERT, M., HARMISON, G. G., RICHARDSON, C. D., and MEIER, E. (1985). Expression of a cDNA encoding a functional 241 kilodalton vesicular stomatitis virus RNA polymerase. Proc. Natl. Acad. Sci. USA **82:** 7984-7988.

STEC, D. S., HILL, M. G., and COLLINS, P. L. (1991). Sequence analysis of the polymerase L gene of human respiratory syncytial virus and predicted phylogeny of nonsegmented negative-strand viruses. Virology **183:** 273-287.

TUAZON, P. T., and TRAUGH, J. A. (1991). Casein kinase I and II--Multipotential serine protein kinases: Structure, function, and regulation. Adv. Sec. Messenger and Phosphoprotein Res. **23:** 123-164.

RNA SYNTHESIS AND mRNA EDITING IN PARAMYXOVIRUS INFECTIONS

Joseph Curran, Thierry Pelet, Jean-Philippe Jacques and Daniel Kolakofsky

Department of Genetics and Microbiology, University of Geneva School of Medicine CMU, 9 Ave Champel CH-1211 Geneva Switzerland

A BRIEF OVERVIEW OF PARAMYXOVIRUS RNA SYNTHESIS

Paramyxoviruses contain nonsegment ssRNA genomes of around 15 kB and negative polarity, i.e., complementary to the viral mRNAs. These genomes (and negative-strand viral genomes in general) function not as free nucleic acid, but as helical nucleocapsids (NC) assembled with the viral nucleoprotein (NP) in which the genome RNA represents only 4% by weight. It is these nucleocapsids which are the templates for RNA synthesis. The vrial polymerase is composed of two subunits, the phosphoprotein P and the large protein L, but its overall structure is unclear.

The genome contains about 6 genes or transcriptional units, preceeded by a short non-coding leader region. Between the genes are highly conserved sequences thought to function as the mRNA polyadenylation/termination and the mRNA start signals, separated by a short non-transcribed spacer. A central feature of (–)RNA genomes is that they are templates for both mRNA synthesis (transcription) and that of the full length antigenomes (the complementary intermediates in genome replication). During transcription, the polymerase starts at the 3' end of the template and sequentially produces the leader RNA and mRNAs by stopping and restarting at each of the junctions.

The mRNAs contain a poly(A) tail of several hundred nucleotides and the start of this tail maps to the U_{5-7} stretch within the conserved termination sequence of the (–) template (e.g., 3'AUUC<u>UUUUU</u> for Sendai virus). The poly(A) tail is presumably made by polymerase slippage and reiterative copying of the U_{5-7} stretch (McGeoch, 1979; Rose, 1980; Schubert *et al.*, 1980; Robertson *et al.*, 1981; Giorgi *et al.*, 1983; Gupta and Kingsbury, 1984), and this phenomenon can be refered to as "stuttering". The same polymerase is thought to start the next mRNA only after terminating poly(A) synthesis. During antigenome synthesis or replication, the polymerase reads through all junctions to make an exact full length complementary copy. Unlike the mRNAs but like genomes, the antigenomes are assembled into NCs and this takes place concomitant with their synthesis. (Antigenomes presumably also function as templates only in the assembled form). This coupling of

Regulation of Gene Expression in Animal Viruses, Edited by
L. Carrasco *et al.*, Plenum Press, New York, 1993

antigenome synthesis and assembly is thought to play a key role in whether the junctional stop/restart signals are obeyed. For a more complete review, see Banerjee, 1987, and Kolakofsky and Roux, 1987.

Except for their poly (A) tails, most mRNAs are faithfull copies of their genes. This is not so, however, for the P gene mRNA(s), the subject of this review.

THE DISCOVERY OF G INSERTIONS WITHIN THE P GENE mRNAs

Even before the realization that a form of mRNA editing operated in paramyxoviruses, their P genes were considered unusual. In every case multiple proteins were expressed from this gene, in contrast to all the other

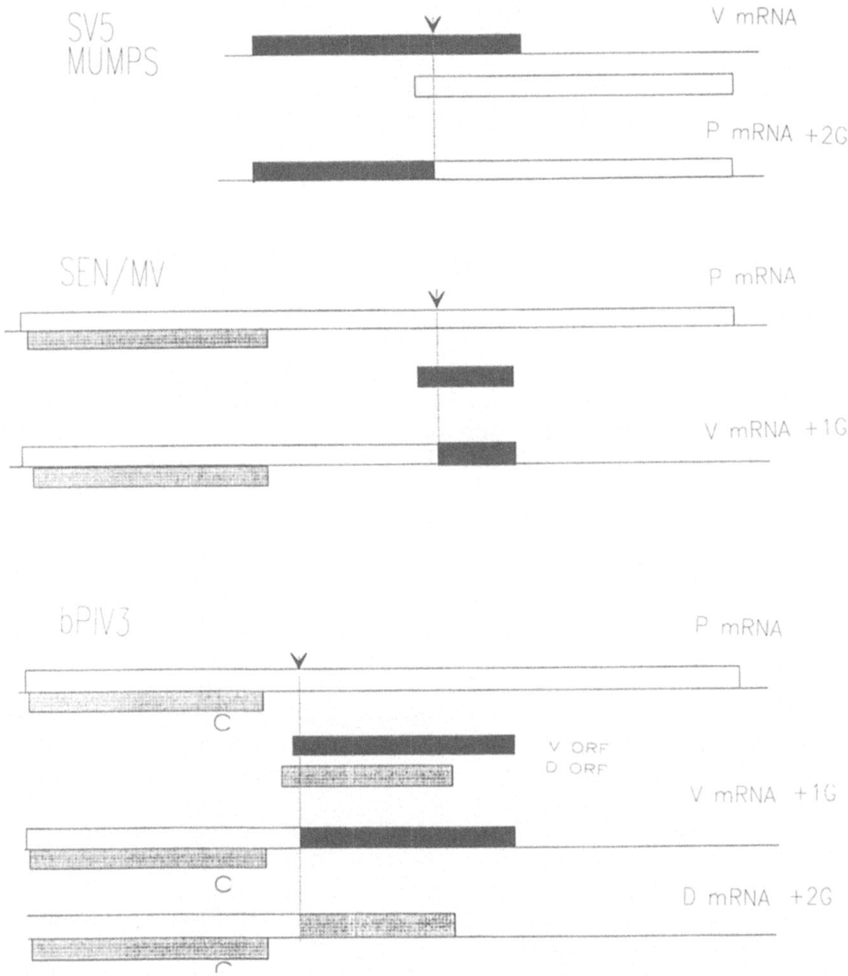

Fig 1. Schematic representation of the paramyxovirus P gene mRNAs.

The mRNAs are indicated as lines and the ORFs as boxes. For each group, the upper line shows the mRNA which is an exact copy of the gene, and the beginning of the ORF box indicates the ribosomal start codon. When more than one ORF box is attached to the line, they are accessed by alternate initiation codons. The boxes below indicate alternate downstream ORFs which are accessed by G insertions in the mRNAs. The positions of the insertions are shown by the arrowheads and the dotted vertical lines. The three possible ORFs are indicated by open, hatched, and black boxes.

viral genes in which only a single primary translation product was expressed. The Paramyxoviridae is a relatively large family of viruses, and their P genes can be devided into basically two groups (see fig 1):

i) the Sendai/measles virus or SEN/MV group, which express longer P proteins (568-603 aa) and which also contain an alternate ORF to express the C protein(s). The C protein ORF overlaps the amino-terminal sequences of the P ORF in the +1 frame, and

ii) the simian virus 5 (SV5)/mumps virus group, which express shorter P proteins (245-400 aa) and do not contain the overlapping C ORF.

For the SEN/MV group, there was evidence that the P and C proteins were translated from the same mRNA by ribosomal choice, via leaky ribosomal scanning and sometimes by what appears to be a scanning independent mechanism (Curran and Kolakofsky, 1988, 1989). However, even for the SV5/mumps group which did not contain the alternate C ORF, these P genes also expressed at least two proteins. For example, the mumps virus P gene also expresses the nonstructural proteins NS1 and NS2 (since renamed V and I, with tongue in cheek, for consistency with SV5) (Herrler and Compans, 1982; Paterson and Lamb, 1990). However, until the sequence of the first P gene in this group was determined (correctly), there was no way of knowing what mechanisms might be operating to produce these multiple proteins.

The first such gene determined was that of SV5 (Thomas *et al.*, 1988) (fig 1). The sequence was determined both from three separate cDNA clones (to mRNA) and then confirmed on the viral genome. This yielded the surprising result that there was no single ORF large enough to code for the SV5 P protein. Rather, the gene contained two separate ORFs at each end, which briefly overlapped in the middle. mRNA transcribed from these cDNAs *in vitro* could be translated into the shorter V protein encoded by the 5' proximal ORF, but not into the longer P protein. After a period of frustration, the solution to this dilemma became clear when Thomas *et al* prepared a second cDNA library with a larger number of P gene clones, and sequenced 22 of these. Twelve of these were identical to the genome and the 3 mRNA clones sequenced initially. The ten other sequences were identical to these, except that they had 2 G residues inserted within a short run of Gs in a region where the 2 ORFs overlapped (arrowhead in fig 1). This fused the separate ORFs at each end into a single continuous unit, and mRNAs prepared from these latter clones translated into P (but not V) proteins *in vitro*.

A second unexpected finding emerged from this study. Whereas the P protein sequences are the least conserved of the viral proteins between different paramyxoviruses, that region of V which is not shared with P (i.e., the C-terminal region) is cysteine rich and among the best conserved of all the viral sequences. This cys-rich V domain was in fact present in all the P genes of the SEN/MV group previously sequenced, but had not been noticed because it was relatively short (ca. 75 aa), in the middle of the genes, and generally lacked a start codon. Thomas *et al* predicted that these other virus would also express V proteins by a similar mechanism, except that in the SEN/MV group IG would need to be inserted to switch from the P to the V ORF.

This prediction was verified, and in fact, before the paper of Thomas *et al* appeared in print. M. Billeter's laboratory in Zurich, who were studying the mutations generated in the various MV genes during persistent infections of human brain (known as subacute sclerosing panencephalitis or SSPE), independently found that some P gene mRNAs generated from cDNA clones made normal sized P proteins whereas others made proteins of only half this size. When the various cDNAs were sequenced, the only real difference between tham was that those that made the shorter V protein contained a single extra G within a short run of Gs, and which was also absent in the genome (Cattaneo *et al.*, 1989). Once it became clear what to look for, other examples of G insertions in P gene mRNAs then followed shortly (Sendai virus, Vidal *et al.*, 1990a; mumps virus, Paterson and Lamb, 1990; bPIV3, Pelet *et al.*, 1991; hPIV3, Galinski *et al.*, 1992; LMPV, Berg *et al.*, 1992).

THE MECHANISM OF THE G INSERTIONS

As mentioned above, poly(A) tail formation of these mRNAs is thought to take place by the viral polymerase reiteratively copying the $U_{5\text{-}7}$ stretch within the termination sequence at the end of each gene. Since the G insertions appear to take place only in the P gene mRNAs and not in the P genes (i.e., during transcription but not during genome replication), and because there was also some homology between the sequences which precede the $U_{5\text{-}7}$ stretch and those of the G insertion site, it seemed reasonable that the G insertion could arise by reiterative copying of the short C stretch on the genome template at the insertion site (Thomas *et al.*, 1988; Cattaneo *et al.*, 1989).

Support for this notion was obtained with Sendai virus (Vidal *et al.*, 1990a). All paramyxovirions contain a template bound-polymerase and purified virions can make mRNA *in vitro*, however, only for SEN had the conditions for efficient mRNA synthesis been worked out. When P gene mRNAs made with purified SEN virions were examined and compared to those made *in vivo*, they were found to be "edited" very similarly; approximately 25% had a single G insertion, and about 2% had insertions which ranged from 2 to 8 Gs. This suggested that the insertiona were due to viral proteins. Equally of importance, when the P gene mRNAs were expressed from recombinant vaccinia viruses in cells doubly infected with SEN, those made from the SEN genome were modified as before whereas those made from the vaccinia genome were not modified at all. The inability of the insertion activity to act *in trans* supports the idea that the insertions occur during mRNA synthesis, rather than on preformed mRNA. The G insertions also do not occur when the mRNA is made *in vivo* from plasmid DNA via T7 polymerase (Curran *et al*, 1991), nor from an RNA vector which uses the Sindbis virus polymerase (U. Gegenmuller and S. Schlesinger, unpublished). Sindbis virus, however, is a plus strand RNA virus, uses "naked" RNA as template, and does not form poly(A) tails by a stuttering mechanism. All these data, of course, are consistent with a co-transcriptional mechanism.

On the other hand, it is entirely possible that when a portion of the SEN P gene containing the editing site is placed in an artificial vector system based on another minus-strand RNA virus, such as influenza virus or a rhabdovirus, G insertion at this site in the mRNAs will take place. Although influenza virus and rhabdoviruses are not known to edit their own mRNAs, they nevertheless use a nucleocapsid as template for RNA synthesis, and also form poly(A) tails on their mRNAs by stuttering on an oligo(U) stretch.

A STUTTERING MODEL FOR THE G INSERTIONS

How, then, would this stuttering or reiterative copying proposed for the G insertion take place? The mechanism implies that the viral polymerase reiteratively copies one of the three C residues on the Sendai genome template (nt 1051-1053). This presumably occurs when the 3' end of the nascent mRNA at the insertion site, which is base-paired to the template over a few

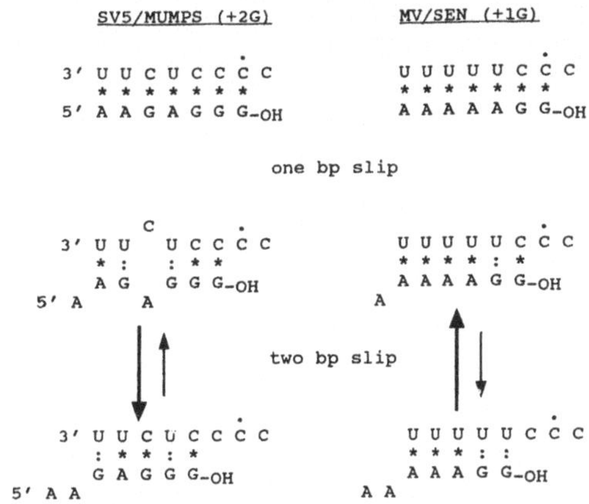

Fig 2. Polymerase and nascent chain slippage at the editing sites.

The proposed events at the insertion site for each virus group are shown. The top line in each case is the genome, and only the consensus sequence is shown, with a dot above marking the proposed pause site. The bottom line is the nascent mRNA, with its 3' –OH end indicated. Only 7 bp between the mRNA and the template are shown, with normal pairs indicated with asterisks, and U:G pairs with 2 dots. The top duplexes show the nascent chains at the pause site before slippage, the middle and bottom duplexes after a one and two bp slippage respectively. The strength of the arrows indicates the preference of a one or two base initial slippage for each group, based on the stabilities of the misalignment intermediates. Incorporation after slippage which brings the polymerase back to the pause site leads to the G insertions, and incorporation beyond the pause site fixes the events.

nucleotides, slips backwards or upstream on the template together with the polymerase before the next base is incorporated (right side, fig 2). The frequency with which the P mRNAs are inserted appears to be tightly controlled. For Sendai virus *in vivo*, 31 +/- 2% of the mRNAs contain a 1G insertion (5 determinations, Vidal *et al.*, 1990a). It seems reasonable that the polymerase would pause at the insertion site, otherwise there would not be time for the slippage to occur. RNA polymerases are known to pause during transcription on both procaryotic and eucaryotic DNA not only at termination sites, but well within the mRNA (von Hippel *et al.*, 1984; Platt, 1986; Reines *et al.*, 1987), and pause times of 10 sec to several minutes have been estimated (Chamberlin, 1976; Krakow *et al.*, 1976). The reasons for the polymerase pausing are unclear, but pausing occurs *in vitro* even in the presence of high NTP concentrations (von Hippel *et al.*, 1984), and for RNA polymerase II the pausing is independent of the structure of the nascent mRNA (Reines *et al.*, 1987). These findings suggest that pausing is an intrinsic property of the template at these sites.

Assuming that the length of the pause is described by a bell-shaped curve (fig 3), a plausible mechanism to explain how the frequency of inserted mRNAs is controlled can then be based on this distribution. If, for example, the average pause time was 10 sec and the minimun time necessary for slippage were slightly longer (e.g. 13 sec), then insertions would occur at a fixed frequency of less than half. The fraction of mRNAs with G insertions is then determined by the fraction of the polymerases which pause longer that the time required for slippage, and changes in either of these two parameters would alter the frequency of G insertions. Conditions which increase the pause, for example, by so limiting the NTP required to continue elongation that further incorporation becomes rate limiting (Ruteshouser and Richardson, 1989), might then increase the frequency of G insertions.

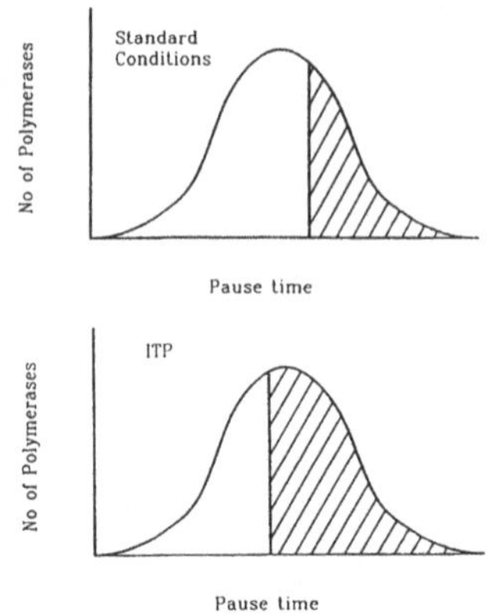

Fig 3. Polymerase pausing and the frequency of G insertions.

The distribution of the length of the pause at the insertion site (before slippage/insertion) for individual polymerases is shown as a bell shaped curve. The vertical line within the curve indicates the minimun time necessary for slippage to occur, and the shaded area to the right indicates the fraction of polymerases which will insert one or more Gs. The difference between the fraction of inserted mRNAs *in vivo* (31 +/– 2%) and under standard conditions *in vitro* (20 +/–2%) could be due to a difference in either of these parameters. The effect of replacing guanosine with inosine (below) is shown as lowering the minimun time required for slippage.

We therefore examined the effect of limiting NTP concentration during Sendai virus mRNA synthesis on the frequency of G insertions, using virion polymerase reactions. CTP, UTP, or GTP were individually lowered from 1 mM to 25 μM (and mRNA synthesis decreased about 9-fold), and the frequency of G insertions were determined by a cloning/oligonucleotide-typing method (Vidal *et al.*, 1990a) (Table I). ATP was not lowered, since all viral RNAs start with ATP and its relatively high Km for synthesis (300 μM) reflects initiation rather that internal incorporation. Neither the frequency of single G insertions nor that for multiple G insertions were basically different from the standard reaction when either CTP or UTP were decreased to 25 μM (Table I). When treated as a group, 19.9 +/- 1.7% contained +1G, and 2.6 +/- 0.5% contained >1G. Only when GTP was lowered to 25 μM were significant differences apparent, but in an unexpected fashion. The frequency of single G insertions was decreased by half, whereas those with multiple G insertions were increased 3-fold. The net result, however, was that the total fraction of inserted mRNAs varied little, if at all.

In terms of the stuttering model, we did expect any differences at 25 μM UTP, as the first U after the G run is 3 nt downstream on the mRNA (5' GGGCAU 3'). Causing the polymerase to wait longer before it can incorporate this base should have no effect because it is too far from the insertion site. By the same criteria, the lack of any effect at 25 μM CTP (5'GGGCAU 3') would argue that if the stuttering mechanism applies, stuttering would not take place at the third C on the template, but presumably on the first or second C (3' CCCGUA 5'). If this were so, at low GTP there might be more time for slippage before the 2nd or 3rd G was incorporated (5' GGGCAU 3'), and so the frequency of insertions should increase. This is in essence what has occurred, even though the total fraction of inserted mRNAs is unchanged. Under standard condition, 1G is aded 18.5% of the time, and an average of 6 Gs are added at 2.7%, such that overall an average of 1.6 Gs are added when insertions occur. The same calculation when 25 μM GTP is

TABLE I. MRNA SYNTHESIS IN VITRO

Reaction conditions	Total Colonies[a]	Colonies examined	No. Gs inserted at nt 1051-1053			Total freq. insertions
			0	1	>1	
Exp # 1						
Mock	34	34	22 (65%)	9 (26%)	3 (9%)	35%
Standard	2688	224	174 (78%)	44 (18.5%)	6 (2.7%)	21.2%
25 uM CTP	408	170	135 (79%)	32 (18.8%)	3 (1.8%)	20.6%
25 uM UTP	362	151	112 (74%)	34 (22.5%)	5 (3.3%)	25.8%
			average + mean	19.9 · 1.7%	2.6 · 0.5%	22.5%·2.2%
25 uM GTP	192	160	128 (80%)	19 (11.8%)	13 (8.1%)	19.9%
Exp # 2						
Mock	10	0				
Standard	2688	176	130 (73.8%)	43 (24.4%)	3 (1.7%)	26.1%
40% Br UTP	952	167	133 (79.6%)	32 (19.1%)	3 (1.8%)	20.9%
40% ITP	317	176	74 (42%)	67 (38%)	35 (19.8%)	57.8%

a : neutral oligo$^+$ colonies

used shows that an average of 3.6 Gs added when insertions occur. In terms of the stuttering model, it would appear that 25 uM GTP is not low enough to significantly lengthen the initial pause at the insertion site, and so the total fraction of inserted mRNA is unchanged. However, 25 μM GTP appears to extend the pause after the 1st insertion has occurred, hence a greater fraction of the inserted mRNAs contain multiple insertions. These results also support the stuttering model in another way. That the frequency of insertions is similar in all respects at low CTP or UTP even though 9-fold less mRNA is made, is inconsistent with the insertions occurring on mature mRNAs; the ratio of insertion activity (viral proteins) to mRNA would be 6 to 9-fold higher here.

Stuttering implies that during the polymerase pause at the insertion site, the base pairing between the 3' end of the nascent mRNA and its template will be broken transiently, for slippage to occur. The precise number of these base pairs and their composition would then be important. We therefore examined the effect of incorporating base analogs into the mRNA chain which alter the strength of these interactions. Inosine (I) incorporation in place of guanosine will decrease the stability of the base pairing (and there must be at least 1 G:C pair available here for substitution), whereas 5-Br-uridine incorporation will have the opposite effect. Parallel transcription reactions were carried out under standard conditions, and in which 40% of the UTP was replaced with Br-UTP, or GTP with ITP.

When the frequency of G insertions in the mRNA were examined (Table I), Br-U substitution for U was found to have little or no effect (a 1G insertion frequency of 19.1% vs 24.4% for the parallel control, and a >1G insertion frequency of 1.8% vs 1.7%). In contrast, inosine substitution for G increase the frequency of 1G insertions to 38%, and the multiple G insertions were 10-fold more frequent now (19.8%). The incorporation of the analogs would have two effects; they would alter the stability of the intramolecular pairing upstream of the 3' end of the nascent chain, as well as that to the template at the 3' end. The absence of an effect when U is substituted with Br-U argues that the folding of the nascent chain has little effect on the insertion frequency. The effect of substituting inosine for G is then more likely to be due to the pairing with the template. In terms of the model in figs 2 and 3, this can be viewed as a decrease in the minimun time necessary for the initial slippage to occur, as well as subsequent slippages, because it would be easier to break the I:C pairs within the polymerase domain.

THE DIFFERENCE BETWEEN THE SEN/MV GROUP AND THE SV5/MUMPS GROUP mRNA EDITING

All attempts to show that the insertions can occur on preformed mRNAs were negative, and manipulations of the *in vitro* system can be interpreted in a coherent way for the predicted insertion site in terms of a stuttering model. The model can account for how the fraction of 1G-inserted mRNAs is tightly controlled both *in vivo* (31 +/-2%), and *in vitro* under standard conditions (20 +/-2%, 4 determinations). Can it also offer any clues why 2Gs are inserted at high frequency in SV5 and mumps virus when insertions occur, rather than the 1G inserted in Sendai and measles virus?

An alignment of the demosntrated (or suspected) insertion regions of 11 paramyxovirus templates is shown in fig 4. They all contain a minimun of 3 Cs on which the polymerase could pause and slip, but interestingly, the length of this C stretch is not well conserved; it varies from 3 to 5 Cs for the SEN/MV group, and from 4 to 7 Cs for the SV5/mumps group. In addition, there is little homology in these alignments downstrean of the C stretch. However, the 1st, 3rd and 4th bases upstream of the C stretch are all uridines. The consensus sequences 3' UUUUUCCC 5' (SEN/MV group) and 3' AAAUUCUCCCC 5' (SV5/mumps group) are evident. The meaning of the longer consensus sequence for the SV5/mumps group is unclear, and may only reflect the fact that the viruses in this group are more closely related to each other than those in the SEN/MV group.

A global consensus sequence 3' UUYUCCC 5' is also evident, in which the Y is a U in those viruses where 1G is inserted, and a C where 2Gs are inserted, and it is this difference which may be involved. Fig 2 shows a schematic representation of the insertion mechanism for each virus group. We do not know which template C is the pause site, nor the number of base

```
SEN/MV Group
                      *
SEN         g U U U U U U C C C       G U a U C C 5'
hPIV3       A U U U U U U C C C C C   u U u U C C 5'
bPIV3       A U U U U U U C C C C     u U c c u U 5'
(CDV)       A A U U U U U C C C C     G U G U C U 5'
MV          A A U U U U U C C C       G U G U C U 5'
(NDV)       g A U U U U U C C C       G g G U a C 5'
                      *
consensus   A U U U U U U C C C       G U G U C C 5'
            g A                       u g n c n U

      ------------------------------------------------

SV5/Mumps Group
                      *
MUMPS       A A A U U C U C C C C C   G g C c c U 5'
SV5         A A A U U C U C C C       G u C c c U 5'
(PIV2)      A A A U U C U C C C C C C u c g a u U 5'
(PIV4)      A A A U U C U C C C C C   u u a u a U 5'
LPMV        A A A U U C U C C C C C   G g C u g g 5'
                      *
consensus   A A A U U C U C C C C     G n C n n U 5'
                                      u   n     g

            ----------------------
global
consensus         U U Y U C C C
```

Fig 4. Paramyxovirus P gene editing sites.

The sequence of the various sites on the genomes or template strands are listed in the 3' to 5' direction, with the C stretch on which the insertions occur in the middle. Lower case letters indicate poor conservation, upper case letters indicate better conservation, and upper case and bold letters indicate complete conservation. The asterisk shows the pyrimidine residue which differentiates the SEN/MV and SV5/mumps groups; n refers to any base.

pairs between the nascent chain and the template. However, we would expect the latter to be limited to the minimum in this region, to facilitate the transient melting required for slippage. The model was originally demonstrated (Vidal *et al.*, 1990b) with 4 bp starting at the middle C of the sequence 3' UUYUCCC 5' to accomodate the proposed NDV editing sequence (3' UUCUUCCC 5', (Cattaneo *et al.*, 1989; Paterson *et al.*, 1989)) which is likely to be the incorrect one. The NDV sequence 3' UUUUUCCC 5' also exists nearby, and is more likely to be the corect one. The model in fig 2 usesthe extended consensus sequence of each group, with 7 bp in each case and the polymerase paused on the penultimate C of the C stretch. There is presumably some pressure following the pause to displace the nascent chain upstream. For SV5/mumps, displacement upstream by 1 bp is unstable as it includes and A:C pair, whereas displacement by 2 bp creates a more stable intermediate. For the measles/Sendai group, on the other hand, an A:U rather than an A:C pair occurs at this position upon displacement by a single bp, and this intermediate is more stable than that obtained on a shift of 2bp. Thus, the relative stabilities of the 1 and 2 bp misalignment intermediates for each virus could determine whether 1 or 2 Gs are inserted at high frequency when insertions occur. In the case of the SV5/mumps virus group, we note that once the initial 2 base slippage has occurred to avoid the unfavorable A:C pair, further rounds of slippage (when they occur) would now proceed one base pair at a time, as there are no further possiblities for forming A:C pairs. The distribution of G insertions in the mumps virus mRNAs has recently been determined (Paterson and Lamb, 1990, see fig 5) and fits nicely with this prediction.

A mechanism with many similar features has recently been proposed to account for the repeate TTGGGG sequences in Tetrahymena telomers (Greider and Blackburn, 1989). These are added by a telomerase which contains a RNA template for this sequence within a larger chain. What is similar here is that the enzyme is proposed to pause on the template and then slip 6 nt upstream, and this distance is determined by the base pairing to the template.

DIFFERENCES BETWEEN THE INITIAL AND SUBSEQUENT ROUNDS OF SLIPPAGE

Once the initial one or two base slippage/insertion has occurred, a second round is somehow avoided so that the 1 or 2 G insertions predominate. This could occur by the initial insertion(s) relieving the polymerase pause, in contrast to what would happen during polyadenylation. However, for Sendai virus, the initial insertion apparently fails to relieve the pause some of the time, as 7% of the mRNAs *in vivo* contain multiple G insertions, as do 2% of those made *in vitro*. When twelve >1G inserted mRNAs were first examined (Vidal *et al.*, 1990a), those with +2Gs did not predominate as expected, but mRNAs with +2 to +8 Gs inclusive were found. If the multiple G insertions arose by simple repetition of single G insertions, one would expect a distribution in which +2Gs > +3Gs > +4Gs, etc. A larger number of these mRNAs made *in vitro* has also been examined, and their distribution is not dissimilar (Vidal *et al.*, 1990b). What is striking here is that 12 Gs are as likely to be incorporated as 2 Gs when multiple insertions occur. This distribution suggests that in these cases the initial insertion has not only failed to relieve the pause, but may rather have increased it. In this respect, the G insertion mechanism would approach the polyadenylation process, where the pause is presumably much stronger, and can only be relieved by chain termination. However, we have not seen any evidence for chain termination at the insertion site under any conditions, using RNase mapping.

BOVINE PARAINFLUENZA VIRUS TYPE 3, THE EXCEPTION THAT STRENGTHENS THE RULE

A remarkable feature of the paramyxovirus editing mechanism is that for each virus group, the requisite number of Gs are added at high frequency to provide access to the alternate downstream ORF, i.e., 2Gs are added in the SV5/mumps group to go from V to P and mRNAs with only 1G inserted are very rare, whereas +1G is the predominant event for the SEN/MV group to go from P to V. This is best illustrated by comparing figs 1 and 5.

PIV3 (parainfluenza virus type 3) is closely related to SEN and MV, but is slightly more complicated. The P gene sequence of both the human (h) (Spriggs and Collins, 1986; Galinski *et al.*, 1986; Luk *et al.*, 1986) and bovine (b) (Sakai *et al.*, 1986) viruses are known, and these contain one more ORF (called D). The D ORF follows shortly after that of C, in the same reading frame as C (fig 1). On the basis of sequence comparisons, Cattaneo *et al* (1989) suggested 2 editing sites for hPIV3, both of which would favor single base slippages according to the above rule. Insertions at the first site (site a) cannot produce a functional V mRNA for hPIV3, as there are several stop codons between this site and the cys-rich domain. However, as 2G insertion here would give rise to a mRNA which has fused the N-terminus of P to 131 amino acids of the D ORF. A second insertion site (site b), ca. 300 nt downstream and at the same relative location as the site which operates in MV and SEN, lies downstream of the stop codons and can produce a functional V mRNA. hPIV3 would then appear to use two insertion sites; the first to express the D ORF exclusively, and the second for V.

The P gene of bPIV3 is organized identically to that of hPIV3 and contains the same two putative insertion sites (fig 1). However, there are no stop codons between the first site and the cys-rich domain of the V ORF. This creates the very unusual situation where a block of 320 nt can potentially be expressed in all 3 reading frames. As it is unlikely that such a situation would exist by chance, this suggests that the first editing site will be used to express both the V and D proteins (fig 1). Moreover, if the second site also operates in bPIV3, there could potentially be 3 different forms of V as well as multiple forms of P.

A primer extension method was used to quantitate the fraction of mRNAs with insertions, and the distribution of the number of bases inserted (Pelet *et al.*, 1991). Examination of the bPIV3 mRNAs indicated that site b was non-functional, whereas site a operates frequently. Moreover, the distribution of G insertions at site a were remarkably broad; from 1 to 6 Gs were added with virtually equal frequency (fig 5). Insertions at site a can then give rise to both V and D protein mRNAs, because a very broad distribution of Gs are added when insertions occur. This broad distribution is very different from the editing sites of Sendai virus or SV5, where predominantly one form of edited mRNA containing either a 1 or 2G insertion respectively is created, to access the single overlapping ORF of these latter viruses.

Although SEN, MV, and bPIV3 all insert Gs within the common sequence 5'AAAAAGGG, bPIV3 does this somewhat differently, and there is a plausible mechanism for how this could occur in the stuttering model. In this model, the polymerase would pause after incorporating the second G at the editing site (top line, fig 6). In some cases, the pause would be too short for the base pairing between the 3' end of the nascent mRNA and the template to be broken. Here the polymerase would continue on, whitout creating insertions. When the pause is long enough, the base pairing is transiently interrrupted, and there is some pressure to move the 3' end of the mRNA upstream. For bPIV3 the pause would presumably be longer than for SEN, as 65% of the bPIV3 mRNAs are inserted versus 25% for SEN. However, once

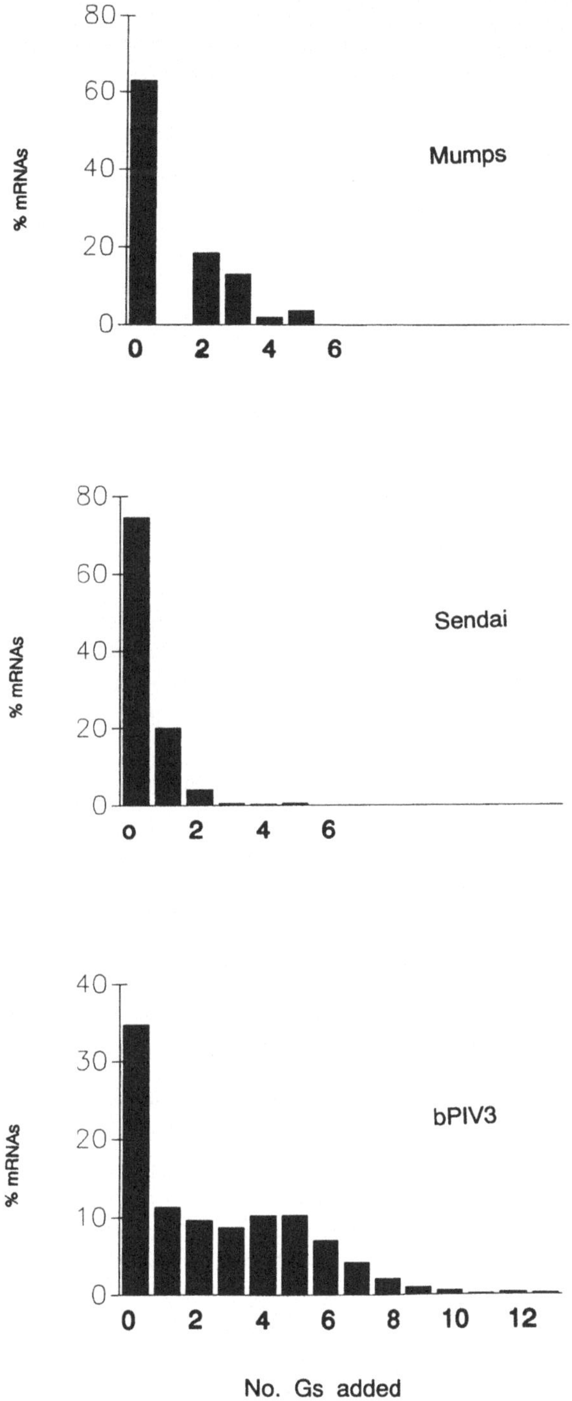

Fig 5. Distribution of G insertions in various paramyxoviruses.

The distribution of the number of Gs inserted in the indicated paramyxovirus P gene mRNAs is shown as a bar graph. The mumps virus data is from Paterson and Lamb, 1990, and was determined by sequencing 54 mRNA clones. The Sendai and bPIV3 data is from Pelet *et al.*, 1991, and was determined by a primer extension method directly on the mRNA population. Zero Gs inserted refers to the unedited mRNA.

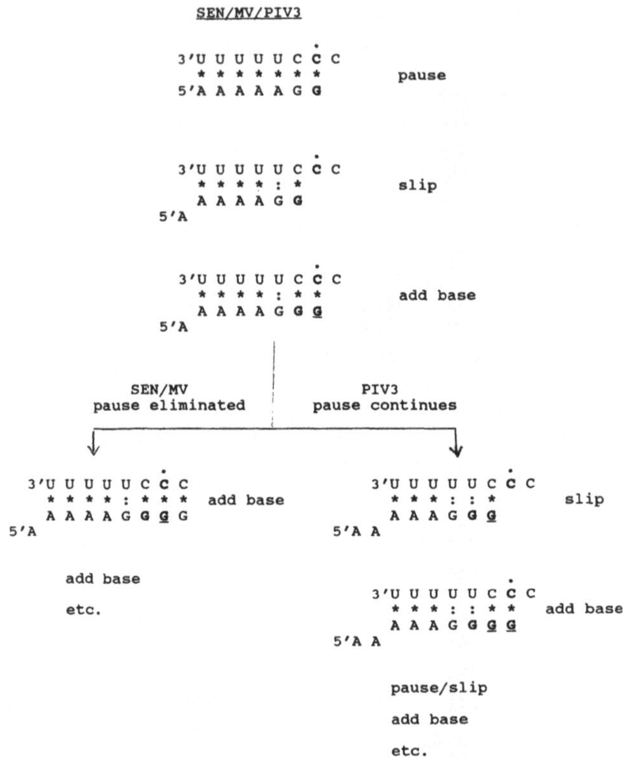

Fig 6. A model to explain the different frequencies of multiple G insertions for SEN/MV and PIV3.

The polymerase pauses after adding the second G to the mRNA (in bold, bottom line of top duplex), and the mRNA slips one base upstream, creating a single G:U pair (double dots). The length of the pause determines the total fraction of mRNAs with insertions, and presumably the pause is stronger for bPIV3 than for SEN. Addition of the pseudotemplated G brings the polymerase back to the pause site (the template C with dot above), creating a single base insertion (underlined). For SEN/MV, the pause is presumably eliminated at this point, the next truly templated G is added, and the polymerase continues in a strictly processive mode. For PIV3, the pause is not eliminated. Because the G:U pair makes it easier to break the bonds between the mRNA and the template for slippage to reoccur, multiple rounds of slippage and insertion are favored, and a large majority of the mRNAs with insertion contain multiple insertions. After about 6 rounds, the probability that the pause is eliminated eventually increases, or the process ceases as oligo G formation becomes self-limiting.

the initial round of slippage/insertion(s) has occurred (2n and 3rd lines, fig 6), the polymerase is back to the same position as before, except that one G:U pair has been formed in the misalignment. A second round is then favored if nothing else has changed, since it should be easier to break the new base pairs for slippage to recur. Nevertheless, both *in vivo* and *in vitro*, 80% or more of SEN mRNAs which contain insertions have inserted only a single G. For 1G insertions to predominate, something else must have changed when the polymerase returns to the pause site after the initial events, to prevent a the polymerase returns to the pause site after the initial events, to prevent a second round. For example, protein modification of the polymerase or template could have occurred during the first round, which eliminates the pause. It is apparently this something else wich has failed to occur in large part during bPIV3 mRNA editing, because hare insertions of >1G are now more than 5 times as frequenct as those with only 1G. Somewhat similar situations can be created with SEN *in vitro*, by altering reaction conditions which are thought to either extend the pause, or decrease the time required for slippage to occur (Vidal *et al.*, 1990b). For example, lowering the GTP concentration during mRNA synthesis does not appear to extend the initial pause at the insertion site, but does appear to extend the subsequent pauses. This suggests that the initial and subsequent pauses operate somewhat differently. Moreover, partial replacement of GTP with ITP during synthesis only doubles the frequency of mRNAs with 1G insertions, whereas those with >1 G insertion increases 10-fold.

CONCLUSIONS

Paramyxovirus mRNA editing, like ribosomal frameshifting of retroviruses (Jacks and Varmus, 1985) or coronaviruses (Brierley *et al.*, 1987, 1989), appears to be composed of two parts; a pause, and a slippery sequence which allows for alternate base pairing (Brierley *et al.*, 1989). The slippery sequence of paramyxoviruses (3'UUYUCCC) is fairly clear. However, the counterpart to the downstream template structure which causes ribosomal pausing is difficult to even guess at. DNA dependent RNA polymerases have long been known to pause in the middle of genes (Chamberlin, 1976; von Hippel *et al.*, 1984; Platt, 1986), and although there is some evidence that this is due to an intrinsic property of the template (Reines *et al.*, 1987), its nature is equally unclear. Further progress in this form of mRNA editing will require some understanding of the nature of the pause. Moreover, the reasons why some viruses have chosen to express alternate downstream ORFs via ribosomal frameshifting, whereas other have chosen "polymerase frameshifting", remains obscure.

The G insertions do not represent mRNA editing in a strict sense, as the mRNA has not been altered after its synthesis. However, it is a form of editing in that it results in a specific change of the mRNA relative to its template which is important for translation. Compared to the "true" forms of editing which take place after synthesis of the transcript, the stuttering mechanism is clearly different.

REFERENCES

Banerjee, A.K., 1987, Transcription and replication of rhabdoviruses, Microbiol Rev. 51:66-87.

Berg, M., Hjertner, B., Moreno López, J., and Linné, T., 1992, The P gene of the porcine paramyxovirus LPMV encodes three possible polypeptides P, V and C: the P protein mRNA is edited, J Gen Virology. In press.

Brierley, I., Boursnell, M.E.G., Binns, M.M., Bilimoria, B., Block, V.C., Brown, T.D.K. and Inglis, S.C., 1987, An efficient ribosomal frameshifting signal in the polymerase encoding region of the coronavirus IBV.

Brierley, L., Digard, P., and Inglis, S.C., 1989, Characterization of an efficient Coronavirus ribosomal frame shifting signal: requirement for an RNA pseudoknot, Cell. 57:537-547.

Cattaneo, R., Kaelin, K., Baczko, K. and Billeter, M.A., 1989, Measles virus editing provides an additional cysteine-rich protein, Cell. 56:759-764.

Chamberlin, M.J., 1976, In Chamberlin, M.J. and Losick. R.,eds., RNA Polymerase. Cold Spring Harbor Laboratory Press, Cold Spring Harbor, NY, pp. 159-192.

Curran, J. and Kolakofsky, D., 1988, Ribosomal initiation form an ACG codon in the Sendai virus P/C mRNA, EMBO J. 7:245-251.

Curran, J. and Kolakofsky, D., 1989, Scanning independent ribosomal initiation of the Sendai virus Y proteins *in vitro* and *in vivo*, EMBO J. 8:521-526.

Curran, J., Boeck, R. and Kolakofsky, D., 1991b, The Sendai virus P gene expresses both an essential protein and an inhibitor of RNA synthesis by shuffling modules via mRNA editing, EMBO J. 10:3079-3085.

Galinski, M.S., Mink, M.A., Lambert, D.M., Wechsler, S.L., and Pons, M.W., 1986, Molecular cloning and sequence analysis of the human parainfluenza 3 virus mRNA encoding the P and C proteins, Virology. 155:46-60.

Galinski, M.S., Troy, R.M. and Banerjee, A.K., 1992, RNA editing in the phosphoprotein gene of the human parainfluenza virus type 3, Virology. 186:543-550.

Giorgi, C., Blumberg, B.M. and Kolakofsky, D., 1983, Sendai virus contains overlapping genes expressed from a single mRNA, Cell. 35:829-836.

Greider, C.W., and Blackburn, E.H., 1989, A telomeric sequence in the RNA of Tetrahymena telomerase required for telomere repeat synthesis, Nature. 337:331-336.

Gupta, M. and Kingsbury, D., 1984, Complete sequences of the intergenic and mRNA start signals in the Sendai virus genome: Homologies with the genome of Vesicular Stomatitis virus, Nucleic Acids Res. 12:3829-3841.

Herrler, G. and Compans, R.W., 1982, Synthesis of mumps virus polypeptides in infected vero cells, Virology. 199:430-438.

Jacks, T. and Varmus, H.E., 1985, Expression of the Rous Sarcoma virus pol gene by ribosomal frameshifting, Science. 230:1237-1242.

Kolakofsky, D., and Roux L., 1987, The molecular biology of paramyxoviruses, in: "The Molecular basis of viral replication," P. Bercoff, ed., Plenum Publishing, New York. pp. 277-297.

Krakow, J.S., Rhodes, G. and Jovin, T.M., 1976, In Chamberlin M.J. and Losick, R. (eds). RNA Polymerase. Cold Spring Harbor Laboratory Press. Cold Spring Harbor, NY pp. 127-158.

Luk, D., Sánchez, A., and Banerjee, A.K., 1986, Messenger RNA encoding the phosphoprotein (P) gene of human parainfluenza virus 3 is bicistronic, Virology. 153:318-325.

McGeoch, D.J., 1979, Structure of the gene N: Gene NS intercistronic junction in the genome of Vesicular Stomatitis virus, Cell. 71:673-681.

Paterson, R.G., Thomas, S.M. and Lamb, R.A., 1989, Specific nontemplated nucleotide addition to a simian virus 5 mRNA: Prediction of a common mechanism by which unrecognized hybrid P-cystein-rich proteins are encoded by paramyxovirus "P" genes. In Kolakofsky. D. and Mahy, B.W.J. (eds). Genetics and Pathogenicity of Negative Stranded Viruses. Elsevier/North-Holland Publishing Co., Amsterdam, pp. 232-245.

Paterson, R.G., and Lamb, R.A., 1990, RNA editing by G nucleotide insertion in mumps virus "P" gene mRNA transcripts, J Virol. 64:4137-4145.

Pelet, T., Curran, J. and Kalokofsky, D., 1991, The P gene of bovine parainfluenza virus 3 expresses all three reading frames from a single mRNA editing site, EMBO J. 10:443-448.

Platt, T., 1986, Transcription termination and the regulation of gene expression, Annu Rev Biochem. 55:339-372.

Reines, D., Wells, D., Chamberlin, M.J. and Kane, C.M., 1987, Identification of intrinsic termination sites *in vitro* for RNA polymerase II within eukaryotic gene sequences, J Mol Biol. 196:229-312.

Robertson, J.S., Schubert, J.S. and Lazzarini, R.A., 1981, Polyadenylation sites for influence virus mRNA, J Virol. 38:157-163.

Ruterhouser, E.C., and Richardson, J.P., 1989, Identification and characterization of transcription termination sites in the Escherichia coli lacZ gene, J Mol Biol. 208:23-43.

Sakai, Y., Suzu, S., Shioda, T., and Shibuta, H., 1987, Nucleoitide sequence of the bovine parainfluenza 3 virus genome: its 3' end and the genes of NP P, C and M proteins, Nucleic Acids Res. 15: 2927-2944.

Schubert, M., Keene, J.D., Herman, R.C. and Lazzarini, R.A., 1980, Site on the Vesicular Stomaititis Virus genome specifying polyadenylation and the end of the L gene mRNA, J Virol. 34:550-559.

Springgs, M.K. and Collins, P.L., 1986, Sequence analysis of the P and C protein genes of human parainfluenza virus type 3: Patterns of amino acid sequence homology among paramyxovirus proteins, J Gen Virol. 67:2705-2719.

von Hippel, P.H., Bear, D.G., Morgan, W.D. and McSwiggen, J.A., 1984, Protein-nucleic acid interactions in transcription: A molecular analysis.JOURNAL

Thomas, S.M., Lamb, R.A., and Paterson, R.G., 1988, Two mRNAs that differ by two nontemplated nucletides encode the amino coterminal proteins P and V of the paramyxovirus SV5, Cell. 74:891-92.

Vidal, S., Curran, J. and Kolakofsky, D., 1990a, Editing of the Sendai virus P/C mRNA by G insertion occurs during mRNA synthesis via a virus-encoded activity, J Virol. 64:239-246.

ANTIGENIC VARIATION OF HUMAN RESPIRATORY SYNCYTIAL VIRUS G GLYCOPROTEIN: GENETIC MECHANISMS AND EVOLUTIONARY SIGNIFICANCE

José A. Melero, Paloma Rueda and Blanca García-Barreno

Centro Nacional de Microbiología
Instituto de Salud "Carlos III"
Majadahonda, 28220 Madrid
Spain

INTRODUCTION

Human respiratory syncytial (RS) virus is the major cause of lower respiratory tract infections in infants and young children (McIntosh and Chanock, 1990). This virus is a member of the *Pneumovirus* genus of the family *Paramixoviridae*. As other members of this family, RS virus has two major surface glycoproteins (G and F) which play essential roles during the initial stages of the infectious cycle. The G protein mediates virus binding to the cell surface receptor (Levine et al., 1987) and the F (fusion) protein promotes the fusion of the viral and cell membranes (Walsh and Hruska, 1983). Polyclonal and monoclonal antibodies directed against either glycoprotein inhibit virus infectivity (Walsh et al., 1987; Anderson et al., 1988).

Early attempts of vaccination with formalin-inactivated virus failed to protect young vaccinates against natural RS virus infections (Kim et al., 1969). Moreover, vaccinated children experienced more frequently an exacerbated disease after a natural infection than control children. Retrospectively, it was found that the failure of the inactivated vaccine to protect children was due, at least in part, to modification of epitopes on the F and/or G proteins recognized by neutralizing antibodies (Murphy et al., 1986). These results highlight the importance of a detailed characterization of these two major RS virus antigens.

Another relevant aspect of RS virus immunobiology is the antigenic variation of clinical isolates. RS virus reinfections are common and antigenic variation could be a factor favouring repeated infections of the same individual (Mufson et al., 1987; Waris, 1991). Monoclonal antibodies (MAbs) have been used to classify RS virus strains into two antigenic subgroups (A and B) (Anderson et al., 1985; Mufson et al., 1985). Major

Regulation of Gene Expression in Animal Viruses, Edited by
L. Carrasco *et al.,* Plenum Press, New York, 1993

antigenic changes, both inter- and intra-subgroups, are found mainly in the G protein (García-Barreno et al., 1989).

To characterize the neutralizing epitopes of the RS virus antigens and to understand the genetic basis of their antigenic variation, viruses that escaped neutralization by either anti-F or anti-G MAbs were isolated. The genetic alterations selected in the escape mutants were then identified by sequencing the corresponding F or G genes. The results obtained indicated that mutants resistant to anti-F antibodies contained nucleotide substitutions that replaced certain amino acids located in two regions of the F primary structure. These genetic changes altered a limited set of epitopes on the F molecule (López et al., 1990; Arbiza et al., 1992). In contrast, most mutants selected with anti-G antibodies had major antigenic changes generated by three alternative types of genetic alterations (García-Barreno et al., 1990; Rueda et al., 1991). We summarize here the results obtained with the last mutants and discuss their relevance for the natural history of RS virus.

ANTIGENIC AND GENETIC CHANGES SELECTED IN RS VIRUS MUTANTS RESISTANT TO ANTI-G ANTIBODIES

The RS virus G protein is a type II glycoprotein with a single hydrophobic region, between amino acids 38 and 66, which serves as both signal sequence and transmembrane anchor (Wertz et al., 1989). The C-terminal ectodomain of the G protein contains three to eight potential sites for N-linked oligosaccharide side chains in different strains, and >70 potential sites for O-linked sugars (Johnson et al., 1987). Monoclonal antibodies identify three types of epitopes onto the RS virus G glycoprotein: i) conserved epitopes shared by all human isolates, ii) subgroup-specific epitopes shared by viruses of the same antigenic subgroup and iii) variable epitopes present in a limited set of strains from the same subgroup.

As stated before, immune pressure might be a major driving force in RS virus evolution by selecting natural variants able to reinfect the same individual. Strain-specific epitopes would be the main target of this immune selection. Consequently, we undertook

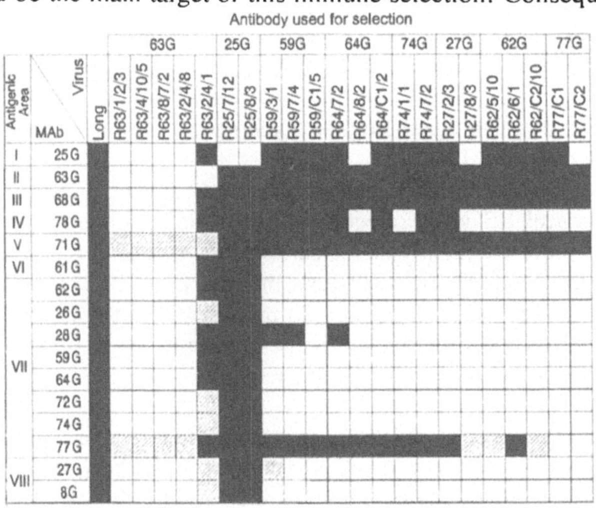

Figure 1. Reactivity of escape mutants with anti-G MAbs. Each mutant (selected with the antibodies indicated above) was tested in an ELISA using the MAbs indicated at the left. Symbols: more than 50% (■), 25 to 50% (▨) and less than 25% (□) of the values obtained for each antibody with the Long strain.

the characterization of the variable epitopes of the G molecule. To this end, escape mutants of the Long strain were isolated after serial passage of the virus in the presence of strain- specific MAbs. Most escape mutants had lost the majority of the variable epitopes (Figure 1), including those classified in separate antigenic areas by virus binding competitions of antibody pairs (García-Barreno et al., 1989). These mutants, however, retained the conserved and subgroup-specific epitopes of the G molecule (not shown).

Sequence analysis of the viruses resistant to anti-G antibodies revealed three types of genetic changes, compared to the parental Long strain: i) deletions and insertions of single adenosine residues that introduced frameshift changes in the G protein gene, ii) nucleotide substitutions that introduced premature in-frame stop codons and iii) nucleotide substitutions leading to amino acid replacements.

The frameshift mutations -all selected in viruses resistant to MAb 63G- originated from deletions of a single adenosine after residues 588 or 623 (Table I). One mutant (R63/2/4/1), however, had an adenosine inserted after nucleotide 648 which restored the original reading frame. The last mutant encoded a protein with seven amino acids changed, between residues 205 and 211, and had lost only the epitope of the antibody (63G) used for its selection (Figure 1). In contrast, the other frameshift mutants encoded G proteins which were 25 amino acids shorter than wild type and with the last 81 or 69 amino acids changed. These mutants had lost the majority of the variable epitopes on the G protein (Figure 1).

The deletion and insertion of adenosines in the mutant G protein genes were always found within runs of six or seven As. A slippery polymerase that shifts on the RNA template during replication has been postulated to be responsible of these changes (García-Barreno et al., 1990). This mechanism may be similar to the stuttering described during polyadenylation of mRNA (Gupta and Kingsbury, 1984) or to the mRNA editing of the paramyxovirus P protein gene (Thomas et al., 1988; Cattaneo et al., 1989; Vidal et al., 1990). Interestingly, frameshift changes in runs of As or Gs have also been observed in a retrovirus shuttle vector after a single replication cycle (Pathak et al., 1990).

The mutants with premature stop codons (Table II) contained transitions (A-G, C-U) or transversions (A-U) that generated alternative in-frame nonsense triplets (Table II). These changes truncated the G polypeptide between 1 and 42 amino acids at the C-terminal end and led to the loss of multiple variable epitopes (Figure 1). The mutants with the largest truncations lacked the unique epitopes recognized by MAbs 25G and 78G, in addition to those included in the antigenic areas VI, VII and VIII of the G molecule.

Tablel. Frame-shift mutations.

Virus	*Adenosines after nucleotide*			*Amino acids changed (residues)*	*Amino acids lost (residues)*	*Scheme of the G protein*
	588	*623*	*648*			
Long	*6*	*7*	*6*			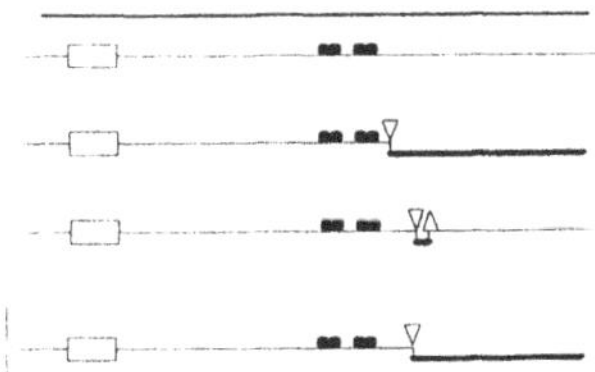
R63/1/2/3	*5*	*7*	*6*	*81(193-273)*	*25(274-298)*	
R63/2/4/1	*6*	*6*	*7*	*7(205-211)*		
R63/2/4/8	*6*	*6*	*6*	*69(205-273)*	*25(274-298)*	
R63/4/10/5	*6*	*6*	*6*	*69(205-273)*	*25(274-298)*	
R63/8/7/2	*6*	*6*	*6*	*69(205-273)*	*25(274-298)*	

The location of frameshift changes in each mutant G protein are illustrated on the right, denoting the hydrophobic region (▭) and the cysteine residues (•) of the G protein ectodomain.

Table 2. Premature stop codons.

Mutant	Nucleotide changed	Amino acid changed	Amino acids lost (residues)	Scheme of the G protein
R27/8/3	784(A-U)	257(Lys-stop)	42(257-298)	
R64/8/2	426(A-G)	NO CHANGE		
	796(C-U)	261(Gln-stop)	38(261-298)	
R77/C2	796(C-U)	261(Gln-stop)	38(261-298)	
R77/C1	847(C-U)	278(Gln-stop)	21(278-298)	
R62/6/1	847(C-U)	278(Gln-stop)	21(278-298)	
R62/C2/10	847(C-U)	278(Gln-stop)	21(278-298)	
R74/1/1	876(C-U)	288(Gln-stop)	11(278-298)	
R74/7/2	907(C-U)	298(Gln-stop)	1(298)	

The location of premature stop codons in each mutant protein is indicated at the right (∇).

Among the mutants with amino acid substitutions (Table III), those selected with MAb 25G had two amino acid changes (residues 265 and 274) that eliminated only the epitope recognized by this antibody. In contrast, the mutants with single amino acid replacements within the last three residues of the G protein had lost multiple overlapping epitopes (Figure 1).

The premature stop codons and the amino acid replacements resulted from nucleotide substitutions, likely originated by misincorporations of the viral polymerase during replication. This type of errors are generated at high frequency by RNA polymerases that lack proofreading activities (Steinhauer and Holland, 1987) being responsible, at least in part, of the extreme genetic heterogeneity found in populations of RNA viruses (Domingo et al., 1985).

The antigenic and genetic analysis of the escape mutants described above indicated that the C-terminal end of the G molecule is essential to determine the structure of its variable epitopes. In addition, the C-terminal third of the G protein can accommodate multiple sequence changes generated by different genetic mechanisms. It is then possible that this plasticity of the G molecule is used by RS virus to evade the immune barrier of

Table 3. Amino acid substitutions.

Mutant	Nucleotide changed	Amino acid changed	Scheme of the G protein
R27/2/3	902(C-U)	296(Thr-Ile)	
R64/C2	901(A-G)	296(Thr-Ala)	
R59/3/1	905(G-A)	297(Arg-His)	
R25/7/12	808(U-C)	265(Phe-Leu)	
	836(U-C)	274(Leu-Pro)	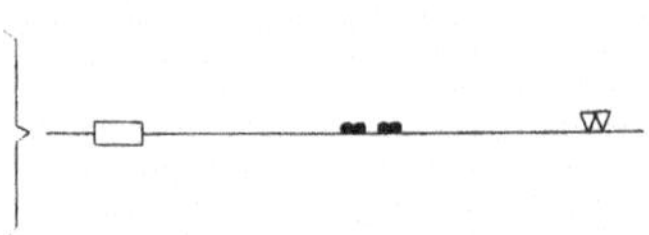
R25/8/3	808(U-C)	265(Phe-Leu)	
	836(U-C)	274(Phe-Leu)	

The location of the amino acid substitutions in the G mutant proteins is indicated at the right (∇).

pre-existing antibodies, allowing reinfections of the same individual. If this were the case, the antigenic and genetic differences observed in natural isolates should resemble the changes found in the escape mutants. We will review now the evidence in support of this hypothesis.

ANTIGENIC AND GENETIC CHANGES FOUND IN NATURAL RS VIRUS ISOLATES

The antigenic differences between RS virus isolates parallel those found in the escape mutants. Conserved and subgroup specific epitopes were retained in the viruses resistant to the strain-specific anti-G antibodies of Figure 1, indicating that the variable epitopes are restricted to the C-terminal third of the G primary structure. More conserved epitopes must be then defined by amino acid sequences included in the N-terminal two thirds of the G protein. In fact, mapping of certain epitopes with synthetic peptides (Norrby et al., 1987) and selection of escape mutants with MAbs recognizing conserved epitopes (P. Rueda et al., unpublished) corroborate this conclusion.

The available sequence data of natural isolates reveal genetic differences which might have originated by the mechanisms described below, similar to those that generated the changes observed in the escape mutants.

Point Mutations Leading To Amino Acid Replacements

Comparison of the G protein sequences from several isolates of the A (Cane et al., 1991) and B (Sullender et al., 1991) antigenic subgroups have been reported. Multiple nucleotide substitutions leading to amino acid sequence differences within each subgroup were found in two hypervariable regions of the G protein ectodomain. The first region preceded the cluster of four cysteines located in the middle of the G ectodomain. The second region was placed in the C-terminal third of the G molecule, including the amino acid residues changed in the escape mutants and that were essential for the integrity of variable epitopes. The high level of nucleotide changes (>50%) among viral isolates that resulted in amino acid changes of the G protein indicated a selective pressure for change (Cane et al., 1991; Sullender et al., 1991). However, neither variable epitopes nor changes in escape mutants selected with strain-specific antibodies have been mapped in the first hypervaribale region of the G molecule. It is then possible that this region is antigenically relevant only in humans but not in the mice strain used to raise the MAbs. Alternatively, there may be structural constrains for the G protein when RS virus replicates in the respiratory tract which are not observed when the virus replicates in tissue culture. Then, the amino acid changes found in the first hypervariable region might be needed only in natural isolates to compensate the changes in the C-terminal hypervariable end.

Alternative Stop Codons

The presence of alternative termination codons in the G protein genes of natural isolates is also inferred from sequence comparison of subgroup A (Cane et al., 1991) and B (Sullender et al., 1991) viruses. The nucleotide length of the G gene is essentially the same for all human isolates; however, the stop codons in subgroup A viruses are alternatively placed at two positions that change the protein length by one amino acid. The position of alternative termination codons introduces still wider variations in the G protein length of subgroup B viruses, from 292 to 299 amino acids. An extreme example of the use of alternative termination codons is observed in bovine RS virus where the G protein is reduced to 257 amino acids (Lerch et al.,1990).

The use of alternative stop codons may not be exclusive of the RS virus attachment protein. The counterpart proteins of other paramyxovirus, although do not share sequence homology with the RS virus G protein, are also type II glycoproteins. Sequence comparison of the HN protein gene of several Newcastle Disease Virus (NDV) isolates have revealed differences in the protein length that originated from the use of alternative termination codons (Sakaguchi et al., 1989). Interestingly, the differences in HN length correlated with evolutionary lineages that differed in their degree of virulence. Since the C-terminal end of type II glycoproteins is located in the external domain, it is likely that structural constrains are less restrictive at this site of the molecule. Thus, paramyxovirus might take advantage of the genetic plasticity at the C-terminal end of their attachment proteins to generate variants with differences in biological properties, such as antigenicity or virulence.

Frame shift mutations

Among the G gene sequences from subgroup B strains, reported by Sullender et al. (1991), two viruses isolated from the same child two years apart had only 11 differences at the amino acid level, providing a 96% total amino acid identity. Of the 11 amino acid changes, 7 were consecutive (positions 226 and 232). The last changes were introduced by two reading frame shifts generated by deletion and insertion of single adenosines in runs of 6 and 7 As. Although the site of these changes did not correlate with the position of the frameshit changes in the escape mutants of the Long strain (belonging to subgroup A), in both cases insertions and deletions occurred in clusters of adenosine residues, likely by the same mechanism. Interestingly, the two natural isolates showed also antigenic differences detected with MAbs.

NATURAL HISTORY OF RS VIRUS

The results mentioned before provide circumstantial evidence that immune pressure is important for RS virus evolution through the selection of new antigenic variants. The two antigenic subgroups of RS virus cocirculate among the human population (Hendry et al., 1986; Akerlind and Norrby, 1986). Within the same subgroup, viruses from separate evolutionary lineages are isolated during the same outbreak (Cristina et al., 1991) and closely related viruses are found in widely separated countries (Cane et al., in press). Thus, immune selection would be only a factor contributing to RS virus diversity. The contribution of other factors -bottle-necks, virulence, etc.- to RS virus propagation remains to be evaluated.

What is then the relative significance of the different genetic mechanisms, that generated the escape mutants of the G protein, to the natural history of RS virus? Three points should be considered in this respect: i) the mutation rates for the different genetic lessions, ii) the tolerance of RS virus to the sequence changes generated by each genetic mechanism and iii) the selective advantage of the mutant proteins.

Amino acid substitutions and premature stop codons are both consequences of nucleotide misincorporations occurring during replication of the viral genome. Although the frequency of this type of errors for RS virus polymerase is unknown, it may be assumed similar to that of other RNA-dependent RNA polymerases (Domingo and Holland, 1988) (4-5x10^{-4} mutations/nucleotide and replication round). Misincorporations generating alternative stop codons are likely to have more profound effects in the biological properties of the G protein, and relatively less tolerated, than those leading to amino acid substitutions. Thus, as it is found among natural isolates, viruses containing a limited set of amino acid replacements are likely to be selected more frequently than variants with alternative termination codons. The last type of viruses might be selected, however, under special circumstances; for instance, when amino acid replacements in an evolutionary lineage are saturated to yield functional G proteins that are antigenically different. In this way, alternative stop codons in the G protein would create new lineages and/or subgroups, in a maner similar to the lineages of NDV virus with alternative stop codons in the HN gene (Sakaguchi et al., 1989). However, whereas no evidence has been provided of immune pressure being the selective force for NDV lineages, the selection with anti-G MAbs of RS virus mutants with alternative termination codons indicates that immune pressure might be the driving force generating new RS virus lineages and/or subgroups.

Regarding the frameshift mutations, there is limited data on the frequency with which this type of changes are generated during replication of RNA viruses. In an elegant experimental model of a spleen necrosis virus based shuttle vector, the rate of frameshift mutations by the reserve transcriptase was estimated 1x10^{-6} per base pair per replication cycle. In the same experimental system, the rate of base pair substitutions was 7x10^{-6} per base pair and replication cycle (Pathak and Temin., 1990). The co-transcriptional addition of untemplated Gs for editing of the paramyxovirus P protein gene occurs also at high frequency (Vidal et al., 1990). The addition of extra Gs was also observed at high frequency during "in vivo" replication of a Sendai virus synthetic RNA which included the sites of mRNA editing (Park et al., 1991). Thus, it seems that the addition or deletion of nucleotides by viral RNA polymerases is a frequent event while copying certain templates and that the RS virus polymerase is also proned to this type of mutations.

Frameshift mutations are likely to have profound effects upon the biological properties of the RS virus G glycoprotein. This would explain the high frequency of reversion to the wild type phenotype of escape mutants with single reading frameshifts, after passage in the absence of antibody (García-Barreno et al., unpublished). In contrast, mutants with two reading frameshifts are stably maintained without antibody. Interestingly, only natural isolates of RS virus with two reading frame shifts have been reported and they were distinguishable with MAbs (Sullender et al., 1991). Although the introduction of two reading frameshifts in the G gene occurs probably at low frequency during genome replication, the new viruses may have a strong selective immune advantage. It should be pointed out, however, that the actual frequency of RS viruses generated by two reading frameshifts may be underestimated from sequence analysis of natural isolates, since nucleotide substitutions accumulated in the progeny genomes, between the limits of the two reading frame changes, can obscure their origin.

In summary, although definitive data are still missing, it seems that the different genetic mechanisms that generate the antigenic variants selected "in vitro" with MAbs also operate during natural propagation of RS virus. Point mutations in the G protein gene, leading to amino acid replacements, are likely generated at high frequencies during genome replication, creating new antigenic variants susceptible of immune selection by pre-existing antibodies. Mutant genomes with alternative stop codons, or adenosine insertion/deletions in the G protein gene, may be also generated at high frequencies during RS virus replication; however, these changes are probably less tolerated than single amino acid substitutions. Nevertheless, frameshift and stop codon mutants may be selected under certain circumstances to originate new RS virus lineages and/or subgroups with major antigenic changes in their G proteins.

Acknowledgements

We thank Juan Ortín and Agustín Portela por critical reading of the manuscript and helpful suggestions. This work was supported by grants from CICYT, FISS and WHO.

REFERENCES

Akerlind, B., and Norrby, E., 1986, Ocurrence of respiratory syncytial virus subtypes A and B strains in Sweden, *J. Med. Virol.* 19:241 .

Anderson, L.J., Gingham, P., and Heirholzer, J.C., 1988, Neutralization of respiratory syncytial virus by individual and mixtures of F and G protein monoclonal antibodies, *J. Virol.* 62:4235.

Anderson, L.J., Heirholzer, J.C., Tson, C., Hendry, R.M., Fernie, B.N., Stone, Y. and McIntosh, K., 1985, Antigenic characterization of respiratory syncytial virus strains with monoclonal antibodies, *J. Infect. Dis.* 151:626.

Arbiza, J., Taylor, G., López, J.A., Furze, J., Wyld, S., Whyte, P., Stott, J., Wertz, G., Sullender, W., Trudel, M., and Melero, J.A., Characterization of two antigenic sites recognized y neutralizing monoclonal antibodies directed against the F glycoprotein of human respiratory syncytial virus, *J. Gen. Virol.* 73:2225.

Cane, P.A., Matthews, D.A., and Pringle, C.R., 1991, Identification of variable domains of the attachment (G) protein of subgroup A respiratory syncytial virus. *J. Gen. Virol.* 72:2091.

Cane, P.A., Matthews, D.A., and Pringle, C.R. Analysis of relatedness of subgroup A respiratory syncytial viruses isolated worldwide, *J. Infect. Dis.* in press.

Cattaneo, R., Kaelin, K., Baczko, K., and Billeter, M.A., 1989, Measles virus editing provides and additional cysteine-rich protein, *Cell.* 56:759.

Cristina, J., Moya, A., Arbiza, J., Russi, J., Hortal, M., Albo, C., García-Barreno, B., García, O., Melero, J.A., and Portela, A., 1991, Evolution of the G and P genes of human respiratory syncytial virus (subgroup A) studied by the RNase A mismatch cleavage method, *Virology.* 184:210.

Domingo, E., Martinez-Salas, E., Sobrino, F., de la Torre, J.C., Portela, A., López-Galindez, C., Perez-Breña, P., Villanueva, N., Nájera, R., VandePol, S., Steinhauer, D., and Holland, J., 1985, The quasispecies (extremely heterogeneous) nature of viral RNA genome populations: biological relevance -a review, *Gene.* 40:1.

Domingo, E., and Holland, J.J., 1988, High error rates, population equilibrium, and evolution of RNA replication systems, *in* "RNA Genetics vol.3", E. Domingo, J.J. Holland, and P. Ahlquist, ed., CRC Press Inc., Boca Raton, Fl.

García-Barreno, B., Palomo, C., Peñas, C., Delgado, T., Perez-Breña, P., and Melero, J.A., 1989, Marked differences in the antigenic structure of human respiratory syncytial virus F and G glycoproteins, *J. Virol.* 63:925 (1989).

García-Barreno, B., Portela, A., Delgado, T., López, J.A., and Melero, J.A., 1990, Frame shift mutations as a novel mechanism for he generation of neutralization resistant mutants of human respiratory syncytial virus, *EMBO J.* 12:4181.

Gupta, K.C., and Kingsbury, D.W., 1984, Complete sequence of the intergenic and mRNA start signals in the Sendai virus genome: homologies with the genome of vesicular stomatitis virus. *Nucl. Acids. Res.* 12:3829.

Hendry, R.M., Talis, A.L., Godfrey, E., Anderson, L.J., Fernie, B.F., and McIntosh, K., 1986, Concurrent circulation of antigenically distinct strains of respiratory syncytial virus during community outbreaks, *J. Infect. Dis.* 153:291.

Johnson, P.R., Spriggs, M.K., Olmsted, R.A., and Collins, P.L., 1987, The G glycoprotein of human respiratory syncytial viruses of subgroups A and B: extensive sequence divergence between antigenically related proteins, *Proc. Nat. Acad. Sci. USA.* 84:5625.

Kim, H.W., Canchola, J.G., Brandt, C.D., Pyles, G., Chanock, R.M., Jensen, K., and Parrott, R.H., 1969, Respiratory syncytial virus diseases in infants despite prior administration of antigenic inactivated vaccine, *Am. J. Epidemiol.* 89:422.

Levine, S., Klaiser-Franco, R., and Paradiso, P.R., 1987, Demonstration that glycoprotein G is the attachment protein of respiratory syncytial virus, *J. Gen. Virol.* 68:2521.

Lerch, R.A., Anderson, K., and Wertz, G.W., 1990, Nucleotide sequence analysis and expression from recombinant vectors demonstrate that the attachment protein G of bovine respiratory syncytial virus is distinct from that of human respiratory syncytial virus, *J. Virol.* 64:5559.

López, J.A., Peñas, C., García-Barreno, B., Melero, J.A., and Portela, A., 1990, Location of a highly conserved neutralizing epitope in the F glycoprotein of human respiratory syncytial virus, *J. Virol.* 64:927.

McInstosh, K., and Chanock, R.M., 1990, Respiratory syncytial virus, *in* "Virology", B.N. Fields and D.M. Knipe, ed., Raven Press, New York.

Mufson, M.A., Belshe, R.B., Örvell, C., and Norrby, E., 1987, Subgroup characteristics of respiratory syncytial virus strains recovered from children with two consecutive infections, *J. Clin. Microbiol.* 25:1535.

Mufson, M.A., Örvell, C., Rafnar, B., and Norrby, E., 1985, Two distinct subtypes of human respiratory syncytial virus. *J. Gen. Virol.* 66:2111.

Murphy, B.R., Prince, G.A., Walsh, E.E., Kim, H.W., Parrott, R.H., Hemming, V.G., Rodriguez, W.J., and Chanock, R.M., 1986, Dissociation between serum neutralizing and glycoprotein antibody responses of infants and children who received inactivated respiratory syncytial virus vaccines, *J. Clin. Microbiol.* 24:197.

Norrby, E., Mufson, M.A., Alexander, H., Houghten, R.A., and Lerner, R.A., 1987, Site-directed serology with synthetic peptides representing the large glycoprotein G of respiratory syncytial virus, *Proc. Natl. Acad. Sci. USA.* 84:6572.

Park, K., Huang, T., Correia, F., and Krystal, M., 1991. Expression and rescue of synthetic RNA by Sendai virus, in "Negative-strand viruses, 8th International Conference". Charleston. S.C.

Pathak, V.K., and Temin, H.M., 1990, Broad spectrum of "in vivo" forward mutations, hypermutations, and mutational hotspots in a retroviral shuttle vector after a single replication cycle: substitutions, frameshifts, and hypermutations, *Proc. Natl. Acad. Sci. USA.* 87:6019.

Rueda, P., Delgado, T., Portela, A., Melero, J.A., and García-Barreno, B., 1991, Premature stop codons in the G glycoprotein of human respiratory syncytial viruses resistant to neutralization by monoclonal antibodies, *J. Virol.* 65:3374.

Sakaguchi, T., Toyoda, T., Gotoh, B., Inocencio, N.M., K. Kuma, K., Miyata, T., and Nagai, Y., 1989, Newcastle disease virus evolution. I. Multiple lineages defined by sequence variability of the hemagglutinin-neuraminidase gene, *Virology.* 169:260.

Steinhauer, D., and Holland, J.J., 1987, Rapid evolution of RNA viruses, *Ann. Rev. Microbiol.* 41:409.

Sullender, W.M., Mufson, M.A., Anderson, L.J., and Wertz, G.W., 1991, Genetic diversity of the attachment protein of subgroup B respiratory syncytial viruses, *J. Virol.* 65:5425.

Thomas, S.M.., Lamb, R.A., and Patterson, R.G., 1988, Two mRNAs that differ by two non-templated nucleotides encode the amino terminal proteins P and V of the paramyxovirus SV5, *Cell.* 54:891.

Vidal, S., Curran, J., and Kolakofsky, D., 1990, A stuttering model for paramyxovirus P mRNA editing, *EMBO J.* 9:2017.

Walsh, E.E., Hall, C.B., Briselli, M., Brandriss, M.W., and Schlesinger, J.J., 1987, Immunization with glycoprotein subunits of respiratory syncytial virus to protect cotton rats against viral infection, *J Infect. Dis.* 155:1198.

Walsh, E.E., and Hruska, J., 1983, Monoclonal antibodies to respiratory syncytial virus proteins: identification of the fusion protein, *J. Virol.* 47:171.

Waris, M., 1991, Pattern of respiratory syncytial virus epidemics in Finland: Two-year cycles with alternative prevalence of groups A and B, *J. Infect. Dis.* 163:464.

Wertz, G.W., Krieger, M., and Ball, L.A., 1989, Structure and cell surface maturation of the attachment glycoprotein of human respiratory syncytial virus in a cell line deficient in O-glycosylation, *J. Virol.* 63:4767.

TRANSLATION REGULATION BY REOVIRUS STRUCTURAL PROTEINS

Aaron J. Shatkin, Ramona M. Lloyd, Laurie S. Seliger and Lucia Tillotson

Center for Advanced Biotechnology and Medicine
679 Hoes Lane
Piscataway, NJ 08854-5638

SUMMARY

Mammalian reoviruses, like many other animal viruses including adeno-, influenza and vaccinia viruses, have evolved mechanisms to counteract the translational shut-off resulting from double-stranded (ds) RNA activation of the interferon-induced eIF-2α kinase, DAI. Consequently, viral mRNA translation can proceed in infected cells concomitantly with a decline in host protein synthesis. This effect in adenovirus infections depends on VA1 RNA blocking DAI and sparing translational initiation capacity for the production of late viral proteins. In reovirus infections, the host translational shut-off phenotype segregates with the S4 gene which codes for polypeptide σ3. This major outer structural component of the virus binds dsRNA, suggesting that it influences translation by regulating the extent of DAI activation. Transient gene expression in DNA transfected cells can also activate DAI, e.g. decreasing the translation of plasmid-encoded dihydrofolate reductase (DHFR) reporter mRNA in a process that is reversed by co-expression of adenovirus VA1. The reovirus S4 gene similarly stimulated CAT mRNA translation in transfected cells. S4 also replaced the VA1 requirement for transient expression of DHFR from DAI-activating plasmids. In addition, σ3 provided in *trans* by S4 transfection restored late viral protein synthesis in 293 cells infected with the VA1-negative adenovirus mutant dl331. Lysates of reovirus-infected mouse L cells and S4 gene-transfected COS cells required more dsRNA to activate DAI, as measured by ^{32}P-labeling, than either uninfected cells or cells transfected with inverted S4 sequences, consistent with the dsRNA binding activity of σ3. In reovirus-infected cells, virus assembly involves proteolytic cleavage of the M2 gene-encoded polypeptide μ1 and association of the cleavage product, μ1C, with σ3 to form the outer virion capsid. Since protein complex formation decreases or prevents σ3 dsRNA binding, any DAI-mediated translational effects of σ3 are likely to be modulated by the presence of μ1C (and perhaps other viral

Regulation of Gene Expression in Animal Viruses, Edited by
L. Carrasco *et al.*, Plenum Press, New York, 1993

products). This was demonstrated by transient co-expression of M2 with S4 DNA which resulted in structural protein complexes and loss of the translational stimulatory effects of σ3.

INTRODUCTION

The persistence in nature of a diverse multiplicity of viruses provides clear evidence that these ubiquitous parasites have evolved measures to evade cellular antiviral defense mechanisms. The interferon response to virus infection includes induction of a protein kinase that is activated by dsRNA and phosphorylates the α subunit of eukaryotic initiation factor eIF-2 (Hovanessian, 1989; Samuel, 1991; Sen and Ramsohoff, 1992). Consequently, the catalytic action of eIF-2 in translation initiation is blocked, and viral as well as cellular protein synthesis ceases. Thus, the infected cell is sacrificed in an effort to spare the organism by restricting the spread of the virus. Many viruses can elude these defensive tactics by producing inhibitors of the eIF-2α kinase, also referred to as DAI. Among the best studied of the DAI inhibitors is adenovirus VA1 RNA, a polIII transcript that prevents DAI activation by out-competing dsRNA for binding to the enzyme (Mathews and Shenk, 1991). Vaccinia virus codes for proteins that prevent kinase activation (Davies et al., 1992), including a dsRNA binding polypeptide (Akkaraju et al., 1989). In influenza virus infected cells a pre-existing cellular protein that inhibits the kinase apparently is unmasked during virus infection (Lee et al., 1990).

Reoviruses are segmented dsRNA viruses that replicate successfully in a variety of plants, insects and vertebrates including humans. In reovirus-infected mammalian cells, protein synthesis is down-regulated and redirected from host to viral mRNA translation. This virus-specified phenotype segregates with the smallest viral genome segment, S4 (Sharpe and Fields, 1982). It codes for polypeptide σ3 which is intriguingly pleiotropic, e.g. it binds dsRNA (Schiff et al., 1988) or alternatively associates with viral polypeptide μ1C in structural complexes that do not bind dsRNA (Huismans and Joklik, 1976). These properties suggested that translation in reovirus-infected cells is modulated by the availability of polypeptide σ3 for sequestering dsRNA activators of DAI. This possibility was explored in virus-infected cells and in COS cells co-transfected with viral genes and reporter expression vectors (Giantini and Shatkin, 1989; Seliger et al., 1992; Tillotson and Shatkin, 1992).

MATERIALS AND METHODS

Vectors

A DNA copy of the reovirus serotype 3 S4 open reading frame was inserted into pBC12BI and the resulting pt3S4 was used for transient expression of polypeptide σ3 under control of the Rous sarcoma virus (RSV) long terminal repeat (LTR) (Giantini and Shatkin, 1989). A control vector, pt3S4inv contains the same S4 sequences but with the 5' two-thirds inverted to eliminate the σ3 coding capacity of the S4 transcripts. Other plasmids used in this study have also been described; they include pRSVCAT (Gorman et al., 1982; Giantini and Shatkin, 1989), DHFR expression vectors p91023 and pD61 (Kaufman et al., 1989), pMHVA (Mellits and Mathews, 1988), pCMVS4 (Lloyd-Italia, 1992), and pBC12BI-based constructs for expressing the reovirus type 3 polypeptide $\mu 1$ from the viral M2 gene (Tillotson and Shatkin, 1992).

Transfections and expression assays

COS-1 and COS-7 cells were transfected and analysed for bacterial chloramphenicol acetyltransferase (CAT) activity, DHFR expression and the synthesis of reovirus polypeptides $\sigma 3$, $\mu 1$ and $\mu 1C$ as described (Giantini and Shatkin, 1989; Tillotson and Shatkin, 1992; Lloyd-Italia, 1992).

Adenovirus Infections

Human embryonic kidney 293 cells were grown and infected with 20 PFU/cell of adenovirus type 5 or the VA1-negative dl331 mutant as described previously (Thimmappaya et al., 1982). To test for trans-complementation of late viral protein synthesis by VA1 RNA or polypeptide $\sigma 3$, 293 cells were transfected with pMHVA or pCMVS4 DNA by the calcium phosphate precipitation method 24 hours before virus infection and then radiolabeled with ^{35}S-methionine 20 hours after infection. Samples were analyzed by SDS-polyacrylamide gel electrophoresis and autoradiography. Relative units were determined by densitometry (Lloyd-Italia, 1992).

RESULTS

Stimulation of CAT mRNA translation by reovirus polypeptide $\sigma 3$ in transfected COS cells

Translation in reovirus-infected mouse L cells is regulated as demonstrated by the concomitant increase in viral protein synthesis and decrease in host mRNA utilization (Zweerink and Joklik, 1970). This phenotype segregates with the viral S4 gene (Sharpe and Fields, 1982). The same gene is also involved in the establishment of persistent infections (Ahmed and Fields, 1982), a finding consistent with a role of the S4 gene product, polypeptide $\sigma 3$, in host translation inhibition and viral cytopathogenesis.

To study its translational effects, the serotype 3 S4 gene was cloned, sequenced and inserted into a vector suitable for the transient co-expression of polypeptide $\sigma 3$ in mammalian cells with a reporter CAT gene (Giantini et al., 1984; Giantini and Shatkin, 1989). In contrast to the inhibition predicted from early studies (Sharpe and Fields, 1984), the presence of polypeptide $\sigma 3$ increased CAT levels several-fold relative to RCAT plus inverted S4 which is transcribed but does not encode $\sigma 3$ (Figure 1). Stimulation depended on the presence of $\sigma 3$ and not on flanking sequences in the vector since S4 constructs containing the SV40 early promoter in place of the LTR also increased CAT levels. Similar results were obtained when pt3S4 was co-transfected with pBC12BI, indicating that $\sigma 3$ can also stimulate bovine preproinsulin expression. These S4-mediated increases appeared to be confined to plasmid-derived genes, as observed by others studying the stimulatory effects of the adenovirus VA1 gene in transfected cells (Akusjarvi et al., 1987; Kaufman and Murtha, 1987).

Northern analyses demonstrated that the steady-state CAT RNA levels differed by less than two-fold in cells co-transfected with pRSVCAT plus pt3S4 or ptS4inv, i.e. differences too small to account for the ~10-fold CAT enhancement. In addition, the stability of the reporter protein, as determined by pulse-chase and immunoprecipitation experiments, was not altered in COS cells by S4 DNA co-transfection. Taken together, these results indicate that reovirus polypeptide σ3 stimulates CAT mRNA translation.

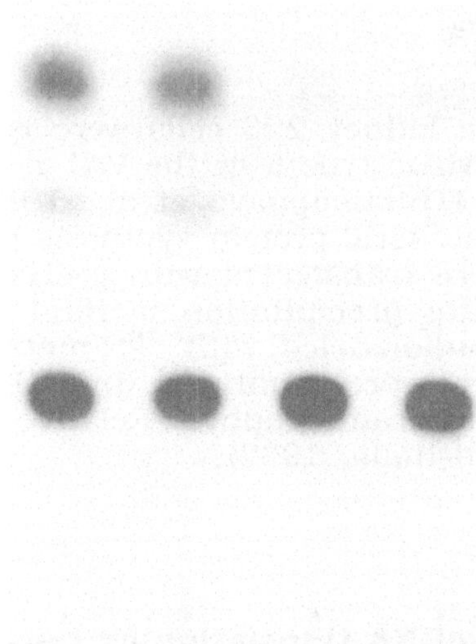

Figure 1. Stimulation of CAT expression by polypeptide σ3. COS cells were co-transfected with 0.2 μg of pRSVCAT (RCAT) and 2 μg of either pt3S4 (S4) or pt3S4 inv (inv). Extracts prepared 48 hr later were assayed for CAT activity (from Giantini and Shatkin, 1989).

Functional replacement of VA1 RNA by polypeptide σ3 in transfected COS cells and in adenovirus dl331 infected 293 cells

Since DAI can be activated by plasmid transfection (Kaufman et al., 1989), it seemed possible that the observed CAT stimulation resulted from σ3 sequestering dsRNA activators of the eIF-2α kinase. Adenovirus VA1 RNA provides a well-studied precedent for a DAI-based mechanism of enhancement because transient transfection of the VA1 gene can increase the expression of reporter DHFR by ten-fold or greater in COS cells (Kaufman and Murtha, 1987). In addition, the effective synthesis of late viral proteins in adenovirus infected 293 cells requires VA1 gene transcripts. This small RNA blocks the activation of DAI, preventing the phosphorylation of the α subunit of eIF-2 and sparing the translation initiation capacity of the cell (Mathews and Shenk, 1991).

To determine if polypeptide σ3 can substitute for VA1 RNA, DHFR expression was assayed in COS cells transfected with the VA1-negative pD61 or VA1-producing p91023, together with pt3S4 or the inverted

Table 1. Replacement of adenovirus VA1 RNA by reovirus polypeptide σ3.

A: DHFR expression in transfected COS cells

DNA (1 μg each)	DHFR (relative units)
pD61 + pt3S4inv	1.0
p91023 + pt3S4inv	4.9
pD61 + pt3S4	4.4
p91023 + pt3S4	6.0

B: Adenovirus late protein synthesis in infected and transfected 293 cells

Virus	DNA (0.5 μg)	100 kd protein (relative units)
dl 331	salmon sperm	1.0
dl 331	pMHVA	4.2
dl 331	pCMVS4	3.5
wild type	salmon sperm	17.7

S4 construct. As shown in Table 1 - part A, the expression of DHFR from the VA1-negative plasmid was increased 4.4-fold by S4 DNA, a level similar to that obtained with the VA1-producing plasmid either in the presence or absence of polypeptide σ3.

Consistent with this result, polypeptide σ3 provided in *trans* from pt3S4 replaced the VA1 RNA requirement and complemented the synthesis of late protein synthesis (as measured by 100 kd nonstructural protein synthesis) in 293 cells infected with the VA1-negative dl331 adenovirus type 5 mutant (Table 1 - part B). At 0.5 μg of plasmid DNA per 22 mm plate, the increase of 3.5-fold was similar to the value of 4.2 obtained with VA1 RNA produced from pMHVA. Higher pMHVA concentrations were more effective, e.g. 2.0 μg yielded 80% of the amount of 100 kd protein obtained in wild type virus-infected cells. With pCMVS4, the level of 100 kd protein increased slightly at 1.0 μg (5.0-fold) but was not elevated further at higher plasmid input levels, although the amount of σ3 was increased. These results support the suggestion that dsRNA (from intramolecular base-pairing or in transfected cells from plasmid bidirectional transcription) activates DAI as in reovirus-infected cells (Imani and Jacobs, 1988; Lloyd-Italia, 1992), and that the resulting eIF-2α phosphorylation and translational shut-down can be reversed by inhibitors of the DAI kinase, including adenovirus VA1 RNA and reovirus polypeptide σ3. In other experiments, it was shown that transfection with pt3S4, like pMHVA, inhibits the extensive phosphorylation of eIF-2α that occurs in dl331-infected 293 cells (Lloyd-Italia, 1992).

Prevention of S4 stimulated CAT synthesis by reovirus M2 gene expression

Reovirus polypeptide σ3 binds dsRNA or alternatively forms complexes with μ1/μ1C. These two events appear to be mutually exclusive and may correspond to early, genome assembly and later, virion maturation stages of reovirus replication. If the stimulation of CAT by σ3 in COS cells occurs via DAI as suggested, the co-expression of M2 would be expected to diminish the enhancing effect of S4 transfection. As shown in Figure 2, the presence of the S4 gene increased CAT expression several-fold as compared to the M2 gene; transfection of both viral genes together resulted in CAT levels that were similar to M2 alone (or to inverted S4 DNA). This finding implies that the M2-encoded polypeptide(s) bind σ3 and allow dsRNA activation of DAI, with accompanying down-regulation of CAT mRNA translation.

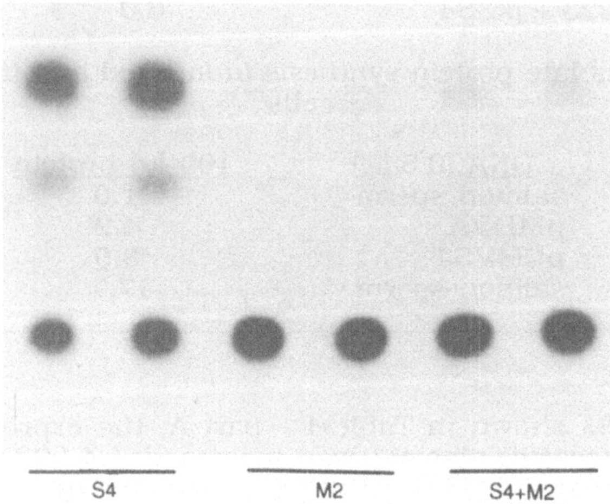

Figure 2. M2 gene co-expression prevents the stimulation of CAT by polypeptide σ3 in COS cells. COS-7 cells were transfected with 0.2 μg of pRSVCAT plus 2 μg of pt3S4 or the same vector containing the reovirus M2 gene in place of the S4 open reading frame, or 1 μg each of the S4 and M2 expression vectors. CAT activity was assayed at 48 hr after transfection (from Tillotson and Shatkin, 1992).

Cleavage of μ1 and formation of σ3/μ1C complexes in COS cells

To demonstrate that the M2 gene was expressed in COS cells and to test if the primary translation product, polypeptide μ1, could be proteolytically processed to μ1C in the absence of virus infection, cells were transfected with M2, S4 or both DNAs and analyzed by antibody precipitation and gel electrophoresis. As shown in Figure 3, the S4 and M2 genes yielded the corresponding σ3 (lane 1) and μ1 polypeptides (lane 2). When cells were expressing both genes together, another viral polypeptide band was apparent (lane 6), and it migrated in the position of authentic μ1C (lane 5). In addition, M2 genes with mutations encompassing the amino acid 42-43 processing site, i.e. N42T/P43A (lane 3) and N42H/P43A (lane 4), were expressed essentially as well as the wild type M2 gene. However, the mutant μ1 primary products were not processed to μ1C (lanes 7 and 8).

These results demonstrated that site-specific processing of μ1 to μ1C can occur in COS cells if σ3 is also being expressed. The specificity of processing was further documented with single amino acid mutants of M2: P43A (Figure 4, lane 5) and N42H (lane 6). They were expressed well compared to wild type M2 DNA (lane 2) but, in contrast to wild type μ1 (lane 8), were not processed in the presence of σ3 (lanes 11, 12). Both Asn42 and Pro43 apparently are important determinants for conversion of μ1 to μ1C, and this processing event clearly involves polypeptide σ3 but no other reovirus proteins.

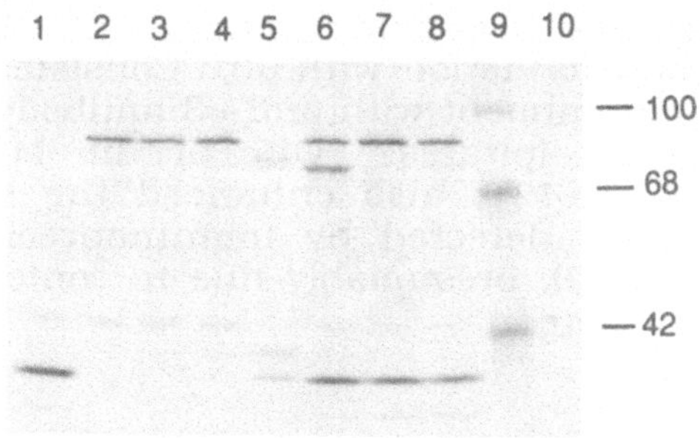

Figure 3. Site-specific proteolytic processing of polypeptide μ1 depends on co-expression of σ3. COS-1 cells were transfected with 5 μg of pt3S4 (lane 1), pt3M2 (lane 2) or the M2 double mutants N42T/P43A (lane 3) and N42H/P43A (lane 4). In addition, co-transfections were done with 2.5 μg each of S4 DNA plus M2 DNA (lane 6) or the TA (lane 7) and HA (lane 8) M2 mutants. Cells were radiolabeled with ^{35}S-methionine 48 hr later, and extracts were immunoprecipitated with antibody to reovirus-infected cells and analyzed by SDS-polyacrylamide gel electrophoresis. Lane 5 contains ^{35}S-methionine labeled viral proteins immunoprecipitated from reovirus infected L cells, and lane 10 corresponds to mock-transfected cells.

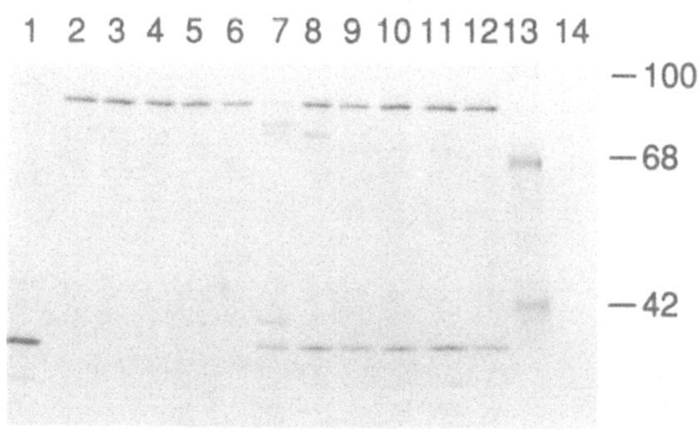

Figure 4. M2 point mutations at amino acid positions 42 or 43 eliminate μ1 processing to μ1C. COS-1 cells were transfected with 5 μg of S4 DNA (lane 1), wild type M2 DNA (lane 2) or M2 mutant DNAs N42T/P43A (lane 3), N42H/P43A (lane 4), P43A (lane 5) and N42H (lane 6). Lanes 8-12 correspond to transfections with 2.5 μg of S4 DNA plus 2.5 μg of the same M2 DNAs as described for lanes 2-6. Lane 7 contains ^{35}S-labeled viral proteins from reovirus-infected L cells; lane 13, ^{14}C-labeled marker proteins and lane 14, mock-transfected cells. Samples in all but lane 13 were immunoprecipitated with antibody to reovirus-infected cells before SDS-polyacrylamide gel electrophoresis and autoradiography(from Tillotson and Shatkin, 1992).

The data in Figure 2 implied that σ3 is sequestered in COS cells by M2 expression products μ1, μ1C or both. To assess this possibility directly, extracts of transfected, reovirus polypeptide-producing cells were also immunoprecipitated with σ3-directed antibody (Figure 5). S4 transfection alone yielded σ3 (lane 1), and the antibody did not cross react with μ1 products (lanes 2-6 as compared to Figure 4, lanes 2-6). However, cells transfected with S4 and M2 DNA yielded σ3 and immunoprecipitable M2 products, mainly μ1C (Figure 5, lane 8), demonstrating complex formation with σ3. Consistent with this result, extracts boiled before treatment with anti-σ3 antibody yielded σ3 but no μ1C (or μ1) in the precipitates. One of the M2 mutations that prevented processing, N42T, also enhanced the stability of μ1/σ3 complexes. They were detected by immunoprecipitation with σ3-directed antibody (lane 9), presumably due to conformational changes that remain to be elucidated.

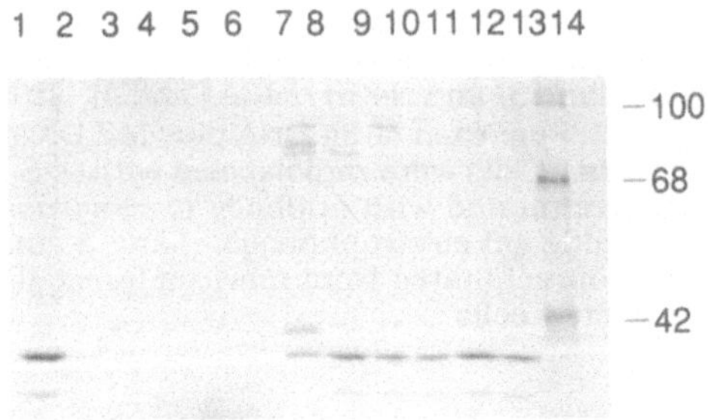

Figure 5. Formation of σ3 complexes with μ1C and a threonine 42 mutant of polypeptide μ1. Samples of the extracts used to obtain the data in Fig. 4 were immunoprecipitated with antibody specific for polypeptide σ3. Lanes are the same as in Fig. 4 (from Tillotson and Shatkin, 1992).

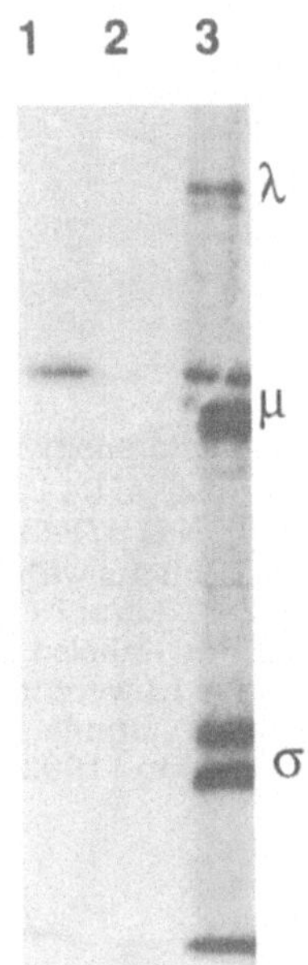

Figure 6. N-myristoylation of polypeptide μ1 in M2 DNA transfected cells. COS-1 cells received 5 μg of pt3M2 (lane 1) or the G2A mutant (lane 2) and were labeled with ^{3}H-myristic acid for 4 hours 48 hr later. Extracts were immunoprecipitated with antibody to reovirus-infected cells and analyzed by SDS-polyacrylamide gel electrophoresis and autoradiography. Lane 3: ^{35}S-methionine labeled reovirus proteins (from Tillotson and Shatkin, 1992).

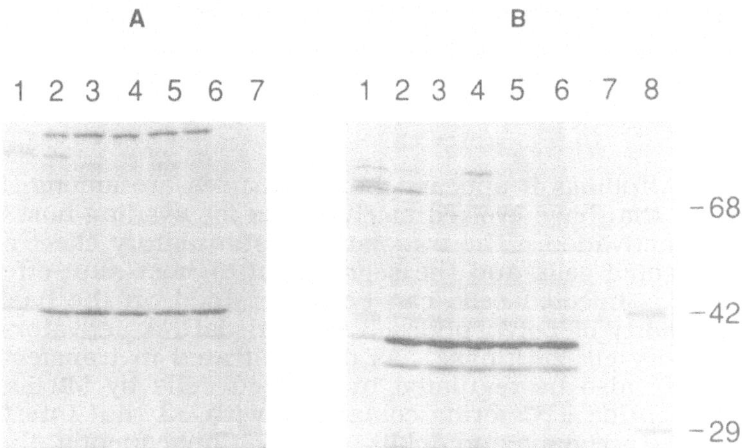

Figure 7. N-terminal myristoylation of μ1 is required for σ3 binding and processing to μ1C. Extracts of transfected COS-1 cells were immunoprecipitated with rabbit antibody to reovirus-infected cells (panel A) or with antibody specific for polypeptide σ3 (panel B). Cells were co-transfected with 2.5 μg each of S4 DNA and either wild type M2 DNA (lane 2) or M2 mutants G2A (lane 3), N42T (lane 4) G2A/N42T (lane 5) and N42Q (lane 6). Lane 1: ^{35}S-labeled reovirus proteins immunoprecipitated from infected L cells; lane 7: immunoprecipitated extract of mock-transfected COS cells; and lane 8: ^{14}C-labeled protein markers (from Tillotson and Shatkin, 1992).

Requirement of μ1 myristoylation for σ3 binding and cleavage to μ1C

Recently it was shown that the type 3 reovirus μ1 polypeptide, like many other proteolytically processed viral structural proteins (Bryant and Ratner, 1990; Kräusslich et al., 1990; Marc et al., 1990; Persing et al., 1987; Streuli and Griffin, 1987), is myristoylated in virus-infected mammalian cells (Nibert et al., 1991). COS cells transfected with M2 DNA also myristoylate μ1 (Figure 6, lane 1). The modification is at the N-terminal glycine, and its conversion to alanine abolished myristoylation (lane 2). Strikingly, the M2 G2A mutation also eliminated the processing of μ1 to μ1C, despite the continued presence of the wild type N42-P43 processing sequence in the mutant (Figure 7A, lane 3).

As shown in Figure 5, the N42T/P43A double mutant of μ1 forms relatively stable and immunoprecipitable complexes with σ3. The single N42T mutant does the same (Figure 7B, lane 4). However, when the myristoylation-minus G2A mutation was superimposed on the N42T mutant, the resulting double-mutant lost the ability to form stable complexes with σ3 (Figure 7B, lane 5). These results suggest that myristoylation of the N-terminus may promote attachment of μ1 to a membrane site as a first step in processing, before σ3 binding. This hypothesis is supported by differential extraction studies of wild type μ1 and the nonmyristoylated G2A mutant. The mutant μ1 and wild type μ1C (which also lacks the modification) were more readily extractable with non-denaturing buffers than the myristoylated wild type μ1 polypeptide (Tillotson and Shatkin, 1992). Myristoylation of the N-terminus of μ1 may anchor the protein for σ3 binding and subsequent cleavage. This order of events is also suggested by the finding that myristoylation is required for σ3 binding of the N42T mutant μ1, which remained uncleaved even in association with σ3.

DISCUSSION

From these findings it appears that reoviruses are among the many animal viruses that have evolved mechanisms for evading host defenses based on DAI activation. The translational stimulatory effect of the S4 gene in transfected cells and the segregation of host shut-off with the S4 gene in virus-infected cells can be understood on the basis of the relative availability of polypeptide σ3 to bind dsRNA activators of DAI, i.e. the inhibitory eIF-2α kinase. As demonstrated in transfected cells, translation may also be regulated in infected cells by M2 expression because polypeptide μ1C forms complexes with σ3 that interfere with dsRNA binding (Huismans and Joklik, 1976). Consequently, the ratio of these two viral structural proteins is important and may explain, at least partly, the reported serotype differences re host inhibition (Sharpe and Fields, 1982). Indeed, DAI activation in extracts of infected L cells required addition of lower amounts of dsRNA with type 2 as compared to type 1 or 3 reovirus (Lloyd-Italia, 1992), implying the presence of less σ3 available for dsRNA binding and correspondingly greater host inhibition, as reported for type 2 infected cells (Sharpe and Fields, 1982).

An intriguing aspect of reovirus maturation is the formation of stable outer capsid complexes of σ3 and μ1C, a process apparently requiring myristoylation of μ1 before its association with σ3 and subsequent specific cleavage. The proteolytic cleavage of μ1 between amino acids 42 and 43 also depends on the presence of the sequence Asn42-Pro43 (which is repeated at positions 587-588 and may correspond to another processing site in μ1). It remains to be determined if σ3 has protease activity or only provides a folding scaffold for presentation of the μ1 cleavage site to a host protease. Alternatively, σ3 may alter the conformation of μ1 to one favorable for autocatalytic proteolysis.

Previously it was noted that σ3 contains sequences homologous to the conserved catalytic sites in the picornavirus 2A and 3C proteases (Schiff et al., 1988). However, it is striking that a sequence in μ1, GDSG, is conserved in several serine proteases. Sindbis virus core protein is a chymotrypsin-like serine proteinase that uses a catalytic triad consisting of residues His141, Asp163 and Ser215 (Choi et al., 1991). The critical serine occurs in the sequence GDSG as in chymotrypsin and bacterial α-lytic proteinase (Choi et al., 1991). Although absent from all three serotypes of reovirus polypeptide σ3 (Seliger et al., 1992), the GDSG sequence appears in μ1 at positions 683-686 in serotypes 2 and 3 and apparently as GDGS in the same position in the serotype 1 polypeptide μ1 (Wiener and Joklik, 1988; Tarlow et al., 1988). These sequence considerations favor an autoproteolytic mechanism of μ1 processing.

It is also striking that the herpes simplex virus type 1 U_L26 gene codes for a serine protease that cleaves itself and a more abundant, N-terminally truncated version of the same protein, both at a site ~20 residues from the C-terminus (Liu and Roizman, 1991). The smaller protein is a capsid component but not a protease. The protease catalytic site in the overlapping, non-structural larger protein has been localized to the N-terminal region (Liu and Roizman, 1992). Examination of the published HSV-1 DNA sequence (McGeoch et al., 1988) has revealed the presence of the conserved GDSG at positions 33-36 of the deduced 635-amino acid U_L26 polypeptide. In light of these observations, it will be of interest to determine if mutagenizing the serine in the GDSG sequence of reovirus polypeptide μ1 alters processing and similarly if the maturation of HSV-1, and possibly other members of the herpes family, is dependent on the serine in the GDSG conserved sequence.

REFERENCES

Ahmed, R., and Fields, B.N., 1982, Role of the S4 gene in the establishment of persistent reovirus infection in L cells, *Cell* 28:605-612.

Akkaraju, G., Whitaker-Dowling, P., Youngner, J., and Jagus, R., 1989, Vaccinia specific kinase inhibitory factor prevents translational inhibition by double-stranded RNA in rabbit reticulocyte lysate, *J. Biol. Chem.* 264:10321-10325.

Akusjarvi, G., Svensson, C., and Nygard, O., 1987, A mechanism by which adenovirus-associated RNAI controls translation in a transient expression assay, *Mol. Cell. Biol.* 7:549-551.

Bryant, M., and Ratner, L., 1990, Myristoylation-dependent replication and assembly of human immunodeficiency virus 1, *Proc. Natl. Acad. Sci. USA* 87:523-527.

Choi, H-K., Tong, L., Minor, W., Dumas, P., Boege, U., Rossmann, M.G., and Wengler, G., 1991, Structure of Sindbis virus core protein reveals a chymotrypsin-like serine proteinase and the organization of the virion, Nature 354:37-43.

Davies, M., Elroy-Stein, O., Jagus, R., Moss, B., and Kaufman, R.J., 1992, The vaccinia virus K3L gene product potentiates translation by inhibiting double-stranded RNA-activated protein kinase and phosphorylation of the alpha subunit of eukaryotic initiation factor 2, *J. Virol.* 66:1943-1950.

Giantini, M., and Shatkin, A.J., 1989, Stimulation of chloramphenicol acetyltransferase mRNA translation by reovirus capsid polypeptide σ3 in cotransfected COS cells, *J. Virol.* 63:2415-2421.

Giantini, M., Seliger, L., Furuichi, Y., and Shatkin, A.J., 1984, Reovirus type 3 genome segment S4: nucleotide sequence of the gene encoding a major virion surface protein, *J. Virol.* 52:984-987.

Gorman, C.M., Merlino, G.T., Willingham, M.C., Pastan, I., and Howard, B.H., 1982, The Rous sarcoma virus long terminal repeat is a strong promoter when introduced into a variety of eukaryotic cells by DNA-mediated transfection, *Proc. Natl. Acad. Sci. USA* 79:6777-6781.

Hovanessian, A.G., 1989, The double-stranded RNA-activated protein kinase induced by interferon: dsRNA-PK, *J. Interferon Res.* 67:641-647.

Huismans, H., and Joklik, W.K., 1976, Reovirus-coded polypeptides in infected cells: isolation of two native monomeric polypeptides with affinity for single-stranded and double-stranded RNA, respectively, *Virology* 70:411-424.

Imani, F., and Jacobs, B.L., 1988, Inhibitory activity for the interferon-induced protein kinase is associated with the reovirus serotype 1 σ3 protein, *Proc. Natl. Acad. Sci. USA* 83:7887-7891.

Kaufman, R., Davies, M., Pathak, V., and Hershey, J., 1989, The phosphorylation state of eucaryotic initiation factor 2 alters translational efficiency of specific mRNAs, *Mol. Cell. Biol.* 9:946-958.

Kaufman, R.J., and Murtha, P., 1987, Translational control mediated by eucaryotic initiation factor 2 is restricted to specific mRNAs in transfected cells, *Mol. Cell. Biol.* 7:1568-1571.

Kräusslich, H.G., Hölscher, C., Reuer, Q., Harber, J., and Wimmer, E., 1990, Myristoylation of the poliovirus polyprotein is required for proteolytic processing of the capsid and for viral infectivity, *J. Virol.* 64:2433-2436.

Lee, T.G., Tomita, J., Hovanessian, A.G., and Katze, M.G., 1990, Purification and partial characterization of a cellular inhibitor of the interferon-induced protein kinase of Mr 68,000 from influenza virus-infected cells, *Proc. Natl. Acad. Sci. USA* 87:6208-6212.

Liu, F., and Roizman, B., 1991, The herpes simplex virus 1 gene encoding a protease also contains within its coding domain the gene encoding the more abundant substrate, *J. Virol.* 65:5149-5156.

Liu, F., and Roizman, B., 1992, Differentiation of multiple domains in the herpes simplex virus 1 protease encoded by the U_L26 gene, *Proc. Natl. Acad. Sci. USA* 89:2076-2080.

Lloyd-Italia, R., 1992, Molecular mechanism of translational control mediated by the reovirus S4 gene, Ph.D. Thesis.

Marc, D., Masson, G., Girard, M., and van der Werf, S., 1990, Lack of myristoylation of poliovirus capsid polypeptide VP0 prevents the formation of virions or results in the assembly of noninfectious virus particles, *J. Virol.* 64:4099-4107.
Mathews, M., and Shenk, T., 1991, Adenovirus virus-associated RNA and translational control, *J. Virol.* 65:5657-5662.
McGeoch, D.J., Dalrymple, M.A., Davison, A.J., Dolan, A., Frame, M.C., McNab, D., Perry, L.J., Scott, J.E., and Taylor, P., 1988, The complete DNA sequence of the long unique region in the genome of herpes simplex virus type 1, *J. gen. Virol.* 69:1531-1574.
Mellits, K., and Mathews, M., 1988, Effects of mutations in stem and loop regions on the structure and function of adenovirus VA RNAI, *EMBO J.* 7:2849-2859.
Nibert, M.L., Schiff, L.A., and Fields, B.N., 1991, Mammalian reoviruses contain a myristoylated structural protein, *J. Virol.* 65:1960-1967.
Persing, D.H., Varmus, H.E., and Ganem, D., 1987, The pre-S1 protein of hepatitis B virus is acylated at its amino terminus with myristic acid, *J. Virol.* 61:1672-1677.
Samuel, C.E., 1991, Antiviral actions of interferon. Interferon-regulated cellular proteins and their surprisingly selective antiviral activities, *Virology* 183:1-11.
Schiff, L.A., Nibert, M.L., Co, M.S., Brown, E.G., and Fields, B.N., 1988, Distinct binding sites for zinc and double-stranded RNA in the reovirus outer capsid protein σ3, *Mol. Cell. Biol.* 8:273-283.
Seliger, L.S., Giantini, M., and Shatkin, A.J., 1992, Translational effects and sequence comparisons of the three serotypes of the reovirus S4 gene, *Virology* 187:202-210.
Sen, G.C., and Ransohoff, R.M., Interferon-induced antiviral actions and their regulation, *Advances in Virus Research*, in press.
Sharpe, A.H., and Fields, B.N., 1982, Reovirus inhibition of cellular RNA and protein synthesis: role of the S4 gene, *Virology* 122:381-391.
Streuli, C., and Griffin, B.E., 1987, Myristic acid is coupled to a structural protein of polyoma virus and SV40, *Nature (London)* 326:619-622.
Tarlow, O., McCorquodale, J.G., and McCrae, M.A., 1988, Molecular cloning and sequencing of the gene (M2) encoding the major virion structural protein (μ1-μ1C) of serotypes 1 and 3 of mammalian reovirus, *Virology* 164:141-146.
Thimmappaya, B., Weinberger, C., Schneider, R.J., and Shenk, T., 1982, Adenovirus VAI RNA is required for efficient translation of viral mRNAs at late times after infection, *Cell* 31:543-551.
Tillotson, L., and Shatkin, A.J., 1992, Reovirus polypeptide σ3 and N-terminal myristoylation of polypeptide μ1 are required for site-specific cleavage to μ1C in transfected cells, *J. Virol.* 66:2180-2186.
Wiener, J.R., and Joklik, W.K., 1988, Evolution of reovirus genes: a comparison of serotype 1, 2, and 3 M2 genome segments, which encode the major structural capsid protein μ1C, *Virology* 163:603-613.
Zweerink, J.J., and Joklik, W.K., 1970, Studies on the intracellular synthesis of reovirus-specified proteins, *Virology* 44:501-518.

THE REGULATION OF CORONAVIRUS GENE EXPRESSION

Stuart G. Siddell

Institute of Virology
University of Würzburg
Germany

INTRODUCTION

The coronaviruses are a monogeneric group of enveloped, RNA viruses that infect vertebrates, including man. In the main, coronavirus infections are associated with respiratory and gastrointestinal illness (Table 1, for a recent review of coronavirus biology see Spaan et al., 1990).

Table 1. Coronaviruses and their hosts

	Acronym	Name	Host
Group I	MHV	Murine hepatitis virus	Mouse
	HCVOC43	Human coronavirus OC43	Man
	HEV	Heamagglutinating encephalitis virus	Pig
	BCV	Bovine coronavirus	Cattle
	TCV	Turkey coronavirus	Turkey
Group II	IBV	Infectious bronchitis virus	Chicken
	HCV229E	Human coronavirus 229E	Man
	TGEV	Transmissible gastroenteritis virus	Pig
	CCV	Canine coronavirus	Dog
	FIPV	Feline infectious peritonitis virus	Cat

In the last few years, our knowledge of coronavirus molecular biology has increased significantly. The basic principles of genomic organization and, at least, some aspects of the coronavirus replication strategy have been elucidated. Also, a number of mechanisms which regulate coronavirus gene expression at the translational level have been identified. This article briefly reviews this information.

Regulation of Gene Expression in Animal Viruses, Edited by
L. Carrasco *et al.*, Plenum Press, New York, 1993

GENOME ORGANIZATION

The coronavirus genome is large, between 27 and 32 kilobases (kb) (Boursnell et al., 1987; Lee et al., 1991). At the present time, the complete genomic sequences of three coronaviruses (IBV, MHV and HCV229E) are available and partial, but quite extensive genomic sequences have been obtained for BCV, TGEV, CCV and FIPV. When this data is compared, it becomes apparent that the genetic organization of all coronaviruses is very similar but that two quite distinct lineages exist.

The overall genomic organization is illustrated in figure 1. The genes encoding the major structural proteins, N (nucleocapsid), M (membrane) and S (surface) are located towards the 3' end of the genome. These structural genes are interspersed by a number of open reading frames (ORFs) (e.g. IBV, 5; MHV, 3 and HCV229E, 3) which encode small (approx. 5-15 kilodalton, KDa) proteins of unknown function. The variable number and occurrence of deletion mutations in these ORFs (Yokomori and Lai, 1991; Wesley et al., 1990) suggests that they encode proteins that are not essential. One exception is the ORF located immediately upstream of the M protein gene (ORF 3c for IBV, ORF 4 for TGEV, etc) which encodes a protein that is incorporated into virions (Liu and Inglis, 1991; Godet et al., 1992). This protein displays striking structural similarities with the M2 protein of influenza virus.

Towards the 5' end of the genome are two large ORFs which overlap by 40-75 bases. These ORFs constitute the coronavirus RNA polymerase locus. The expression of gene products from this locus is complex and appears to involve mechanisms such as (-1) ribosomal frameshifting and autoproteolytic processing of one or more polyproteins (Brierley et al., 1989; Denison and Perlman, 1986; Denison et al., 1991, 1992; Baker et al., 1989). The locus is polycistronic and 5 complementation groups of ts mutants with an RNA-minus phenotype have been mapped within this area by recombination analysis (Baric et al., 1990). The size of the locus (approximately 20 kb) leads one to suspect that this region of the genome may also encode functions which are not related to RNA synthesis.

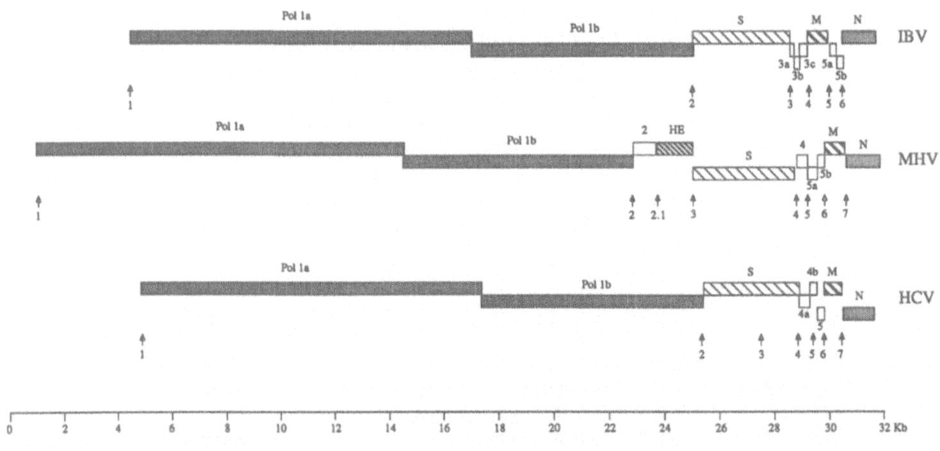

Figure 1. The organization of coronavirus genomes. The major open reading frames in the genomes of IBV, MHV and HCV229E are shown. The sizes of the ORFs are to scale.

Coronavirus genomes fall into two lineages: Group I, represented by MHV, HCVOC43, HEV, BCV and TCV and Group II, represented by IBV, HCV229E, TGEV, CCV and FIPV. These lineages manifest themselves as a clear distinction in the degree of genetic relatedness and the occurrence of an additional genetic element, comprised of two ORFs, ORF 2 and HE, which is found only in the Group I viruses (fig. 1). The HE gene encodes a fourth structural protein, the haemagglutinin-esterase, which shows a striking similarity with the HA1 region of the influenza C virus surface glycoprotein HEF (Luytjes et al., 1988) Spontaneous deletion mutants of MHV which do not express the HE protein or ORF 2 gene product (Luytjes et al., 1988, Schwarz et al., 1990) are still capable of efficient replication, at least in tissue culture.

TRANSCRIPTION

The expression of the coronavirus genome is mediated by genomic and subgenomic mRNAs. Together these mRNAs represent about 95% of the viral RNA present in the infected cell. T_1 oligonucleotide analysis and sequencing of these mRNAs shows that they are structurally related as a 3' coterminal (nested) set (fig. 2). Furthermore, at the 5' end of each mRNA, a common sequence at about 70 nucleotides, the leader RNA, is present. This sequence is found only once in the genomic RNA, at the 5' end, and the synthesis of the subgenomic mRNAs must, therefore, involve a process of discontinuous transcription. A specific motif, UCUAAAC, is found at those positions in the genome which define the 5' end of the unique region at each mRNA, i.e. the region not found in the next smallest RNA. This motif is thought to have an important function in mRNA synthesis, however, its precise role is still uncertain (for a review of coronavirus mRNA structure see Lai, 1990).

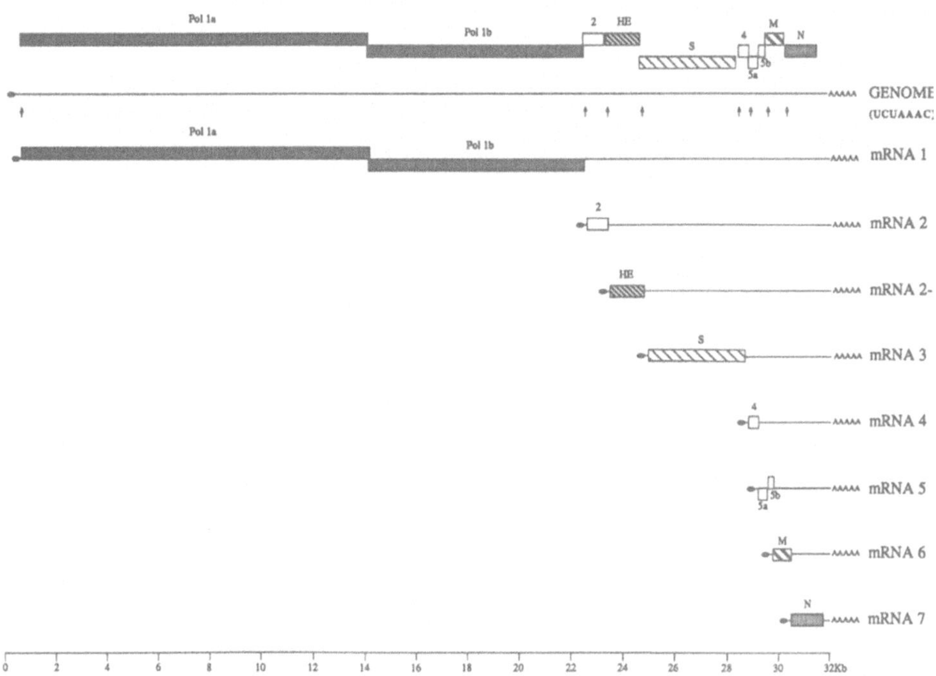

Figure 2. Expression of the coronavirus (MHV) genome. The arrangement of ORFs in the genome, their relationship to the 5′unique regions of the subgenomic mRNAs and the positions of the UCUAAAC and leader RNA sequences in the genomic RNA are shown.

At the present time, it is not possible to present a definitive model for the genesis of coronavirus mRNAs. The following information is, however, relevant;

1. The dosage of UV-irradiation required to block the synthesis of the subgenomic mRNAs is proportional to their individual lengths (Jacobs et al., 1981, Stern and Sefton, 1982).
2. Infected cells contain subgenomic length negative strand RNAs which correspond to each of the subgenomic mRNAs (Sethna et al., 1989, Hofmann et al., 1990).
3. RNAse treatment of infected cell RNA yields one genomic and several subgenomic RF structures, the number of which corresponds to the number of subgenomic mRNAs (Hofmann et al., 1990, Sawicki and Sawicki, 1990).

4. The sizes of the RF structures are proportional to the expected sizes of the negative-strand templates for each of the mRNA species (Sawicki and Sawicki, 1990).
5. The subgenomic RF structures are derived from transcriptionally active subgenomic RI structures (Sawicki and Sawicki 1990).
6. The RI structures synthesize both positive and negative strand RNA. The genomic and subgenomic RF structures and mRNAs are labelled coincidentally throughout this infection (Sawicki and Sawicki, personal communication).
7. The subgenomic negative-strand RNAs have a 3' antileader sequence (Sethna et al., 1991).

With this information, at least, two quite different models for the genesis of coronavirus mRNA can be proposed. Both models assume that the genomic RNA is infectious (see Jeong and Makino, 1992) and both models propose that a process of "leader-primed" discontinuous transcription, (Spaan et al., 1983, Baric et al., 1983) is used to generate, at least the initial subgenomic mRNA molecules. However, they differ from each other in the role postulated for the UCUAAAC motif. In the first model , which is termed "internal initiation" in figure 3, this motif is an element which regulates the frequency of leader-primed initiation on genome length negative strand templates. Lai and coworkers (see Lai et al., 1990) originally proposed that this regulation was achieved by the degree of complementarity between the 3' end of the leader RNA and template sequences flanking the UCUAAAC motif. Later, Konings et al. (1988) argued that additional mechanisms had to be considered, for example, the role of transcriptional blockage at the initiation site. In any case, by analogy to other positive-strand RNA viruses, the model predicts subgenomic RFs corresponding to the unique regions of each mRNA and these structures have not yet been found (see Sawicki and Sawicki, 1990)

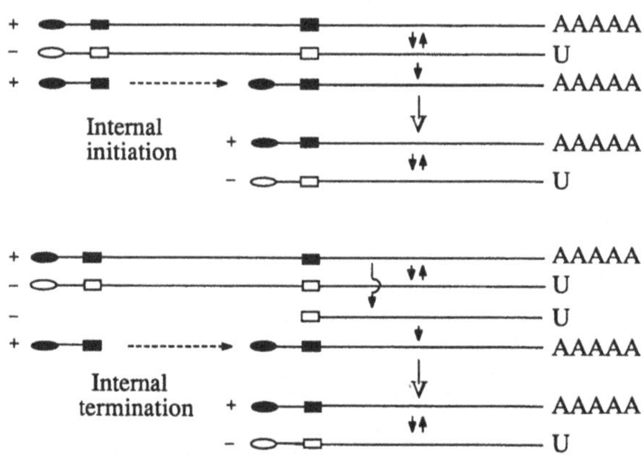

Figure 3. Two models for the genesis of coronavirus subgenomic mRNA.The leader RNA and UCUAAAC motifs are represented as oval and box structures respectively, in plus (closed) and minus (open) orientation. The molecules are not drawn to scale and only one subgenomic RNA is illustrated.

In the second model, termed "internal termination" in fig. 3, the role of the UCUAAAC motif is postulated to be a termination signal for negative strand RNA synthesis (Sawicki and Sawicki, 1990). In this way a (nested) set of 5' coterminal negative strand RNAs are generated and these serve as templates for "leader-primed" transcription to produce subgenomic mRNA. It is possible that the UCUAAAC motif may also have a role in this event. In both models the subgenomic mRNAs can subsequently be copied to subgenomic negative-strand templates which possess antileader RNA, and subgenomic replicons are amplified independently of RNA synthesis involving genomic sized templates. The internal termination model is compatible with all of the available data to date, but important predictions, for example the presence of subgenomic negative-strand RNA lacking antileader sequences have not yet been fulfilled. Clearly, because of the exceedingly low amounts of viral RNA in the earliest stages of infection, it will be very difficult to distinguish the two models. Most probably, PCR technology and reverse genetics will be needed.

TRANSLATION

The basic premise of coronavirus translation is that only the information encoded in the 5' unique region of each mRNA is expressed from that RNA. This conclusion is based upon the in vitro translation of mRNAs (Rottier et al., 1981; Leibowitz et al., 1982, Siddell, 1983) and is consistent with the general requirements of translation in eucaryotic cells. For the majority of coronavirus mRNAs the 5' unique region encompasses only a single ORF. Thus, for example, the mRNAs encoding the structural proteins N, M, S and, if present, HE are functionally monocistronic. For some mRNAs, however, the 5' unique region contains more than one ORF. In these cases, additional strategies are used to regulate gene expression.

Expression of the RNA Polymerase Locus (mRNA 1)

As has been mentioned, the coronavirus RNA polymerase locus is polycistronic and is comprised of two large, overlapping ORFs, ORF 1a and ORF 1b. There is now substantial evidence that ORF b is expressed by a (-1) ribosomal frame-shifting event mediated by a sequence specific element, the "slip-sequence" and an RNA structure, the RNA pseudoknot, both located at the ORF 1a/ORF 1b junction (Brierley et al., 1987, Brierley et al., 1989, Bredenbeek et al., 1990). In the absence of proteolytic processing the polymerase locus could, therefore, encode a 400 KDa OFR 1a product or a 700 KDa ORF 1a-ORF 1b product. Brierley and his colleagues have performed a detailed mutational analysis of the pseudoknot component of the IBV frameshifting signal (Brierley et al., 1991) and conclude that the efficiency of frameshifting, at least in vitro, can approach 30%. Because of the small amounts of polymerase gene products synthesized in infected cells, it has not yet been possible to provide direct evidence for coronavirus ribosomal frame shifting in vivo.

The second level at which the expression of the coronavirus polymerase locus is regulated is proteolytic processing. Structural analysis of the IBV, MHV and HCV229E polymerase region has revealed sequence motifs characteristic of papain-like proteases and the poliovirus 3c-like protease (Boursnell et al., 1987, Lee et al., 1991, Herold and Siddell, unpublished). These "protease domains" are located in ORF 1a and an autoproteolytic activity has been mapped to the middle of this ORF by in vitro transcription/translation (Baker et al., 1989). Recently, the first steps in the analysis of proteolytic processing in relation to polymerase gene expression in vivo have been reported (Denison et al., 1991, 1992), but it is too early to propose a processing pathway.

Functionally Polycistronic Subgenomic mRNAs

Whereas most of the coronavirus subgenomic mRNAs contain only a single ORF in their 5' unique region, there is inevitabley at least one mRNA with two or three ORFs at this position. The best characterized examples are the mRNA 3 of IBV and the mRNA 5 of MHV. The IBV mRNA 3 has three ORFs in its 5' unique region and in infected cells all three translation products, 3a, 3b and 3c are expressed (Liu et al., 1991). The MHV mRNA 5 has two ORFs, 5a and 5b, and the downstream ORF product has been found in vivo (Leibowitz et al., 1988). A striking structural feature of both mRNAs is that the upstream ORFs (3a and 3b in IBV mRNA 3 and 5a in MHV mRNA 5) are, with the exception of the initiation codon, devoid of AUG triplets in any reading frame. This lead to the idea that expression of the downstream ORFs might be mediated by initiation events involving leaky scanning (Skinner et al., 1985, Budzilowicz and Weiss, 1987). This idea was, however, never confirmed experimentally and very recently Inglis and coworkers (Liu and Inglis, personal communication) have provided evidence that the IBV 3c protein may be expressed by a mechanism involving the internal entry of ribosomes in a cap-independent fashion.

CONCLUSIONS

The regulation of coronavirus gene expression appears to be complex. At the transcriptional level, coronavirus mRNAs are synthesized in a tightly regulated, non-equimolar fashion throughout the infection cycle. However, we do not yet have a clear picture of the mechanism of mRNA synthesis nor the regulatory controls which are used. At the translational level, coronaviruses appear to exploit a remarkable variety of strategies, including ribosomal frame-shifting, proteolytic processing and functionally polycistronic subgenomic mRNAs. The size and complexity of the coronavirus genome suggests that additional regulatory mechanisms have yet to be discovered.

Acknowledgments

I thank Drs S.G.and D.L. Sawicki, I. Brierley and S.C. Inglis for communicating results before publication. The authors work is supported by the DFG and the BMFT.

References

Baker, S.C., Shieh, C.-K., Soe, L.H., Chang, M.-F., Vannier, D.M., and Lai, M.M.C., 1989, Identification of a domain required for autoproteolytic cleavage of murine coronavirus gene A polyprotein, J. Virol. 63, 3693.

Baric, R.S., Fu, K.S., Schaad, M.C., and Stohlman, S.A., 1990, Establishing a genetic recombination map for MHV-A59 complementation groups, Virology 177:646.

Baric, R.S., Stohlman, S.A., and Lai, M.M.C., 1983, Characterization of replicative intermediate RNA of mouse hepatitis virus: presence of leader RNA sequences on nascent chains, J. Virol. 48:633.

Boursnell, M.E.G., Brown, T.D.K., Foulds, I.J., P.F., Tomley, F.M., and Binns, M.M., 1987, Completion of the sequence of the genome of the coronavirus avian infections bronchitis virus, J. Gen. Virol. 68:57.

Bredenbeek, P.J., Pachuk, C.J., Noten, A.F.H., Charité, A., Luytjes, W., Weiss, S.R., and Spaan, W.J.M., 1990, The primary structure and expression of the second open reading frame of the polymerase gene of the coronavirus MHV-A59; a highly conserved polymerase is expressed by an efficient ribosomal frame-shifting mechanism, Nucl. Acids Res. 18:1825.

Brierley, I., Boursnell, M.E.G., Binns, M.M., Bilimosia, B., Blok, V.C., Brown, T.D.K., and Inglis, S.C., 1987, An efficient ribosomal frame-shifting signal in the polymerase encoding region of the coronavirus IBV, EMBO J. 6:3779.

Brierley, I., Digard, P., and Inglis, S.C., 1989, Characterization of an efficient coronavirus ribosomal frameshifting signal: requirement for an RNA pseudoknot, Cell 57:537.

Brierley, I., Rolley, N.J., Jenner, A.J., and Inglis, S.C., 1991, Mutational analysis of the RNA pseudoknot component of a coronavirus ribosomal frame-shifting signal, J. Mol. Biol. 220:889.

Budzilowicz, C.J., and Weiss, S.R., 1987, In vitro synthesis of two polypeptides from a non-structural gene of coronavirus mouse hepatitis virus strain A59, Virology 157:509.

Denison, M.R., and Perlman, S., 1986, Translation and processing of mouse hepatitis virus virion RNA in a cell-free system, J. Virol. 60:12.

Denison, M.R., Zoltick, P.W., Hughes, S.A., Giangreco, B., Olson, A.L., Perlman, S., Leibowitz, J.L., and Weiss, S.R., 1992, Intracellular processing of the N-terminal ORF 1a proteins of the coronavirus MHV-A59 requires multiple proteolytic events, Virology 189:274.

Denison, M.R., Zoltick, P.W., Leibowitz, J.L., Pachuk., C.J., and Weiss, S.R., 1991, Identifiction of polypeptides encoded in open reading frame 1b of the putative polymerase gene of the murine coronavirus mouse hepatitis virus A59, J. Virol 65:3076.

Godet, M., L'Haridon, R., Vautherot, J.-F., and Laude, H., 1992, TGEV corona virus ORF 4 encodes a membrane protein that is incorporated into virions, Virology 188:666-675.

Hofmann, M.A., Sethna, P.B., and Brian, D.A., 1990, Bovine coronavirus mRNA replication continues throughout persistent infection in cell culture, J. Virol. 64:4108.

Jacobs, L., Spaan, W.J.M., Horzinek, M.C., and van der Zeijst, B.A.M., 1981, Synthesis of subgenomic mRNAs of mouse hepatitis virus is initiated independently: evidence from UV transcription mapping. J. Virol. 39:401.

.Jeong, Y.S., and Makino, S., 1992, Mechanism of coronavirus transcription: duration of primary transcription initiation activity and effects of subgenomic RNA transcription on RNA replication, J. Virol. 66, 3339.

Konings, D.A.M., Bredenbeek, P.J., Noten, J.F.H., Hogeweg, P., and Spaan, W.J.M., 1988, Differential premature termination of transcription as a proposed mechanism for the regulation of coronavirus gene expression, Nucl. Acids Res. 16:10849.

Lai, M.M.C., 1990, Coronavirus: organization, replication and expression of genome, Ann. Rev. Microbiol. 44:303.
Lee, H.-J., Shieh, C.-K., Gorbalenya, A.E., Koonin, E.V., La Monica, N., Tuler, J., Bagdzhadzhyan, A., and Lai, M.M.C., 1991, The complete sequence (22 kilobases) of murine coronavirus gene encoding the putative proteases and RNA polymerase, Virology 180:567.
Leibowitz, J.L., Perlman, S., Weinstock, G., DeVries, J.R., Budzilowicz, C., Weissemann, J.M., and Weiss, S.R., 1988, Detection of a murine coronavirus non-structural protein encoded in a downstream open reading frame, Virology 164:156.
Leibowitz, J.L., Weiss, S.R., Paavola, E., and Bond, C.W., 1982, Cell free translation of murine coronavirus RNA, J. Virol. 43:905.
Liu, D.X., Cavanagh, D., Green, P., and Inglis, S.C., 1991, A polycistronic mRNA specified by the coronavirus infectious bronchitis virus, Virology 184:531.
Liu, D.X., and Inglis, S.C., 1991, Association of the infectious bronchitis virus 3c protein with the virion envelope, Virology 185:911.
Luytjes, W., Bredenbeek, P.J., Noten, A.F.H., Horzinek, M.C., and Spaan, W.J.M., 1988, Sequence of mouse hepatitis A59 mRNA 2: indications for RNA recombination between coronaviruses and influenza C virus." Virology 166:415.
Rottier, P.J., Spaan, W.J.M., Horzinek, M.C., and van der Zeijst, B.A.M., 1981, Translation of three mouse hepatitis virus strain A59 subgenomic RNAs in Xenopus laevis oocytes, J. Virol. 38:20.
Sawicki, S.G., and Sawicki, D.L., 1990, Coronavirus transcription: subgenomic mouse hepatitis virus replicative intermediates function in RNA synthesis, J. Virol. 64:1050.
Schwarz, B., Routledge, E., and Siddell, S.G., 1990, Murine coronavirus non-structural protein ns2 is not essential for virus replication in transformed cells, J. Virol. 64:4784.
Sethna, P.B., Hofmann, M.A., and Brian, D.A., 1991, Minus-strand copies of replicating coronavirus mRNAs contain antileaders, J. Virol. 65:320.
Sethna, P.B., Hung, S.-L., and Brian, D.A., 1989, Coronavirus subgenomic minus-strand RNAs and the potential for mRNA replicons, Proc. Natl. Acad. Sci. USA 86:5626.
Siddell, S., 1983, Coronavirus JHM: coding assignments of subgenomic mRNAs, J. Gen. Virol. 64:113.
Skinner, M.A., Ebner, D., and Siddell, S.G., 1985, Coronavirus MHV-JHM mRNA 5 has a sequence arrangement which potentially allows translation of a second, downstream open reading frame. J. Gen. Virol. 66:581.
Spaan, W., Cavanagh, D., and Horzinek, M.C., 1990, Coronavirus, in "Immunochemistry of Viruses. II. The Basis for Serodiagnosis and Vaccines," M.H.V. van Regenmortel and A.R. Neurath, eds., Elsevier, Amsterdam.
Stern, D.F., and Sefton, B.M., 1982, Synthesis of coronavirus mRNAs: kinetics of inactivation of IBV RNA synthesis by UV light, J. Virol. 42:755.
Wesley, R.D., Woods, R.D., and Cheung, 1990, Genetic basis for the pathogenesis of transmissible gastroenteritis virus, J. Virol. 64:4761.

INTERACTION OF INITIATION FACTORS AND CAPSID PROTEIN WITH THE CAP STRUCTURE OF CHIMAERIC MRNAS CONTAINING THE 5' UNTRANSLATED REGIONS OF THE RNAS OF SEMLIKI FOREST VIRUS

Gerda Berben-Bloemheuvel and Harry O. Voorma

Department of Molecular Cell Biology, Utrecht University
Padualaan 8, 3584 CH Utrecht, The Netherlands

INTRODUCTION

A considerable effort has been put into the understanding how Semliki Forest virus overrules the protein synthetic machinery of the host, promoting the translation of its own late 26S mRNAs while suppressing those of the host cell and its own genomic 42S RNA (van Steeg *et al.*, 1981a; van Steeg *et al.*, 1981b). Two major virus-specific RNAs are found in alphavirus-infected cells: the genomic 42S RNA and the subgenomic 26S RNA, comprising about one third of the 3' part of the 42S RNA (Kääriäinen and Söderlund, 1978). Both the 42S viral mRNA, which is utilized as an early mRNA for the synthesis of non structural proteins (Kääriäinen and Söderlund, 1978; Wengler *et al.*, 1979) and the subgenomic late 26S mRNA, which encodes the structural proteins later in infection (Kääriäinen and Söderlund, 1978), are capped (Petterson *et al.*, 1980; Wengler *et al.*, 1979). Late in infection proteins are synthesized from 26S RNA in SFV-infected cells, despite the fact that 50% of the mRNA population is of cellular origin (van Steeg *et al.*, 1981b; Tuomi *et al.*,1975). In an effort to explain how host shut off takes place, it has been suggested that viral mRNAs compete with cellular mRNAs for components of the translational apparatus of the host cell (Tuomi *et al.*, 1975; Kozak, 1986; Schneider and Shenk, 1987).

It has been established that 26S mRNA, which is translated late in infection when the expression of other mRNAs is suppressed, displays a low need for eIF-4B and eIF-4E as measured in an *in vitro* translation system (van Steeg *et al.*, 1981a). In addition, optimal translation of 42S mRNA requires a 2-4 fold larger amount of these cap binding proteins (van Steeg *et al.*,1981a). Since both viral RNAs depend completely on a cap structure for efficient translation, a decreased importance of the cap structure of 26S RNA can be excluded as an explanation of its decreased requirement for eIF-4B and eIF-4E (van Steeg *et al.*, 1981a; van Duijn *et al.*, 1986). In order to determine whether the 5' untranslated regions of 42S

Regulation of Gene Expression in Animal Viruses, Edited by
L. Carrasco *et al.*, Plenum Press, New York, 1993

RNA and 26S RNA are responsible for the difference in translational behaviour, we studied the translation of chimaeric CAT mRNAs containing the leader sequences of early and late RNA of SFV. The leader sequence of AMV RNA 4 is used as a positive control. The requirements for the cap structure and initiation factors were determined in *in vitro* translation assays.

Furthermore, we studied the interaction of cap binding proteins to these chimaeric mRNAs. Special attention is given to the interactions of eIF-4B and eIF-4E, factors known to play an antagonistic role in the translational inhibition caused by the SFV capsid protein during viral infection (van Steeg *et al.*,1984). 26SCAT mRNA and AMVCAT mRNA both have a high affinity, whereas 42SCAT mRNA has a low affinity for these cap binding proteins, and the interactions appear to be correlated to the accessibility of the cap structure. The interaction of eIF-4E is influenced by structures present downstream of the cap structure, indicating that this part of the 5' UTR is involved in mRNA selection during the initial steps of translation.

Moreover, SFV capsid protein impairs efficient translation of host mRNAs both *in vitro* (van Steeg *et al.*,1984) and *in vivo* (Elgizoli *et al.*, 1989), indicating that a general mechanism is involved in the overall repression of protein synthesis by the viral capsid protein. The multiple interactions of SFV capsid protein with viral and cellular components can not account for the observed host shut off (Glanville and Ulmanen, 1976; Jalanko and Söderlund, 1985; Ulmanen *et al.*,1976). Nevertheless, the molecular mechanism underlying this translational inhibition by SFV capsid is not understood yet. It has also been reported that the viral capsid protein inhibits *in vitro* translation of 42S SFV RNA, while the translation of 26S SFV RNA is much less sensitive to the addition of capsid protein (van Steeg *et al.*,1984). Therefore, we studied the translational behaviour of these chimaeric mRNAs, containing the leader sequences of both RNAs of SFV in the presence of purified capsid protein as well.

EXPERIMENTAL PROCEDURES

Materials

As a source of initiation factors a crude ribosomal wash was used. Subfractionation of the ribosomal wash into a 0-40% ammonium sulphate fraction (A-wash) and a 40-70% ammonium sulphate fraction (BC-wash) was as described (Voorma *et al.*, 1979). Eukaryotic initiation factor eIF-3,-4B and -4F (nomenclature of initiation factors is according to the International Union of Biochemistry (Safer, 1989) were purified from rabbit reticulocyte lysates according to procedures described earlier (Voorma *et al.*, 1979; Amesz *et al.*, 1979; Buckley and Ehrenfeld, 1987).

Construction of DNA clones

General DNA techniques were used as described by Maniatis *et al.*, (1982). The constructs p26SCAT, p42SCAT, and pAMVCAT were made in pEMBL19 (Dente and Cortese, 1987) and contain the T7 promoter sequence derived from pET (Rosenberg *et al.*, 1987).They contain the prokaryotic reporter gene encoding CAT (Alton and Vapnek, 1979) downstream of the leader sequences of the subgenomic 26S RNA of SFV (Ou *et al.*, 1982; Riedel *et al.*, 1982), the genomic 42S RNA of SFV (Wengler *et al.*, 1979; Lehtovaara *et al.*, 1982) and RNA 4 of AMV (Brederode *et al.*, 1980). Due to the cloning procedures the RNAs produced *in vitro* have 2 additional G-residues at their 5' end (see Fig. 1). Construction of p42SCAT-TTG was accomplished by using single stranded p42SCAT DNA. Subsequently, the upstream ATG of p42SCAT, underlined in Fig. 1, was changed into TTG. In addition three nucleotides, i.e. ATC were inserted into p26SCAT, 43 nucleotides downstream of the transcription start site of the T7 promoter sequence, using Amersham site-directed mutagenesis system RPN.1523. All plasmids were checked by restriction enzyme analysis and the primary sequence of the different leaders was checked by DNA sequence analysis (see Fig. 1).

26S (GG)AUUGGUGCGUUAAUACACAGAAUUCUGAUUAUAGCGCACUAUUAUAGCACC**AUG**

42S (GG)UGGCGG<u>AUG</u>UGUGACAUACACGACGCCAAAAGAUUUUGUUCCAGCUCCUGCCACCUCCGCUACGCGAGAGAUUAACCACCCACC**AUG**

AMV (GG)UUUUUAUUUUUAAUUUUCUUUCAAAUACUUCCACC**AUG**

Fig. 1A

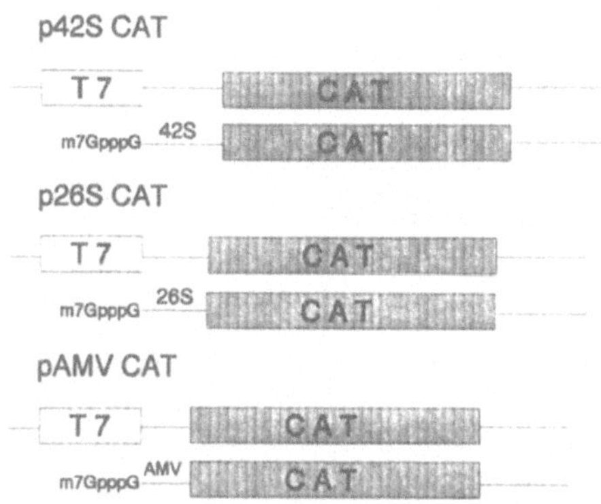

Fig. 1B

Figure 1. Sequence of the 5' untranslated regions of 26S-, 42S-, and AMV-CAT mRNAs.
Panel A: Nucleotides added to the 5' end during cloning procedures are in parentheses. The upstream AUG in the 5' UTR of 42SCAT mRNA is underlined, startcodons are in bold.
Panel B:The DNA constructs, 26S CAT, 42S CAT and AMV CAT, see Experimental procedures.

Preparation of capped and uncapped mRNA

Capped mRNAs were prepared in a 200 µl reaction mixtures containing 40 mM Tris/HCl pH 7.5, 10 mM dithiothreitol, 0.05 mM EDTA, 0.1 µg of BSA/µl, 11 mM $MgCl_2$, 0.5 mM cap analog m^7GpppG, 50 µM GTP, 1 mM UTP, 1 mM CTP, 1 mM ATP, 2 mM spermidine, 1 U of RNAsin/µl, 0.7-1 U of T7 RNA polymerase/µl and 0.08 µg of BamHI linearized DNA/µl. To prepare uncapped mRNAs, 100 µl assays were used, capanalog was omitted and the concentration of GTP was raised to 1 mM. After incubation for 2 hours at 37°C, 0.2 U of DNAse /µl was added and incubation was continued for 15 min. Free nucleotides and capanalog were removed using a 2-ml Sephadex G50-fine column. The mRNA present in the pooled fractions was precipitated with ethanol and after centrifugation the dried pellet was dissolved in sterile water. The amount of RNA recovered was determined by measuring the absorbance at 260 nm (Maniatis et al, 1982). Non-denaturing agarose-gel electrophoresis indicated that all mRNAs were intact. In order to check the capping efficiencies, internally labelled leaders were made which were examined as described (Konarska *et al.*, 1984).

Translation *in vitro*

1. Translation in reticulocyte lysate was performed as described (Pelham and Jackson, 1976), in a total volume of 10 µl containing 2 µCi of [^{35}S]-methionine (1000 Ci/mmol), 0.04 µg of tRNA/µl, 0.05 mM amino acids minus methionine, 144 mM potassium acetate and 0.4 mM magnesium acetate. Incubation was for 1 hour at 30°C.
2. Translation in the pH5 system was performed as described (Thomas *et al.*, 1980). In short, 10 µl assays contained 20 mM HEPES/KOH pH 7.6, 1 mM ATP(Tris)$_3$, 0.4 mM GTP(Tris)$_3$, 5 mM creatine phosphate(Tris)$_2$, 0.05 unit of creatine kinase, 2 mM dithiothreitol, 120 mM potassium acetate, 2 mM magnesium acetate, 100 µM spermine, 50 µM amino acids minus methionine, 2 µCi of [^{35}S]-methionine, 0.4 µg of tRNA, 2.5 µg of 'pH5 enzymes', 0.12 A_{260} units of ribosomal subunits and initiation factors as indicated. Incubation was for 1 hour at 37°C. After translation, 2 µl aliquots were taken to determine hot-trichloroacetic-acid-precipitable radioactivity. The products were analysed on 12.5% SDS-polyacrylamide gels (Laemmli, 1970).

Cross-linking of initiation factors to mRNA

1. Ultraviolet light-induced cross-linking of factors to mRNA
Labelled mRNAs (about 0.11 µg≈10,000-20,000 cpm) were incubated with initiation factors (ribosomal salt wash or purified eIFs) for 10 min at 30°C in 10 µl incubation mixtures containing 12 mM Tris/HCl pH 7.6, 6% glycerol, 1 mM magnesium acetate, 4.2 mM 2-mercaptoethanol, and 2 mM ATP. Ultraviolet light-induced cross-linking was carried out as described (van Heugten *et al.*, 1991). Proteins labelled by cross-linking to the radioactive RNA were located by autoradiography. Quantification was performed by scanning the autoradiogram with a laser scanning densitometer (LKB Ultroscan, Gelscan XL 2.0) or using the PhosphorImager Type 400A (Molecular Dynamics).
2. Chemical cross-linking of factors to mRNA
In this assay ^{32}P-cap-labelled and $NaIO_4$ oxidized mRNAs were used. Because of the reactivity of the oxidized cap structure with primary amines, initiation factors were dialyzed against 20 mM HEPES/KOH pH 7.6, 100 mM potassium acetate, 10% glycerol and 1 mM dithiothreitol. The incubation mixture is identical to the one described above except that 12 mM Tris was replaced by 20 mM HEPES/KOH pH 7.6. Following incubation cross-linked complexes formed between protein and mRNA were stabilized by reduction with 10 mM $NaBH_3CN$ for 3 hr at 0°C (Sonenberg and Shatkin, 1977). Treatment with RNAse A and analysis of radioactive labelled proteins was performed as described above.

Isolation of SFV capsid protein

Semliki Forest virus was propagated in BHK-21 cells and purified as described (van Steeg *et al.*, 1979). Virus containing gradient fractions were pooled, centrifuged for 2 hours at 27,000 rpm in a SW27 rotor and the viral pellet was used to isolate the capsid protein by cation-exchange chromatography using S Sepharose Fast Flow (Pharmacia) according to procedures described by Omar and Koblet (1985). The protein was resuspended in water and protein concentration was determined according to the method of Layne (Layne, 1957).

RESULTS

Previous studies with the mRNAs of SFV showed that the genomic 42S RNA is a rather poor template for *in vitro* translation, in contrast to the subgenomic 26S viral mRNA (van Steeg *et al.*, 1981a) In order to define sequences responsible for this differential behaviour, we fused the 5' untranslated regions of both mRNAs to the bacterial reporter gene encoding CAT. Using the same reporter gene, problems concerning differences in the methionine content of the various proteins are avoided (Wengler and Wengler, 1975). The nucleotide sequences of the 5' UTRs of the chimaeric transcripts, together with the constructs, are shown in Figs. 1A and 1B.

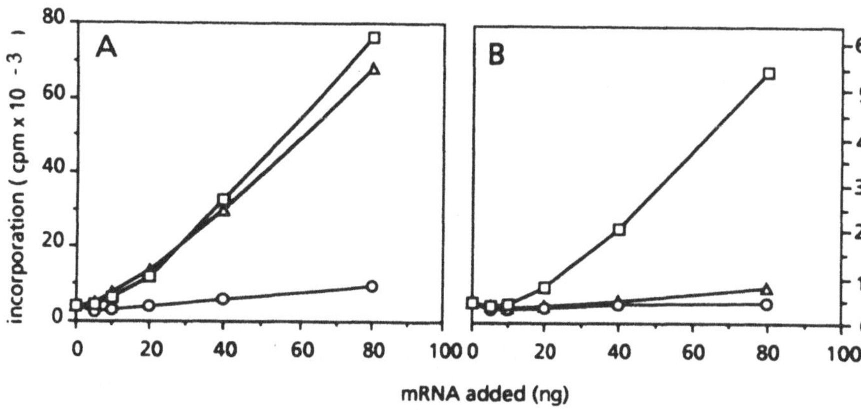

Figure 2. Translation of mRNAs in reticulocyte lysate.
Panel A: capped and Panel B: uncapped transcripts, Reticulocyte lysates were supplemented with 26SCAT mRNA: (Δ), 42SCAT mRNA: (O) or AMVCAT mRNA: (□), (amounts indicated) [^{35}S]-methionine incorporation was determined in 2 μl aliquots.

To gain further insight into the translational behaviour of capped and uncapped 26SCAT mRNA, 42SCAT mRNA and AMVCAT mRNA, protein synthesis was carried out in a micrococcal nuclease-treated reticulocyte lysate (Pelham and Jackson, 1976). The translational efficiency of 26SCAT mRNA is comparable to that of AMVCAT mRNA (Fig. 2A), however, capped 42SCAT mRNA is profoundly impaired in protein synthesis, irrespective of the presence of a cap structure (Fig. 2B). In addition, the translation of 26SCAT mRNA appears to be fully dependent on the presence of a cap structure (c.f. Figs. 2A and 2B), which is in line with observations obtained with chemically decapped viral 26S RNA (van Steeg *et al.*, 1981a).

Uncapped AMVCAT mRNA is translated almost equally efficient compared to capped AMVCAT transcripts (Figs. 2A+B), which is in agreement with data presented by Gehrke *et al.*, (1983). All three mRNAs direct only the synthesis of CAT-protein. In order to find out whether the SFV viral leader-CAT-fusion transcripts resemble their corresponding viral counterparts, the sensitivity of translation for capanalog was determined. The results presented in Fig. 3 show that in the presence of m^{7}GpppG capped 26SCAT mRNA is translated efficiently, whereas 42SCAT mRNA translation is sensitive to the addition of capanalog. Thus we are inclined to conclude that the different translational behaviour of the fusion transcripts resembles the situation when authentic SFV RNAs are translated (van Steeg *et al.*, 1981a; van Steeg *et al.*, 1981b; van Steeg *et al.*, 1979) and therefore may function as model messengers to study the underlying mechanisms of SFV-induced host shut off.

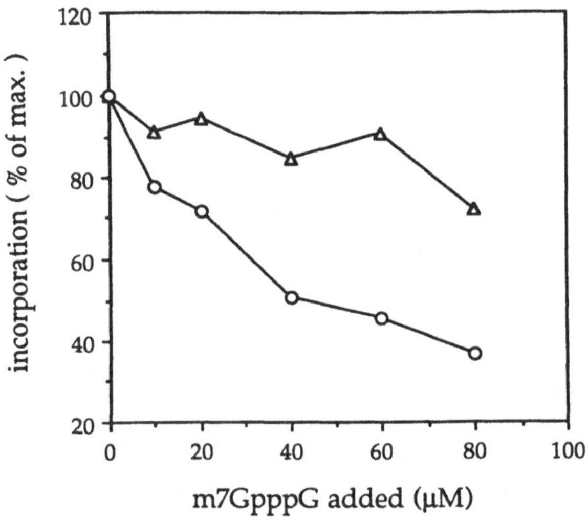

Figure 3. Effect of capanalogue on the translation of different mRNAs
pH5 systems were supplemented with 40 ng of mRNA, 0.38 μg of ribosomal wash fraction A, 3.88 μg of ribosomal wash fraction fraction BC and increasing amounts of capanalogue as indicated. Incorporation of [^{35}S]-methionine without added capanalogue was arbitrarily set at 100%, i.e. 26SCAT mRNA: (Δ) 186,571 cpm/2μl and 42SCAT mRNA: (O) 92,609 cpm/2μl.

Because the reticulocyte lysate is not a suitable system for determining factor dependence, translation was therefore performed with purified components (pH5 system). In Fig. 4 it is shown that 42SCAT mRNA is unable to support translation with the same efficiency as that observed with 26SCAT mRNA, even when relatively high amounts of eIF-3, eIF-4F and eIF-4B are added. These data suggest that the poor translation of 42SCAT mRNA is a feature of this mRNA, most likely determined by the 5' UTR of 42S SFV RNA.

One possible reason for the observed inefficient translation of 42SCAT mRNA may be the presence of an upstream ORF. The leader sequence of 42S RNA contains an open reading frame of 22 triplets including one encoding methionine and three encoding cysteine (Lehtovaara *et al.*, 1982).

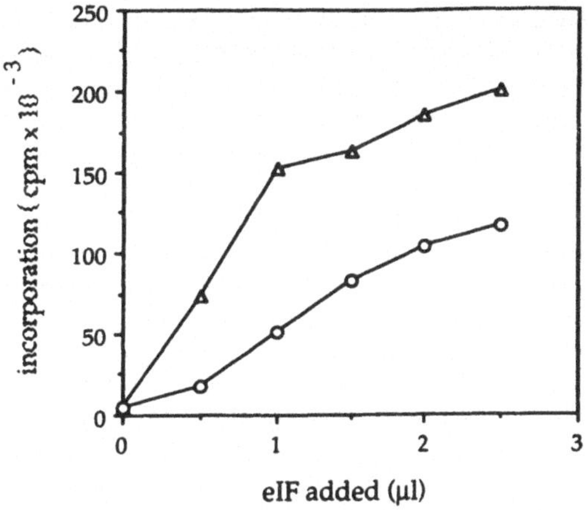

Figure 4. eIF-dependence of different transcripts in pH5 systems
pH5 systems were supplemented with 20 ng capped of mRNA. eIF-3, -4B, -4F-dependence. The reaction mixtures contain 7.8 µg of ribosomal salt wash fraction BC and amounts of eIF-3 plus eIF-4F and eIF-4B as indicated. 1 µl contains 0.8 µg of eIF-3 plus eIF-4F and 0.1 µg of eIF-4B.After translation was stopped 2 µl aliquots were taken to determine hot TCA-precipitable radioactivity. Symbols: 26SCAT mRNA: (Δ), 42SCAT mRNA: (O).

We wanted to establish whether this ORF impedes the translation of the downstream CAT cistron. Consequently, the 5' proximal AUG was changed into an UUG codon, which appears to be a very inefficient startcodon (Peabody, 1989). Together with 42SCAT mRNA and 26SCAT mRNA this mutated 42SCAT-UUG mRNA was translated in the reticulocyte lysate. The results presented in Fig. 5 show that the inefficient translation of 42SCAT mRNA is not improved at all by inactivating this 5' AUG codon. Apparently, another mechanism governs the reduced rate of translation of 42SCAT mRNA.

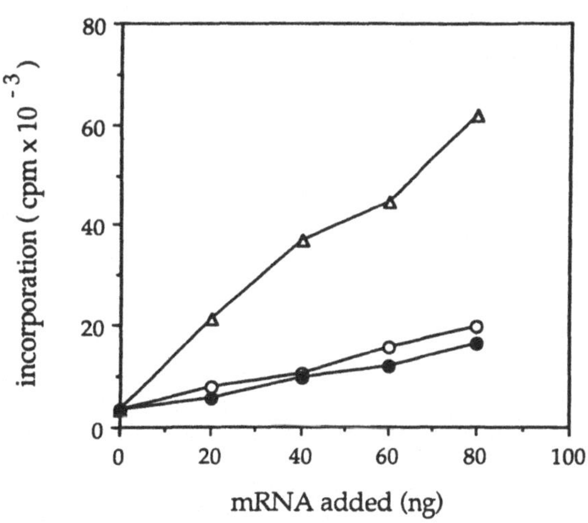

Figure 5. Translation of 26SCAT, 42SCAT and 42SCAT-UUG mRNA in reticulocyte lysate.
A) Translation was performed in 10 µl assay as described in the experimental section. Lysates were programmed with amounts as indicated: 26SCAT mRNA: (Δ), 42SCAT mRNA: (O), and 42SCAT-UUG mRNA: (●)

In order to establish whether the low factor need of 26SCAT mRNA is caused by an efficient binding of eIFs to the mRNA and in particular to the 5' cap structure, we performed mRNA-protein cross-linking experiments. A typical result using crude eIFs and 26SCAT mRNA versus 42SCAT mRNA (AMVCAT mRNA is used as control) is depicted in Fig. 6. Cap-specific cross-linking occurred primarily to eIF-4B (80 kD), whereas the eIF-4E band (24 kD) is less pronounced (van Heugten *et al.*, 1991), both features well-known for the photochemical cross-linking technique (Pelletier and Sonenberg, 1985). The addition of increasing amounts of crude initiation factors enhances the extent of cross-linking of proteins to all mRNAs. However, at non-saturating amounts of proteins, cross-linking of eIF-4B to 26SCAT mRNA and AMVCAT mRNA is more efficient as compared to cross-linking of eIF-4B to 42SCAT mRNA (c.f. Fig. 6, lanes 1 and 9 with 5). To investigate whether this is also valid under competitive conditions, the extent of cross-linking of cap binding proteins to mRNAs was also determined in the presence of increasing amounts of unlabelled globin mRNA (Fig. 7). Evidently, 42SCAT mRNA which interacts inefficiently with eIF-4B and eIF-4E, appears also to be a far weaker competitor with globin mRNA as compared to mRNAs which interact very efficiently with these factors. We, therefore, conclude that 26SCAT mRNA and AMVCAT mRNA have a high affinity for cap binding proteins, lending further support to our previous observations that 26SCAT mRNA and AMVCAT mRNA are efficient templates in in vitro translation assays.

Additional proteins can interact with AMVCAT mRNAs, which are not detected in cross-linking assays containing 26SCAT mRNA, 42SCAT mRNA (Figs. 6 and 7). Although we did not investigate these proteins

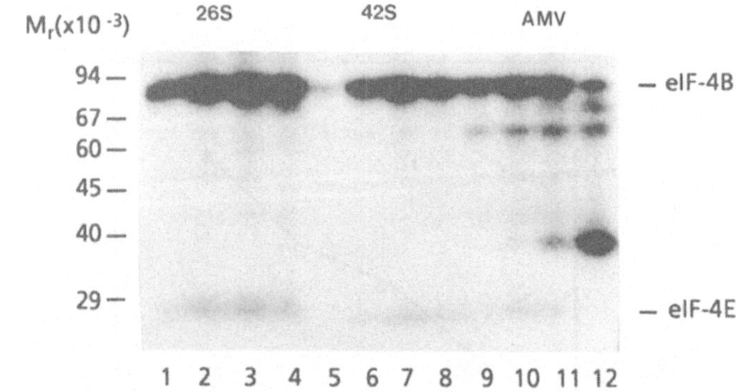

Figure 6. UV cross-linking of crude initiation factors to different mRNAs
Messenger RNA, labelled in the cap structure (17,000 cpm≈0.11 μg), was incubated for 10 min at 30°C with increasing amounts of ribosomal wash subfractions in the presence of 2 mM ATP, 1 mM magnesium acetate, and 60 mM potassium acetate. Based on yields of A- and BC-wash upon isolation, normal A-/BC-wash ratio was taken to be 1:0.7. UV light-induced cross-linking and analysis of the cross-linked samples was performed as described in the experimental section. Lanes 1, 5, and 9: 2.2 μg ABC-wash, lanes 2, 6, and 10: 5.6 μg ABC-wash, lanes 3, 7, and 11: 14 μg ABC-wash, and lanes 4, 8, and 12: 33.6 μg ABC-wash. The mRNAs used are: 26SCAT mRNA: lanes 1-4; 42SCAT mRNA: lanes 5-8; AMVCAT mRNA: lanes 9-12.

further, we assume that they are specific for RNAs containing the leader sequence of RNA 4 of AMV. Especially, the protein band with an apparent molecular weight of 67 kD interacts also specifically with other chimaeric mRNAs containing the 5' UTR of RNA 4 of AMV (results not shown).

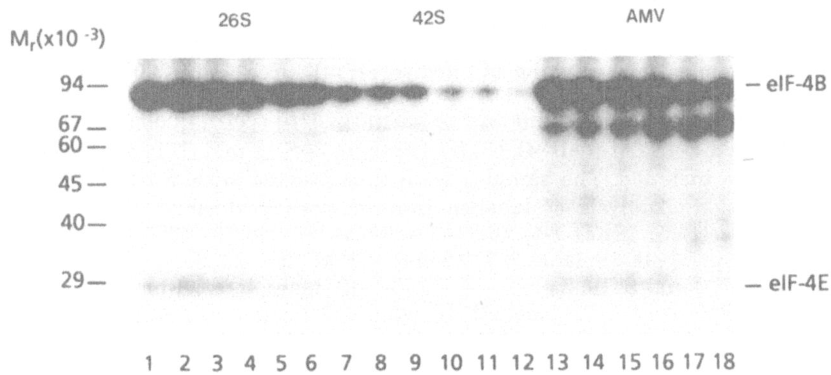

Figure 7. UV cross-linking of crude eIFs to mRNA labelled in their cap structure: competition with unlabelled globin mRNA

Cap labelled messengers (10,000 cpm≈0.05 μg≈0.2 pmol) were incubated with 3 μg A-wash, 2.6 μg BC-wash, 2 mM ATP, 1 mM magnesium acetate, 60 mM potassium acetate, and increasing amounts of competitor RNA. Cross-linking and analysis of labelled proteins was performed as described in Experimental Procedures. The autoradiogram shows in lanes 1, 7 and 13: no competitor RNA added, lanes 2, 8 and 14: 0.1 pmol globin mRNA, lanes 3, 9 and 15: 0.2 pmol globin mRNA, lanes 4, 10 and 16: 0.4 pmol globin mRNA, lanes 5, 11 and 17: 0.6 pmol globin mRNA, and lanes 6, 12 and 18: 0.8 pmol globin mRNA. Lanes 1-6: 26SCAT mRNA, lanes 7-12: 42SCAT mRNA, lanes 13-18: AMVCAT mRNA.

It has been shown that eIF-4E plays an antagonistic role in the translational inhibition caused by the SFV-encoded capsid protein (van Steeg *et al.*, 1984). Furthermore, it has been reported that most of eIF-4E is not complexed with p220 (Buckley and Ehrenfeld, 1987; Hansen *et al.*, 1982). To establish whether eIF-4E, either free or bound to p220, binds differentially to mRNAs, we studied the interaction of this protein with 26SCAT mRNA and 42SCAT mRNA. It is known that the efficiency of UV light-induced cross-linking of the 24 kD protein is significantly lower than that with chemical cross-linking (Pelletier and Sonenberg, 1985). Hence, to increase the eIF-4E signal, the chemical cross-linking technique was used to establish whether both eIF-4E and -4F can discriminate between the mRNAs containing the different leader sequences of Semliki Forest virus. Consistent with a discriminatory function of eIF-4F (Ray *et al.*, 1983; Sarkar *et al.*, 1984), the extent of cross-linking of its smallest subunit, eIF-4E, to different mRNAs varies: 26SCAT mRNA has a high affinity, whereas 42SCAT mRNA has a low affinity for the 24 kD subunit of eIF-4F (c.f. Fig. 8, lanes 1-5 with lanes 6-10). These data confirm the role of eIF-4E either free or complexed in eIF-4F in mRNA discrimination.

It is known that mRNA translation in vitro strongly depends on ionic conditions, especially the K^+-concentration. Furthermore, it has been established that different species or classes of mRNAs respond differently to

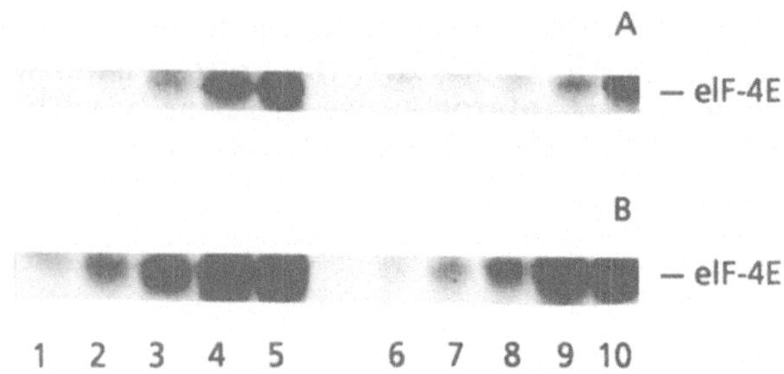

Figure 8. Chemical cross-linking of purified eIF-4E, free and p220 bound, to ^{32}P cap-labelled transcripts.
Oxidized mRNA labelled at the 5' cap site with [a-^{32}P]-GTP (15,000 cpm≈0.11 μg) was incubated with eIF-4E (A) or eIF-4F (B) for 10 min at 30°C; 2 mM CTP, 1 mM $MgAc_2$, and 60 mM KAc were included. The samples were then treated with $NaBH_3CN$, followed by RNAse digestion. Analysis of labelled proteins was performed as described in the experimental section. Lanes 1-5: 26SCAT mRNA; lanes 6-10: 42SCAT mRNA. Fig. 8A: lanes 1, 6: 0.25 pmol eIF-4E, lanes 2, 7: 0.5 pmol eIF-4E, lanes 3, 8: 1 pmol eIF-4E, lanes 4, 9: 1.5 pmol eIF-4E, lanes 5, 10: 2 pmol eIF-4E. Fig. 8B: lanes 1, 6: 0.1 pmol eIF-4F, lanes 2, 7: 0.2 pmol eIF-4F, lanes 3, 8: 0.4 pmol eIF-4F, lanes 4, 9: 0.6 pmol eIF-4F, lanes 5, 10: 0.8 pmol eIF-4F.

these ionic conditions (Brendler *et al.*, 1981a; Brendler *et al.*, 1981b; Carrasco *et al.*, 1979). It is also known that increasing amounts of K^+ stabilize the secondary structure of mRNAs (Fukushima *et al.*, 1988; Holder and Lingrel, 1975; Kozak, 1989). Ribosome binding depends probably on prior interaction of CBPs with an accessible cap structure on mRNA. Hence, it is reasonable to assume that the secondary structure in the 5' UTR of mRNAs is able to prevent the interaction of cap recognition factors with the cap structure and consequently impedes binding. To explore this the effect of increase in the secondary structure by K^+-ions on the binding of eIF-4B and eIF-4E to different mRNAs, labelled exclusively in their 5' cap structures, was monitored. From the autoradiogram given in Fig. 9 it seems that cross-linking of eIF-4B to all tested cap-labelled mRNAs decreases with increasing potassium acetate concentrations, which is feasible when secondary structure is enhanced by higher salt concentrations (Fukushima *et al.*, 1988; Holder and Lingrel, 1975; Kozak, 1989). Quantification of the radioactivity present in the eIF-4B bands indicates that the deleterious effect of the increased salt concentration on eIF-4B cross-linking is more pronounced with 42SCAT mRNA and reovirus mRNA than with 26SCAT mRNA and AMVCAT mRNA. This reinforces previous findings that the cross-linking of eIF-4B is dependent on the degree of secondary structure present in the 5' UTR (Pelletier and Sonenberg, 1985). The data presented in Fig. 9 also suggest that the 5' proximal sequence of 26SCAT mRNA contains almost no secondary structure as has been already reported for RNAs containing the leader sequence of AMV RNA 4 (Gehrke *et al.*, 1983; Gehrke,1986).

In conclusion, the observations presented here, support the idea that the leader sequence of 26S RNA of SFV is devoid of secondary structure at its extreme 5' end, leading to an efficient recognition of this mRNA by cap binding proteins. In contrast, the cap structure present at the 5' end of

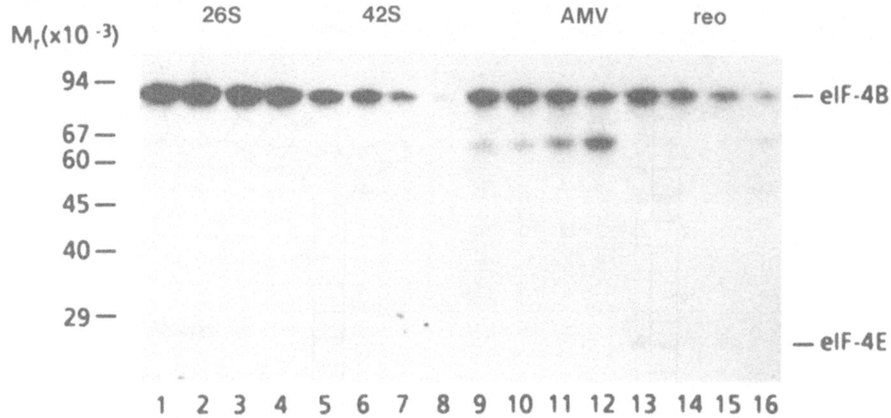

Figure 9. Effect of K^+-concentration on UV light-induced cross-linking of proteins to different mRNAs
^{32}P-mRNA (cap labelled) (17,000 cpm≈0.11 μg) was incubated with 3 μg A-wash and 2.6 μg BC-wash in the presence of 2 mM ATP and 1 mM $MgAc_2$ for 10 min at 30°C. The mixtures were UV irradiated for 30 min at 4°C. KAc was added to give final concentrations of 20 mM (lanes 1, 5, 9 and 13), 40 mM (lanes 2, 6, 10 and 14), 80 mM (lanes 3, 7, 11 and 15), and 160 mM (lanes 4, 8, 12 and 16). The mRNAs used are: 26SCAT mRNA: lanes 1-4; 42SCAT mRNA: lanes 5-8; AMVCAT mRNA: lanes 9-12, and reovirus mRNA: lanes 13-16.

42SCAT mRNA appears to be masked by secondary structure and therefore is rather unaccessible to cap binding proteins. The results presented here are in accordance with computer predicted secondary structures of the leader sequences of 26S RNA and 42S RNA of SFV, see Fig. 10.

The SFV-encoded capsid protein seems to play a crucial role in the SFV-induced shut off of cellular protein synthesis (van Steeg *et al.*, 1984; Elgizoli *et al.*, 1989). This structural polypeptide inhibits host protein synthesis *in vitro* (van Steeg *et al.*, 1984). and *in vivo* (Elgizoli *et al.*, 1989), as well as translation of 42S RNA. The mechanism by which the capsid protein inhibits translation of host mRNAs and 42S RNA, leaving the late 26S RNA directed translation intact, has not been elucidated yet.The inhibitory effect on translation was examined in translation systems directed with 26SCAT mRNA, 42SCAT mRNA or AMVCAT mRNA. Fig. 11 shows a typical result of such an experiment. The inhibitory effect of capsid protein is only pronounced when 42SCAT mRNA is used as template. The results are consistent with previous findings in which a similar differential inhibition of translation of viral 26S RNA and 42S RNA was noticed, together with restoring activities of eIF-4E and eIF-4B in similar assay systems (van Steeg *et al.*, 1984). Therefore, we are inclined to conclude that the inhibitory effect of the capsid protein on *in vitro* translation, of host as well as early 42S SFV mRNAs, leading to SFV-induced host shut off, is exerted at the level of the 5' UTR.

Since eIF-4F and eIF-4B play a crucial role in cap recognition (Sonenberg, 1988), it is tempting to speculate that the capsid protein interferes with these initiation factors for binding to capped mRNAs. To explore this further, the chemical cross-linking technique was used. The capsid protein interacts with cap-labelled mRNA in a specific manner: identical to the behaviour of eIF-4E (Pelletier and Sonenberg, 1985), the capsid protein interacts with m^7GpppG...-ended mRNA and not with GpppG...-ended mRNA (Fig. 12A). The affinity of capsid protein for the cap structure of mRNA was investigated by chemical cross-linking of capsid

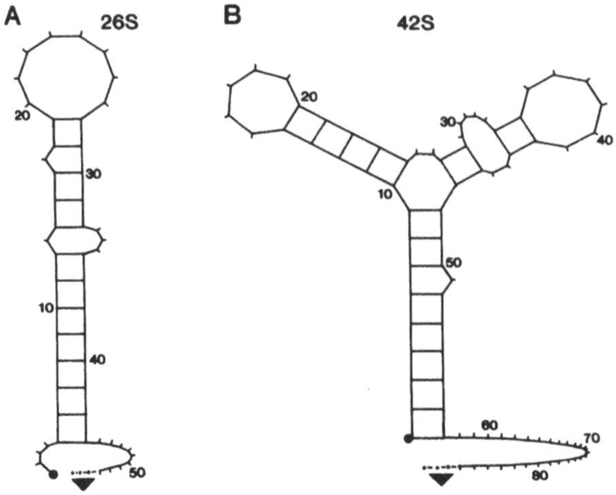

Figure 10. Computer predicted secondary structure of the 5' UTR of 26S and 42S RNA of SFV.
The 5' UTRs of 26S RNA (A) and 42S RNA (B) were subjected to secondary structure analysis using the computer program FOLD (Devereux *et al.*, 1984). The predicted free energy (ΔG) of the structures is -43.3 kJ/mol and -72.7 kJ/mol for the 5' UTR of 26S SFV RNA and 42S SFV RNA, respectively.

protein in the presence of increasing amounts of the capanalog m^7GpppG. Fig. 12B shows the autoradiogram from which can be concluded that binding of capsid protein to mRNA is inhibited by the addition of m^7GpppG (lanes 1-6), but is not affected by 0.6 mM m^7GTP (lane 7). These results indicate that capsid protein interacts with mRNA provided that an intact cap structure is present and also that the capsid protein binds with high affinity.

It is concluded that the general inhibitory action of the capsid protein is caused by its high affinity for methylated cap structures, implying that all capped (host) mRNAs become inhibited in their translation. However, earlier findings (van Steeg *et al.*, 1984) and the results presented in Fig. 11 indicate that SFV 26S RNA, 26SCAT mRNA and AMVCAT mRNA are relatively resistant to the inhibitory action of the capsid protein. We showed already that these messages, in contrast to 42SCAT mRNA, display a high affinity for cap binding proteins as well. Therefore, it was of interest to see whether a 26SCAT-derivative, which interacts inefficiently with CBPs, becomes sensitive to the translational inhibition by the capsid protein. To this end, the three proximal 5' nucleotides (excluding the cap structure) in 26SCAT mRNA, which most likely are single stranded , were forced to become part of the small hairpin structure of the leader by inserting a ATC-sequence 43 nucleotides downstream from the cap structure. In this derivative of 26SCAT mRNA the cap structure is most probably sequestered because of the proximity to the base-paired region and are not available for interaction with CBPs . The translational efficiency of this 26SCAT-AUC mRNA is dramatically decreased compared to that of 26SCAT mRNA: its translational efficiency is comparable to that of 42SCAT mRNA. These data confirm the prediction made earlier that mRNAs that interact efficiently with CBPs i.e. 26SCAT mRNA and AMVCAT mRNA, are almost insensitive to the inhibitory action of the capsid protein compared to mRNAs which interact inefficiently with CBPs i.e. 42SCAT mRNA and 26SCAT-AUC mRNA.

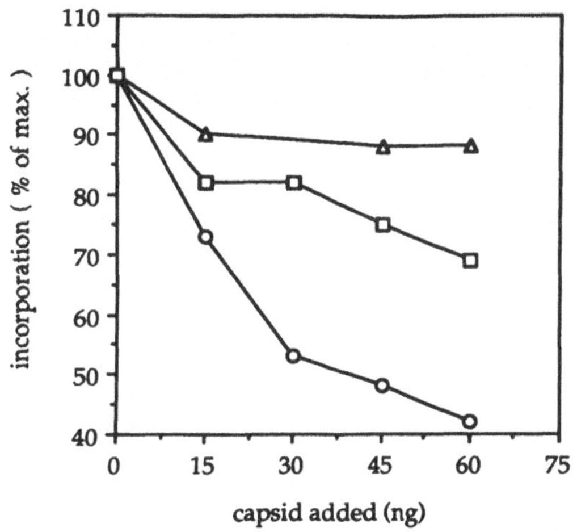

Figure 11. Selective inhibitory effect of capsid protein on translation of different mRNAs Increasing amounts of pure capsid protein (as indicated) were added to pH5 systems (10 µl) containing 0.38 µg of A-wash and 3.88 µg of BC-wash. Protein synthesis was directed by 40 ng of 26SCAT mRNA (Δ), 42SCAT mRNA (O), or AMVCAT mRNA (□). The effect on protein synthesis was determined by analysis of 4 µl samples by SDS-PAGE and quantification of the radioactivity incorporated into CAT protein, was performed with the PhosphorImager. For each RNA the CAT protein band in the absence of capsid was taken as reference point (100%).

DISCUSSION

In this study we show that the translation of 42SCAT mRNA is inefficient, sensitive to capanalog, and needs much higher amounts of eIFs. In contrast 26SCAT mRNA translation is comparable to the efficient translation of AMVCAT mRNA, its translation is relatively insensitive to capanalog, and needs low amounts of eIF-3, -4B and -4F (Figs. 2, 3 and 4). All these features resemble those found for the intact viral SFV RNAs (van Steeg *et al.,* 1979; 1981a; 1981b). Therefore, the intrinsic properties of 26S and 42S viral RNA most likely reside in the untranslated regions of the two messengers, implying that the chimaeric RNAs may be used to unravel the mechanisms of SFV-induced host shot off.

The question of address is what is the difference between the 5' UTR of 26S RNA and 42S RNA, making 26SCAT mRNA an efficient messenger and 42SCAT mRNA an inefficient template. The upstream AUG present in the 5' UTR of 42SCAT mRNA seem to be irrelevant in this respect (Fig.5). Differences in secondary structure, however, may account for the observed differences in translational behaviour of 26SCAT mRNA and 42SCAT mRNA. Stable hairpin structures in 5' leader sequences are thought to inhibit translation by preventing scanning of 40S ribosomal subunits (Pelletier and Sonenberg, 1985) or impeding the recognition by initiation factors (Lawson *et al.*, 1986). Fig. 10 shows the computer predicted secondary structure of the leaders of 26S RNA and 42S RNA. It is clear that those computer predicted secondary structures yield low minimal energies, i.e. -43.3 kJ/mol for the 5' UTR of 26S RNA and -72.7 kJ/mol for the leader sequence of 42S RNA. However, the extreme 5' nucleotides are thought to be single stranded in 26SCAT mRNA, whereas the extreme 5' terminal

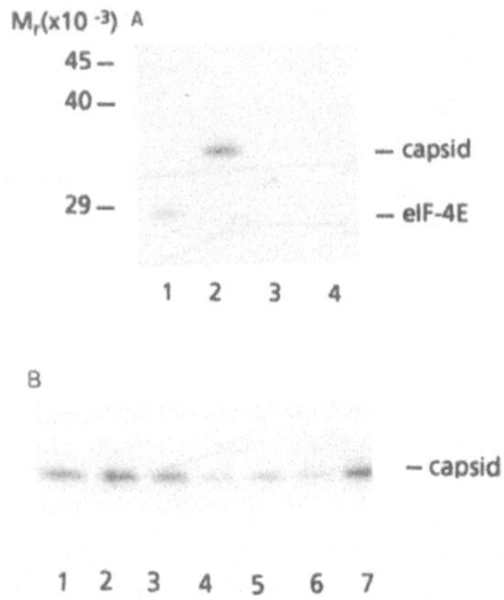

Figure 12. Chemical cross-linking of the capsid protein
A : Chemical cross-linking of capsid protein to ^{32}P-labelled 42SCAT mRNA
Oxidized mRNA labelled at their 5' end with [a-^{32}P]-GTP (10,000 cpm≈0.11 μg) was incubated with 80 ng eIF-4F (lanes 1, 3) or 30 ng capsid protein (lanes 2, 4) for 10 min at 30°C in the presence of 2 mM CTP, 1 mM magnesium acetate and 60 mM potassium acetate. The samples were then treated with NaBH3CN, followed by RNAse digestion and analysis by polyacrylamide gelelectrophoresis. Lanes 1+2: m^{7}GpppG-42SCAT mRNA, lanes 3+4: GpppG-42SCAT mRNA.
B: Effect of capanalog on chemical cross-linking of capsid protein to mRNA
^{32}P-labelled 42SCAT mRNA (10,000 cpm≈0.11 μg) was incubated in the presence of different amounts of capanalogues and cross-linked under essentially the same conditions as described under section A. Lanes 1+2: no capanalog added, lane 3: 0.1 mM m^{7}GpppG, lane 4: 0.2 mM m^{7}GpppG, lane 5: 0.4 mM m^{7}GpppG, lane 6: 0.6 mM m^{7}GpppG, lane 7: 0.6 mM m^{7}GTP.

sequences of 42S RNA may be capable to mask the cap structure, as was suggested by Ou *et al.*, (1983). This finding together with other data presented in this study support the hypothesis that RNA folding at the extreme 5' end of 42SCAT mRNA impedes its translation.

It is proposed that mRNAs with little potential for the formation of hairpins in their 5' UTR, have a decreased dependence on a cap structure and cap binding proteins for efficient translation (Gehrke *et al.*, 1983; Jobling and Gehrke, 1987; Altmann *et al.*, 1990; Edery *et al.*, 1984).The results obtained with AMVCAT mRNA are fully consistent with this hypothesis.

Therefore, the presence of a cap structure together with a low amount of secondary structure may be a combined prerequisite for late 26S RNA to escape SFV-induced cellular protein synthesis shut off. To substantiate this further the interaction of eIFs with the cap structures of 26SCAT mRNA and 42SCAT mRNA, the interference of the SFV capsid protein in translation initiation, and changes in the secondary structure of the leader of 26S RNA has been studied.

In order to get further insight into the mechanisms of SFV-induced host shut off, we investigated the direct interactions between the cap structure of transcripts containing the 5' UTRs of the RNAs of SFV and different cap recognition factors. We report here that 26SCAT mRNA and AMVCAT mRNA react very efficiently with eIF-4B and eIF-4E, most likely by their lack of secondary structure at their extreme 5' end (Figs. 6 and 7). This indicates that the observed difference in eIF recognition between 26S and 42S mRNA results from differences in their 5' UTR. Furthermore, under *in vitro* competitive conditions (Fig. 7), simulating *in vivo* competition between SFV mRNAs and host mRNAs, 26SCAT mRNA, as well as AMVCAT mRNA, still interacts efficiently with cap binding proteins, this is in contrast to 42SCAT mRNA. We, therefore, conclude that 26S SFV RNA is efficiently recognized by CBPs and that this may be the main reason for its preferential translation late in infection.

Several times it has been suggested that a lack of secondary structure at the extreme 5' end of viral mRNAs may reduce the requirement for cap binding proteins (Dolph *et al.*, 1988; Dolph *et al.*, 1990). Unfortunately, only indirect proof has been obtained by using polio-infected lysates (Lee and Sonenberg, 1982) or polio-infected cells (Dolph *et al.*, 1988; Dolph *et al.*, 1990). Here, we have shown unambiguously, by studying the direct interactions between the cap structures of the mRNAs and cap binding proteins, that the most efficient messages are also the ones which interact most efficiently with CBPs, similar conclusions were reported recently by Goss *et al* ., 1990; Carberry *et al.*, 1991.

Noteworthy is that the translational inhibition is more pronounced with 42SCAT mRNA as compared to 26SCAT mRNA and AMVCAT mRNA, Fig. 11. Viral 42S RNA translation is also sensitive to the inhibitory action of the capsid protein (van Steeg *et al.*, 1984). Although quite different locations have been found for the encapsidation signal present on 42S RNA (Jalanko and Soderlund, 1985; van Duijn, 1987; Weiss *et al.*, 1989), it is not clear whether the translational inhibition of 42SCAT mRNA by the capsid protein is due to this signal, of which a more downstream position is proposed. Hardly any translational inhibition is observed for 26SCAT mRNA and AMVCAT mRNA; apparently these messages are able to escape by the single stranded nature of their 5' leader the interaction with the capsid protein. Provided the presence of a structured 5' UTR the binding of the capsid protein to capped mRNAs is strong and this is consistent with its regulatory function in host shut off. The addition of eIF-4F and eIF-4B can relieve inhibition caused by the capsid protein. The first factor to interact with an initiating mRNA is thought to be eIF-4E, the smallest subunit of eIF-4F and after hydrolysis of ATP a stable complex of initiation factors and mRNA can be formed which may involve eIF-4A and eIF-4B (Lawson *et al.*, 1986). The restoring activity of eIF-4F may, therefore, be explained by virtue of its function (Ray *et al.*, 1983): it is the first factor to interact with mRNA, so probably capsid protein must compete with eIF-4F for available cap structures. Addition of eIF-4B can also relieve the translational inhibition caused by the capsid protein. This factor stimulates the ATPase and RNA-helicase activities of eIF-4A and eIF-4F (Pelletier and Sonenberg, 1985) and it also functions in the recycling of the p24 component of eIF-4F (Ray *et al.*, 1986) At present it is unknown how eIF-4B relieves the inhibition, although from the experiments presented a picture emerges that the capsid protein displays a high affinity for the 5' UTR, the capstructure included, provided the presence of secondary structure.

Acknowledgements

The support in the preparation of the manuscript and in the preparation of factors by Adri Thomas, Harry van Steeg, Marcellé Kasperaitis and Cor van der Mast is gratefully acknowledged.

REFERENCES

Altmann, M., Blum, S., Wilson, T.M.A. and Trachsel, H., 1990, The 5'-leader sequence of tobacco mosaic virus RNA mediates initiation-factor-4E-independent, but still initation-factor-4A-dependent translation in yeast extracts, Gene. 91:127.

Alton, N.K. and Vapnek, D., 1979, Nucleotide sequence analysis of the chloramphenicol resistance transposon Tn9, Nature (London). 282:864.

Amesz, H., Goumans, H., Haubrich-Morree, T.H.O., Voorma, T. and Benne, R., 1979, Purification and characterization of a protein factor that reverses the inhibition of protein synthesis by the heme-regulated translational inhibitor in rabbit reticulocyte lysate, Eur J Biochem. 98 :513.

Brederode, F.T., Koper-Zwarthoff, E.C., and Bol, J.F., 1980, Complete nucleotide sequence of alfalfa mosaic virus RNA 4, Nucl Acids Res. 8:2213.

Brendler,T., Godefroy-Colburn, T., Carlill, R.D., and Thach, R.E., 1981, The role of mRNA recognition in regulating translation, II. Development of a quantitative in vitro assay, J Biol Chem. 256:11747.

Brendler, T., Godefroy-Colburn, T., Yu, S., and Thach, R.E., 1981, The role of mRNA recognition in regulating translation, III. Comparison of in vitro and in vivo results, J Biol Chem. 256:11755.

Buckley, B., and Ehrenfeld, E., 1987, The cap-binding protein complex in uninfected and poliovirus-infected HeLa cells, J Biol Chem. 262:13599.

Buckley, B., and Ehrenfeld, E., 1987, The cap-binding protein complex in uninfected and poliovirus-infected HeLa cells, J Biol Chem. 262:13599.

Carberry, S.E., Friedland, D.E., Rhoads, R.E., and Goss, D.J., 1991, Binding of protein synthesis initiation factor 4E to oligoribonucleotides: Effects of cap accessibility and secondary structure, Biochemistry 31:1427.

Carrasco, L., Harvey, R., Blanchard, C., and Smith, A.E., 1979, Regulation of translation of eukaryotic virus mRNAs, in: Modern Trends In Human Leukemia III, Springer Verlag, 277.

Dente, L., and Cortese, R, 1987, [9] pEMBL : a new family of single-stranded plasmids for sequencing DNA, Methods Enzymol. 155:111.

Devereux, J., Haeberli, P., and Smithies, O., 1984, A comprehensive set of sequence analysis programs for the VAX. Nucl Acids Res. 12:387.

Dolph, P.J., Huang, J., and Schneider, R.J., 1990, Translation by the adenovirus tripartite leader: elements which determine independence from cap-binding complex, J Virol. 64:2669.

Dolph, P.J., Racaniello, V., Villamarin, A., Palladino, F., and Schneider, R.J., 1988, The adenovirus tripartite leader may eliminate the requirement for cap-binding protein complex during translation initiation,J Virol. 62:2059.

Edery, I., Lee, K.A.W., and Sonenberg, N., 1984, Functional characterization of eukaryotic mRNA cap binding protein complex: effects on translation of capped and naturally uncapped RNAs, Biochemistry 23:2456.

Elgizoli, M., Dai, Y., Kempf, C., Koblet, H., and Michel, M.R., 1989, Semliki Forest virus capsid protein acts as a pleiotropic regulator of host cellular protein synthesis, J Virol. 63:2921.

Fukushima, T., Nitta, K., and Sugai, S., 1988, Conformation and conformational fluctuation of Bacillus stearothermophilus 5S ribosomal RNA, Int. J Biol Macromol. 10:25.

Gehrke, L., Auron, P.E., Quigley, G.J., Rich, A., and Sonenberg, N., 1983, 5'-Conformation of capped alfalfa mosaic virus ribonucleic acid 4 may reflect its independence of the cap structure or a cap-binding protein for efficient translation, Biochemistry 22:5157.

Gehrke, L., 1986, Preparation of mRNA transcripts for secondary structure analysis using SP6 polymerase, guanyltransferase, and preparative gel electrophoresis, Gene Anal Techn. 3:45.

Glanville, N., and Ulmanen, I., 1976, Biological activity of in vitro synthesised protein: binding of Semliki Forest virus capsid protein to the large ribosomal subunit, Biochem Biophys Res Comm. 71:393.

Goss, D.J., Carberry, S.E., Dever, T.E., Merrick, W.C., and Rhoads, R.E., 1990, Fluorescence study of the binding of m^7GpppG and rabbit globin mRNA to protein synthesis initiation factors 4A, 4E and 4F, Biochemistry 29:5008.

Hansen, J.L., Etchison, D.O., Hershey, J.W.B., and Ehrenfeld, E. (1982). Localization of cap-binding proteins in sub-cellular fractions of HeLa cells, Mol Cell Biol. 2:1639.

Holder, J.W., and Lingrel, J.B., 1975, Determination of secondary structure in rabbit globin messenger RNA by thermal denaturation, Biochemistry 14:4209.

Jalanko, A., and Söderlund, H., 1985, The repeated regions of Semliki Forest virus defective-interfering RNA interferes with the encapsidation process of the standard virus. Virology 141:257.

Jobling, S.A., and Gehrke, L, 1987, Enhanced translation of chimaeric messenger RNAs containing a plant viral untranslated leader sequence, Nature (London). 325:622.

Kääriäinen, L., and Söderlund, H., 1978, Structure and replication of a-viruses, Curr Top Microbiol Immunol. 82:15.

Konarska, M.M., Padgett, R.A., and Sharp, P.A., 1984, Recognition of cap structure in splicing in vitro of mRNA precursors, Cell 38:731.

Kozak, M., 1989, Circumstances and mechanisms of inhibition of translation by secondary structure in eucaryotic mRNAs, Mol Cell Biol. 9:5134.

Kozak, M., 1986, Regulation of protein synthesis in virus-infected animal cells, Adv Virus Res. 31:229.

Laemmli, U.K., 1970, Cleavage of structural proteins during the assembly of the head of bacteriophage T4, Nature (London). 227:680.

Lawson, T.G., Ray, B.K., Dodds, J.T., Grifo, J.A., Abramson, R.D., Merrick, W.C., Betsch, D.F., Weith, H.L., and Thach, R.E., 1986, Influence of 5' proximal secondary structure on the translation efficiency of eukaryotic mRNAs and their interaction with initiation factors, J Biol Chem. 261:13979.

Layne, E., 1957, Protein estimation by ultraviolet absorption, Methods Enzymol. 3:451.

Lee, K.A.W. and Sonenberg, N. 1982, Inactivation of cap-binding proteins accompanies the shut-off of host protein synthesis by poliovirus, Proc Natl Acad Sci USA. 79:3447.

Lehtovaara, P., Söderlund, H., Keränen, S., Pettersson, R.F., and Kääriäinen, L., 1982, Extreme ends of the genome are conserved and rearranged in the defective interfering RNAs of Semliki Forest Virus, J Mol Biol. 156:731.

Maniatis, T., Fritsch, E.F., and Sambrook, J., 1982, Molecular cloning , Cold Spring Harbor Laboratory, Cold Spring Harbor, New York.

Omar, A., and Koblet, H., 1985, One-step separation of the components of Semliki Forest virus by cation exchange chromatography, J Virol Methods 12:71.

Ou, J.-H., Rice, C.M., Dalgarno, L., Strauss, E.G., and Strauss, J.H., 1982, Sequence studies of several alphaviruses genomic RNAs in the region containing the start of the subgenomic RNA, Proc Natl Acad Sci USA. 79:5235.

Ou, J.H., Strauss, E.G., and Strauss, J.H., 1983, The 5'-terminal sequences of the genomic RNAs of several alphaviruses, J Mol Biol. 168:1.

Peabody, D.S., 1989, Translation initiation at non-AUG triplets in mammalian cells, J Biol Chem. 264:5031.

Pelham, H.R.B., and Jackson, R.J., 1976, An efficient mRNA-dependent translation system from reticulocyte lysates, Eur J Biochem. 67:247.

Pelletier, J., and Sonenberg, N., 1985, Insertion mutagenesis to increase secondary structure within the 5' noncoding region of a eukaryotic mRNA reduces translational efficiency, Cell. 40:515.

Pelletier, J., and Sonenberg, N., 1985, Photochemical cross-linking of cap binding proteins to eucaryotic mRNAs: effect of mRNA 5' secondary structure, Mol Cell Biol. 5:3222.

Pettersson, R.F., Söderlund, S., and Kääriäinen, L., 1980, The nucleotide sequences of the 5'-terminal T1 oligonucleotides of Semliki-Forest-Virus 42-S and 26-S RNAs are different, Eur J Biochem. 105:435.

Ray, B.K., Brendler, T.G., Adya, S., Daniels-McQueen, S., Miller, J.K., Hershey, J.W.B., Grifo, J.A., Merrick, W.C., and Thach, R.E., (1983). Role of mRNA competition in regulating translation: further characterization of mRNA discriminatory initiation factors. Proc Natl Acad Sci USA. 80:663.

Ray, B.K., Lawson, T.G., Abramson, R.D., Merrick, D.F. and Thach, R.E., 1983 Recycling of messenger RNA cap-binding proteins mediated by eukaryotic initiation factor 4B. J Biol Chem 261:11466

Riedel, H., Lehrach, H., and Garoff, H., 1982, Nucleotide sequence at the junction between the nonstructural and structural genes of the Semliki Forest Virus genome, J Virol. 42:725.

Rosenberg, A.H., Lade, B.N., Chui, D., Lin, S.-W., Dunn, J.J. and Studier, F.W., 1987, Vectors for selective expression of cloned DNAs by T7 RNA polymerase, Gene. 56:125.

Safer, B., 1989, Nomenclature of initiation, elongation and termination factors for translation in eukaryotes, Eur J Biochem. 186:1.

Sarkar, G., Edery, I., Gallo, R., and Sonenberg, N., 1984, Preferential stimulation of rabbit a globin mRNA by a cap-binding protein complex, Biochim Biophys Acta. 783:122.

Schneider, R.J., and Shenk, T., 1987, Impact of virus infection on host cell protein synthesis, Ann Rev Biochem. 56:317.

Sonenberg, N., and Shatkin, A.J., 1977, Reovirus mRNA can be covalently crosslinked via the 5' cap to proteins in initiation complexes, Proc Natl Acad Sci USA. 74:4288.

Sonenberg, N., 1988, Cap-binding proteins of eukaryotic messenger RNA: Functions in initiation and control of translation, Prog. Nucleic Acids Res Mol Biol. 35:173.

Thomas, A., Spaan, W., van Steeg, H., Voorma, H.O., and Benne, R., 1980, Mode of action of protein synthesis initiation factor eIF-1 from rabbit reticulocytes, FEBS Lett. 116:67.

Ulmanen, I., Söderlund, H., and Kääriäinen, L., 1979, Role of protein synthesis in the assembly of Semliki Forest virus nucleocapsid,Virology. 99:265.

Ulmanen, I., Söderlund, H., and Kääriänen, L., 1976, Semliki Forest virus capsid protein associates with the 60S ribosomal subunit in infected cells. J Virol. 20:203.

van Duijn, L.P., Kasperaitis, M., Ameling, C., and Voorma, H.O., 1986, Additional methylation at the N(2)-position of the cap of 26S Semliki Forest Virus late mRNA and initiation of translation,Virus Res. 5:61.

van Duijn, L.P., 1987, Ph.D Thesis, Utrecht University, Utrecht.

van Heugten, H.A.A., Kasperaitis, M.A.M., Thomas, A.A.M., and Voorma, -H.O., 1991, Evidence that eukaryotic initiation factor (eIF) 2 is a cap-binding protein that stimulates cap recognition by eIF-4B and eIF-4F, J Biol Chem. 266:7279.

van Steeg, H., Thomas, A., Verbeek, S., Kasperaitis, M., Voorma, W.O., and Benne, R., 1981, Shut off of neuroblastoma cell protein synthesis by Semliki Forest virus : loss of ability of crude initiation factors to recognize early Semliki Forest Virus and host mRNAs, J Virol. 38:728.

van Steeg, H., Kasperaitis, M., Voorma, H.O. and Benne, R., 1984, Infection of neuroblastoma cells by Semliki Forest virus. The interference of viral capsid protein with the binding of host messenger RNAs into initiation complexes is the cause of the shut-off of host protein synthesis, Eur J Biochem. 138:473.

van Steeg, H., van Grinsven, M., van Mansfeld, F., Voorma, W.O., and Benne, R., 1984, Initiation of protein synthesis in neuroblastoma cells by Semliki Forest Virus: a decreased requirement of late viral mRNA for eIF-4B and cap-binding protein, FEBS Lett. 129:62.

van Steeg, H., Pranger, M.H., van der Zeyst, B.A.M., Benne, R., and Voorma, H.O., 1979, In vitro translation of Semliki Forest Virus 42S RNA. Initiation at two different sites, FEBS Lett. 108:292.

Voorma, H.O., Thomas, A., Goumans, H., Amesz, H., and van der Mast, C., 1979, Isolation and purification of initiation factors of protein synthesis from rabbit reticulocyte lysate, Methods Enzymol. 60:124.

Weiss, B., Nitschko, H., Ghattas, I., Wright, R., and Schlesinger, S., 1989, Evidence for specificity in the encapsidation of Sindbis virus RNAs, J Virol. 63:5310.

Wengler, G., and Wengler, G., 1975, Comparative studies on polyribosomal, non polysome-associated and viral 42S RNA from BHK 21 cells infected with Semliki Forest Virus, Virology. 65:601.

Wengler, G., Wengler, G., and Gross, H.J., 1979, Replicative form of Semliki Forest Virus RNA contains an unpaired guanosine, Nature (London). 282:754.

K. Tuomi, L. Kääriäinen, and H. Söderlund, 1975 Quantitation of Semliki Forest virus RNAs in infected cells using ^{32} P equilibrium labelling, Nucl Acids Res. 2:555

Wengler, G., Wengler, G., Boege, U., and Wahn, K., 1984, Establishment and analysis of a system which allows assembly and disassembly of alphavirus core-like particles under physiological conditions in vitro,Virology. 132:401.

Wengler, G., Boege, U.,Wengler, G., Bischoff, H., and Wahn, K., 1982, The core protein of the alphavirus Sindbis virus assembles into core-like nucleoproteins with the viral genome RNA and with other single-stranded nucleic acids in vitro, Virology. 118:401.

ASPECTS OF THE MOLECULAR BIOLOGY OF POLIOVIRUS REPLICATION

James Harber and Eckard Wimmer
State University of New York at Stony Brook
Stony Brook, New York, 11794

I. INTRODUCTION

The Picornaviridae are divided into five genera: Enterovirus, Rhinovirus, Hepatovirus, Cardiovirus, and Aphthovirus. The first three of these genera are important in causing human disease; the remainder are of significance in causing animal disease. The virulent Enterovirus members include three poliovirus serotypes which cause paralytic poliomyelitis and meningitis; the coxsackie A viruses (types A1-22, 24), whose symptoms upon infection include herpangina and hand-foot-mouth disease; the coxackievirus B viruses types 1-6, which are responsible for meningitis, pleurodynia, myocarditis, pancreatitis, and hepatitis; and the echoviruses types 1-9, 11-27 and 29-34, which cause meningitis and rash. Rhinoviruses, which are divided into the major and minor groups depending on the receptor they use, are the cause of rhinitis or the common cold (Melnick, 1990). The genus Hepatovirus includes the human pathogen hepatitis A virus (HAV). Representative Cardioviruses include mengovirus, which has been implicated in insulin-dependent diabetes mellitus in mice, and the Theilers murine encephalomyelitis viruses (Yoon et al., 1987). The members of the Aphthovirus genus are foot and mouth disease viruses (FMDV), which are highly infectious in cloven-hoofed animals (Wittman, 1990). This brief summary of the disease potential of the Picornaviridae underscores the importance of the research summarized in this book. While we recognize the fact that there are important differences among the various Picornaviridae, for the purposes of this chapter poliovirus will serve as an archetype to demonstrate recent advances in the following topic areas: virion structure and antigenicity; receptor mediated uptake of the virus by the cell; translation of viral RNA; dicistronic analysis of the polyprotein; and *de novo* synthesis of poliovirus via *in vitro* translation. The knowledge gained from this research refines the understanding of the viral life cycle and provides insights into the mechanisms by which the Picornaviridae cause highly specialized diseases.

II. VIRION STRUCTURE

Geometry and Components of the Virion

The three dimensional structures of representatives of four genera of the Picornaviridae have now been solved. The underlying geometry of all of these structures

Regulation of Gene Expression in Animal Viruses, Edited by
L. Carrasco *et al.*, Plenum Press, New York, 1993

is the same and they share identical strategies for virion formation. The crystallographic structures known include poliovirus type 1 Mahoney (Hogle et al., 1985), poliovirus type 3 (Sabin) (Filman et al., 1989) , a poliovirus type 1 type 2 antigenic hybrid [which is neurovirulent in mice (Yeates et al., 1991)], rhinovirus type 14 (Rossmann et al., 1985), rhinovirus type 1A (Kim et al., 1989), FMDV (Acharya et al., 1989), mengovirus (Luo et al., 1987) and two strains of Theilers virus (Grant et al., 1992; Luo et al., 1992). The polio capsid proteins are geometrically ordered as an icosahedral dodecahedral shell. It consists of 60 repeats of a subunit protein complex (called the protomer) which encases the genomic RNA (Hogle et al., 1985). During assembly each protomer consists of three protein subunits [termed VP0, VP1 and VP3 for the order of these viral proteins seen by electrophoretic separation (Maizol, 1963; Maizol and Summers, 1968)] which are derived from the poliovirus polyprotein by an ordered cascade of proteolytic cleavages (see section V). VP0 is myristoylated (Chow et al., 1987; Paul et al., 1987) by the addition of this fatty acid to a glycine residue at the second position of VP0 (Krausslich et al., 1990; Marc et al., 1989; the N-terminal initiating methionine is removed during this reaction). Virus infectivity is influenced by mutation to VP0 amino acids at positions 2 and 5, which is consistent with the substrate requirements of the N-myristoyl transferase, which is the likely catalytic enzyme for VP0 myristoylation (Marc et al., 1989). Coincident with the packaging of the viral RNA is the as yet unexplained autocatalytic maturation cleavage of VP0 (to yield myristoylated VP4 and VP2), which is not mediated by a conserved serine residue found proximal to the cleavage site in the rhinovirus and poliovirus crystal structures (Harber et al., 1991). Compton et al. (1990) isolated a virus that is temperature sensitive for the VP0 cleavage which contains a point mutation distal to the cleavage site. At the nonpermissive temperature, VP0 is not cleaved and a RNA containing particle resembling a proposed virion assembly intermediate (the provirion) is observed (Compton et al., 1990). Although the maturation cleavage is regarded as the final step in the assembly of a virus, it is important to note that the mature virion contains one to two copies of VP0.

Antigenicity and Hybrid Viruses

Sequencing of escape mutants to neutralizing monoclonal antibodies identified the major antigenic sites of the virus (Diamond et al., 1985) (Minor et al., 1983) before the crystal structure of the virion was reported. A large body of evidence now favors the designation of six neutralization epitopes (termed neutralization antigenic or Nag) to which antibodies can bind: Nag IA, Nag IB, Nag IIA, NagIIB, NagIIIA and NagIIIB (Hogle et al., 1986; Minor, 1990; Minor et al., 1986; Page et al., 1988). Only the NagIA and NagIIA neutralization sites are continuous (linear amino acid sequences). These sequences have been exchanged with those of other polioviruses (Burke et al., 1988; Evans et al., 1989; Jenkins et al., 1990; Martin et al., 1988; Minor et al., 1990; Murdin et al., 1991; Murdin et al., 1992; Murdin et al., 1989; Murray et al., 1988a; Murray et al., 1988b) or with that of human rhinovirus type 14 (Altmeyer et al., 1991). A trivalent poliovirus carrying neutralization antigenic sites from all three serotypes has been constructed in a similar manner (Lu et al., 1992; Murdin et al., 1992). Indeed, the linear antigenic sites can be exchanged by entirely heterologous sequences (Evans et al., 1989; Mirzayan et al., 1991), but the use of poliovirus as a carrier of foreign antigens for vaccines has proved to be less promising than anticipated because of poor growth characteristics of the hybrid viruses or weak immunogenicity of the inserted epitope (Murdin et al., 1990). Interestingly, the exchange of N-AgIA of PV1(M) with that of the mouse neurovirulent poliovirus type 2 Lansing revealed that determinants of mouse neurovirulence are found in this sequence (Martin et al., 1988; Murray et al., 1988a). The type 1/type 2 hybrid virus is altered with respect to the distribution of surface charges in the region of the five fold axis (Yeates et al., 1991). How this charge redistribution relates to the potential adaptation of the virus to mice may only be understood in the context of the elusive murine poliovirus receptor (this problem remains intractable; see section II). Genetic manipulation of the poliovirus antigenic regions has provided much insight into the mechanisms of antibody mediated virus neutralization, virus host range, antigenic structure and potential vaccine strategies [reviewed in (Lu et al., 1992; Murdin et al., 1990)].

The Canyon

A remarkable surface feature of both poliovirus and rhinovirus is a conspicuously deep cleft (called the canyon) between the broad plateau of the threefold axis and the star-shaped mesa of the fivefold axis. Mengovirus has a deep pit region in approximately the same position as the deepest part of the rhinovirus canyon. The observation that amino acid residues of the rhinovirus "canyon" and mengovirus "pit" were more conserved than other surface residues led to the hypothesis that they are the putative sites of specific virus-receptor interactions (Rossmann et al., 1988). The canyon hypothesis (as extended to poliovirus, rhinovirus and mengovirus) postulates that the receptor binding site is located in a region of the virion which is inaccessible to antibodies (Rossmann, 1989a). Several sets of data support the canyon hypothesis: i) the mutation of rhinoviral canyon residues alters the attachment affinity of the virus to cell membranes (Colonno et al., 1988) and ii) inhibition of viral attachment is observed when a conformational change in the canyon floor of rhinovirus is induced by antiviral drugs that bind in a hydrophobic pocket beneath the canyon floor (Badger et al., 1989; Pevcar et al., 1989). Furthermore, viral mutants which are resistant to neutralization by soluble poliovirus receptor have been isolated (Kaplan et al., 1990). Clearly, the further definition of the site for interaction of poliovirus with the cellular receptor will assist in deciphering how virus attachment to the receptor induces conformational changes in the virion (this topic is mentioned again at the end of section III). Finally, as the canyon residues responsible for receptor binding are assigned, it will become possible to determine the interrelationship of the canyon and antigenic amino acids (as in the case of the mouse neurovirulent type 1/type 2 Nag-I antigenic hybrid). [For a thorough review of issues related to picornaviral architecture, refer to the chapter by Dr. Rossmann in this publication and the following reviews: (Casjens, 1985; Harrison, 1989; Rossmann, 1989b).].

III. THE POLIOVIRUS RECEPTOR

Identification of cDNA Clones and the Chromosomal Gene

Viral infection of susceptible cells occurs as a result of virion attachment, penetration, eclipse and subsequent viral replication. Many models have been developed which explain the strategies viruses use to enter cells and provide a necessary theoretical framework for research [reviewed in Marsh and Helenius (1989)]. A poliovirus receptor paradigm was established some thirty years ago by the experiments of Holland and collaborators. They were able to show that the affinity of the polio virion to the surface of susceptible cells was higher than for nonsusceptible cells, and in other experiments, demonstrate that the RNA extracted from virions was infectious when applied (transfected) onto nonsusceptible cells (reviewed in Holland, 1964). From this they concluded that the resistance to poliovirus infection is a cell surface phenomenon, that is, dependent on the availability of a receptor that mediates viral uptake. Many studies showed that the poliovirus receptor is localized to the plasma membrane of the cell

(Holland and McLaren, 1961; Guttman et al., 1977; Krah et al., 1982). However, the identity of the poliovirus receptor remained a mystery until potent molecular cloning techniques supplemented traditional biochemical methodologies for the identification of rare cellular molecules (Mendelsohn et al, 1986, Mendelsohn et al, 1989).

In its natural context *in vivo*, it is now known that poliovirus propagation is restricted to primate cells which bear the poliovirus receptor protein (PVR) [for a review see Zibert et al., (1992c)]. For the purposes of this chapter, the PVR will be discussed only with respect to its role in poliovirus infection; that is, there will be no attempt to speculate on its natural function. The human receptor which we will denote hPVR is encoded by a gene that resides on chromosome 19 (Miller et al., 1974) [Recently, the locus of the hPVR gene has been determined more precisely to be 19q 13.1-13.2 [(Koike et al., 1990; Schonk et al., 1990)]. The strategy for identification of the coding region of the receptor gene involved a rosetting assay, in which the anti-hPVR D171 monoclonal antibody (Nobis et al., 1985) was used to aggregate sheep red blood cells to mouse L cells transformed with human DNA (Mendelsohn et al., 1986). A cDNA lambda phage library was generated from the hPVR rosetting L- cell clones and subsequently two cDNAs which differed in the length and sequence of the carboxy terminal cytoplasmic tail (termed pUCH20A and pUCH20B; see figure 2) were identified (Mendelsohn et al., 1989). Expression of these cDNAs in mouse L cells allowed the propagation of poliovirus in the transformants, a result that proved that both the cDNAs carry genetic information for a poliovirus receptor. At the same time, a cosmid clone bearing a functional genomic copy of the hPVR was isolated from a chromosome 19 library. Mouse L cells were successfully transformed with the cosmid to yield a poliovirus-permissive host cell (termed the JA-1 cell line; J. Bradley, A. Kameda and E. Wimmer, unpublished results). Using similar strategies, both hPVR cDNAs (and other hPVR forms), and genomic DNA encompassing the hPVR gene were independently isolated by Koike et al. (1990).

Sequence and Putative Structure of the hPVR Proteins

The open reading frame of the cDNA clones revealed that hPVR is a molecule with regularly repeated cysteine-spacer-cysteine motifs punctuated by tryptophane residues, potential N-glycosylation sites and a hydrophobic sequence likely specifying a membrane spanning region. This, together with other properties of the deduced polypeptide chain, suggested that hPVR is a member of the immunoglobulin (Ig) superfamily (Mendelsohn et al., 1989). hPVR contains three extracellular Ig-like domains [an amino terminal variable (V) type and two constant (C) type) domains], a hydrophobic membrane spanning region and a cytoplasmic tail. The closest relative with a similar Ig-like domain structure of V-C-C is the *Drosophilia* protein amalgam (Hellen et al., 1989). The V domain of hPVR is of particular interest since it mediates binding of the polio virion (see figure 3 and accompanying section). A three dimensional model of the V-domain has been constructed and is currently being used as a tool to assist in understanding the phenotype of mutations made in putative binding regions of both the virus and the receptor (A. Zibert, G. Bernhardt, M. DeCrombrugghe, J. Harber, J. Bibb and E. Wimmer; unpublished data).

Splice Variants of the hPVR pre-mRNA

Koike et al. (1990) determined that the human gene coding for hPVR contains eight exons that are distributed over 20kb of chromosomal DNA (Koike et al., 1990). Significantly, these investigators found splice variants of hPVR pre-mRNA [named: α, β, γ and δ; see figure 2; (Koike et al., 1990)]. The previously identified cDNA clones H20A and H20B correspond to the hPVR α and hPVRδ respectively. H20B is identical to H20A from exons 1-7. The open reading frame of H20B and H20A are different past amino acid 384 (a serine) in the cytoplasmic tail.

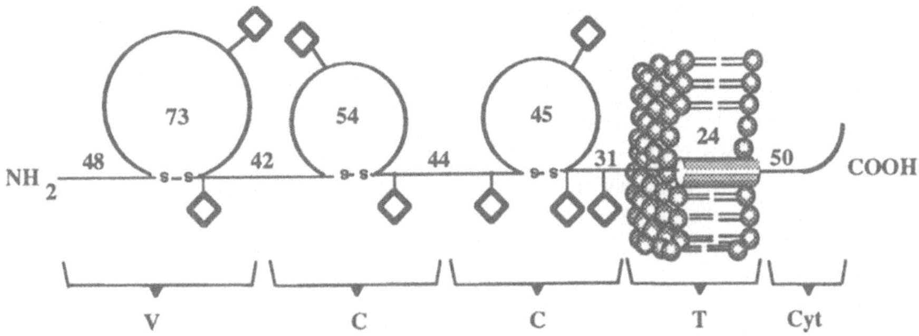

Figure 1. Structure of the hPVR and its orientation relative to the plasma membrane.
The amino and carboxy terminus of the hPVR H20A (PVRα ; see figure 2) are indicated by NH_2 and COOH respectively. The borders of the three Ig-like domains are indicated with brackets and labeled V, C and C. Lengths of amino acid segments of the protein dissected by disulfide bonds (S-S) and the plasma membrane are numbered. The putative N-linked glycosylation sites are indicated by diamonds on the exterior section of the molecule. The transmembrane region (illustrated as a cylinder) passes through the lipid bilayer in this model (adapted from an illustration provided by J. Bibb).

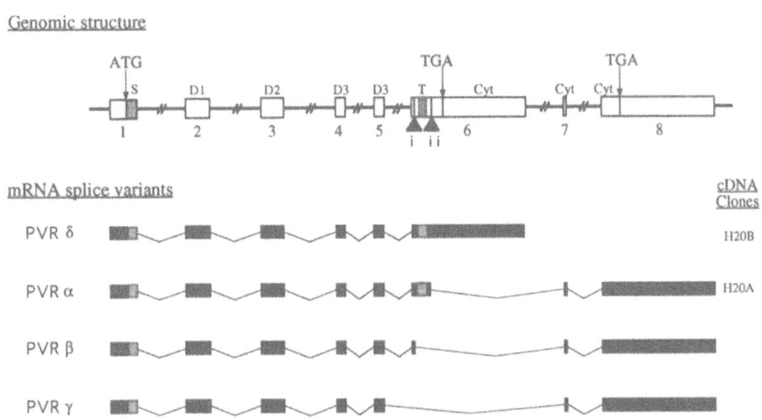

Figure 2. The hPVR gene and its mRNA splice variants.

The genomic arrangement of exons is indicated by numbered boxes. Indicated are the exon 1 ATG start codon for all hPVR protein species and two TGA stop codons located in exons 6 and 8. S (light shading) is the proposed signal polypeptide of the hPVR species which is presumably removed after the molecule is translated across the endoplasmic reticulum; D1, D2 and D3 are the three Ig-like domains; T=transmembrane region (also lightly shaded); Cyt= cytoplasmic region of the hPVR. The four mRNA splice variants are hPVR δ (= pUCH20B), hPVR α (= pUCH20A), hPVR β, and hPVR γ. Two alternative splice donor sites (labeled i and ii) exist within exon 6. hPVRα is spliced at i and hPVR β at ii. with the consequence that only the hPVR δ and hPVR α mRNAs contain the transmembrane coding region. (illustration courtesy of G. Bernhardt; based on data of Koike et al. 1990).

In addition to the four known hPVR cDNAs, two different homologs from a African green monkey kidney cell line have been isolated (Koike et al., 1992). The monkey cDNA clones AGMα1 and AGMδ1 are splice variants which are very similar to hPVRα and δ. Actually three forms of AGMα (AGMα,1, AGMα,1', AGMα,2) were isolated. This revealed that two alleles exist for monkey PVR (AGMα,1 and AGMα,1'). It is likely that in contrast to man, AGMα,1 and AGMα,2 map to two genetic PVR loci in monkeys because they are only 92.5% identical (Koike and Nomoto, 1992). In addition to the human and monkey PVR clones, the cDNA of a PVR mouse homolog has been isolated (Morrison et al., 1992). Interestingly, the mouse homolog does not allow infection by poliovirus presumably because its V-domain contains too many amino acid changes compared to hPVR. Genomic clones have meanwhile been used to generate transgenic mice carrying multiple copies of the hPVR transgene. These mice are susceptible to infection with all three poliovirus serotypes and thus are ideal models for the examination of the progression of poliovirus infection (Koike et al., 1991a; Ren and Racaniello, 1992).

hPVR expression and Viral Tissue Tropism

Poliovirus infection was thought to be restricted to cells of the gastrointestinal tract and the central nervous system, and this was believed to be a consequence of expression of hPVR in these tissues. Indeed, the specific destruction of motoric neurons by poliovirus causes the major poliomyelitis symptom: loss of limb function (reviewed in Holland, 1964) The discovery that hPVR specific mRNA can be found in numerous other human tissue cells (Koike et al., 1991b; Mendelsohn et al., 1989), and that hPVR-specific antigens can be detected on the surface of many cell types came as a surprise (Freistadt et al., 1990). A problem inherent in these studies is that the antibodies used in immunoblots cannot differentiate between functional or non-functional hPVR-related antigens. Moreover, not all hPVR-related polypeptide chains may be at the surface of the cell [as was observed following expression of hPVR using a vaccinia virus vector which did not bind the quantity of virus predicted by the overexpression system (Zibert et al., 1991)] or in the case of secreted PVR molecules (see figure 2: PVRβ and PVRγ) Tissue specific differences in the electrophoretic mobility of hPVR are possibly the result of alternative splicing or post-translational modification (Freistadt et al., 1990). In addition, hPVR expression seems to be regulated by posttranscriptional mechanisms as exemplified in the case of the JA-1 cell line (established by J. Bradley, A. Kameda and E. Wimmer, unpublished). This cell line contains multiple copies of the hPVR gene. But the increased amount of PVR mRNA found in JA-1 is not paralleled by a concommitant increase in surface expression of the hPVR protein (G. Bernhardt, unpublished).

A single monoclonal antibody directed against the receptor can block infection by all three poliovirus serotypes (Minor et al., 1984; Nobis et al., 1985). This observation invokes one molecule capable of mediating the attachment, conformational change and eclipse of poliovirus. Indeed, the expression of hPVR alone in L-cells renders these cells susceptible to poliovirus infection. While it is true that the hPVR is required for virion eclipse, a residual level of poliovirus attachment to non primate cells (McLaren et al., 1959) suggests other molecules can potentially bind the virus. It is not known if attachment of poliovirus to cells not bearing the hPVR is physiologically relevant. A monoclonal antibody capable of partially inhibiting attachment and infection of Hela cells with poliovirus type 2 recognizes a 100-kDa membrane protein on the surface of the human spinal cord and HeLa tissue culture cells whose relationship to the hPVR is not known (Shepley et al., 1988). The recent identification of 50 and 23/25 kDa proteins as low affinity binding sites on Hela cells for poliovirus by monoclonal antibodies (Barnert et al., 1992) appears to reinforce the emerging view that multiple proteins capable of mediating attachment of poliovirus exist on the surface of cells. Visualization of the poliovirus receptor by immunoelectron microscopy with a mAb D171 antibody probe revealed that the receptor is arranged in multivalent protein complexes (Mannweiler et al., 1990). The composition of these complexes might provide some insight into native cellular hPVR function and could answer other questions about the early stages (attachment and tight binding) of poliovirus interaction with the cell surface.

Domain of the hPVR involved in Virus Binding

The amino-terminal V-domain of hPVR is necessary and sufficient to bind poliovirus and effect infection (Freistadt et al., 1991; Koike et al., 1991; Selinka et al., 1991; Selinka et al., 1992). Numerous experiments have revealed that the second domain also contributes towards receptor function (Freistadt et al., 1991; Koike et al., 1991; Selinka et al., 1991). Expression of the first domain alone with the hPVR transmembrane region provides a poor substrate for viral interaction (Selinka et al., 1991). The context in which the virus requires the first domain, however, is not very specific. Hybrids of the hPVR first domain conjugated to I-CAM or CD4 allow virus binding and infection (Selinka et al., 1991; Selinka et al., 1992). Deletion of the carboxy terminus does not destroy hPVR function for poliovirus infection (Koike et al., 1991). Examples of constructs used by H-C. Selinka, A. Zibert and E. Wimmer to localize the domain involved in poliovirus receptor function are illustrated in Figure 3. Point mutations of amino acids in domain 1 of the hPVR important for virus binding and infection initiation are currently being analyzed.

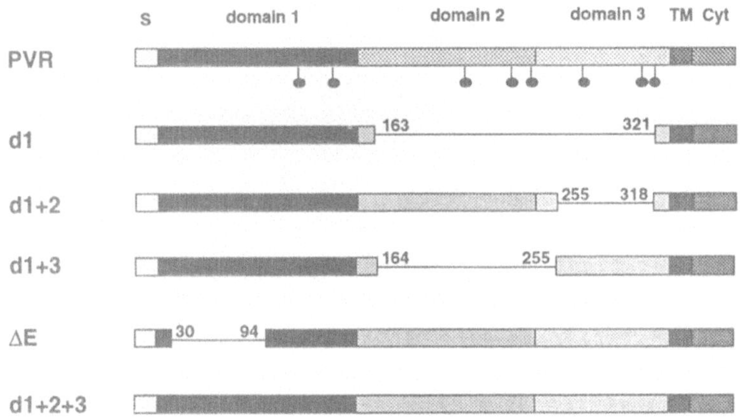

Figure 3. Domains of the hPVR important for viral receptor function.

Schematic overview of constructs coding for poliovirus receptor deletion mutants. The structure of the native PVR with a signal peptide (S), three extracellular domains, the transmembrane (TM) and cytoplasmic domain (Cyt) is illustrated at the top. Dots show potential N-glycosylation sites. Numbers inside the scheme refer to the first and last amino acid deleted in each construct. These constructs, cloned into transient expression vectors pMT2 and pMT1, were expressed in mouse MOP8 and OST7-1 cells. In a single step growth experiment, the yield of infectious virus from mouse cells expressing the complete hPVR (d1+2+3) or the receptor with the first two domains of hPVR (d1+2) was almost identical. The receptor consisting of the hPVR domains 1 and 3 (d1+3) resulted in lower virus titers. Domain 1 (d1) alone with a transmembrane region functioned poorly as a receptor. Finally, partial deletion of the first domain (ΔE) eliminated all receptor function (text and figure from: Selinka et al., 1991).

Glycosylation of the hPVR

Glycosylation of hPVR at multiple sites was demonstrated by *in vitro* translation of transcripts of the pUCH20A clone of hPVR, producing a protein product of 46 kDa in the rabbit reticulocyte lysate system that shifted to 67 kDa when microsomal membranes (which support posttranslational glycosylation) were added (Zibert et al., 1991). This result was confirmed when the overexpression of the hPVR using a vaccinia virus system was studied (see figure 4). Sugar residues had been thought to be important at some stage of viral binding to hPVR because lectins could inhibit this process. Attachment of poliovirus to Hela cells was reduced by the D-mannose binding lectin Concavalin A (Lonberg-Holm, 1975). However, the two potential N-glycosylation sites present in domain 1 of the human PVR can be mutated to abrogate glycosylation without destroying the function of the human receptor in viral infectivity (Zibert et al., 1992b; Koike et al., 1992). Similar to hPVR, the first immunoglobulin-like domain of the rhinovirus receptor ICAM interacts with the virus. Interestingly, the abolition of first domain ICAM glycosylation by site directed mutagenesis does not destroy receptor function (McClelland et al., 1991). Finally, no glycosylation sites exist in the native first domain of the monkey hPVR (Koike and Nomoto, 1992). These results collectively indicate that while the glycosylation sites of hPVR are close to the binding site for the virus, the carbohydrate moieties of the first domain are not essential for the poliovirus-hPVR interaction.

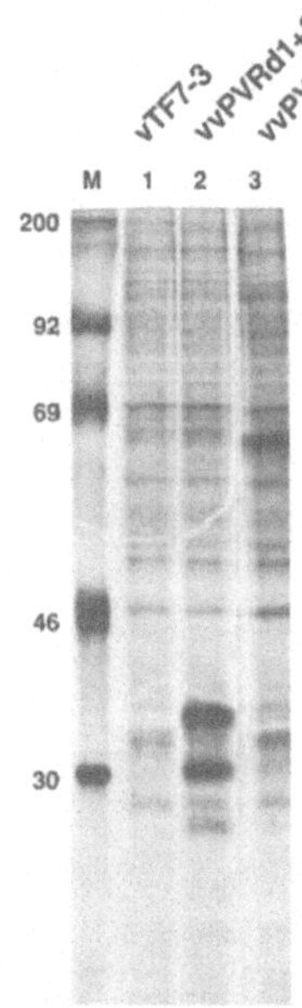

Figure 4. Expression of hPVR constructs via recombinant vaccinia viruses.
Immunoprecipitation of from vaccinia virus infected cell lysates. $5x10^6$ OST7-1 cells were infected with vTF7-3, which is a recombinant vaccinia virus expressing the T7 RNA polymerase protein (lane 1) or coinfected (both moi of 10) with the indicated hPVR recombinant viruses (lane 2 and 3). vvPVRd1+2 contains the first two domains of the hPVR. vvPVR contains the entire hPVR H20A insert. At 14 h.p.i. the medium was changed and cells were labeled with S35-methionine (100uCi). The medium (lane 1 and 2) or the cell lysates were subjected to immunoprecipitation (20 h.p.i) using a polyclonal anti-PVR antiserum (#625) and analyzed by 12.5% SDS-PAGE. M indicates the marker proteins in kdal. A band corresponding to full length hPVR is visible at 67 kdal in lane 3 and two bands of the hPVR domain 1+2 construct are visible in lane 2 near 31 and 35 kdal (presumably the doublet is due to two glycosylated forms of hPVRd1+2) [text and figure adapted from Zibert et al., (1992a)].

Function of the hPVR in A-particle Formation

A mechanism of invasion of the cell by poliovirus has been described as proceeding through the stages of attachment, tight binding and eclipse (Lonberg-Holm et al., 1976 and references therein). The early events in infection coincide with the conformational rearrangement of the virion that is reflected by the loss of antigenic sites on VP1 and VP2 (Wiegers et al., 1985) and the shift in sedimentation coefficient of the virion from 150S to 135S due to the complete loss of VP4 (Guttman and Baltimore, 1977; Lonberg-Holm et al., 1975). This drastically altered particle is termed the "A-particle" (Crowell et al., 1983). Furthermore, extrusion of the hydrophobic amino terminus of VP1 that accompanies the conformational rearrangements of the virion renders the 135S particle capable of attachment to a lipid surface (Fricks et al., 1990; Lonberg-Holm et al., 1976). On the whole, the transition from virion attachment to tight binding is not very efficient because significant numbers of altered particles are found sloughed off into the medium early in infection, before entry can occur (Fenwick et al., 1962; Joklik et al., 1961). These rejected 135S particles are no longer infectious. The generation of 135S poliovirus particles by soluble hPVR mimics the stage of infection in which the virion is conformationally altered. Recombinant forms of the hPVR produced in baculovirus expression vectors (Kaplan et al., 1990) and vaccinia virus expression vectors (Zibert et al., 1992) are capable of inactivating poliovirus. Soluble receptors as small as the first two domains (V-C) inactivate the virus by promoting the eclipse-like transition from 150S to 135S particle in solution (Zibert et al., 1992a). Viral mutants which escaped inactivation (called srr for soluble receptor resistant) were identified in assays with a three domain version of the hPVR (Kaplan et al., 1990). It is thought that sequencing of srr mutants will reveal the location of amino acids involved in structural rearrangements [see the chapter by Rossmann in this volume for further information about structural changes in the capsid proteins following receptor binding].

Entry of the Virion into the Cell

The viral RNA is released from the capsid into the cytoplasm at an unknown cellular location. For the purposes of this chapter, we will first assume that poliovirus enters the cell by hPVR mediated endocytosis during infection. As different picornaviruses use different receptors, the function of the cognate receptors in the process of entry is not the same. In general, viral endocytosis is known to occur as receptor mediated endocytosis or as receptor mediated phagocytosis. The subcellular structures responsible for the two mechanisms are not identical. Endocytosis is a constitutive process common to all cell types that involves receptor mediated internalization of bound ligands into clathrin coated pits (which subsequently bud into internal vesicles). The coated vesicles become endosomes which are acidified by ATPases to a final pH of less than 5.3. At this stage, low pH serves to dissociate the receptor from the ligand, allowing the sorting and recycling of the receptor to the cell surface. The ligand remains in the endosome, which is fused with a lysosome rendering the vesicle competent for low pH proteolysis. Phagocytosis, occurring in cells like macrophages, involves activation and induction of a region of the cell membrane through ligand signaling. The cell then utilizes an actin-based mechanism to internalize a section of the cell membrane which binds a large particle, bacterium, yeast or opsonized particle. These internalized vesicles are smooth, but like endosomes, are acidified and eventually fuse with lysosomes (the cell biology of endocytosis and phagocytosis and its relationship to viral entry and infection are explained in detail in Marsh and Helenius (1989), and in references therin to research by Brown and Goldstein).

The uncoating system which internalizes poliovirus is a constituitive component of Hela cells. At 25 degrees centigrade (when cellular processes are maintained but incapable of induction) poliovirus penetrates Hela cells (Mandel 1967). This experiment utilized a light sensitive virion which had encapsidated RNA intercalated with neutral red. Light sensitive virus applied to cells at 25 degrees centigrade can be inactivated before virion uncoating when it is no longer possible to neutralize the virus with antibody (Mandel, 1962; Mandel, 1967). Furthermore, Kronenberger et al. (1992) showed that at 25 degrees centigrade, it is possible to recover internalized 150S virions from cell lysates by sucrose gradient sedimentation Additional insight into the cellular nature of the poliovirus uncoating system came from studies of cold synchronized infection by Everaert et al. (1989). A formerly unseen 110S eclipse product (in which VP2 was cleaved) appeared following the 135S (presumably A-particle) species upon analysis of cell lysates. Following the emergence of the 110S species, the capsid species were further degraded, suggesting that they had encountered a lysosomal organelle (Everaert, et al., 1989).

The exact cellular location of the virion at the time of eclipse and uncoating has been the subject of much investigation. When ingested, poliovirus (unlike rhinovirus) can pass through the stomach, which has an extremely low pH. Following attachment to the hPVR of an appropriate cell in the gut, the virion's biochemical properties are in a state of transition as the conformational change to A-particle ensues. If the virus were to infect the cell by entering clathrin coated pits (characteristic of receptor mediated endocytosis), then the acidification of the internalized vesicle might be critical to infection by mediating uncoating by dissociating the virus and receptor and/or by directing the release of the RNA from the virion. Initial reports indicated that the ionophore monensin, which raises the pH of endosomal vesicles, affected poliovirus uptake (Madshus, et al., 1984; Zeichhardt, et al., 1985). The experiments of Madshus et al. involved the use of light-sensitive virus, which was inactivated with light following application of the drug monensin. It now appears that monensin simply delays the onset of infection by about 30 minutes (Gromeier and Wetz, 1990). Additional evidence that uncoating is not occurring due to an decrease in pH comes from a system in which A-particles are produced from virions at 37 degrees celcius in a low ionic strength buffer containing calcium chloride (Wetz and Kuzinski, 1991). Wetz and Kucinski also found that in low ionic strength buffer without calcium chloride, complete degradation of the virion was faster at alkaline pH.

Deciphering the relationship between the virus and the cell surface is complicated by the fact that "it is more difficult to obtain a good signal at reasonable multiplicities of infection because the early events take place, by definition, before amplification in signal due to replication" (Marsh and Helenius, 1989). Despite this problem, some methodologies for the examination of early events are available. The light sensitive virus is an excellent tool for examining the penetration of the virus into the cell, but light inactivation does not provide information about the biochemical nature (proteolytic?) of the cellular location (or organelle?) where uncoating is occurring. The basic conflict between the evidence for uncoating inside the cell on the one hand and why so many A-particles are nonproductively produced and sloughed off outside the cell on the other hand has not yet been adequately explained either.

Other tools for investigating the early events of infection involve the attachment of picornaviruses to cell surface molecules other than the native receptor (this avoids problems with the background of sloughed off, nonproductive particles). One ingenious methodology involves binding a streptavidin conjugated poliovirus to a biotinylated murine L-cell to simulate viral attachment to a receptor (Aarnes et al. 1987). This system did not allow for poliovirus replication in L-cells. Another method, detailed below, uses antibodies to mediate the cellular attachment of poliovirus and FMDV to Fc receptors on

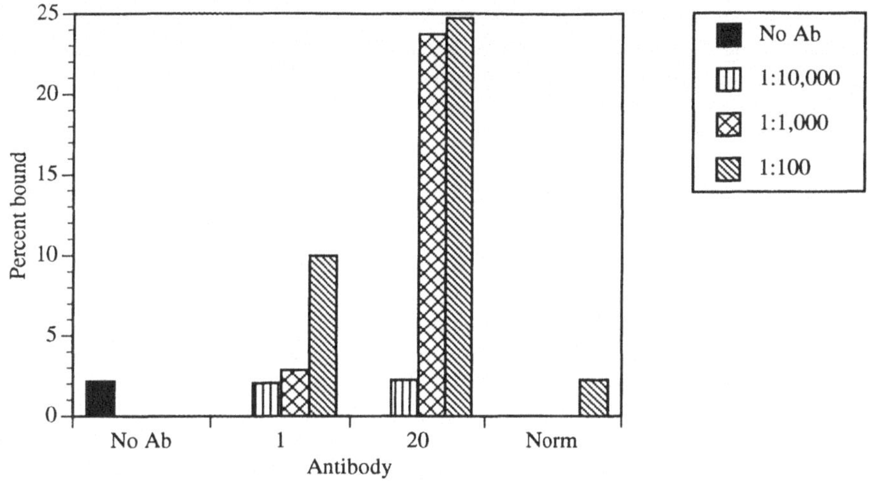

Figure 5 . Binding of the poliovirion to an Fc receptor via antibody
^{35}S-methionine labeled poliovirus hybrid PV1/HRV (Altmeyer et al., 1991) was mixed with anti-Nim I rhinovirus ascities #1 or #20 (Sherry et al., 1986; Sherry et al., 1985), normal mouse ascites (norm) or medium alone (MEM + antibiotics + 25 mM Hepes pH 7.5) at the dilutions indicated for 1 hour at room temperature. Monolayers of CHO cells containing a stable integrated plasmid encoding the murine FcRII-B2 receptor (K. Matter and I. Mellman, Yale University) and treated for 18 hours with 10mM Na Butyrate were incubated for 1 hour at 37°C with the appropriate aliquot of the virus antibody mixture at an MOI of 10 pfu/cell. The virus innoculum was then removed and the remaining virus associated with the cells was determined by the method of B. Baxt (Baxt et al., 1984). These experiments were repeated with numerous monoclonal and polyclonal sera and although binding was always detected, no poliovirus replication was subsequently observed (Mason et al., 1992).

the surface of cells. In both of these systems, any replication in the absence of receptor would be registered as the amplification of the input virus.

Opsonized poliovirus or FMDV were attached to the surface of mouse L-cells or hamster CHO cells (both of which are normally resistant to infection by these viruses) that were transformed to express the B2 isoform of the murine antibody receptor, the Fc molecule (see figure 5). The attachment of FMDV to the Fc receptor of mouse L-cells via a neutralizing antibody mediates infection. In contrast, poliovirus which was bound to the surface of mouse L-cells by antibody-Fc interactions did not initiate infection (Mason, et al., 1992). These experiments imply that the receptor for FMDV is any molecule capable of attaching and internalizing the virus. The receptor for poliovirus, however, must provide additional functions beyond simple attachment (that is, it must destabilize the virion). FMDV, unlike poliovirus, rhinovirus and coxackievirus, produces no conformationally altered particle prior to eclipse (Baxt, et al., 1980; Lonberg-Holm, et al., 1972). Whereas FMDV dissociates below pH 7.0 (Randrup, 1954), the poliovirion is stable in an acidic environment (physiologically isotonic) but low strength ionic buffers can uncoat the poliovirion (Wetz and Kucinski, 1991). FMDV is perhaps the least stable picornavirus, and is capable of severe persistent infection. The possibility that Fc mediated replication is involved in the endurance of the disease is being investigated.

The poliovirus-hPVR relationship is complex. Attempts at mimicry of the uncoating process have revealed that the virus is initially internalized intact, does not require low pH for uncoating (and uncoats more efficiently under basic conditions), is delayed in infection by monensin, and is eventually proteolytically degraded after becoming a 110S species. We and others have developed systems to explore the issue of picornaviral internalization independent of the native cellular receptor. We are actively investigating the possibility that poliovirus infection can be initiated by a combination of virus conjugated to Fc-expressing cells and soluble hPVR. This work is intended to extend the knowledge of the early stages of infection and of the commencement of translational initiation.

IV. INITIATION OF TRANSLATION

A Comparison of the Uncoated Viral RNA and Cellular mRNA

The genome of poliovirus is plus-stranded RNA and serves as mRNA in the cytoplasm of the infected cell. Several unique features of picornaviral translational initiation can be attributed to the structure of the virion RNA. These are the presence of the genome linked protein, VPg, a long 5' non-translated region (NTR) and the lack of a cap structure. VPg, 22 amino acids in length (Adler et al., 1983; Kitamura et al., 1980) is covalently linked to the 5' end of the viral RNA via a tyrosine residue (Lee et al. 1977; Nomoto et al. 1977; Flanegan et al., 1977). VPg is removed following uncoating in the cell (Ambros, 1978) leaving pUp at the 5' end of the RNA (Hewlett et al., 1976; Nomoto et al., 1976). The function of VPg is likely to be at a viral stage other than translational initiation because RNA produced *in vitro* via T7 RNA polymerase lacks VPg but is infectious (van der Werf et al., 1986) The unique 5' structure of the picornaviral RNA thus differs from cellular messages which contain the 7mGpppG cap structure. Moreover, picornavirus RNAs have 5' nontranslated regions that are unusually long [between 624 and 1199 nt.(Palmenberg, 1986)] compared with the non-translated regions of typical cellular messages [between 20 and 100 nt (Kozak, 1987)]. The Picornavirus RNAs contain as many as 13 non-initiating AUGs (Palmenberg, 1986), whereas fewer than 10% of mammalian messages contain AUGs upstream of the initiating codon (Kozak, 1987).

The Shutoff of Host Translation

Picornaviral RNA and eukaryotic mRNA differ with respect to the mechanism used for translational initiation. A ribosomal scanning model for the initiation of translation of eukaryotic mRNAs proposes that the binding of the 43S ribosomal subunit can only occur at the 5' end of the mRNA [through the interaction of the cap and the cap-binding protein complex eIF-4F (Kozak, 1978)]. The ribosomal 43S subunit then scans the mRNA in the 5' to 3' direction until an initiation codon is found preferentially in the context: RXXAUGR where R=purine and X=any nucleotide (Kozak, 1989b). One component of the cap binding complex, eIF-4F, is a cellular protein of molecular weight 220kd referred to as p220. Inhibition of cap dependent protein synthesis during poliovirus infection correlates with the proteolysis of p220 (Etchison et al., 1982). Cellular protein p220 is not cleaved directly by poliovirus encoded proteinases 2A or 3C (Lee et al., 1985; Lloyd et al., 1988), but the activity of 2A is required for p220 cleavage (Bernstein et al. 1985; Kraüsslich et al., 1987). Furthermore, the autocatalytic activity of 2A is ultimately responsible for the cleavage of eukaryotic initiation factor 4F polypeptide p220 (Hellen et al., 1991). The mechanism of cleavage of p220 is not understood but, surprisingly, involves eiF-3 (Wykoff et al. 1990). The unusual properties of picornavirus viral mRNAs, and the fact that they do not need an intact cap binding complex suggests that picornavirus translation occurs by an alternative mechanism. In the following section, data will be discussed which indicate that picornavirus translation is initiated following direct internal entry of ribosomes into a genetic element within the 5' NTR which we have termed the internal ribosomal entry site or IRES.

The Discovery of the IRES Element of the Picornaviridae

Three kinds of experiments support the existence of the IRES. These are: (i) experiments with dicistronic RNAs which demonstrated the phenomenon of internal ribosomal entry, (ii) deletions and mutations in the 5'NTR which define nucleotide borders for the IRES and (iii) construction of an infectious dicistronic virus which confirms that internal ribosomal entry does occur *in vivo*. The deletion and mutation experiments will be discussed later in this section, following presentation of the 5' NTR secondary structure models in Figure 7, and the dicistronic virus experiments will be presented in section VI of this chapter.

Initiation of picornaviral translation was observed by Jang et al. (Jang et al., 1988) in the E. Wimmer laboratory and also by N. Sonenberg and colleagues (Pelletier and Sonenberg, 1988) to occur as a result of internal entry of ribosomes into a segment of the 5' NTR. The genetic element which mediates this internal initiation event was termed the internal ribosomal entry site (IRES) (Jang et al., 1989; Jang et al., 1988), or the ribosomal landing pad (RLP) (Pelletier and Sonenberg, 1988; Pelletier et al., 1989). Both groups used a similar strategy to demonstrate internal entry of ribosomes: the construction of clones that can be transcribed to yield dicistronic RNAs. Those built by Jang and Wimmer will be used in this discussion for the purpose of illustrating the discovery of internal initiation.

In Figure 6 panel A, the poliovirus and EMCV 5'NTRs are shown in a dicistronic construct where each NTR is followed by a reporter gene (X and Y respectively). The rationale for using these constructs was that the poliovirus IRES functions very inefficiently in rabbit reticulocyte lystate whereas the EMCV IRES is very active. If RNA of the type shown in Fig 6A were translated in RRL, ribosomes should only enter at the 5' end as mandated by the scanning model. The level of protein X expression would be low, but higher than the level of protein Y expression. Contrary to this prediction, translation of protein Y was always found to be expressed more efficiently than protein X. It should be noted that, in general, a reporter gene translates better when linked to the EMCV 5' NTR than to the poliovirus 5'NTR in the rabbit reticulocyte lysate (RRL). In addition, the second cistron was expressed earlier than the first, and small deletions in the EMCV 5' NTR abolished expression of protein Y. These results are consistent with a mechanism of translational initiation whereby the ribosome enters the EMCV 5'NTR independently of translational initiation of protein X. Since the initiation codon for protein Y is ~3000 nt from the 5' end of the RNA, the ribosome could not have initiated translation of protein Y by entry at the RNA in a position resembling an authentic 5' terminus. The existence of the IRES was further demonstrated *in vivo* using construct B in which, as in construct A, gene Z under control of the EMCV 5'NTR was translated preferentially over reporter gene Y. Construct C was designed such that gene X was under the control of the globin 5' NTR whereas Y, as before, was controlled by the EMCV IRES. Expression of the mRNA in poliovirus-infected cells yielded only product Y, a result consistent with the shut-off of the cap-dependent translation of X and internal ribosomal entry for the translation of Y.

These results were clearly not compatible with the conventional scanning model (Kozak, 1978). IRES dependent translation mechanisms have been postulated for other picornaviruses including FMDV (Kühn et al. 1990), HAV (Brown et al., 1991), and for the flavivirus-like hepatitis C virus (Tsukiyama-Kohara et al. 1992). Remarkably, a cellular mRNA encoding the glucose regulated immunoglobulin heavy chain binding protein (BIP), is actively translated following picornaviral shutoff of cap dependent translation, and has also been shown to have an IRES function for translational initiation (Macejak et al., 1991; Sarnow, 1989). The repeated demonstration that ribosomes could enter the RNA at an IRES, in a location other than the 5' terminus, has prompted a re-evaluation of the conventional model for eukaryotic translation initiation.

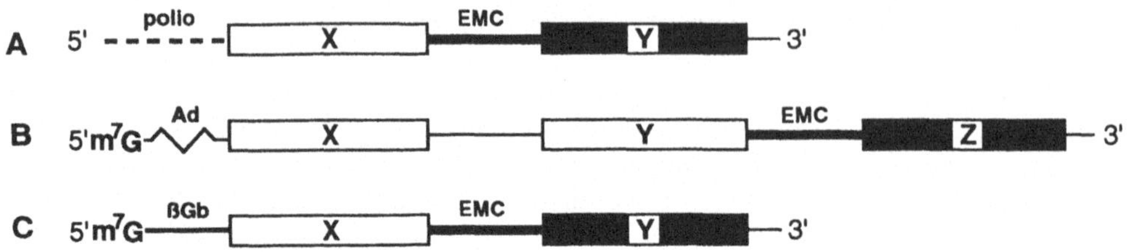

Figure 6. Schematic representation of di- and tricistronic mRNAs whose cistrons are controlled by different 5'NTRs. Polio = The entire 5'NTR of poliovirus; EMC = the 5'NTR of EMCV from nt 260 to the initiating AUG (the sequence preceding nt 260 contains a heteropolymeric segment and poly(C) which have no apparent influence on the efficiency of translation; note that 21 nt of poly C follows nt. 260); Ad and ßGb = the tripartite leader of late mRNA of adenovirus type 2 and the 5'NTR of ß-globin mRNA, respectively. X, Y, and Z denote reporter genes. Expression of construct A was tested in RRL and in a HeLa cell-free system. Expression of constructs B and C were tested in COS cells. Text and illustration from Jang et al (1990). Further details can be found in: (Jang et al., 1989; Jang et al., 1988).

A Unifying Working Hypothesis for the Selection of the Initiating AUG codon

Picornavirus 5' NTRs contain multiple AUGs preceeding the polyprotein initiation site. For example, EMCV initiates polyprotein synthesis at AUG #11 although four AUGs are found in this immediate vicinity (Witherell and Wimmer 1992), and FMDV can initiate protein synthesis at either of two AUGs (Beck et al., 1983). The initiating AUG codon in poliovirus is located 154 nucleotides from the 3' border of the IRES, which has led to suggestions that translational initiation occurs by the internal binding of ribosomes and subsequent scanning of this spaced sequence to the AUG codon (Jackson et al., 1990; Pelletier and Sonenberg, 1988; Pelletier and Sonenberg, 1989; Sonenberg, 1990). With the complexity of observations in the various picornaviral systems mounting, Jang et al. postulated that internal initiation, ribosomal scanning and the selection of the AUG for translation initiation were linked (Jang et al. 1990a). Contained within the sequence involved in internal initiation (the IRES), is a specific motif (YnXmAUG) which is proposed to be responsible for ribosomal binding and depending on the virus genus, contains the site of AUG translational initiation. Y is a pyrimidine stretch of length n (approximately 7 nt long) which is separated from an AUG by a non conserved nucleotide sequence called Xm (of length 10-20 nt.) The polypyrimidine tract (Yn) was originally identified in FMDV (Beck et al., 1983). EMCV and FMDV initiate translation at the AUG of the YnXmAUG motif (FMDV has two such motifs and produces two polyproteins) while poliovirus and rhinovirus have an additional spacer region between the YnXmAUG motif and the initiating AUG. Jang et al. reasoned that direct ribosomal binding and initiation at the AUG of the YnXmAUG motif occurs in the Cardiovirus and Aphthovirus systems, but binding of the ribosome near YnXmAUG and translocation to a downstream AUG occurs in the Enterovirus and Rhinovirus systems. In all cases the Kozak context of the initiating codon (Kozak, 1987) is assumed to be of some importance. Altogether, available data from four picornaviral genera focus on the YnXmAUG motif as a necessary element of translational initiation. Jang et al do not dispute the possible competency of the ribosome for scanning, but have concluded that scanning is not necessary for translational initiation. Agol and his colleagues (Pilipenko et al. 1992) have come to the same conclusion (see discussion later in this section).

Several tenets of the unified theory of picornaviral translation have been verified experimentally. The Kozak context of the initiating AUG, the requirement for correct spacing (Xm) between the polypyrimidine and the AUG, the necessary presence of the polypyrimidine tract Yn and the interaction of the IRES with cellular factors involved in translational initiation have been tested experimentally. The dispensability of all poliovirus AUGs upstream of the initiation codon was demonstrated by mutating each AUG to UUG. Viral viability was not influenced by these mutations with one exception: mutation of the "silent AUG" (at nt. #586) of the YnXmAUG motif yielded a small plaque phenotype, an observation underscoring the importance of this triplet (Pelletier et al. 1988). Hellen et al. explored the possibility whether this triplet can be "activated" to serve as an initiation codon by changing its context (Hellen, Pestova and Wimmer, unpublished data). Accordingly, they changed the sequence of the motif from CUU<u>AUG</u>G to ACC<u>AUG</u>G. As expected, the AUG of the YnXmAUG motif of poliovirus was now used. Since translation from AUG #743 was not inhibited, it is possible that scanning does not occur and thus ribosomes may bind the IRES and then jump to the initiating AUG. Alternatively, ribosomal binding to the IRES and initiation at AUG 743 of poliovirus could occur simultaneously as is thought to occur in the case of EMCV (Jang and Wimmer, 1990; Kaminski et al. 1990; Witherell and Wimmer, 1992). In poliovirus, Iizuka et al. found that a deletion between Yn and AUG 743 (nt 564 to 726) does not have an effect on viral viability, although differences in neurovirulence of Mahoney virus were detected (Iizuka et al., 1989). In this interesting construct, a YnXmAUG motif was restored in that AUG 743 was moved upstream through the deletion. Conservation of distance between the polypyrimidine tract and the downstream AUG could play an important role in the mechanism of internal initiation.(Nicholson et al., 1991; Pilipenko et al., 1992). Particularly interesting is the study of Agol and his colleagues in which the YnXmAUG was extensively modified by insertions and deletions. The mutants were tested *in vitro* and *in vivo*. Generally, polioviruses with changes in the motif grew poorly, and revertants had the motif restored (Pilipenko et al., 1992). Figure 7 illustrates the proposed structural motifs used for ribosomal binding to initiate translation in the EMCV and poliovirus 5'NTR regions.

In the EMCV IRES, the sequences surrounding the YnXmAUG motif that contains the initiating codon AUG#11 have been altered as well. In some cases, the AUGs upstream of AUG#11 were detected to initiate translation (Davis et al. 1992). Moreover, deleting the Xm (m=18) by 9 nt activates an AUG downstream of AUG#11 such that a suitable length of the Xm (m=21 in the mutant) is restored (Jang and Wimmer, unpublished results). Finally, mutations to Yn alone which dramatically reduce translation support the application of the unified theory not only to poliovirus (Nicholson et al. 1991; Pestova et al. 1991) and EMCV (Jang and Wimmer, 199b), but also to FMDV (Kühn et al. 1991). Invariably, experiments in many other picornaviral systems will be necessary to determine the subtleties of the unified theory of picornaviral translational initiation.

IRES Binding Proteins

Several lines of evidence indicate that protein factors which interact with the 5' NTR may play a role in IRES-dependent translation. These include differences in the capacity of various cell-free extracts to support polioviral translation, and the identification of mutations within the poliovirus 5'NTR that both attenuate the neurovirulent phenotype and impair the ability of the RNA to initiate translation. Various techniques have been used to identify factors which interact with the IRES, including mobility shift assay, UV cross-linking [^{32}P] labeled probes to cellular proteins, and biochemical fractionation.

Dorner et al. noted that initiation of translation of monocistronic poliovirus RNA was aberrant in rabbit reticulocyte lysate (RRL), but found that the deficiency could be corrected by supplementing RRL with a crude mixture of translation factors from Hela cells (Dorner et al., 1984). Subsequently, Svitkin et al. purified a complex ("initiation correction factor" or ICF) from Krebs-2 cells that contained eIF-2/2B and which corrected

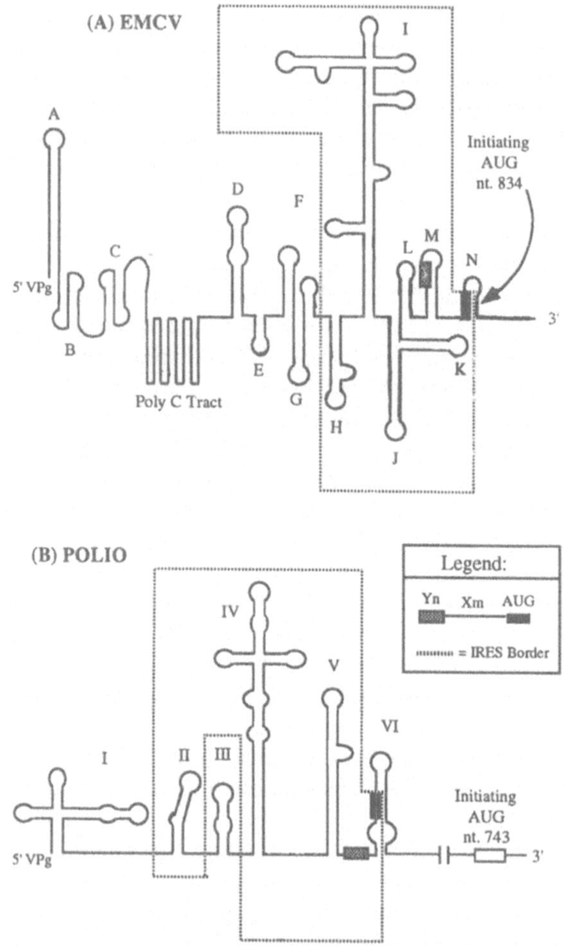

Figure 7. The structure of the IRES

Schematic representations of predicted secondary structures for the EMCV (A) (Duke et al. 1992; Pilipenko et al. 1990) and poliovirus (B) (Pilipenko et al., 1989) IRES elements. Yn-Xm AUG motifs are indicated by shaded boxes (where the light shaded box = Yn, the dark shaded box = AUG and the space between these boxes = Xm). The Yn-Xm AUG motifs contain the start codon (AUG #11) of the EMCV IRES and non-initiating AUG-586 of the poliovirus type 1 IRES. Stem loop structures of the EMCV IRES and poliovirus IRES elements are labeled with letters and roman numerals, respectively. The poliovirus AUG start codon at nt 743 is labeled with an open box, and the spacer distance between the YnXmAUG motif and the initiating AUG is 154 nucleotides. The borders of the IRES are indicated by dashed boxes. The EMCV IRES 5' border is stem loop H and the 3' border is the initiating AUG (as defined by deletion analysis) (Jang and Wimmer, 1990b). The poliovirus IRES is bordered by stem loop III and the YnXmAUG motif (Jackson et al., 1990; Meerovitch et al., 1991; Nicholson et al., 1991). The approximate numerical nucleotide positions of the bases near stem loop border are for EMCV: A. 1-84 ; B. 85-106; C. 107-129; D. 286-319 ; E. 321-336 ; F. 346-371 ; G. 372-392 ; H 403-551 ; I. 552-679 ; J. 680-730 ; K. 731-767 ; L 786-800 ; M 801-825 ; N 831-850 and for Polio: I. 1-78 ; II. 124-162 ; III. 186-220 ; IV. 234-440 ; V. 448-556 ; VI 578-618. Please note that publications from this laboratory prior to 1992 had a different nomenclature for EMCV which did not include the A loop, and the B or C pseudo-knots which preceeded the poly C tract (i.e.- 1992 nomenclature stem loop E = 1993 nomenclature stem loop H). Figure adapted from Duke et al. 1992; Jackson et al, 1990; Jang et al., 1990a; Witherell and Wimmer, 1992.

aberrant translation of poliovirus RNA in RRL (Svitkin et al., 1985). Other components of this complex have not yet been characterized.

Neurovirulent strains of poliovirus have been found to be translated more efficiently than their attenuated counterparts (Svitkin et al., 1990; Svitkin et al., 1985). The expression of these attenuating mutations is tissue-specific *in vivo* (Agol et al., 1989; La Monica et al., 1989) and *in vitro* (Svitkin et al., 1988), indicating that the impairment of translation may reflect a reduction in binding of a factor or factors to domain V of the 5' NTR (Figure 7), possibly caused by alteration of the secondary structure of this domain (Evans et al., 1985; Muzychenko et al., 1991). Taken together, these data strongly suggest that RNA binding proteins modulate translation of picornaviral genomes, and that mutations within the 5' NTR that attenuate the neurovirulent phenotype cause structural or sequence changes in the RNA motifs recognized by such proteins.

Although many proteins are being investigated because they interact specifically with a picornaviral IRES, the following discussion will focus only on those proteins which have been specifically implicated in translational initiation. At the present time, there are only three such proteins: p57 or PTB, p52 and eIF-2A. The first experimental method used to identify IRES specific RNA-binding proteins was the RNA mobility shift. In this assay, proteins from a cell extract are crosslinked to radiolabeled RNA and applied to SDS-PAGE gels to observe the difference in migration due to the addition of a protein to the RNA. Using this method, eukaryotic translational initiation factor EIF-2A was found to be associated with different protein complexes on two sites of the poliovirus 5'NTR. One binding site (nt 510-629) is a sequence inside the IRES which overlaps the poliovirus YnXmAUG motif (Del Angel et al., 1989). Competition for cellular factors occurs between picornaviral and capped messages. In a competition experiment between the 5' nontranslated region of EMCV and globin mRNA for translational initiation complexes in rabbit reticulocyte lysate, eIF-2/2B was identified as the limiting factor. The specific EMCV region responsible for the competition effect was found by mutagenesis and lies 100 to 140 nucleotides upstream of the initiation codon (Scheper et al., 1991). However, the role of eIF-2A in the mechanism of IRES-dependent translation remains obscure.

Nucleotide substitution within the AUG component of the YnXmAUG motif of poliovirus reduced the efficiency of translation (Pelletier et al., 1988) *in vitro* and *in vivo*, and this has been found to be correlated with the ability of a 52 KDa protein to interact with the appropriate RNA probes (Meerovitch et al., 1989). However, the absolute requirement of this factor for internal initiation remains unproved, and it is notable that its putative binding site can be deleted by *in vitro* manipulation (Iizuka et al., 1989; Kuge et al., 1987) or as a result of pseudo-reversion *in vivo* (Haller et al., 1992; Pilipenko et al., 1992).

IRES elements specifically associate with the p57 component of a protein complex. p57 was identified as binding to the H-loop of EMCV (Borovjagin et al. 1991; Jang and Wimmer, 1990b), to regions of domain V of the poliovirus IRES [M. Litterst and E. Wimmer, unpublished results ; (Pestova et al., 1991)], and to the IRES of FMDV (Luz and Beck, 1990; Luz and Beck, 1991). Substitutions within stem loop H of EMCV which disrupt its secondary structure abolish both UV crosslinking of this element to p57, and the ability of the IRES to direct initiation of translation. Compensatory substitutions which restore the structure of stem loop H result in re-establishment of both properties. This provides strong evidence for the direct involvement of p57 in IRES dependent translation of EMCV RNA. Subsequent experiments provide evidence for the involvement of p57 in poliovirus and FMDV translation (Borovjagin et al. 1991; Luz and Beck, 1991; Pestova et al., 1991). Multiple p57 binding sites have been identified within EMCV and FMDV IRES elements (Luz and Beck, 1991;Witherell and Wimmer, submitted), but it is not clear whether this reflects binding of several p57 molecules to a single IRES element or binding of several domains of a single p57 moiety to elements of a complex tertiary structure.

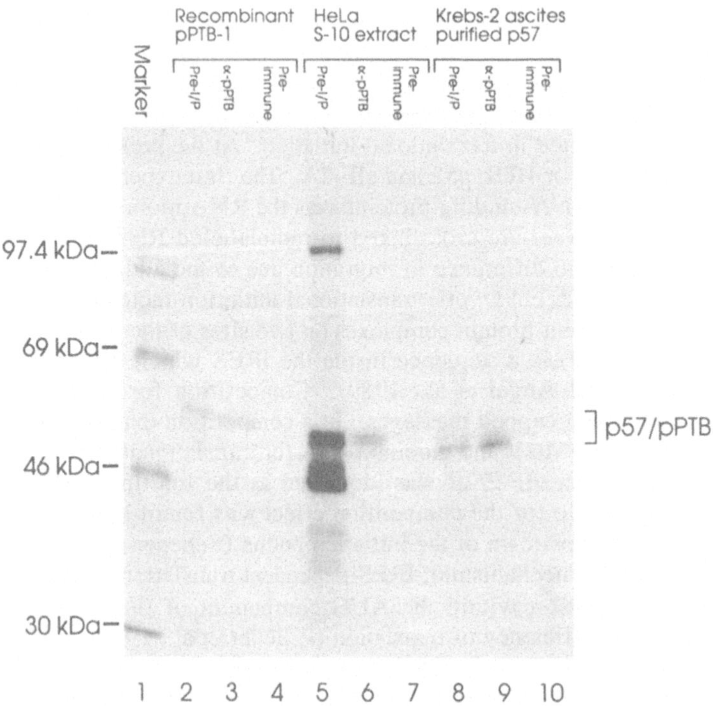

Figure 8. Antigenic relationship between p57 and pPTB

Immunoprecipitation of p57 by anti-pPTB serum. Purified recombinant pPTB-1 (lanes 2-4), a HeLa S10 cytoplasmic extract (lanes 5-7) and partially purified murine p57 from Krebs-2 ascites (lanes 8-10) were cross-linked by UV irradiation to a [32P]-UTP-labeled EMCV 5'NTR segment from nt. 260-488 (Meerovitch et al., 1989; Pestova et al., 1991). 5ng pPTB and 1μg Hela cytoplasmic proteins were used in each 25 μl UV cross-linking reaction; the amount of murine p57 was not determined. Samples prior to addition of antiserum (lanes 2,5 and 8) and those immunoprecipitated with preimmune (lanes 4,7 and 10) or immune serum (lanes 3,6 and 9) were resolved by electrophoresis in a 10-20% PAGE-SDS gradient gel. The positions of [14C]- labeled methylated protein standards (Amersham) are indicated on the left hand side [text and figure from Hellen et al., 1992d).

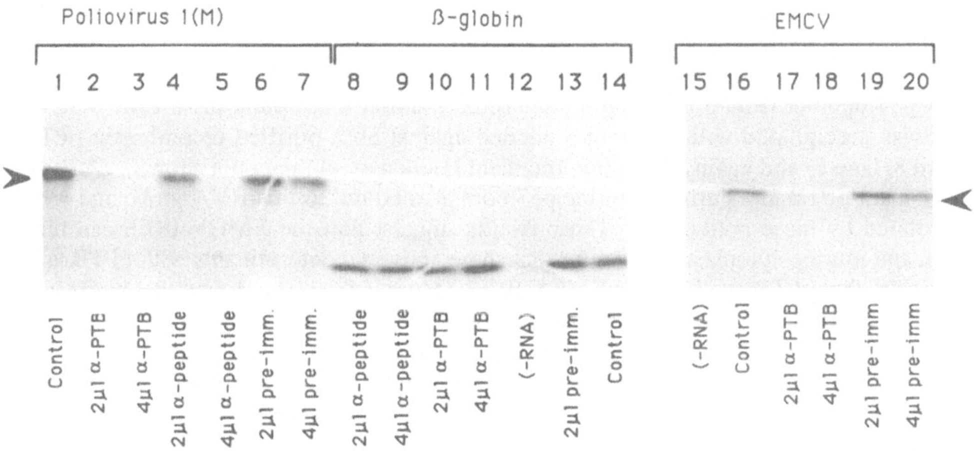

Figure 9. Inhibition of IRES-dependent mRNA translation by anti-pPTB serum.

Aliquots (1 µg) of mRNA transcribed from pBS+ (polioIRES- ΔPl) (lanes 1-7), pBS- (ß-globin) (lanes 8-14) and pBS-ECAT (lanes 15-20) were translated in 25 µl aliquots of HeLa cell extracts (Pestova et al., 1991) in the presence of 2 µl (lanes 2,10,17) or 4 µl (lanes 3,11,18) anti-pPTB serum, 2 ul (lanes 4,8) or 4 µl (lanes 5,9) pPTB-specific antipeptide serum, 2 µl (lanes 6,13,19) or 4 µl (lanes 7,20) preimmune serum, or without addition of serum (lanes 1,14,16). Lanes 12 and 15 correspond to lysate incubated without exogenous mRNA. The poliovirus ΔPl, ß-globin and ECAT translation products are indicated by arrowheads. pBS+(polioIRES- ΔPl) consists of nt. 70-742 of the 5'NTR of poliovirus type 1 (M) and part of the Pl capsid protein precursor cloned into pBS+(KS) (Stratagene) and placed under the transcriptional control of a T3 polymerase promoter. pBS-(ß-globin) contains cDNA corresponding to the complete ß-globin mRNA derived from pOGß 34, (Schwemmle et al., 1991) placed under the transcriptional control of a T3 promoter. pBS (ECAT) contains the EMCV IRES (Jang et al., 1988). Text and figure from Hellen et al. (1992d).

After UV cross-linking to picornaviral IRES elements, p57 was resolved as a doublet or triplet of bands by SDS-PAGE [(Borovjagin, 1991; Borovjagin et al., 1990; Jang and Wimmer, 1990; Luz, 1990; Luz and Beck, 1991; Pestova et al., 1991; Pestova et al., 1989) and Figure 8, lanes 5 and 6]. This and other biochemical properties of p57 closely resemble those of a nuclear RNA binding protein previously identified as polypyrimidine tract binding protein (pPTB) (Garcia-Blanco et al., 1989; Gil et al., 1991; Hellen et al., 1992d; Patton et al., 1991) or hnRNPI (Ghetti et al., 1992). Recently, it has been demonstrated that at least one of the isoforms of pPTB and p57 is the same protein (Hellen et al., 1992d). pPTB is a predominantly nuclear protein which has been implicated in pre-spliceosomal assembly; its binding to pyrimidine-rich tracts located near 3' termini of splice sites is thought to play a role in splice-site selection. However pPTB has also been identified as hnRNP I, a component of the ribonucleoprotein complex that surrounds hnRNA, and has been shown to be present in the cytoplasm in significant amounts (Ghetti et al., 1992). Indications are that p57 is a ribosome-bound protein, and as such it may mediate binding of ribosomes to the IRES by virtue of its affinity for specific sequences or structural elements (Jang and Wimmer, 1990b; Pestova et al., 1991).

A p57 doublet from a Hela S10 cytoplasmic fraction that bound to an EMCV RNA probe was precipitated with antibodies elicited against both purified recombinant pPTB (Figure 8, lane 6) and against a peptide fragment [Hellen et. al., data not shown]. Purified recombinant pPTB and purified murine p57 both also bound to this RNA probe and were precipitated by these antibodies. These results suggest that the EMCV IRES can bind human and murine cytoplasmic proteins that share antigenic determinants with pPTB and, conversely, that pPTB can bind the EMCV IRES. Moreover, it was found that binding of pPTB to the mutant and pseudorevertant EMCV RNA probes, described in Jang and Wimmer (1990b), parallels binding of p57 to these probes (Hellen et al., 1992d). Additional strong supporting evidence for the identity of p57 and pPTB is provided by the close similarity in the patterns of cleavage of these proteins by V8 protease and cyanogen bromide (Hellen et al., 1992d).

V. DICISTRONIC ANALYSIS OF THE POLYPROTEIN

While the analysis of RNA motifs and protein factors of picornaviral internal initiation has provided a framework for dicistronic experiments *in vitro*, a related approach has proven insightful *in vivo*. This theoretical leap was the construction of a viable dicistronic poliovirus (Molla et al., 1992). Whereas poliovirus is functionally monocistronic (Kitamura et al., 1981), in contrast the dicistronic virus construct contains IRES elements from poliovirus and EMCV, which functionally separate the genome of poliovirus into two independent cistrons (Figure 11). The virus is constructed so that the single poliovirus polyprotein is divided by the insertion of the IRES of EMCV. Thus in order for progeny dicistronic virus to be produced, two independent ribosomal internal initiation events must occur to translate the two segments of the polyprotein (Molla et al., 1992). The dicistronic virus provides an important tool for further dissection of functional aspects of the polioviral genome. What follows here is a short summary of the organization of the polioviral genome and the proteolytic processing of the polyprotein Further detail of the role of the dicistronic virus construct in the dissection of polioviral gene function in proteolytic processing will then be addressed.

The electrophoretic separation of a sample from a poliovirus infected Hela cell lysate reveals more than a dozen different virus specific polypeptides. The correlation of amino acid sequences of the termini of major polypeptides with genomic RNA sequences has yielded a map of the genomic organization of these electrophoretically separable polypeptides in the context of the polyprotein (Emini et al., 1982; Hanecak et al., 1982;

Kitamura et al., 1980, 1981; Larsen et al., 1982; Pallansch et al., 1984; Semler et al., 1983). In addition, protein sequencing shows that the polyprotein reading frame is initiated at the AUG at nt. 743 (Dorner et al., 1982). Translation terminates at a stop codon at nt. 7361 (Emini et al., 1982), thus revealing the polyprotein border relative to the 3' untranslated region (Kitamura et al., 1981) and the genomically coded 3' polyadenine (Yogo and Wimmer, 1972). The nomenclature of the genomic map divides poliovirus into three main sections (P1, P2 and P3), which are subdivided by letters in the nonstructural region (P2 and P3) (Rueckert and Wimmer 1984). The capsid proteins of the P1 section are called 1A, 1B, 1C and 1D, but they usually retain their original designations (VPO, VP1, VP2, and VP3), which are based on their migration patterns (and ultimately molecular weight) on SDS-PAGE gels (Rueckert and Wimmer, 1984).

Picornaviral encoded proteinases serve to process the polyprotein by a cascade of proteolytic cleavages into functional capsid and structural protein components. In poliovirus, these proteases are encoded by the 2A and 3CD regions of the genome. The first proteolytic cleavage of the cascade occurs cotranslationally after the ribosome passes the 2A region of the polyprotein. Autocatalytic cleavage by 2Apro at its own amino-terminus (likely in cis) separates the P1 capsid precursor region from the non-structural proteins of the P2 and P3 regions (Toyoda et al., 1986). Later, the catalytic activity contained in 3Cpro likely separates the P2-P3 region into P2-3AB and 3CD by a similar autocatalytic mechanism at its amino-terminus. 3CD is an unusual case of a protein in that it is both an enzyme and precursor substrate. 3CD is specifically competent for cleavage of the P1 region into three functional capsid proteins: VP0, VP3 and VP1 (Ypma-Wong et al., 1988; Jore et al. 1988; for a more detailed discussion of the proteolysis cascade see the chapter of Dr. Semler in this volume). Cleavage of 3CD yields the proteinase 3C, which is active in non-structural region cleavages, and the replication enzyme 3D, an RNA-dependent, primer-dependent RNA polymerase. Research into protein processing of the polyprotein in this laboratory over the last ten years has revealed many details regarding the function, substrate recognition and purification of the polio proteinases (Clark et al., 1991; Hämmerle et al., 1991; Hanecak et al., 1984; Harris et al., 1993; Hellen et al., 1992a; Kraüsslich et al., 1987; Lee et al., 1988; Nicklin et al., 1988; Toyoda et al., 1986). [For further details of the proteolytic cleavage cascade and the role of proteolytic processing in viral morphogenesis and replication, refer to the following reviews: Harris et al., 1990; Hellen et al., 1992b; Hellen et al., 1992c; and Kräusslich et al., 1988]. A complete cleavage scheme, including the identity and organization of protein products of the polyprotein is illustrated in Figure 10.

Studies of the polioviral proteolytic processing cascade can now be assayed in the context of a dicistronic virus. Gene dosage can now be manipulated easily and so, for example, the effect of two 3CD proteins on viral function can now be assayed. The efficiency of translation in each cistron can be manipulated by changing the characteristics of one IRES independently of the other. Does the most efficient rate of virus production result from a ratio of non-structural proteins to capsids of 1:1? Might a 2:1 ratio be limited by the rate of viral genome synthesis? A formerly intractible question such as the effect of substituting VP4 and VP2 for all VP0 protein can be addressed with a dicistronic construct. Some questions have already been answered; in fact by placing the EMCV IRES between P1 and P2-P3 (Figure 11), it can be ascertained that the capsid region is self sufficient in its capacity to fold correctly (is this true for every cleavage product of the polyprotein?). Novel uses for the dicistronic construct will undoubtedly be explored in the future. Originally, the dicistronic virus was designed to demonstrate that internal initiation of ribosomes occurs *in vivo* and to provide further proof for IRES function (Molla et al., 1992). In accomplishing this objective, it has allowed picornavirologists access to the genetic tools of gene dosage and complementation.

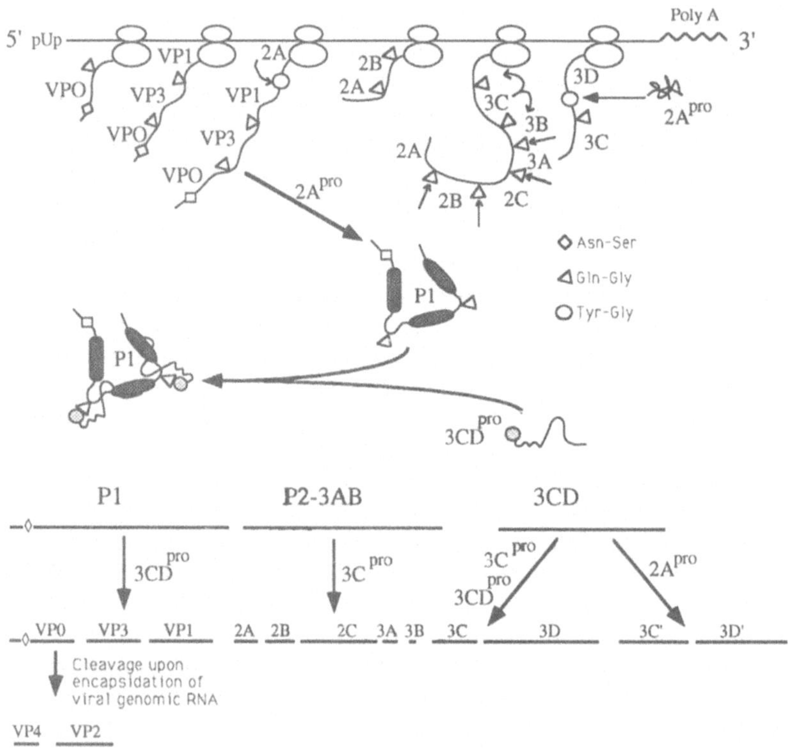

Figure 10. Model of the proteolytic processing of the poliovirus polyprotein
Co-translational intramolecular cleavage by 2Apro occurs at a Y-G site [o] at its amino-terminus. 3Cpro probably cleaves the Q-G bond [Δ] at its amino-terminus in a similar manner, and then cleaves other Q-G sites in the p2-3AB precursor *in trans* . The 3D portion of 3CDpro interacts with the P1 substrate in such a way that the Q-G sites in this precursor can be cleaved by the 3C moiety of the 3CDpro protein. 2Apro can cleave 3CD *in trans* to produce 3C' and 3D'. The mechanism by which the maturation cleavage occurs upon encapsidation of RNA at an N-S site [◊] has not been elucidated (Harber et al., 1991). The text and figure are from Harris et al., (1990).

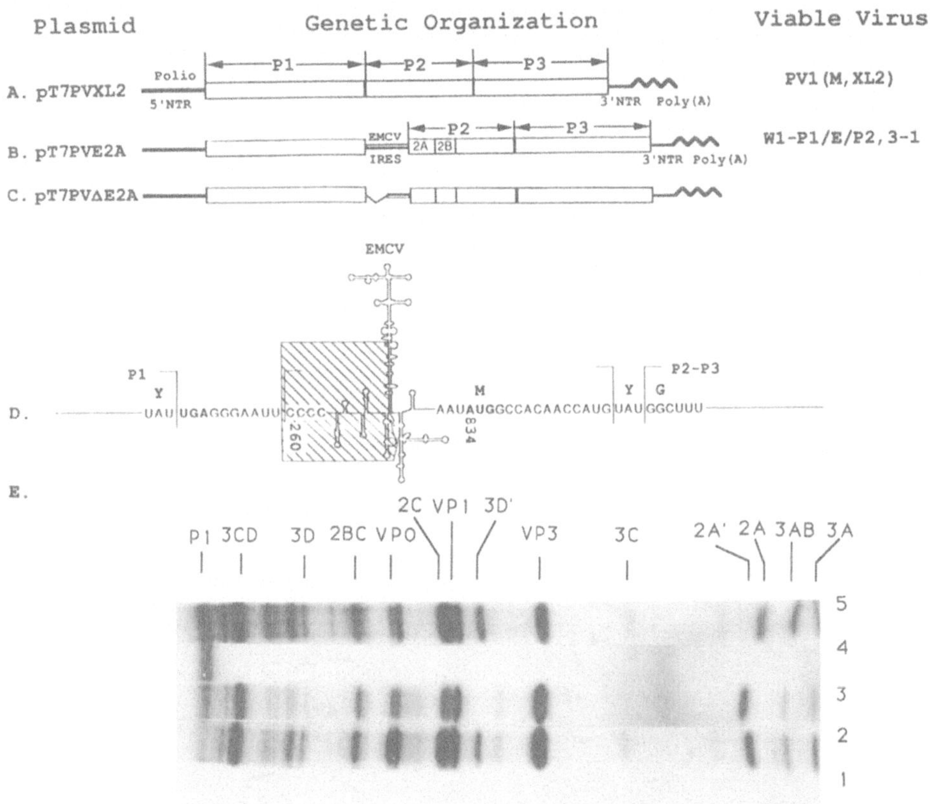

Figure 11. The genetic organization of the dicistronic poliovirus

Genetic organization of poliovirus and its dicistronic variants. Shown is the genetic content of the hypothetical viruses (open bars: reading frames; P1: capsid precursor; P2 and P3: non-structural proteins) that were engineered into plasmids containing the promoter for phage T7 RNA polymerase (van der Werf et al., 1986), and viruses recovered after transfection of HeLa cell monolayers with transcripts spanning the poliovirus cDNAs (minus sign, no virus recovered). a, Parental construct consisting of pT7PV1-5 ((van der Werf et al., 1986)), of which the 3' fragment (BglII site at nucleotide 5,600. to EcoRI, immediately downstream of the poly(A)) had been exchanged with a corresponding fragment from plasmid T7pGEM polio (supplied by P. Sarnow). b, Dicistronic constructs in which the polyprotein was interrupted by 602 nucleotides of non-coding RNA containing the EMCV IRES. 2A and 2B are gene products mapping to the N-terminal portion of P2. 2A is the proteinase normally cleaving P1/P2. c, *As b,* except that a segment of the EMCV IRES was deleted (nucleotide 260 to 434) as shown in d (shaded area). d, Detail of the insert between domains P1 and P2 of the polyprotein. Capital letters Y, M and G (bold) are single-letter abbreviations for amino acids. Numbers refer to the boundaries of the segment of the 5'NTR of EMCV RNA (Jang and Wimmer, 1990b). Line drawing depicts the secondary structure of the EMCV IRES (Jang et al., 1990a). Termination of the P1 ORF is described in (Nicklin et al., 1987). e, Polyacrylamide gel (12.5%) electrophoresis of poliovirus-specific proteins synthesized under the direction of T7 transcripts in a HeLa cell-free extract. Lane 1, no RNA; lane 2, pT7PVXL2 RNA; lane 3, pT7PVE2A RNA; lane 4, pT7PVΔE2A RNA, lane 5, poliovirus proteins synthesized in HeLa cells after infection with PV1(M),XL2 as markers. 2A, Wild-type 2Apro; 2A', 2Apro with additional amino acids at the N terminus originating from the ORF of the insert (see d). [figure and text from Molla et al., (1992)].

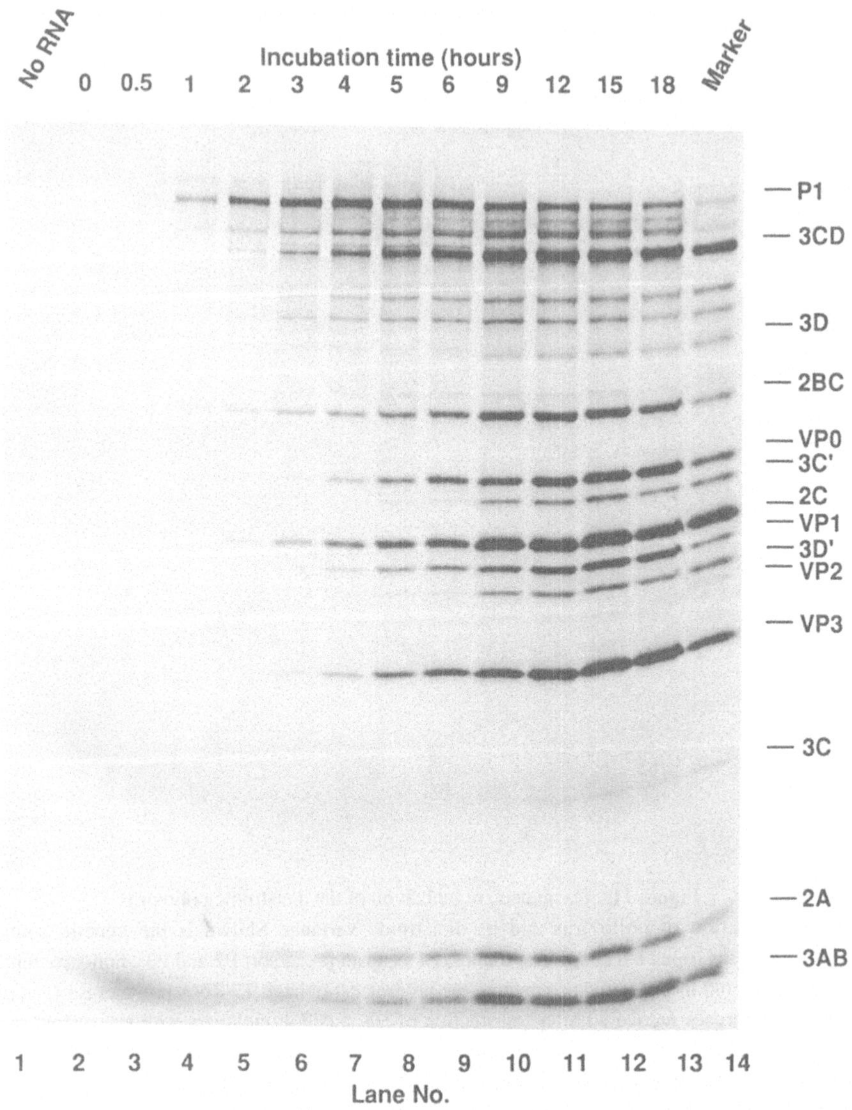

Figure 12. Time course of poliovirus polypeptide synthesis *in vitro*.

The synthesis and proteolytic processing of poliovirus proteins in a HeLa S10 extract (Lee et al., 1982; Molla et al., 1991; Pelham et al., 1976; Pelletier and Sonenberg, 1985; Pestova et al., 1991) complemented with a translation mixture and viral mRNA. The *in vitro* incubation mixtures (250 μl) contained 93 μl of S10 extract with the following additions: 1 mM ATP, 50 μM GDP, 10 mM creatine phosphate, 6 μg creatine phosphokinase, 2 mM DTT, 6 μg of calf liver tRNA, 12 μM each of all amino acids except methionine, 300 μCi of [35S]methionine, 18 mM Hepes (pH 7.4), 240 μM spermidine, 0.1 M K acetate, 0.35 mM Mg acetate, 0.75 mM Mg chloride and 2 μg of PV1(M) RNA [the K+ and Mg2+ ions were added in addition to those contributed from the dialysis buffer (Molla et al, 1991). Incubation was for 18 hours at 30°C. At each time point samples (20 μl) were removed, mixed with RNAse A (to 20 μg/ ml), RNAse T1 (to 100 U/ml), and cyclohexamide (to 5 μg/ml), incubated for 5 min. at 30°C, and frozen immediately. Portions (6 μl) were analyzed on 12.5% SDS-polyacrylamide gels. Gels were fixed, treated with EN3HANCE, dried, and exposed for 15 to 24 hours to x-ray film at 70°C. Lane 1, no RNA; lanes 2 to 13, incubation at 0, 0.5, 1, 2, 3, 4, 5, 6, 9, 12, 15, and 18 hours, respectively; lane 14, cytoplasmic extract of [35S]methionine-labeled PVI(M) infected HeLa cells as marker. [words and text from: (Molla et al., 1991)].

VI. *IN VITRO* SYNTHESIS OF POLIOVIRUS

Many problems in viral research can be solved *in vitro* if suitable assays for the function of proteins or complex molecular aggregates (such as replication complexes) are available. The cell-free translation, replication and packaging of poliovirus, which can be accomplished in an extract of human Hela cells, is the starting point for the development and refinement of such assays (Molla et al., 1991). Briefly, the general events of the *in vitro* poliovirus synthesis are as follows: RNA is first added to the extract, which then translates the polyprotein. The proteolytic processing cascade in turn releases the virus specific structural and non-structural proteins. Subsequent polio specific RNA synthesis then replicates the genome. Packaging of newly synthesized plus stranded genomic RNA inside capsid proteins to create virions *de novo* then ensues (Molla et al., 1991). Documentation of the progression of the *in vitro* poliovirus synthesis reaction is illustrated by the time course of protein synthesis and processing in Figure 12. In addition, the formation of infectious virions can be monitored by plaque assay (see Molla et al. 1992).

The *in vitro* synthesis of poliovirus provides an opportunity to assay reactions with various ingredients so as to determine their effect on virion production. Since these effects can be seen independently of the cell, the possibility for determining the exact step at which inhibition or stimulation occurs is conceivable. In Figure 13 the results of one such experiment are listed.

Since the production of poliovirus virions has now been reduced to a large scale biochemical reaction (Molla et al., 1991), the next step will be to dissect the individual steps in this process. The possibility that a host factor is involved in virion production (Putnak et al., 1982) can now be logically approached by assaying subfractions of the Hela cell lysate for stimulation of the packaging reaction. This is only one example of many conceivable *in vitro* experiments to assay elements involved in the synthesis of poliovirus; a complex macromolecule with the empirical formula of its organic matter: $C_{332,652}H_{492,388}N_{98,245}O_{131,196}P_{7501}S_{2340}$†.

PFU per milliliter

RNA	Guanidine HCL (2 mM)		MAb D171		Anti-PV1(M)		Anti-PV3(L)		Mg^{2+} ∞		Edeine (4 μM)	
	-	+	-	+	-	+	-	+	-	+	-	+
PV1(M)g[s]	24,948	0	24,965	0	21,868	0	15,932	15,924	0	12,000	12,000	0
PV1(M)g[r]	400	2,200	ND	ND	ND	ND	ND	ND	ND	ND	ND	ND

∞ : "-" lysate contained 1.5 mM Mg^{2+}; and "+" lysate contained 2.6 mM Mg^{2+}

Figure 13. Effect of antibodies and inhibitors on the synthesis of poliovirus in the incubation mixture. Conditions were as in Figure 12 except that concentrations of GTP, CTP, and UTP were adjusted to 125μM each; different batches of HeLa extracts were used; and PFUs were measured after 15 hours of incubation. The "-" and "+" refer to the absence or presence of the indicated treatment, except as noted. ND, not determined. mAb D171 is a poliovirus receptor-specific antibody. Anti-PV1(M) and anti-PV3(L) are neutralizing antiserum specific for types 1 and 3 poliovirus, respectively. [figure and text from: (Molla et al., 1991)].

ACKNOWLEDGMENTS

The authors wish to thank Christopher Hellen, Günter Bernhardt, Gary Witherell, and Marion Freistadt for assistance with revisions of this manuscript. Other current members of the lab whose work and suggestions appear in this manuscript include: Sung-Key Jang, Caroline Mirzayan, Kevin Harris, Akteruzzaman Molla, Andree Zibert, Hans-Christoph Selinka, HuiHua Lu, Michael Schmid, Mark Litthurst, Tatyana Pestova, Aniko Paul, XueMei Cao, Sang-Hoon Shin, Lou Alexander, James Bibb, and Marie DeCrombrugghe. We also appreciate advice from Jorge Galan, Joe Lipsick, Peter Tegtmeyer, Bill Bauer, and Stan Fields. Technical assistance from Connie Rafferty, Sue Newlin, Kathy Donnelly and the photography of Chris Helmke was appreciated as was the computer expertise of Mark Galassi, Bob Gallo, Bob Butz, Gert Bolwig and Peter Kissel.

† Calculations for the molecular formula for poliovirus: J. Harber, S.-H. Shin, A. Paul and Eckard Wimmer.

REFERENCES

Aarnes, C. M., Madshus, I. H., Guillemot, J. C., Sandvig, K. and Olsnes, S. (1987). Formation and activity of covalent conjugates of poliovirus and ligands binding to cell surface structures. *Exp Cell Res.* **170**, 483-490.

Acharya, R., Fry, E., Stuart, D., Fox, G., Rowlands, D. and Brown, F. (1989). The three-dimensional structure of foot-and-mouth disease virus at 2-9 Å resolution. *Nature.* **337**, 709-716.

Adler, C. J., Elzinga, M. and Wimmer, E. (1983). The genome-linked protein of picornaviruses. VIII. Complete amino acid sequence of poliovirus VPg and carboxy-terminal analysis of its precursor, P3-9. *J Gen Virol.* **64**, 349-355.

Agol, V. I., Drozdov, S. G., Ivannikova, T. A., Kolesnikova, M. S., Korolev, M. B. and Tolskaya, E. A. (1989). Restricted growth of attenuated poliovirus strains in cultured cells of a human neuroblastoma. *J. Virol.* **63**, 4034-4038.

Altmeyer, R., Murdin, A. D., Harber, J. J. and Wimmer, E. (1991). Construction and characterization of a poliovirus/rhinovirus antigenic hybrid. *Virology.* **184**, 636-644.

Ambros, V., and Baltimore, D. (1978). Protein is linked to the 5' end of poliovirus RNA by a phosphodiester linkage to tyrosine. *J. Biol. Chem.* **253**, 5263-5266.

Badger, J., Minor, I., Oliveira, M. A., Smith, T. J. and Rossmann, M. G. (1989). Structural analysis of antiviral agents that interact with the capsid of human rhinoviruses. *Proteins.* **6**, 1-19.

Barnert, R. H., Zeichhardt, H. and Habermehl, K. O. (1992). Identification of 50- and 23-/25-kDa HeLa cell membrane glycoproteins involved in poliovirus infection: occurrence of poliovirus specific binding sites on susceptible and nonsusceptible cells. *Virology.* **186**, 533-542.

Baxt, B. and Bachrach, H. (1980). Early interactions of foot-and-mouth-disease virus with cultured cells. *Virology.* **104**, 42-55 .

Baxt, B., Morgan, D. O., Robertson, B. H. and Timpone, C. A. (1984). Epitopes on foot-and-mouth disease virus outer capsid protein VP1 involved in neutralization and cell attachment. *J Virol.* **51**, 298-305.

Beck, E., Forss, S., K., S., Cattaneo, R. and Feil, G. (1983). Structure of the FMDV translation initiation site and of the structural proteins. *Nucleic Acids Res.* **11**, 7873-7885.

Bernstein, H. D., Sonenberg, N. and Baltimore, D. (1985). Poliovirus Mutant that does not selectively inhibit host cell protein synthesis. *Molecular and Cellular Biology.* **5**, 2913-2923.

Borovjagin, A. V., Ezrokhi, M.V., Rostapshov, V.M., Ugarova, T.Y., Bystrova, T.F., and Shatsky, I.N. (1991). RNA-protein interactions within the internal translation initiation region of encephalomyocarditis virus RNA. *Nucleic Acids Res.* **19**, 4999-5005.

Borovjagin, A. V., Evstafieva, A. G., Yu. Ugarova, T. and Shatsky, I. N. (1990). A factor that specifically binds to the 5'-untranslated region of encephalomyocarditis virus RNA . *FEBS Lett.* **261**, 237-240.

Brown, E. A., Day, S.P., Jansen, R.W., and Lemon, S.M. (1991). The 5' nontranslated region of hepatitis A virus RNA: secondary structure and elements required for translation *in vitro*. *J. Virol.* **65**, 5828-5838.

Burke, K. L., Dunn, G., Ferguson, M., Minor, P. D. and Almond, J. W. (1988). Antigenic chimeras of poliovirus as potential new vaccines. *Nature.* **332**, 81-82.

Casjens, S. (1985). "Virus Structure and Assembly". Jones and Bartlett Pub. Inc., Boston.

Chow, M., Newman, J. F. E., Filman, D., Hogle, J. M., Rowlands, D. J. and Brown, F. (1987). Myristylation of picornavirus capsid protein VP4 and its structural significance. *Nature.* **327**, 482-486.

Clark, M. E., Hammerle, T., Wimmer, E. and Dasgupta, A. (1991). Poliovirus proteinase 3C converts an active form of transcription factor IIIC to an inactive form: a mechanism for inhibition of host cell polymerase III transcription by poliovirus. *Embo J.* **10**, 2941-2947.

Colonno, R. J., Condra, J. H., Mizutani, S., Callahan, P. L., Davies, M. E. and Murcko, M. A. (1988). Evidence for the direct involvement of the rhinovirus canyon in receptor binding. *Proc Natl Acad Sci U S A.* **85**, 5449-5453.

Compton, S. R., Nelsen, B. and Kirkegaard, K. (1990). Temperature-sensitive poliovirus mutant fails to cleave VP0 and accumulates provirions. *J Virol.* **64**, 4067-4075.

Crowell, R. L. and Landau, B. J. (1983). Receptors in the initiation of picornavirus infections. *In "Comprehensive Virology"* (H. Fraenkel-Conrat and R. R. Wagner, ed.), Plenum Press, New York.

Davies, M. V. and Kaufman, R. J. (1992). The sequence context of the initiation codon in the encephalomyocarditis virus leader modulates efficiency of internal translation initiation. *J Virol.* **66**, 1924-1932.

Del Angel, R. M., Papavassiliou, A. G., Fernandez, T. C., Silverstein, S. J. and Racaniello, V. R. (1989). Cell proteins bind to multiple sites within the 5' untranslated region of poliovirus RNA. *Proc Natl Acad Sci U S A.* **86**, 8299-303.

Diamond, D. C., Jameson, B. A., Bonin, J., Kohara, M., Abe, S., Itoh, H., Komatsu, T., Arita, M., Kuge, S., Nomoto, A. and et, a. l. (1985). Antigenic variation and resistance to neutralization in poliovirus type 1. *Science.* **229**, 1090-3.

Dorner, A. J., Dorner, L. F., Larsen, G. R., Wimmer, E. and Anderson, C. W. (1982). Identification of the initiation site of poliovirus polyprotein synthesis. *J Virol.* **42**, 1017-1028.

Dorner, A. J., Semler, B. L., Jackson, R. J., Hanecak, R., Duprey, E. and Wimmer, E. (1984). *In vitro* translation of poliovirus RNA: utilization of internal initiation sites in reticulocyte lysate. *J Virol.* **50**, 507-514.

Duke, G. M., Hoffman, M.A., and Palmenberg, A.C. (1992). Sequence and structural elements that contribute to efficient encephalomycarditis virus RNA translation. *J. Virol.* **66**, 1602-1609.

Emini, E. A., Elzinga, M. and Wimmer, E. (1982). Carboxy-terminal analysis of poliovirus proteins: termination of poliovirus RNA translation and location of unique poliovirus polyprotein cleavage sites. *J Virol.* **42**, 194-199.

Etchison, D., Milburn, S., Edery, 1., Sonenberg, N. and HersheyJWB (1982). Inhibition of HeLa cell protein synthesis following poliovirus infection correlates with the proteolysis of a 220,000 dalton polypeptide associated with eucaryotic initiation factor 3 and a cap binding protein complex. *J. Biol. Chem.* **257**, 14806-14810 .

Evans, D. J., McKeating, J., Meredith, J. M., Burke, K. L., Katrak, K., John, A., Ferguson, M., Minor, P. D., Weiss, R. A. and Almond, J. W. (1989). An engineered poliovirus chimaera elicits broadly reactive HIV-1 neutralizing antibodies. *Nature.* **339**, 385-388.

Evans, D. M., Dunn, G., Minor, P. D., Schild, G. C., Cann, A. J., Stanway, G., Almond, J. W., Currey, K. and Maizel, J. V. (1985). Increased neurovirulence associated with a single nucleotide change in a noncoding region of the Sabin type 3 poliovaccine genome. *Nature.* **314**, 548-550.

Everaert, L., Vrijsen, R. and Boeye, A. (1989). Eclipse products of poliovirus after cold-synchronized infection of HeLa cells. *Virology.* **171**, 76-82.

Fenwick, M. and Cooper, P. (1962). Early interactions between poliovirus and ERK cells. Some observations on the nature and significance of the rejected particles. *Virology.* **18**, 212-223 .

Filman, D. J., Syed, R., Chow, M., Macadam, A. J., Minor, P. D. and Hogle, J. M. (1989). Structural factors that control conformational transitions and serotype specificity in type 3 poliovirus. *EMBO J.* **8**, 1567-1579.

Flanegan, J., Pettersson, R., Ambros, V., Hewlett, M. and Baltimore, D. (1977). Covalent linkage of a protein to a defined nucleotide sequence at the 5'-terminus of virion and replicative intermediate RNAs of poliovirus. *Proc. Natl. Acad. Sci. USA.* **74**, 961-965 .

Freistadt, M. S., Kaplan, G. and Racaniello, V. R. (1990). Heterogeneous expression of poliovirus receptor-related proteins in human cells and tissues. *Mol Cell Biol.* **10**, 5700-5706.

Freistadt, M. S. and Racaniello, V. R. (1991). Mutational analysis of the cellular receptor for poliovirus. *J Virol.* **65**, 3873-3876.

Fricks, C. E. and Hogle, J. M. (1990). Cell-induced conformational change in poliovirus: externalization of the amino terminus of VP1 is responsible for liposome binding. *J. Virol.* **64**, 1934-1945.

Garcia-Blanco, M. A., Jamison, S. F. and Sharp, P. A. (1989). Identification and purification of a 62,000-dalton protein that binds specifically to the polypyrimidine tract of introns. *Genes and Development.* **3**, 1874-1886.

Ghetti, A., Pinol, R. S., Michael, W. M., Morandi, C. and Dreyfuss, G. (1992). hnRNP I, the polypyrimidine tract-binding protein: distinct nuclear localization and association with hnRNAs. *Nucleic Acids Res.* **20**, 3671-3678.

Gil, A., Sharp, P. A., Jamison, S. F. and Garcia-Blanco, M. A. (1991). Characterization of cDNAs encoding the polypyrimidine tract-binding protein. *Genes-Dev.* **5**, 1224-1236.

Grant, R. A., Filman, D. J., Fujinami, R. S., Icenogle, J. P. and Hogle, J. M. (1992). Three-dimensional structure of Theilers virus. *Proc. Natl. Acad. Sci. U.S.A.* **89**, 2061-2065.

Gromeier, M. and Wetz, K. (1990). Kinetics of poliovirus uncoating in HeLa cells in a nonacidic environment. *J Virol.* **64**, 3590-3597.

Guttman, N. and Baltimore, D. (1977). A plasma membrane component able to bind and altervirions of poliovirus type 1: studies on cell free alteration using a simplified assay. *Virology.* **82**, 25-36.

Haller, A. A. and Semler, B. L. (1992). Linker scanning mutagenesis of the internal ribosome entry site of poliovirus RNA. *J. Virol.* **66**, 5075-5086.

Hämmerle, T., Hellen, C. U. T. and Wimmer, E. (1991). Site-directed mutagenesis of the putative catalytic triad of poliovirus 3C proteinase. *J. Biol. Chem.* **266**, 5412-5416.

Hanecak, R., Semler, B. L., Anderson, C. W. and Wimmer, E. (1982). Proteolytic processing of poliovirus polypeptide P3-7c inhibit cleavage at glutamine-glycine pairs. *Proc. Natl. Acad. Sci. USA.* **79**, 3973-3977.

Hanecak, R., Semler, B. L., Ariga, H., Anderson, C. W. and Wimmer, E. (1984). Expression of a cloned gene segment of poliovirus in E. coli: evidence for autocatalytic production of the viral proteinase. *Cell.* **37**, 1063-1073.

Harber, J. J., Bradley, J., Anderson, C. W. and Wimmer, E. (1991). Catalysis of poliovirus VP0 maturation cleavage is not mediated by serine 10 of VP2. *J Virol.* **65**, 326-334.

Harris, K. S., Hellen, C. U. T. and Wimmer, E. (1990). Proteolytic processing in the replication of picornaviruses. *Seminars in Virology.* **1**, 323-333.

Harris, K. S., Reddigari, S. R., Nicklin, M. J. H., Hämmerle, T. and Wimmer, E. (1993). Purification and characterization of poliovirus polypeptide 3CD, a proteinase and a precursor for RNA polymerase. *submitted.*

Harrison, S. C. (1989). Common Features in the Design of Small RNA Viruses. *In "Concepts in Viral Pathogenesis III"* (A. L. Notkins and M. B. A. Oldstone, ed.), Springer-Verlag, New York.

Hellen, C. U.T, Facke, M., Krausslich, H. G., Lee, C. K. and Wimmer, E. (1991). Characterization of poliovirus 2A proteinase by mutational analysis: residues required for autocatalytic activity are essential for induction of cleavage of eukaryotic initiation factor 4F polypeptide p220. *J Virol.* **65**, 4226-4231.

Hellen, C. U. T., de Crombrugghe, M., Wimmer, E., Seeger, M. A. and Kaufman, T. C. (1989). Amalgamated sequences. *Nature.* **340**, 682.

Hellen, C. U. T., Lee, C. K. and Wimmer, E. (1992a). Determinants of substrate recognition by poliovirus 2A proteinase. *J Virol.* **66**, 3330-3338.

Hellen, C. U. T. and Wimmer, E. (1992b). Maturation of poliovirus capsid proteins. *Virology.* **187**, 391-397.

Hellen, C. U. T. and Wimmer, E. (1992c). The role of proteolytic processing in the morphogenesis of virus particles. *Experientia.* **48**, 201-215.

Hellen, C. U. T., Witherell, G. W., Schmid, M., Shin, S. H., Pestova, T. V., Gil, A. and Wimmer, E. (1992d). pPTB and p57 are identical RNA binding proteins involved in nuclear splicing and cytoplasmic translation. *submitted.*

Hewlett, M. J., Rose, J. K. and Baltimore, D. (1976). 5'-Terminal structure of poliovirus is pUp. *Proc. Natl. Acad. Sci.* **73**, 327-330.

Hogle, J. M., Chow, M. and Filman, D. J. (1985). The three dimensional structure of poliovirus at 2.9 A resolution. *Science.* **229**, 1358-1365.

Hogle, J. M., Chow, M. and Filman, D. J. (1986). The three dimensional structure of poliovirus: Implications for assembly and immune recognition. *In "Postitive Strand RNA Viruses"* (M. A. Brinton and R. R. Rueckert, ed.), Alan Liss, Inc., New York.

Holland, J. (1964). Enterovirus entrance into specific host cells and subsequent alterations of cell protein and nucleic acid synthesis. *Bacteriol. Rev.* **28**, 3.

Holland, J. and McLaren, L. (1961). The location and nature of enterovirus receptors in susceptible cells. *J. Exp. Med.* **114**, 161-171 .

Holland, J. J. (1961). Receptor affinities as major determinants of enterovirus tissue tropisms in humans. *Virology.* **15**, 312-326.

Iizuka, N., Kohara, M., Hagino, Y. K., Abe, S., Komatsu, T., Tago, K., Arita, M. and Nomoto, A. (1989). Construction of less neurovirulent polioviruses by introducing deletions into the 5' noncoding sequence of the genome. *J Virol.* **63**, 5354-5363.

Jackson, R. J., Howell, M. T. and Kaminski, A. (1990). The novel mechanism of initiation of picornavirus RNA translation. *TIBS.* **15**, 477-483.

Jang, S. K., Davies, M. V., Kaufman, R. J. and Wimmer, E. (1989). Initiation of protein synthesis by internal entry of ribosomes into the 5' nontranslated region of encephalomyocarditis virus RNA *in vivo.* *J Virol.* **63**, 1651-1660.

Jang, S. K., Krausslich, H. G., Nicklin, M. J., Duke, G. M., Palmenberg, A. C. and Wimmer, E. (1988). A segment of the 5' nontranslated region of encephalomyocarditis virus RNA directs internal entry of ribosomes during *in vitro* translation. *J Virol.* **62**, 2636-2643.

Jang, S. K., Pestova, T. V., Hellen, C. U. T., Witherell, G. W. and Wimmer, E. (1990a). Cap-Independent translation of picornavirus RNAs: structure and function of the internal ribosomal entry site. *Enzyme.* **44**, 292-309.

Jang, S. K. and Wimmer, E. (1990b). Cap-independent translation of encephalomyocarditis virus RNA: Structral elements of the internal ribosomal entry site and involvement of a cellular 57-kDa RNA-binding protein. *Genes and Dev.* **4**, 1560-1572.

Jenkins, O., Cason, J., Burke, K. L., Lunney, D., Gillen, A., Patel, D., McCance, D. J. and Almond, J. W. (1990). An antigen chimaera of poliovirus induces antibodies against human papillomavirus type 16. *J. Virol.* **64**, 1201-1206.

Joklik, W. and Darnell, J. (1961). The adsorption and early fate of purified poliovirus in HeLa cells. *Virology.* **13**, 439-447.

Jore, J., De-Geus B., Jackson, R. J., Pouwels, P. H. and Enger-Valk, B.E. (1988). Poliovirus protein 3CD is the active protease for processing of the precursor protein P1 *in vitro.* J. Gen Virol **69**, 1627-36.

Kaminski, A., Howell, M. T., and Jackson, R. J. (1990). Initiation of encephalomyocarditis virus RNA translation: the authentic initiation site is not selected by a scanning mechanism. *EMBO J.* **9**, 3753-3759.

Kaplan, G., Peters, D. and Racaniello, V. R. (1990). Poliovirus mutants resistant to neutralization with soluble cell receptors. *Science.* **250**, 1596-1599.

Kim, S. S., Smith, T. J., Chapman, M. S., Rossmann, M. C., Pevear, D. C., Dutko, F. J., Felock, P. J., Diana, G. D. and McKinlay, M. A. (1989). Crystal structure of human rhinovirus serotype 1A (HRV1A). *J Mol Biol.* **210**, 91-111.

Kirkegaard, K. (1990). Mutations in VP1 of poliovirus specifically affect both encapsidation and release of viral RNA. *J Virol.* **64**, 195-206.

Kitamura, N., Adler, C. J., Rothberg, P. G., Martinko, J., Nathenson, S. G. and Wimmer, E. (1980). The genome-linked protein of picornaviruses. VII. Genetic mapping of poliovirus VPg by protein and RNA sequence studies. *Cell.* **21**, 295-302.

Kitamura, N., Semler, B. L., Rothberg, P. G., Larsen, G. R., Adler, C. J., Dorner, A. J., Emini, E. A., Hanecak, R., Lee, J. J., van der Werf, S., Anderson, C. W. and Wimmer, E. (1981). Primary structure, gene organization and polypeptide expression of poliovirus RNA. *Nature.* **291**, 547-553.

Kitamura, N. and Wimmer, E. (1980). Sequence of 1060 3'-terminal nucleotides of poliovirus RNA as determined by a modification of the dideoxynucleotide method. *Proc Natl Acad Sci U S A.* **77**, 3196-3200.

Koike, S., Horie, H., Ise, I., Okitsu, A., Yoshida, M., Iizuka, N., Takeuchi, K., Takegami, T. and Nomoto, A. (1990). The poliovirus receptor protein is produced both as membrane-bound and secreted forms. *EMBO J.* **9**, 3217-3224.

Koike, S. and Nomoto, A. (1992). A second gene for the for the african green monkey poliovirus receptor that has no putative N-glycosylation site in the functional N-terminal immunoglobulin-like domain. *J. Virology* **66**, 7059-7066.

Koike, S., Ise, I. and Nomoto, A. (1991). Functional domains of the poliovirus receptor. *Proc Natl Acad Sci U S A.* **88**, 4104-4108.

Koike, S., Taya, C., Kurata, T., Abe, S., Ise, I., Yonekawa, H. and Nomoto, A. (1991a). Transgenic mice susceptible to poliovirus. *Proc Natl Acad Sci U S A.* **88**, 951-955.

Korant, B. D., Lonberg, H. K., Noble, J. and Stasny, J. T. (1972). Naturally occurring and artificially produced components of three rhinoviruses. *Virology.* **48**, 71-86.

Kozak, M. (1978). How do eukaryotic ribosomes select initiation regions in messenger RNA? *Cell.* **15**, 1109-1123.

Kozak, M. (1987). An analysis of 5'-noncoding sequences from 699 vertebrate messenger RNAs. *Nucl. Acids Res.* **15**, 8125-8133.

Kozak, M. (1989a). Context effects and inefficient initiation at non-AUG codons in eukaryotic cell-free translation systems. *Mol. Cell. Biol.* **9**, 5073-5080.

Kozak, M. J. (1989b). The scanning model for translation: An update. *J. Cell Biol.* **108**, 229-241.

Krah, D. and Crowell, R. (1982). A solid-phase assay of solubilized HeLa cell membrane receptors for binding group B coxsackieviruses and polioviruses. *Virology.* **118**, 148-156.

Kräusslich, H.-G. and Wimmer, E. (1988). Viral proteinases. *Ann. Rev. Biochem.* **57**, 701-754.

Kräusslich, H. G., Holscher, C., Reuer, Q., Harber, J. and Wimmer, E. (1990). Myristoylation of the poliovirus polyprotein is required for proteolytic processing of the capsid and for viral infectivity. *J Virol.* **64**, 2433-2436.

Kraüsslich, H. G., Nicklin, M. J., Toyoda, H., Etchison, D. and Wimmer, E. (1987). Poliovirus proteinase 2A induces cleavage of eucaryotic initiation factor 4F polypeptide p220. *J Virol.* **61**, 2711-2718.

Kräusslich, H. G., Nicklin, M. J. H., Lee, C. K. and Wimmer, E. (1988). Polyprotein processing in picornavirus replication. *Biochimie.* **70**, 119-130.

Kuge, S. and Nomoto, A. (1987). Construction of viable deletion and insertion mutants of the Sabin strain type 1 poliovirus: Function of the 5' noncoding sequence in vial replication. *J. Virol.* **61**, 1478-1487.

Kühn, R., Luz, N. and Beck, E. (1990). Functional analysis of the internal translation initiation site of foot-and-mouth disease virus. *J. Virol.* **64**, 4625-4631.

La Monica, N. and Racaniello, V. R. (1989). Differences in replication of attenuated and neurovirulent polioviruses in human neuroblastoma cell line SH-SY5Y. *J. Virol.* **63**, 2357-2360.

Larsen, G. R., Anderson, C. W., Dorner, A. J., Semler, B. L. and Wimmer, E. (1982). Cleavage sites within the poliovirus capsid protein precursors. *J Virol.* **41**, 340-344.

Lee, Y. F., Nomoto, A., Detjen, B. M. and Wimmer, E. (1977). A protein covalently linked to poliovirus genome RNA. *Proc Natl Acad Sci U S A.* **74**, 59-63

Lee, C. K. and Wimmer, E. (1988). Proteolytic Processing of Poliovirus Polyprotein: Elimination of 2Apro-Mediated, Alternative Cleavage of Polypeptide 3CD by *in vitro* Mutagenesis. *Virology.* **166**, 405-414.

Lee, K. and Sonenberg, N. (1982). Inactivation of cap-binding proteins accompanies the shut-off of host protein synthesis by poliovirus. *Proc. Natl. Acad. Sci. USA.* **79**, 3447-3451 .

Lee, K. A., Edery, I., Hanecak, R., Wimmer, E. and Sonenberg, N. (1985). Poliovirus protease 3C (P3-7c) does not cleave P220 of the eucaryotic mRNA cap-binding protein complex. *J Virol.* **55**, 489-493.

Lloyd, R. E., Grubman, M. J. and Ehrenfeld, E. (1988). Relationship of p220 cleavage during picornavirus infection to 2A proteinase sequencing. *J Virol.* **62**, 4216-23.

Lonberg-Holm, K. (1975). The effects of concanavalin A on the early events of infection by rhinovirus type 2 and poliovirus type 2. *J Gen Virol.* **28**, 313-27.

Lonberg-Holm, K., Gosser, L. B. and Kauer, J. C. (1975). Early alteration of poliovirus in infected cells and its specific inhibition. *J Gen Virol.* **27**, 329-342.

Lonberg-Holm, K., Gosser, L. B. and Shimshick, E. J. (1976). Interaction of liposomes with subviral particles of poliovirus type 2 and rhinovirus type 2. *J Virol.* **19**, 746-749.

Lonberg-Holm, K. and Korant, B. D. (1972). Early interaction of rhinoviruses with host cells. *J. Virol.* **9**, 29-40.

Lonberg-Holm, K. and Whiteley, N. M. (1976). Physical and metabolic requirements for early interaction of poliovirus and human rhinovirus with HeLa cells. *J Virol.* **19**, 857-870.

Lu, H. H., Murdin, A. D., Altmeyer, R., Harber, J. and Wimmer, E., ed. (1992). Antigenic hybrids of poliovirus: trivalent polioviruses, and characterization of a poliovirus/rhinovirus hybrid. *In "Vaccines '92"* (H. Ginsberg, F. Brown, R. A. Lerner and R. M. Chanock, series ed.), 287-93. Cold Spring Harbor Press, Cold Spring Harbor, N.Y.

Luo, M., He, C., Toth, K. S., Zhang, C. X. and Lipton, H. L. (1992). Three-dimensional structure of Theiler murine encephalomyelitis virus (BeAn Strain). *Proc. Natl. Acad. Sci. U.S.A.* **89**, 2409-2413.

Luo, M., Vriend, G., Kamer, G., Minor, I., Arnold, E., Rossmann, M. G., Boege, U., Scraba, D. G., Duke, G. M. and Palmenberg, A. C. (1987). The atomic structure of Mengo virus at 3.0 A resolution. *Science.* **235**, 182-91.

Luz, N., and Beck, E (1990). A cellular 57 kDa protein binds to two regions of the internal translation initiation site of foot-and-mouth disease virus. *FEBS Lett.* **269**, 311-314.

Luz, N. and Beck, E. (1991). Interaction of a cellular 57-kilodalton protein with the internal translation initiation site of foot-and-mouth disease virus. *J Virol.* **65**, 6486-6494.

Macejak, D. G. and Sarnow, P. (1991). Internal initiation of translation mediated by the 5' leader of a cellular mRNA. *Nature.* **353**, 90-94.

Madshus, I. H., Olsnes, S. and Sandvig, K. (1984). Different pH requirements for entry of the two picornaviruses, human rhinovirus 2 and murine encephalomyocarditis virus. *Virology.* **139**, 346-357.

Maizol, J., and Summers, D.F. (1968), Evidence for a difference in size and composition of the poliovirus specific polypeptides in infected Hela cells. Virology **36**, 48.

Mannweiler, K., Nobis, P., Hohenberg, H. and Bohn, W. (1990). Immunoelectron microscopy on the topographical distribution of the poliovirus receptor. *J.Gen.Virol.* **71**, 2737-2740.

Mandel, B. (1962). Early stages of virus-cell interaction as studied by using antibody. *Cold Spring Harbor Symp. Quant. Biol.* **27**, 123-136.

Mandel, B. (1967a). The interaction of neutralized poliovirus with HeLa cells. 1. Adsorption. *Virology.* **31**, 238-247.

Mandel, B. (1967b). The relationship between penetration and uncoating of poliovirus in HeLa cells. *Virology.* **31**, 702-712.

Marc, D., Drugeon, G., Haenni, A.-L., Girard, M. and van der Werf, S. (1989). Role of myristoylation of poliovirus capsid protein VP4 as determined by site-directed mutagenesis of its N-terminal sequence. *EMBO J.* **8**, 2661-2668.

Maizol, J. (1963). Evidence for multiple components in the structural proteins of type 1 poliovirus. Biochemistry and Biophysics Research Communications 13, 483.

Maizol, J. and Summers, D.F. (1968). Evidence for a difference in size and composition of the poliovirus specific polypeptides in infected Hela cells. Virology 36, 48.

Marc, D., Drugeon, G., Haenni, A.-L., Girard, M. and van der Werf, S. (1989). Role of myristoylation of poliovirus capsid protein VP4 as determined by site-directed mutagenesis of its N-terminal sequence. *EMBO J.* **8**, 2661-2668.

Marsh, M. and Helenius, A. (1989). Virus entry into animal cells. *Adv. Virus Res.* **36**, 107-151.

Martin, A., Wychowski, C., Couderc, T., Crainic, R., Hogle, J. and Girard, M. (1988). Engineering a poliovirus type 2 antigenic site on a type 1 capsid results in a chimeric virus which is neurovirulent for mice. *EMBO J.* **7**, 2839-2847.

Mason, P. W., Baxt, B., Brown, F., Harber, J., Murdin, A. and Wimmer, E. (1992). Antibody-Complexed Foot-and-Mouth Disease Virus, but Not Poliovirus, Can Infect Cells Via the Fc Receptor. Virology. in Press.

McClelland, A., deBear, J., Yost, S. C., Meyer, A. M., Marlor, C. W. and Greve, J. M. (1991). Identification of monoclonal antibody epitopes and critical residues for rhinovirus binding in domain 1 of intercellular adhesion molecule 1. *Proc Natl Acad Sci U S A.* **88**, 7993-7.

McLaren, L. C., Holland, J. J. and Syverton, J. T. (1959). The mammalian cell-virus relationship I. Attachment of poliovirus to cultivated cells of primate and non-primate origin. *J. Exp. Med.* **109**, 475-485.

Meerovitch, K., Nicholson, R. and Sonenberg, N. (1991). *In vitro* mutational analysis of cis-acting RNA translational elements within the poliovirus type 2 5' untranslated region. *J Virol.* **65**, 5895-5901.

Meerovitch, K., Pelletier, J. and Sonenberg, N. (1989). A cellular protein that binds to the 5'-noncoding region of poliovirus RNA: implications for internal translation initiation. *Genes Dev.* **3**, 1026-1034.

Melnick, J. L. (1990). "Enteroviruses: Polioviruses, coxsackieviruses, echoviruses, and newer enteroviruses. *in "Virology"* (B. Fields, ed.) 549-604. Raven Press, New York.

Mendelsohn, C., Johnson, B., Lionetti, K. A., Nobis, P., Wimmer, E. and Racaniello, V. R. (1986). Transformation of a human poliovirus receptor gene into mouse cells. *Proc Natl Acad Sci U S A.* **83**, 7845-7849.

Mendelsohn, C. L., Wimmer, E. and Racaniello, V. R. (1989). Cellular receptor for poliovirus: molecular cloning, nucleotide sequence, and expression of a new member of the immunoglobulin superfamily. *Cell.* **56**, 855-865.

Miller, D. A., Miller, O. J., Dev, V. G., Hashmi, S., Tantravahi, R., Medrano, L. and Green, H. (1974). Human chromosome 19 carries a poliovirus receptor gene. *Cell.* **1**, 167-173.

Minor, P., Schild, G., Bootman, J., Evans, D., Ferguson, M., Reeve, P., Spitz, M., Stanway, G., Cann, A., Hauptmann, R., Clarke, L., Mountford, R. and Almond, J. (1983). Location and primary structure of a major antigenic site for poliovirus neutralization. *Nature.* **301**, 674 .

Minor, P. D. (1990). Antigenic structure of picornaviruses. *In "Picornaviruses"* (V. R. Racaniello, ed.), 121-154. Springer-Verlag, Berlin.

Minor, P. D., Ferguson, M., Evans, D. M. A. and Icenogle, J. P. (1986). Antigenic structure of polioviruses of serotypes 1, 2 and 3. *J. Gen. Virol.* **60**, 1283-1291.

Minor, P. D., Ferguson, M., Katrak, K., Wood, D., John, A., Howlett, J., Dunn, G., Burke, K. and Almond, J. W. (1990). Antigenic chimeras of type 1 and type 3 poliovirus involving antigenic site 1. *J. Gen. Virol.* **71**, 2543-2551.

Minor, P. D., Pipkin, P. A., Hockley, D., Schild, G. C. and Almond, J. W. (1984). Monoclonal antibodies which block cellular receptors of poliovirus. *Virus Res.* **1**, 203-212.

Mirzayan, C. M., Pallai, P. and Wimmer, E. (1991). Specificity of the polioviral proteinase 3C towards genetically engineered cleavage sites in the viral capsid. *J. Gen. Virol.* **72**, 1159-1163.

Molla, A., Jang, S. K., Paul, A. V., Reuer, Q. and Wimmer, E. (1992). Cardioviral internal ribosomal entry site is functional in a genetically engineered dicistronic poliovirus. *Nature.* **356**, 255-257.

Molla, A., Paul, A. V. and Wimmer, E. (1991). Cell-free, de novo synthesis of poliovirus. *Science.* **254**, 1647-1651.

Morrison, M. E. and Racaniello, V. R. (1992). Molecular Cloning and Expression of a Murine Homolog of the Human Poliovirus Receptor Gene. *Journal of Virology.* **66**, 2807-2813.

Murdin, A. D., Kameda, A., Murray, M. G. and Wimmer, E. (1991). Phenotypic characterization of antigenic hybrids of poliovirus. *Microb Pathog.* **10**, 39-45.

Murdin, A. D., Lu, H. H., Murray, M. G. and Wimmer, E. (1992). Poliovirus antigenic hybrids simultaneously expressing antigenic determinants from all three serotypes. *J Gen Virol.* **73**, 607-611.

Murdin, A. D., Murray, M. G. and Wimmer, E. (1990). "Novel Approaches to Picornavirus Vaccines" in *Viral Vaccines.* 301-323. Wiley-Liss, Inc., New York.

Murdin, A. D. and Wimmer, E. (1989). Construction of a poliovirus type 1/type 2 antigenic hybrid by manipulation of neutralization antigenic site II. *J Virol.* **63**, 5251-5257.

Murray, M. G., Bradley, J., Yang, X. F., Wimmer, E., Moss, E. G. and Racaniello, V. R. (1988a). Poliovirus host range is determined by a short amino acid sequence in neutralization antigenic site I. *Science.* **241**, 213-215.

Murray, M. G., Kuhn, R. J., Arita, M., Kawamura, N., Nomoto, A. and Wimmer, E. (1988b). Poliovirus type 1/type 3 antigenic hybrid virus constructed *in vitro* elicits type 1 and type 3 neutralizing antibodies in rabbits and monkeys. *Proc Natl Acad Sci U S A.* **85**, 3203-3207.

Muzychenko, A. R., Lipskaya, G., Maslova, S. V., Svitkin, Y. V., Pilipenko, E. V., Nottay, B. K., Kew, O. M. and Agol, V. I. (1991). Coupled mutations in the 5'-untranslated region of the Sabin poliovirus strains during *in vivo* passages: structural and functional implications. *Virus Res.* **21**, 111-122.

Nicholson, R., Pelletier, J., Le, S. Y. and Sonenberg, N. (1991). Structural and functional analysis of the ribosome landing pad of poliovirus type 2: *in vivo* translation studies. *J Virol.* **65**, 5886-5894.

Nicklin, M. J. H., Harris, K. S., Pallai, P. V. and Wimmer, E. (1988). Poliovirus proteinase 3C: Large-scale expression, purification, and specific cleavage activity on natural and synthetic substrates *in vitro*. *J. Virol.* **62**, 4586-4593.

Nicklin, M. J. H., Kräusslich, H. G., Toyoda, H., Dunn, J. J. and Wimmer, E. (1987). Poliovirus polypeptide precursors: Expression *in vitro* and processing by 3C and 2A proteinases. *Proc. Natl. Acad. Sci. USA.* **84**, 4002-4006.

Nobis, P., Zibirre, R., Meyer, G., Kühne, J., Warnecke, G. and Koch, G. (1985). Production of a monoclonal antibody against an epitope on HeLa cells that is the functional poliovirus binding site. *J.Gen.Virol.* **66**, 2563-2569.

Nomoto, A., Kitamura, N., Golini, F. and Wimmer, E. (1977). The 5'-terminal structures of poliovirion RNA and poliovirus mRNA differ only in the genome-linked protein VPg. *Proc Natl Acad Sci U S A.* **74**, 5345-5349.

Nomoto, A., Lee, Y. F. and Wimmer, E. (1976). The 5' end of poliovirus mRNA is not capped with m^7G(5')ppp(5')Np. *Proc. Natl. Acad. Sci. USA.* **73**, 375-380.

Page, G. S., Mosser, A. G., Hogle, J. M., Filman, D. J., Rueckert, R. R. and Chow, M. (1988). Three-dimensional structure of poliovirus serotype 1 neutralizing determinants. *J Virol.* **62**, 1781-1794.

Pallansch, M. A., Kew, O. M., Semler, B. L., Omilianowski, D. R., Anderson, C. W., Wimmer, E. and Rueckert, R. R. (1984). Protein processing map of poliovirus. *J Virol.* **49**, 873-880.

Palmenberg, A. C. (1986). Comparative organization and genome structure in picornaviruses. *UCLA Symp. Mol. Cell. Biol.* **54**, 25-34.

Patton, J. G., Mayer, S. A., Tempst, P. and Nadal-Ginard, B. (1991). Characterization and molecular cloning of polypyrimidine tract-binding protein: A component of a complex necessary for pre-mRNA splicing. *Genes Dev.* **5**, 1237-1251.

Paul, A. V., Schultz, A., Pincus, S. E., Oroszlan, S. and Wimmer, E. (1987). Capsid protein VP4 of poliovirus is N-myristoylated. *Proc Natl Acad Sci U S A.* **84**, 7827-7831.

Pelham, H. R. and Jackson, R. J. (1976). An efficient mRNA-dependent translation system from reticulocyte lysates. *Eur J Biochem.* **67**, 247-256.

Pelletier, J., Flynn, M. E., Kaplan, G., Racaniello, V. and Sonenberg, N. (1988). Mutational analysis of upstream AUG codons of poliovirus RNA. *J Virol.* **62**, 4486-4492.

Pelletier, J. and Sonenberg, N. (1985). Insertion mutagenesis to increase secondary structure within the 5'-noncoding region of eukaryotic mRNA reduces translational efficiency. *Cell.* **40**, 515-526.

Pelletier, J. and Sonenberg, N. (1988). Internal initiation of translation of eukaryotic mRNA directed by a sequence derived from poliovirus RNA. *Nature.* **334**, 320-325.

Pelletier, J. and Sonenberg, N. (1989). Internal binding of eucaryotic ribosomes on poliovirus RNA: translation in HeLa cell extracts. *J Virol.* **63**, 441-444.

Pestova, T. V., Hellen, C. U. T. and Wimmer, E. (1991). Translation of poliovirus RNA: role of an essential cis-acting oligopyrimidine element within the 5' nontranslated region and involvement of a cellular 57-kilodalton protein. *J Virol.* **65**, 6194-6204.

Pestova, T. V., Maslova, S. V., Potapov, V. K. and Agol, V. I. (1989). Distinct modes of poliovirus polyprotein initiation *in vitro*. *Virus Res.* **14**, 107-118.

Pevear, D. C., Fancher, M. J., Felock, P. J., Rossmann, M. G., Miller, M. S., Diana, G., Treasurywala, A. M., McKinlay, M. A. and Dutko, F. J. (1989). Conformational change in the floor of the human rhinovirus canyon blocks adsorption to HeLa cell receptors. *J Virol.* **63**, 2002-2007.

Pilipenko, E. V., Gmyl, A. P., Maslova, S. V., Svitkin, Y. V., Sinyakov, A. N. and Agol, V. I. (1992). Prokaryotic-like cis elements in the cap-independent internal initiation of translation on picornavirus RNA. *Cell.* **68**, 119-131.

Pilipenko, E. V., Blinov, V. M., Romanova, L. I., Sinyakov, A. N., Maslova, S. V. and Agol, V. I. (1989). Conserved structural domains in the 5'-untranslated region of picornaviral genomes: an analysis of the segment controlling translation and neurovirulence. *Virology.* **168**, 201-209.

Pilipenko, E. V., Blinov, V. M., Chernov, B. K., Dmitrieva, T. M. and Agol, V. I. (1989). Conservation of the secondary structure elements of the 5'-untranslated region of cardio- and aphthovirus RNAs. *Nucleic Acids Res.* **17**, 5701-11.

Putnak, J. and Phillips, B. (1982). Poliovirus empty capsid morphogenesis: evidence for conformational differences between self- and extract-assembled empty capsids. *J. Virol.* **41**, 792-800.

Randrup, A. (1954). Ultracentrifugation of the virus of foot-and-mouth disease. IV. Splitting of the greater complement fixed particle. *Acta. path. microbiol. scand.* **34**, 366-374.

Ren, R. E. and Racaniello, V.R. (1992). Human Poliovirus Receptor Gene Expression and Poliovirus Tissue Tropism in Transgenic Mice. *Journal of Virology.* **66**, 296-304.

Rossmann, M. G. (1989a). The canyon hypothesis. *Viral Immunol.* **2**, 143-161.

Rossmann, M. G., and J. E. Johnson (1989b). Icosahedral RNA virus structure. *Annu. Rev. Biochem.* **58**, 533-573.

Rossmann, M. G., Arnold, E., Erickson, J. W., Frankenberger, E. A., Griffith, J. P., Hecht, H. J., Johnson, J. E., Kamer, G., Luo, M., Mosser, A. G., Ruecker, R. R., Sherry, B. and Vriend, G. (1985). Structure of a human common cold virus and functional relationship to other picornaviruses. *Nature.* **317**, 145-153.

Rossmann, M. G. and Palmenberg, A. C. (1988). Conservation of the putative receptor attachment site in picornaviruses. *Virology.* **164**, 373-382.

Rueckert, R. R. and Wimmer, E. (1984). Systematic nomenclature of picornavirus proteins. *J. Virol.* **50**, 957-959.

Sabin, A. B. (1985). Oral poliovirus vaccine: history of its development and use and current challenge to eliminate poliomyelitis from the world. *J. Infect. Dis.* **151**, 420-436.

Salk, J. (1984). One-dose immunization against paralytic poliomyelitis using a noninfectious vaccine. *Rev. Infect. Dis.* **6**, S444-450.

Sarnow, P. (1989). Translation of glucose-regulated protein 78/immunoglobin heavy-chain binding protein mRNA is increased in poliovirus-infected cells at a time when cap-dependent translation of cellular mRNAs is inhibited. *Proc. Natl. Acad. Sci. USA.* **86**, 5795-5799.

Scheper, G. C., Thomas, A. A. and Voorma, H. O. (1991). The 5' untranslated region of encephalomyocarditis virus contains a sequence for very efficient binding of eukaryotic initiation factor eIF-2/2B. *Biochim Biophys Acta.* **1089**, 220-226.

Schonk, D., van Dijk, P., Riegmann, P., Trapman, J., Holm, C., Willcocks, T. C., Sillekens, P., van Venrooij, W., Wimmer, E., and Geurts, K. A. (1990). Assignment of seven genes to distinct intervals on the midportion of human chromosome 19q surrounding the myotonic dystrophy gene region. *Cytogenet Cell Genet.* **54**, 15-19.

Schwemmle, M., Schickinger, J., Bader, M., Sarre, T. F. and Hilse, K. (1991). A 60-kDa protein from rabbit reticulocytes specifically recognizes the capped 5' end of beta-globin mRNA. *Eur J Biochem.* **201**, 139-145.

Selinka, H. C., Zibert, A. and Wimmer, E. (1991). Poliovirus can enter and infect mammalian cells by way of an intercellular adhesion molecule 1 pathway. *Proc Natl Acad Sci U S A.* **88,** 3598-3602.

Selinka, H. C., Zibert, A. and Wimmer, E. (1992). A chimeric poliovirus/CD4 receptor confers susceptibility to poliovirus on mouse cells. *J Virol.* **66**, 2523-2526.

Semler, B. L., Hanecak, R., Dorner, L. F., Anderson, C. W. and Wimmer, E. (1983). Poliovirus RNA synthesis *in vitro*: structural elements and antibody inhibition. *Virology.* **126**, 624-35.

Shepley, M. P., Sherry, B. and Weiner, H. L. (1988). Monoclonal antibody identification of a 100-kDa membrane protein in HeLa cells and human spinal cord involved in poliovirus attachment. *PNAS.* **85**, 7743-7747.

Sherry, B., Mosser, A. G., Colonno, R. J. and Rueckert, R. R. (1986). Use of monoclonal antibodies to identify four neutralization immunogens on a common cold picornavirus, human rhinovirus 14. *J. Virol.* **57**, 246-257.

Sherry, B. and Rueckert, R. (1985). Evidence for at least two dominant neutralization antigens on human rhinovirus 14. *J. Virol.* **53**, 137-143.

Sonenberg, N. (1990). Poliovirus translation. *Curr. Top. Microbiol. Immunol.* **161**, 23-47.

Svitkin, Y. V., Cammack, N., Minor, P. D. and Almond, J. W. (1990). Translation deficiency of the sabin type 3 poliovirus genome: Association with an attenulating mutation C_{472} to U. *Virology.* **175**, 103-109.

Svitkin, Y. V., Maslova, S. V. and Agol, V. I. (1985). The genomes of attenuated and virulent poliovirus strains differ in their *in vitro* translation efficiencies. *Virology.* **147**, 243-252.

Svitkin, Y. V., Pestova, T. V., Maslova, S. V. and Agol, V. I. (1988). Point mutations modify the response of poliovirus RNA to a translation initiation factor: a comparison of neurovirulent and attenuated strains. *Virology.* **166**, 394-404.

Toyoda, H., Nicklin, M. J., Murray, M. G., Anderson, C. W., Dunn, J. J., Studier, F. W. and Wimmer, E. (1986). A second virus-encoded proteinase involved in proteolytic processing of poliovirus polyprotein. *Cell.* **45**, 761-770.

Tsukiyama-Kohara, K., Iizuka, N., Kohara, M., and Nomoto, A. (1992). Internal ribosome entry site within hepatitis C virus RNA. *J. Virol.* **66**, 1476-1483.

van der Werf, S., Bradley, J., Wimmer, E., Studier, F. W. and Dunn, J. J. (1986). Synthesis of infectious poliovirus RNA by purified T7 RNA polymerase. *Proc. Natl. Acad. Sci. USA.* **83**, 2330-2334.

Wetz, K. and Kucinski, T. (1991). Influence of different ionic and pH environments on structural alterations of poliovirus and their possible relation to virus uncoating. *J. Gen. Virol.* **72**, 2541-2544.

Wiegers, K. and Dernick, R. (1985). Evidence for conformational changes of poliovirus precursor particles during virus morphogenesis. *J. Gen. Virol.* **66**, 1037-1044.

Witherell, G. W. and Wimmer, E. (1992). Cap-independent translation of picornavirus RNAs. *submitted.*

Wittman, G. (1990). Virustrager bei Maul- und Klauenseuche. Ubersichtsreferat. *Berliner und Munchener Tierarztliche Wochenschrift.* **103**, 145-150.

Wyckoff, E. E., Hershey, J. W. B. and Ehrenfeld, E. (1990). Eukakryotic initiation factor 3 is required for poliovirus 2A protease-induced cleavage of the p220 component of eukaryotic initiation factor 4F. *Proc. Natl. Acad. Sci. USA.* **87**, 9529-9533.

Yeates, T. O., Jacobson, D. H., Martin, A., Wychowski, C., Girard, M., Filman, D. J. and Hogle, J. M. (1991). Three-dimensional structure of a mouse-adapted type 2/type 1 poliovirus chimera. *Embo J.* **10**, 2331-2341.

Yogo, Y. and Wimmer, E. (1972). Polyadenylic acid at the 3'-terminus of poliovirus RNA. *Proc Natl Acad Sci U S A.* **69**, 1877-82.

Yoon, J. W., Eun, H. M., Essani, K., Roncari, D. A. and Bryan, L. E. (1987). Possible mechanisms in the pathogenesis of virus-induced diabetes mellitus. *Clin Invest Med.* **10**, 450-456.

Ypma-Wong. M. F., Dewalt, P. G., Johnson, V. H., Lamb, J. G. and Semler, B. L. (1988). Protein 3CD is the major poliovirus proteinase responsible for cleavage of the P1 capsid precursor. *Virology.* **166**, 265-70.

Zeichhardt, H., Wetz, K., Willingmann, P. and Habermehl, K. O. (1985). Entry of poliovirus type 1 and Mouse Elberfeld (ME) virus into HEp-2 cells: receptor-mediated endocytosis and endosomal or lysosomal uncoating. *J Gen Virol.* **66**, 483-492.

Zibert, A., Selinka, H. C., Elroy, S. O., Moss, B. and Wimmer, E. (1991). Vaccinia virus-mediated expression and identification of the human poliovirus receptor. *Virology.* **182**, 250-259.

Zibert, A., Selinka, H. C., Elroy, S. O. and Wimmer, E. (1992a). The two N-terminal domains of the poliovirus receptor are sufficient for neutralization of virus. *Vir. Res.* (in press).

Zibert, A. and Wimmer, E. (1992b). N-glycosylation of the virus binding domain is not essential for function of the human poliovirus receptor. *J. Virology* **66**, 7368-7373.

Zibert, A. and Wimmer, E. (1992c). The receptor for poliovirus, and other picornaviruses. *In "Techniques and Applications of Genome Research"* (K. Adolph, ed.), Academic Press, Minneapolis.(in press)

3CD CLEAVAGE OF THE POLIOVIRUS P1 PRECURSOR: A MODEL FOR COMPLEX PROTEINASE/SUBSTRATE INTERACTIONS

Wade S. Blair, Xiaoyu Li, and
Bert L. Semler

Department of Microbiology
and Molecular Genetics
University of California
Irvine, California 92717

INTRODUCTION

Eukaryotic viruses have evolved numerous strategies of replication that allow the propagation of their nucleic acid genomes in a potentially hostile host environment. Such strategies often include the limitation of genomic RNA or DNA sequences to only those essential, the production of multiple gene products from a single viral mRNA species, the use of host-cellular replication and translation machinery, and the selective inhibition of host-cell gene expression. Storage of viral genetic information in the form of ribonucleic acid is an additional replication strategy employed by a subgroup of eukaryotic viruses (RNA viruses). Most RNA viruses (excluding orthomyxoviruses and retroviruses) have eliminated the need for a nuclear phase in their viral life cycles, thus facilitating viral gene expression. An instructive example of a well-studied RNA virus which incorporates all of the above mentioned strategies to orchestrate a productive and efficient replication cycle is poliovirus, a member of the family *Picornaviridae*. Poliovirus is a small, positive-sense RNA virus that encodes at least 14 distinct viral gene products within a monocistronic genome of 7.5 kb (Rueckert, 1985; Lawson and Semler, 1990). This is achieved by translation of the viral mRNA into a single polyprotein that is subsequently cleaved into mature structural and nonstructural viral proteins by viral proteinases 2A and 3C. Proteinase 2A specifically recognizes a tyrosine-glycine

Regulation of Gene Expression in Animal Viruses, Edited by
L. Carrasco, *et al.*, Plenum Press, New York, 1993

(Y-G) amino acid pair and completes the initial cleavage on the polyprotein at the P1/P2 junction (Toyoda et al., 1986). The remainder of the important cleavages within the poliovirus P2 and P3 regions of the polyprotein is mediated by proteinase 3C, which exclusively recognizes glutamine-glycine (Q-G) amino acid pairs (Semler et al., 1981; Emini et al., 1982; Hanecak et al., 1982; Larsen et al., 1982). Cleavage of the P1 precursor requires polymerase sequences (3D) as well as 3C sequences in the form of proteinase 3CD (Jore et al. 1988; Ypma-Wong et al., 1988a). The poliovirus genomic RNA is also characterized by a large untranslated region (~740 nt) located at the 5' end (5' UTR). The 5' UTR facilitates the translation of viral mRNA by host-cell translation machinery via a novel cap-independent mechanism of translation initiation (Pelletier et al., 1988; Trono et al., 1988). The use of a cap-independent translation initiation mechanism allows for the selective inhibition of host-cell cap-dependent translation of eukaryotic mRNAs, an event mediated by proteinase 2A (Etchison et al., 1982; Sun and Baltimore, 1989).

Perhaps one of the most unique features of poliovirus gene expression is the complete dependence on proteolysis for the production of mature gene products. The uniqueness manifests itself in the complicated set of viral proteinase/substrate interactions required to insure proteolytic specificity or fidelity. For example, substrate recognition by proteinase 3C activity seems to depend on recognition of the cleaved Q-G amino acid pair, as well as recognition of other primary amino acid sequence and structural substrate

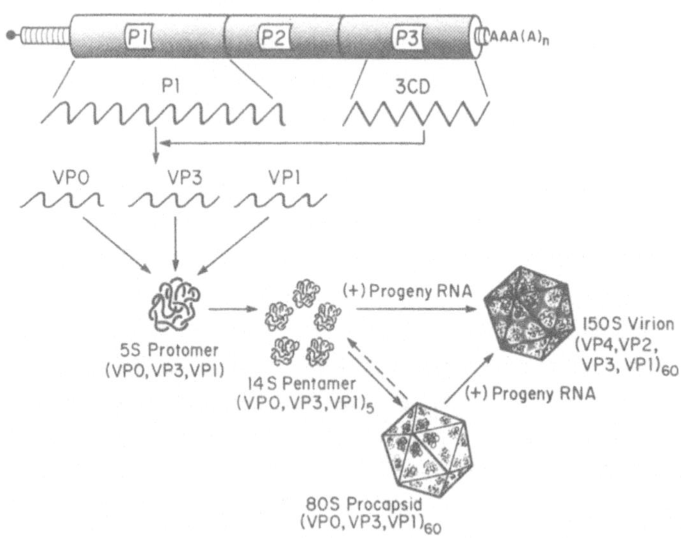

Figure 1. P1 processing cascade. An abbreviated poliovirus processing cascade showing cleavage of the capsid protein precursor (P1) by proteinase 3CD and putative pathways of virion assembly. Poliovirus 5' noncoding and 3' noncoding regions of the genomic RNA are denoted by hatched cylinders. Encoded P1, P2 and P3 regions are denoted by shaded cylinders, and the filled dot corresponds to genome-linked protein (VPg). Wave form and jagged lines denote capsid proteins and proteinase 3CD, respectively.

determinants. This is most apparent for the interaction of proteinase 3CD with its substrate, the P1 precursor (discussed below). Also unique are the poliovirus (or picornavirus) 2A and 3C proteinases themselves. Although these proteinases utilize cysteine nucleophilic residues, 2A and 3C share sequence homology to trypsin-like serine proteinases rather than members of the cysteine proteinase family (Bazan and Fletterick, 1988; Gorbalenya et al., 1989). In addition to sequence homology, poliovirus 3C exhibits functional homology to serine proteases (Lawson and Semler, 1990). It seems, therefore, that poliovirus 2A and 3C comprise, together with other picornaviral proteinases, a novel class of serine-like proteinases.

Presented below is a detailed discussion of one important aspect of the proteolytic regulation of poliovirus gene expression, the determinants required for the interaction of proteinase 3CD with its cognate substrate P1. The 3CD-mediated cleavage of P1 serves a pivotal role in the poliovirus replication cycle by initiating the process of virion assembly and possibly effecting a temporal switch from viral RNA synthesis to virion specific RNA packaging. To insure that only authentically cleaved capsid proteins assemble into virion structural intermediates and associate with newly synthesized viral RNA, the stringent 3CD/P1 interaction incorporates all of the determinants required for 3C recognition of P2 and P3 precursor polypeptides as well as additional enzyme/substrate determinants. 3CD-mediated P1 processing may, therefore, serve as a complete model for the proteinase/substrate interactive complexity evolved to insure the fidelity of poliovirus gene expression. We will examine the primary sequence and structural substrate determinants and briefly describe some enzymatic determinants required for P1 recognition and cleavage by proteinase 3CD in the context of previous and ongoing studies.

THE P1 PROCESSING CASCADE

Processing of the capsid protein precursor is initiated by cotranslational 2A-mediated cleavage at the P1/P2 junction. After liberation from the polyprotein, P1 is subsequently cleaved at Q-G amino acid pairs by proteinase 3CD to yield capsid protein precursor VP0 and capsid proteins VP3 and VP1. Cleavage of P1 into such products allows the assembly of 5S protomers into 14S pentamer structures, which specifically aggregate into 80S procapsids. Newly synthesized viral RNA associates with 14S pentamer structures or 80S procapsids, inducing cleavage of VP0 into capsid proteins VP4 and VP2, by a yet unidentified mechanism, resulting in the formation of mature 150S virions (for review see Putnak and Phillips, 1981; Hellen and Wimmer, 1992). An additional morphogenetic intermediate, the 120S or 150S provirion, which consists of viral RNA encapsidated by virion-like structures comprised of capsid proteins VP0, VP3, and VP1, has been identified (Fernandez-Tomas and Baltimore, 1973; Guttman and Baltimore, 1977; Compton et al., 1990). However, the role of the provirion as a true morphogenetic intermediate is still unclear (Fig. 1).

An additional question about the P1 processing cascade is whether 3CD-mediated cleavage of the P1 precursor occurs sequentially. A recent study suggests that cleavage of P1 is indeed ordered, with cleavage occurring initially at the VP3/VP1 junction followed by cleavage at the VP0/VP3 junction. Reynolds et al., (1992) observed the presence of capsid protein intermediate 1ABC (resulting from initial cleavage of P1 at the VP3/VP1 site) and not intermediate 1CD (resulting from initial cleavage at the VP0/VP3 site) in wild-type poliovirus infected cell extracts. The authors proposed that cleavage of the P1 precursor must occur at the VP3/VP1 cleavage site initially, and that cleavage at the VP3/VP1 site might be required for efficient 3CD-mediated cleavage of the VP0/VP3 site. Contradictory to this study, other investigators have failed to observe significant levels of either intermediate 1ABC or 1CD in PV1 infected cells (Blair and Semler, 1991). However, recent immunoprecipitation analysis of lysates from [^{35}S]-methionine-labeled, PV1-infected cells using antisera directed against VP1 and VP3 demonstrated that both intermediates 1ABC and 1CD can be detected at low levels (Blair and Semler, unpublished observations). In addition, the kinetics of *in vitro* P1 processing suggested the 3CD-mediated generation of both intermediates 1ABC and 1CD from intact P1 precursors (Fig. 2). *In vitro*-generated [^{35}S]-methionine-labeled P1 precursor polypeptides were incubated in the presence of an extract from PV1-infected HeLa cells (a source of proteinase 3CD), and at various times the cleavage products were analyzed on SDS-polyacrylamide gels. Cleavage of P1 resulted in the production of capsid protein intermediates 1ABC and 1CD, which were subsequently cleaved into capsid proteins VP0, VP3, and VP1. The production of both intermediates 1ABC and 1CD *in vivo* and *in vitro* argues against a required initial 3CD-mediated P1 cleavage event at either of the VP0/VP3 or VP3/VP1 cleavage sites, and suggests that P1 processing may not necessarily be sequential. A final argument against a required sequential 3CD/P1 cleavage order comes from the observation that elimination of 3CD-mediated cleavage at either of the VP0/VP3 (discussed below) or VP3/VP1 (Blair and Semler, 1991) poliovirus P1 cleavage sites by substrate amino acid insertion or substitution mutation, has no effect on cleavage of P1 at the remaining wild-type cleavage sites. Similarly, elimination of cleavage at the VP3/VP1 cleavage site in encephalomyocarditis virus (EMCV) P1 precursors has no effect on cleavage at the VP0/VP3 cleavage site (Parks and Palmenberg, 1987). Taken together, these data demonstrate that cleavage at either the VP0/VP3 or the VP3/VP1 cleavage site does not depend on an initial cleavage event. However, cleavage rates at either site may differ *in vivo*, and such differential processing activity may account for the presence of unequal quantities of 1ABC and 1CD in poliovirus infected cells.

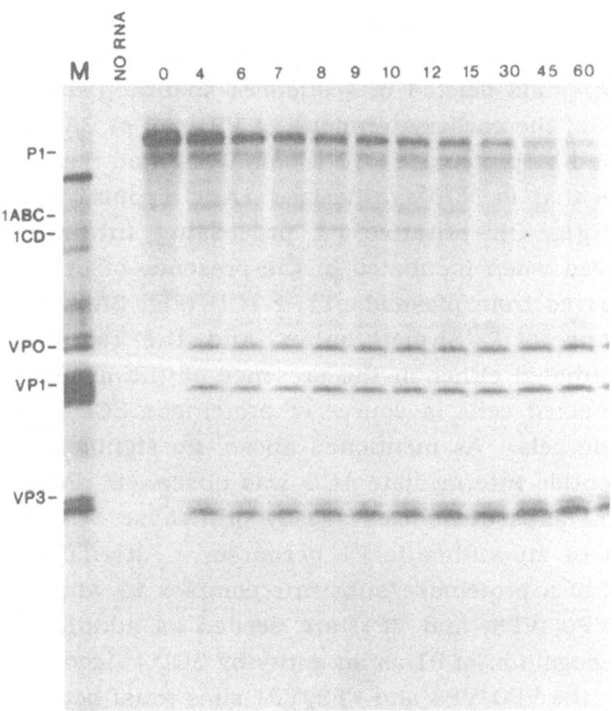

Figure 2. *In vitro* P1 processing kinetics. Subgenomic poliovirus cDNA pT7-P1 was linearized by digestion using restriction endonuclease Hind III at a site located 3' of the poliovirus sequences. Linear poliovirus subgenomic cDNAs were subsequently transcribed *in vitro* using bacteriophage T7 RNA polymerase, and the resulting transcripts were translated (at 0.5 μg ml-1) in the presence of $[^{35}S]$-methionine at 30°C in rabbit reticulocyte lysate supplemented with a HeLa cell S10 cytoplasmic extract (20% of the reaction volume). Translation reactions were terminated after 3h and incubated with RNase A (200 μg ml-1) and cycloheximide (5 μg ml-1) for 20 min. The reaction mixtures were then incubated with an extract from PV1-infected cells (a source of proteinase 3CD) at 30°C and, at various times, the cleavage products were harvested, diluted 10-fold in Laemmli sample buffer, and analyzed by SDS-polyacrylamide gel electrophoresis. NO RNA indicates that translation reactions were incubated in the absence of viral mRNA. Numbers above the lanes designate minutes after the start of the incubation with the extract from infected cells that the cleavage reactions were harvested. M, marker lane of $[^{35}S]$-methionine-labeled proteins isolated from PV1-infected cells.

STRUCTURAL DETERMINANTS REQUIRED FOR 3CD RECOGNITION OF P1 SUBSTRATES

Recognition of the P1 precursor by proteinase 3CD depends on the presentation of a complete and correctly folded substrate. *In vitro* translated P1 precursors truncated at the carboxy-terminus (Ypma-Wong and Semler, 1987b) or containing carboxy-terminal uncleaved 2A sequences (Nicklin et al., 1987) are not cleaved by proteinase 3CD. Additionally, an *in vitro* synthesized P1/P2 fusion polypeptide deleted in sequences spanning the P1/P2 junction (including deletion of the carboxy-terminus of VP1, all of 2A, and a portion of 2B) is not recognized by proteinase 3CD (Ypma-Wong and Semler, 1987a). It is interesting to note that *in vitro* synthesized 1ABC (Ypma-Wong and Semler, 1987b) or 1CD (Fig. 3), putative P1 processing intermediates, remain essentially uncleaved when incubated in the presence of proteinase 3CD. In Fig. 3B, RNAs derived from plasmid pT7P1-1CD (Fig. 3A) were translated *in vitro* in the presence of [^{35}S]-methionine, and the resulting polypeptide products were incubated either in the presence or the absence of an extract from poliovirus-infected cells (a source of proteinase 3CD) and analyzed on SDS-polyacrylamide gels. As mentioned above, no significant 3CD-mediated cleavage of polypeptide intermediate 1CD was observed. Possibly, proper P1 protein folding and subsequent cleavage by proteinase 3CD depends on the initial production of an authentic P1 precursor. Alternatively, P1 may be processed as part of a proteinase/substrate complex in which three complete capsid proteins (VP0, VP3, and VP1) are needed as additional determinants required for the recognition of P1 as an entity by 3CD. According to this latter model, cleavage at the VP0/VP3 and VP3/VP1 sites must occur prior to release of capsid proteins from the complex.

A requirement for intact P1 secondary structures for 3CD recognition of P1 has also been demonstrated. The solved crystal structure of the poliovirion revealed highly ordered capsid protein secondary structures, which were predicted to be present in the capsid protein precursor (Hogle et al., 1985; Ypma-Wong et al., 1988b). *In vitro*-generated P1 precursors containing six amino acid insertion mutations designed to disrupt predicted β-sheet secondary structures were not cleaved by proteinase 3CD, while P1 precursors containing six amino acid insertions in regions predicted to be devoid of defined secondary structure motifs were processed by 3CD (Ypma-Wong et al., 1988b). Surprisingly, disruption of secondary structures in any capsid protein (VP0, VP3, or VP1) interrupted 3CD-mediated cleavage of the entire P1 molecule, suggesting that P1 processing determinants depend on an intact P1 tertiary structure. P1 tertiary structure may be specifically recognized by proteinase 3CD. Alternatively, P1 tertiary structure may be required for correct presentation or to maintain the accessibility of P1 cleavage sites to proteinase 3CD. The above mentioned study (Ypma-Wong et al., 1988) as well as a later study (Mirzayan et al., 1991), demonstrated that introduction of amino acid insertion mutations containing Q-G pairs into regions devoid of secondary

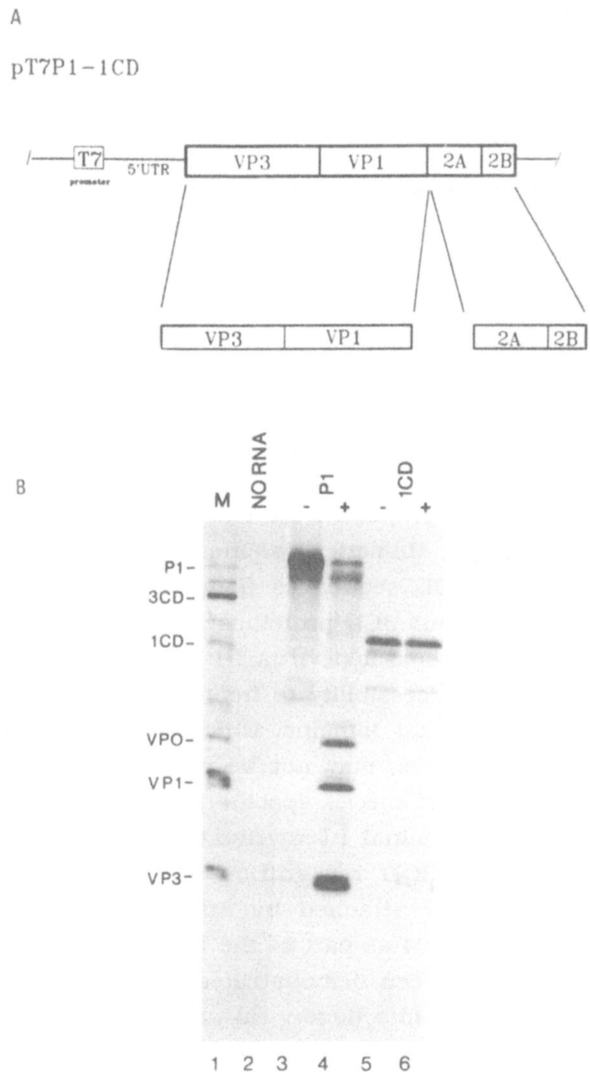

Figure 3. *In vitro* processing profile of 1CD polypeptide intermediates derived from plasmid pT7P1-1CD. (A) Diagram of subgenomic cDNA pT7P1-1CD. pT7P1-1CD contains the bacteriophage T7 promoter, the PV1 5' untranslated region (5' UTR), and sequences encoding authentic PV1 polypeptides VP3, VP1, 2A, and a portion of 2B located directly downstream from the 5' UTR. (B) RNAs were transcribed *in vitro* from plasmids pT7-P1 and pT7P1-1CD and translated *in vitro* in the presence of [^{35}S]-methionine as described in Figure 2. *In vitro* translation products were incubated either in the presence (+) or absence (-) of an extract from PV1-infected cells for 2 h prior to analysis on SDS-polyacrylamide gels.

structure within capsid protein VP1 resulted in 3C or 3CD-mediated cleavage at the inserted Q-G amino acid pairs. These observations suggested that Q-G accessibility determines 3C- or 3CD-mediated recognition, and support the latter model of 3CD/P1 tertiary structure interaction proposed above.

The cleavage profile of picornavirus P1 capsid protein precursors mediated by heterologous 3CD proteinases provides further evidence for the stringent requirements in P1 processing. Poliovirus (PV1), rhinovirus (HRV14), and coxsackievirus (CVB3) non-structural protein precursors are efficiently cleaved by heterologous 3C proteinases (Dewalt et al., 1989; Lawson et al., 1990). However, poliovirus, rhinovirus, and coxsackievirus P1 precursors are only incompletely processed by heterologous 3CD proteinases. CVB3 P1 precursors are cleaved exclusively at the VP3/VP1 cleavage site by poliovirus 3CD proteinase (Lawson et al., 1990), while rhinovirus P1 precursors are apparently cleaved exclusively at the VP0/VP3 cleavage site by poliovirus 3CD proteinase (Li and Semler, unpublished observations). Poliovirus P1 precursors are not cleaved at either the VP0/VP3 or VP3/VP1 cleavage sites by rhinovirus 3CD proteinase (Dewalt et al., 1989; Li and Semler, unpublished observations). Interestingly, poliovirus P1 precursors are cleaved at the VP3/VP1 cleavage site by a chimeric proteinase containing coxsackievirus 3C sequences and poliovirus 3D sequences (Lawson et al., 1990). Incomplete cleavage of P1 by heterologous 3CD proteinases may result form differences in poliovirus, rhinovirus, and coxsackievirus P1 secondary or tertiary structure which limits cleavage site accessibility to heterologous 3CD proteinases. It is also possible that P1 structural domains, which may be specifically recognized by homologous 3C proteinases, may not be recognized by heterologous 3CD proteinases due to the loss of species-specific protein-protein interactions.

The effect of amino-terminal P1 myristate modification is an additional consideration concerning 3CD recognition of P1. N-tetradecanoic acid (myristic acid) is covalently attached by an amide linkage to the amino-terminus of capsid protein VP4 as part of the P1 precursor (Chow et al., 1987; Paul et al., 1987). It has been demonstrated that myristate modification is required for the assembly of infectious viral particles (Marc et al., 1990; Marc et al. 1991; Moscufo et al., 1991; Ansardi et al., 1992), but its role in P1 processing is unclear. Initial studies demonstrated that *in vitro* synthesized P1 precursors containing amino-terminal glycine to alanine substitutions, preventing P1 myristylation, were inefficiently cleaved at the VP0/VP3 site by proteinase 3CD (Krausslich et al., 1990). These studies suggested that myristate modification of P1 was required for efficient cleavage at the VP0/VP3 cleavage site by proteinase 3CD. A later study reported conflicting *in vivo* observations. In cells transfected with full-length poliovirus cDNAs expressing mutant P1 precursors incapable of being myristylated, cleavage of such P1 precursors was observed at both the VP0/VP3 and VP3/VP1 cleavage sites (Marc et al. 1991). It is possible that myristate modification of P1 affects the efficiency of cleavage at the VP0/VP3 site *in vitro*, but it may not be required for 3CD-mediated cleavage of P1 *in vivo*.

PRIMARY AMINO ACID SEQUENCE DETERMINANTS REQUIRED FOR 3CD-MEDIATED P1 PROCESSING

In addition to structural substrate determinants, primary amino acid sequences surrounding cleaved Q-G amino acid pairs also have a role in 3C recognition of viral substrates. Although not strictly conserved, an alanine residue is usually present at the -4 position (four residues proximal) of cleaved Q-G amino acid pairs, and not present at the -4 position of uncleaved Q-G amino acid pairs in the poliovirus polyprotein (Nicklin et al., 1986). Recent experimental evidence demonstrated that a -4 alanine residue was important for the efficiency of cleavage of synthetic peptides containing authentic poliovirus P2 and P3 cleavage sites by purified 3C proteinase. Pallai et al. (1989) observed that synthetic peptides (sixteen residues in length) containing the authentic 3C/3D cleavage site, in which a threonine is present at the -4 position, were cleaved inefficiently compared to other P2 or P3 synthetic peptides containing -4 alanine residues. Substitution of an alanine residue for the -4 threonine in synthetic peptides containing the authentic 3C/3D cleavage site resulted in an increased cleavage rate of such peptides in the presence of purified poliovirus 3C proteinase. Similarly, substitution of an alanine residue for the wild-type threonine at the -4 position of the HRV-14 2C/3A cleavage site in synthetic peptides resulted in an increased cleavage rate of such peptides in the presence of purified HRV-14 3C proteinase (Cordingley et al., 1990). These data demonstrated that a -4 alanine residue is important for efficient cleavage of a Q-G amino acid pair by 3C *in vitro* and suggested that cleavage rates of particular Q-G amino acid pairs in the poliovirus or rhinovirus polyprotein may be influenced by their corresponding -4 residues.

To investigate the role of the -4 alanine residue in the recognition of poliovirus Q-G cleavage sites by 3C activity in the context of an authentic viral precursor (P1), a series of four amino acid insertion and single amino acid substitution mutations was produced at sequences proximal to the VP3/VP1 cleavage site. The four amino acid insertion mutations were designed to disrupt either sequences proximal to the -4 position or the -3 position of the VP3/VP1 cleavage site. Analysis of the substrate properties of such mutations demonstrated that substitution of the -4 alanine residue at the VP3/VP1 cleavage site resulted in a significantly greater defect in 3CD-mediated processing at that site than disruption of sequences proximal to the -4 position *in vitro* and *in vivo* (Blair and Semler, 1991). Additionally, the single amino acid substitution of a serine for the alanine residue at the -4 position of the VP3/VP1 cleavage site resulted in a significant reduction of 3CD-mediated cleavage at that site *in vitro* and *in vivo* (Blair and Semler, 1991). These data directly demonstrated *in vitro* and *in vivo* that a -4 alanine residue is important for recognition and cleavage of the VP3/VP1 site by 3CD.

To determine if a -4 alanine residue is capable of directing cleavage at a particular Q-G amino acid pair, a series of six amino acid insertion mutations was introduced into sequences proximal to the VP3/VP1 cleavage site (Fig. 4A). The initial six amino acid insertion mutation [pT7P1-1C(AG)], which contained a Q-G amino acid pair, resulted in the substitution of a glycine for the alanine residue at the -4 position of the wild-type VP3/VP1 cleavage site and maintained an alanine residue at the -4 position of the inserted Q-G amino acid pair. Oligonucleotide site-directed mutagenesis was then used to construct all pair-wise combinations of alanine and glycine at the -4 positions of the inserted Q-G and wild-type VP3/VP1 cleavage sites (Fig. 4A). [^{35}S]-methionine labeled P1 precursors containing such mutations were synthesized *in vitro*, incubated in the presence of an extract from poliovirus-infected HeLa cells, and the final cleavage products were analyzed on SDS-polyacrylamide gels. Cleavage at the inserted Q-G amino acid pair was determined based on an altered electrophoretic mobility of the resulting VP1 cleavage product (Blair et al., 1990). As demonstrated in Figure 4B, 3CD-mediated cleavage of P1 precursors containing the above described six amino acid insertion mutations at either the inserted Q-G or the wild-type VP3/VP1 cleavage site depended on the presence of an alanine residue at the -4 position. Similar cleavage patterns of such P1 precursors were also observed *in vivo* (Blair and Semler, 1991). These data demonstrate that a -4 alanine residue is sufficient to select a cleavage site from two adjacent Q-G amino acid pairs and suggest that the -4 alanine residue is specifically recognized by 3CD.

Based on the above mentioned data demonstrating the requirement of a -4 alanine residue for efficient cleavage of a Q-G amino acid pair by 3C or 3CD, it is surprising that a proline residue occupies the -4 position of the VP0/VP3 cleavage site. The VP0/VP3 cleavage site appears to be cleaved efficiently by proteinase 3CD (refer to Fig. 2) despite the lack of a -4 alanine residue. To examine the importance of a -4 proline residue in facilitating cleavage at the VP0/VP3 cleavage site, single amino acid substitution mutations were introduced at that position. Threonine and arginine residues were substituted for the -4 proline at the VP0/VP3 cleavage site. The substitution of a threonine residue for the proline residue at the -4 of the VP0/VP3 cleavage site resulted in a reduction in cleavage at that site *in vitro* and *in vivo* (Blair and Semler, 1991). The substitution of an arginine residue for the -4 proline eliminated cleavage at the VP0/VP3 cleavage site *in vitro* (Fig. 5). As shown in Figure 5, [^{35}S]-methionine-labeled P1 precursors translated *in vitro*, which contained a proline to arginine single amino acid substitution at the -4 position of the VP0/VP3 cleavage site, were cleaved only at the VP3/VP1 cleavage site when incubated in the presence of poliovirus 3CD. These data demonstrate that amino acid substitutions at the -4 position of the VP0/VP3 cleavage site affect cleavage at that site by 3CD. Interestingly, recent observations showed that dilution of the 3CD/P1 cleavage reaction resulted in a decreased P1 cleavage rate at the VP0/VP3 cleavage site relative to the VP3/VP1 cleavage site, and that the substitution of an alanine for the -4 proline residue increases the efficiency of 3CD-mediated cleavage *in vitro* at the VP0/VP3 site. Additionally, a cellular cofactor may be required to mediate efficient interaction of 3CD with P1 at the VP0/VP3 cleavage site *in vitro*, possibly compensating for the lack of a -4 alanine residue (Blair, Li, and Semler, submitted for publication).

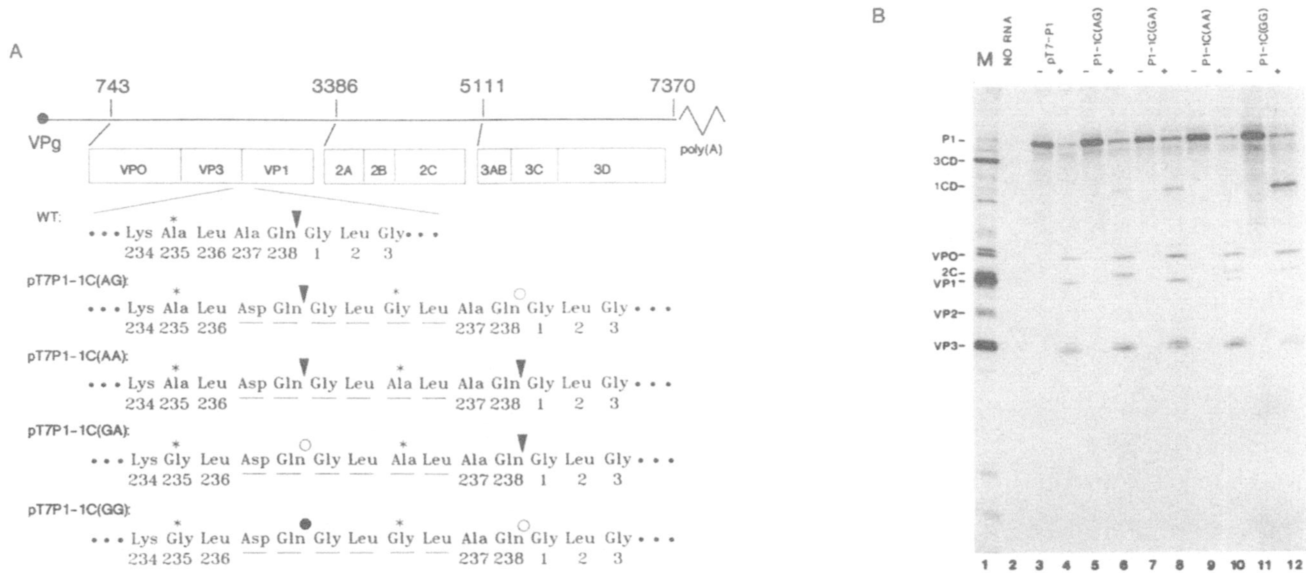

Figure 4. **Processing profile of P1 precursors derived from 6 amino acid insertion mutations.** (A) Map of the poliovirus genome and its translated products showing the 6 amino acid insertion mutations. The amino acid insertions are underlined, and asterisks designate the -4 residues of the wild-type VP3/VP1 cleavage sites and inserted Q-G amino acid pairs. The filled arrows denote Q-G amino acid pairs cleaved efficiently by proteinase 3CD, the open circle denotes a Q-G amino acid pair cleaved inefficiently by proteinase 3CD, and the filled circle denotes Q-G amino acid pairs rarely cleaved by proteinase 3CD. (B) RNAs transcribed *in vitro* from the different plasmids were translated *in vitro* as described in the legend to Figure 2. The translated products were analyzed on SDS-polyacrylamide gels. (+), translated products incubated in the presence of poliovirus-infected HeLa cell extract, the source of proteinase 3CD, for 1 h prior to analysis on SDS-polyacrylamide gels. (-), translated products analyzed on SDS-polyacrylamide gels without incubation in the presence of poliovirus-infected HeLa cell extract. (Reproduced from Blair and Semler, 1991, with permission)

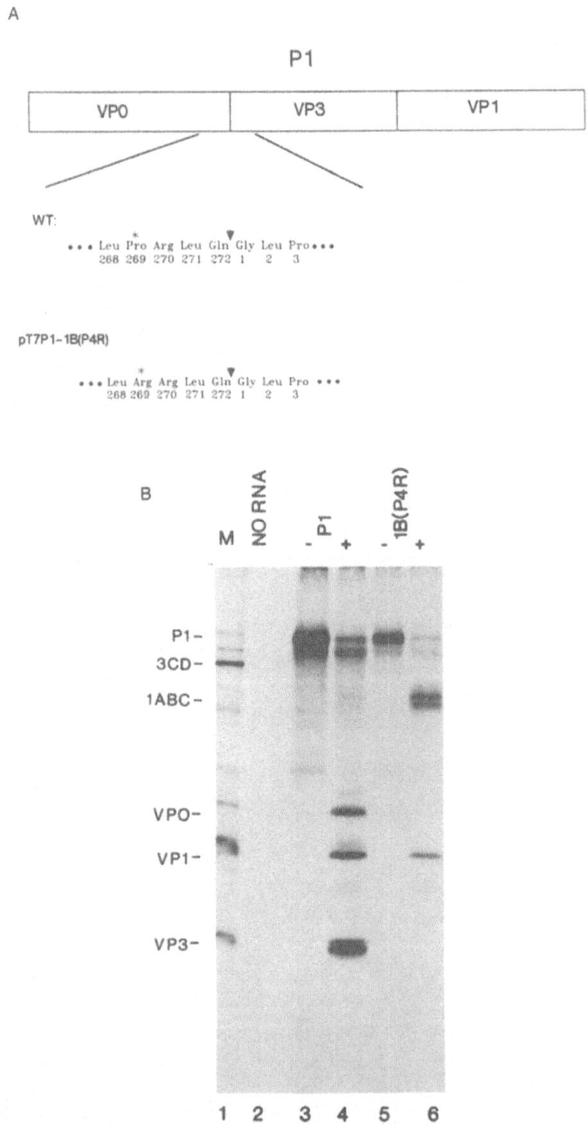

Figure 5. Processing profile of the P1 precursors derived from pT7-P1 and pT7P1-1B(P4R). (A) Simplified diagram of the P1 precursor showing the single amino acid substitution mutation pT7P1-1B(P4R). Amino acid sequences correspond to residues spanning the VP0/VP3 cleavage site for wild-type (WT) and pT7P1-1B(P4R). Asterisks denote residues in the -4 position of the VP0/VP3 cleavage site and filled arrows designate VP0/VP3 cleavage sites. (B) RNAs were transcribed *in vitro* from plasmids pT7-P1 (P1) and PT7P1-1B(P4R) and translated *in vitro* in the presence of [^{35}S]-methionine as described in the legend to Figure 2. Translated products were incubated either in the presence (+) or the absence (-) of an extract from PV1-infected cells for 1 h, boiled in Laemmli sample buffer, and analyzed on SDS-polyacrylamide gels.

THE REQUIREMENT OF PROTEINASE 3CD FOR P1 PROCESSING

As mentioned above, cleavage of the P1 precursor is mediated by proteinase 3CD, depending on both intact polymerase (3D) sequences as well as 3C proteolytic activity. Ypma-Wong et al. (1987a) initially observed that P1 precursors from nearly full-length poliovirus cDNAs, truncated within 3D sequences, remain uncleaved by the endogenously encoded 3C or truncated 3CD proteinases. In such experiments, P2 and P3 precursor polypeptides were efficiently processed. The requirement for 3D sequences in P1 processing was demonstrated by Ypma-Wong et al. (1988a) and Jore et al. (1988). In parallel studies, these two groups of investigators reported that P1 precursors derived *in vitro* from subgenomic poliovirus cDNAs were cleaved by proteinase 3CD synthesized *in vitro* and supplied in trans, but not by *in vitro* synthesized 3C proteinase. Ypma-Wong et al. (1988a) also showed that a four amino acid insertion mutation within 3D sequences resulted in the selective inhibition of P1 cleavage without affecting P2 or P3 processing. Additional evidence for the requirement of 3CD for P1 processing came from later studies, which reported that large molar excesses of purified 3C proteinase were able to cleave *in vitro* synthesized poliovirus P1 precursors only inefficiently and almost exclusively at the VP3/VP1 cleavage site (Nicklin et al., 1988). Finally, P1 precursor polypeptides were cleaved efficiently by bacterially-expressed, wild-type 3CD, but not by bacterially-expressed 3CD containing various mutations within 3D sequences (Burns et al., 1989). A mutant virus containing one such single amino acid insertion mutation in 3D exhibited a temperature-sensitive defect in P1 processing *in vivo* (Charini et al., 1991).

The requirement for 3D sequences in the form of proteinase 3CD for P1 processing seems to be the case for some picornaviruses such as poliovirus, coxsackievirus (Lawson et al., 1990), and human rhinovirus 14 (Dewalt et al., 1989), but not for others such as foot-and-mouth disease virus (FMDV) (Vakharia et al., 1987; Clarke and Sangar, 1988) and possibly EMCV (Parks et al., 1986; Parks et al., 1989). Although it appears that picornaviruses which require either 3CD or 3C proteinase for cleavage of P1 may be subdivided based on evolutionary relatedness, an explanation for such P1 processing differences is not readily apparent. Utilization of proteinase 3CD may have evolved to compensate for P1 capsid protein divergence or to compensate for 3C proteinase divergence. The evolution of more structurally stable enterovirus capsid proteins may have required 3C co-evolution, resulting in the incorporation of additional 3D enzymatic determinants, for cleavage of such enterovirus capsid protein precursors. Alternatively, enterovirus and rhinovirus proteinase divergence may have resulted in 3C proteinases with altered enzymatic properties from those of FMDV or EMCV 3C proteinases, such that additional 3D sequences are required in the form of proteinase 3CD for enterovirus or rhinovirus P1 cleavage.

To further study P1/3CD interactions, we have partially purified bacterially-expressed poliovirus type 1 3CD proteinase. A bacteriophage T7

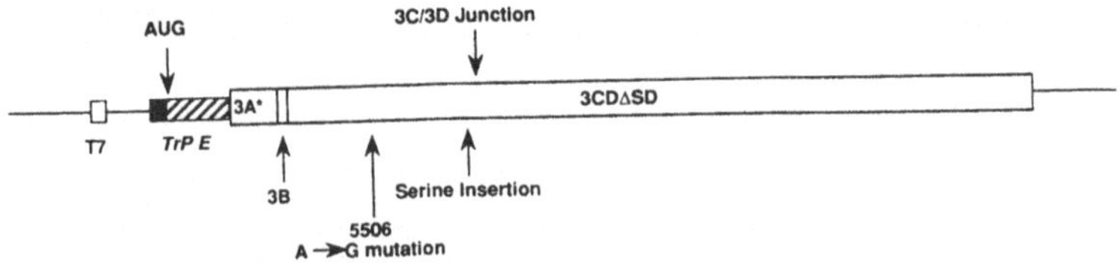

Construction of pet-3-*Trp E*-3CDΔSD

Figure 6. Plasmid pet-3-TrpE-3CDΔSD, used to express proteinase 3CD in *E. coli*. Sequences encoding a portion of 3A, all of 3B and poliovirus proteinase 3CD were fused to bacterial Trp E sequences and introduced into plasmid pet-3 downstream of the bacteriophage T7 promoter. Expres-sion of 3CD was performed by transformation of pet-3-TrpE-3CDΔSD into *E. coli* strain BL21(DE3), which encodes the phage T7 RNA polymerase gene under control of the *lac* promoter, and growth of such transformed *E. coli* in the presence of IPTG. A single amino acid (serine) insertion was introduced into poliovirus sequences proximal to the 3C/3D junction to prevent autocatalysis at the 3C/3D cleavage site. A single nucleotide substitution (A to G) mutation was introduced into sequences encoding poliovirus 3C (at poliovirus nt 5506) to disrupt a fortuitous Shine-Dalgarno sequence and reduce internal translation initiation. The filled box corresponds to Trp E leader sequences, and the cross-hatched box corresponds to translated amino-terminal Trp E sequences. (*) denotes a truncated 3A encoded polypeptide.

RNA polymerase-directed expression plasmid was engineered which encodes a portion of the *E. coli* trp E gene product and poliovirus 3CD proteinase (Fig. 6). The encoded 3CD proteinase contains a single amino acid insertion mutation in sequences proximal to the 3C/3D junction to prevent cleavage at that junction (Semler et al., 1987). This plasmid was transformed into an *E. coli* strain which contains the bacteriophage T7 RNA polymerase gene under control of the *lac* promoter. After induction with IPTG, efficient bacterial expression of 3CD was observed (Fig. 7). Supernatants from lysates of 3CD-expressing bacterial cells were then subjected to DEAE-Sephacel ion exchange chromatography. A fraction containing 3CD was eluted with 0.4 M NaCl in buffer D (50 mM Tris pH 7.6, 1 mM EDTA, 0.5% NP-40, 10% glycerol, and 1 mM DTT) and subsequently subjected to phosphocellulose ion exchange chromatography. Following a wash with 0.3 M NaCl in buffer D, 3CD was obtained by elution of the column with 0.4 M NaCl in buffer D. The 3CD-containing fraction was finally loaded onto a CM-Sepharose ion exchange column, which was eluted with a linear salt gradient (0.1 M to 0.5 M NaCl) in buffer E (50 mM Tris pH 8.5, 2.5 mM EDTA and 2 mM DTT) . The activity of the resulting partially-purified 3CD (Fig. 8, lanes 2 through 5) was demonstrated using an *in vitro* P1 cleavage assay (Fig. 9). An equal amount of total protein (12 μg) from each purification step was incubated with *in vitro* synthesized P1 (labeled with [^{35}S]-methionine) for 3 h, and the resulting

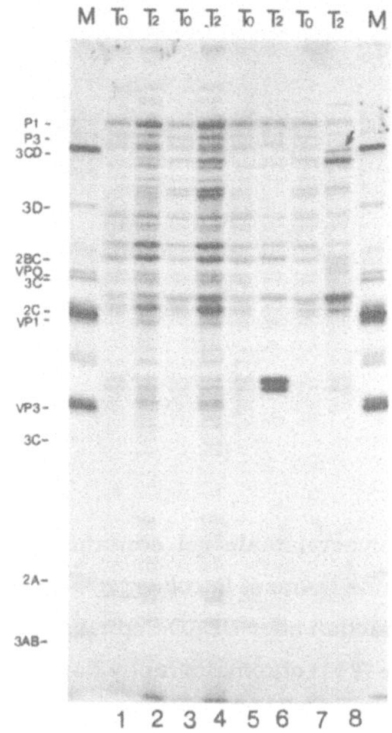

Figure 7. IPTG induction and partial purification of bacterially-expressed 3CD. Plasmid pet-3 (vector control) or pet-3-TrpE-3CDΔSD was transformed into *E. coli* [BL21(DE3)] and grown in minimal media (M9ZB) at 37°C. After growth to an OD_{600} of 0.3, cultures were pulsed with [^{35}S]-methionine for 30 min. then either induced with IPTG (lanes 5-8) or not induced (lanes 1-4) and incubated at 37°C for 2 h. T_0, cultures were harvested after the 30 min. pulse with [^{35}S]-methionine (prior to IPTG induction). T_2, cultures were harvested 2.5 h after the pulse with [^{35}S]-methionine (or 2 h after IPTG induction). Lanes 1, 2, 5, and 6, *E. coli* transformed with pet-3. Lanes 3, 4, 7, and 8, *E. coli* transformed with pet-3-TrpE-3CDΔSD. The arrow denotes proteinase 3CD.

cleavage products were analyzed on SDS-polyacrylamide gels. As shown in Figure 9, an active partially-pure 3CD proteinase was obtained, and this enzyme appeared to cleave with nearly equal efficiency at both the VP0/VP3 and VP3/VP1 cleavage sites. This enzyme preparation will prove useful for studies examining the role of 3D sequences in 3CD-mediated P1 processing. In addition, proteinase 3CD will also be further purified to homogeneity for use in X-ray crystallographic studies aimed at understanding the structure/function relationships of this key picornavirus enzyme.

CONCLUSIONS

The recognition of poliovirus polypeptide P1 by proteinase 3CD appears to be a complex interaction that is defined in terms of at least four parameters. First, P1 cleavage depends on 3CD recognition of the Q-G cleavage sites. Second, 3CD-mediated cleavage of P1 at the VP3/VP1 cleavage site depends on the presence of an alanine residue at the -4 position. The -4 alanine residue may be specifically recognized by the proteinase and may serve to stabilize the interaction of 3CD with P1 at the VP3/VP0 cleavage site. Third, intact P1 secondary structures or an intact P1 tertiary structure is required for

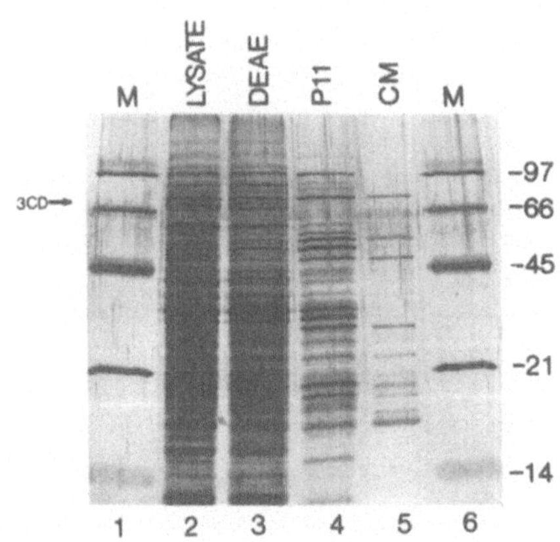

Figure 8. Silver stained SDS-polyacrylamide gel containing partially-pure 3CD proteinase. Samples: 1.5 mg total protein of crude lysate of *E. coli* expressing proteinase 3CD (lane 2), 0.15 mg total protein of a 3CD-containing fraction after DEAE-Sephacel chromatography (lane 3), 40 μg of total protein after phosphocellulose (P11) chromatography (lane 4), and 18 μg of total protein after CM-Sepharose chromatography (lane 5). M, marker lanes of molecular weight standards (molecular weights in kDa). Arrow denotes proteinase 3CD.

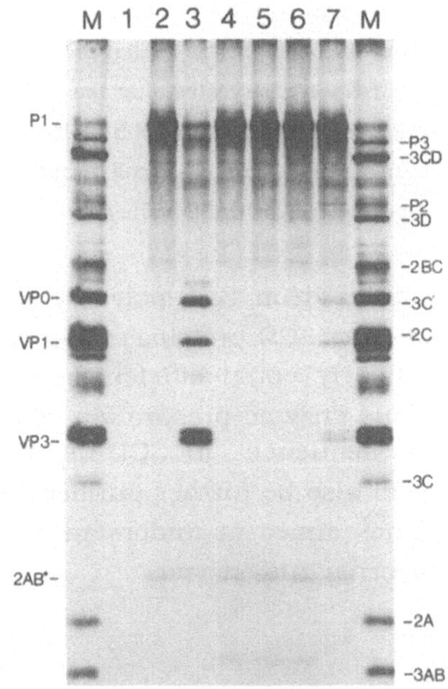

Figure 9. Cleavage activity of partially- purified poliovirus 3CD proteinase. [^{35}S]-methionine labeled P1 precursors synthesized *in vitro* as described in the legend to Figure 2 were incubated for 3 h in the absence of exogenous 3CD proteinase (lane 2), the presence of an extract from PV1-infected cells (lane 3), 12 μg of total protein from a crude lysate of *E. coli* expressing-3CD (lane 4), 12 μg of 3CD containing fractions after DEAE-Sephacel chromatography (lane 5), 12 μg of total protein after phosphocellulose chromatography (lane 6), and 12 μg of total protein after CM-Sepharose chromatography (lane 7) prior to analysis on SDS-polyacrylamide gels.

recognition and cleavage by 3CD. Possibly P1 secondary structures or a P1 tertiary structure is specifically recognized by proteinase 3CD, or perhaps such structural determinants have a more passive role in P1 processing. A correctly folded P1 precursor may be required simply for rendering the Q-G cleavage sites accessible to proteinase 3CD. Fourth, P1 processing depends on polymerase sequences (3D) in addition to 3C proteinase sequences, as part of proteinase 3CD. The 3D domain of proteinase 3CD may interact directly with the P1 substrate to stabilize the proteinase/substrate complex, or may act in *cis* to alter the conformation of proteinase 3C allowing recognition of P1. Preliminary data suggest that in addition to the above four parameters, a cellular factor may be required to facilitate efficient interaction of 3CD with P1 (a putative fifth parameter). Such a complicated set of proteinase/substrate interactions appears to be necessary to ensure efficient cleavage of a structurally rigid substrate specifically at two independent cleavage sites, such that the resulting cleavage products (viral capsid proteins) are authentic, abundant, and enter the virion morphogenetic pathway in equimolar quantities.

LITERATURE CITED

Ansardi, D. C., Porter, D. C., and Morrow, C. D., 1992, Myristylation of poliovirus capsid precursor-P1 is required for assembly of subviral particles, *J. Virol.* 66:4556-4563.

Bazan, J. F., and Fletterick, R. J., 1988, Viral cysteine proteases are homologous to the trypsin-like family of serine proteases: structural and functional implications, *Proc. Natl. Acad. Sci. USA.* 85:7872-7876.

Blair, W. S., Hwang, S. -S., Ypma-Wong, M. F., and Semler, B. L., 1990, A mutant poliovirus containing a novel proteolytic cleavage site in VP3 is altered in viral maturation, *J. Virol.* 64:1784-1793.

Blair, W. S. , and Semler, B. L., 1991, Role for the P4 amino acid residue in substrate utilization by the poliovirus 3CD proteinase, *J. Virol.* 65:6111-6123.

Burns, C. C., Lawson, M.A., Semler, B. L. , and Ehrenfeld, E. , 1989, Effects of mutations in poliovirus $3D^{pol}$ on RNA polymerase activity and on polyprotein cleavage, *J. Virol.* 63:4866-4874.

Charini, W. A., Burns, C. C., Ehrenfeld, E., and Semler, B. L., 1991, Trans rescue of a mutant poliovirus RNA polymerase function, *J. Virol.* 65:2655-2665.

Chow, M., Newman, J. F. E., Filman, D., Hogle, J. M., Rowlands, D. J., and Brown, F., 1987, Myristylation of picornavirus capsid protein VP4 and its structural significance, *Nature* 327:482-486.

Clarke, B. E., and Sangar, D. V. , 1988, Processing and assembly of foot-and-mouth disease virus proteins using subgenomic RNA, *J. Gen. Virol.* 69:2313-2325.

Compton, S. R., Nelsen, B., and Kirkegaard, K., 1990, Temperature-sensitive poliovirus mutant fails to cleave VP0 and accumulates provirions, *J. Virol.* 64:4067-4075.

Cordingley, M. G., Callahan, P. L., Sardana, V. V., Garsky, V. M., and Colonno, R. J., 1990, Substrate requirements of human rhinovirus 3C protease for peptide cleavage *in vitro*, *J. Biol. Chem.* 265:9062-9065.

Dewalt, P. G., Lawson, M. A., Colonno, R. J., and Semler, B., L., 1989, Chimeric picornavirus polyproteins demonstrate a common 3C proteinase substrate specificity, *J. Virol.* 63:3444-3452.

Emini, E. A., Elzinga, M. , and Wimmer, E. , 1982, Carboxy-terminal analysis of poliovirus proteins: termination of poliovirus RNA translation and location of unique poliovirus polyprotein cleavage sites, *J. Virol.* 42:194-199.

Etchison, D., Milburn, S. C., Edery, I., Sonenberg, N., and Hershey, J. W. B., 1982, Inhibition of HeLa cell protein synthesis following poliovirus infection correlates with the proteolysis of a 220,000 Dalton polypeptide associated with eukaryotic initiation factor 3 and a cap binding protein complex, *J. Biol. Chem.* 257:14806-14810.

Fernandez-Tomas, C. B. , and Baltimore, D. , 1973, Morphogenesis of poliovirus II. Demonstration of a new intermediate, the provirion, *J. Virol.* 12:1122-1130.

Gorbalenya, A. E., Dochenko, A. P. , Blinov, V. M. , and Koonin, E. V. , 1989, Cysteine proteases of positive strand RNA viruses and chymotrypsin-like serine proteases: a distinct protein superfamily with a common structural fold, *FEBS Lett.* 243:103-114.

Guttman, N. , and Baltimore, D. , 1977, Morphogenesis of poliovirus IV. Existence of particles sedimenting at 150S and having the properties of provirion, *J. Virol.* 23:363-367.

Hanecak, R., Semler, B. L. , Anderson, C. W. , and Wimmer, E. , 1982, Proteolytic processing of poliovirus polypeptides: Antibodies to polypeptide P3-7c inhibit cleavage at glutamine-glycine pairs, *Proc. Natl. Acad. Sci. USA* 79:3973-3977.

Hellen, C. U. T., and Wimmer, E. , 1992, Maturation of poliovirus capsid proteins, *Virology* 187:391-397.

Hogle, J. M., Chow, M., and Filman, D. J, 1985, Three-dimensional structure of poliovirus at 2.9 Å resolution, *Science* 229:1358-1365.

Jore, J., De Geus, B. , Jackson, R. J. , Pouwels, P. H. , and Enger-Valk, B. E. , 1988, Poliovirus protein 3CD is the active protease for processing of the precursor protein *in vitro*, *J. Gen. Virol.* 69:1627-1636.

Krausslich, H. G., Holscher, C., Reuer, Q., Harber J., and Wimmer, E., 1990, Myristoylation of the poliovirus polyprotein is required for proteolytic processing on the capsid and for viral infectivity, *J. Virol.* 64:2433-2436.

Larsen, G. R., Anderson, C. W. , Dorner, A. J. , Semler, B. L. , and Wimmer, E. , 1982, Cleavage sites within the poliovirus capsid protein precursors, *J. Virol.* 41:340-344.

Lawson, M. A., Dasmahapatra, B., and Semler, B. L. , 1990, Species-specific substrate interaction of picornavirus 3C proteinase suballelic exchange mutants, *J. Biol. Chem.* 265:15920-15931.

Lawson, M. A. and B. L. Semler, 1990, Picornavirus protein processing: enzymes, substrates, and genetic regulation, *Curr. Top. Microbiol. Immunol.* 161:49-87.

Lawson, M. A., and Semler, B. L., 1991, Poliovirus thiol proteinase 3C can utilize a serine nucleophile within the putative catalytic triad, *Proc. Natl. Acad. Sci. USA* 88:9919-9923.

Marc, D., Masson, G., Girard, M., and van der Werf, S., 1990, Lack of myristoylation of poliovirus capsid polypeptide VP0 prevents the formation of virions or results in the assembly of noninfectious virus particles, *J. Virol.* 64:4099-4107.

Marc, D., Girard, M., and van der Werf, S., 1991, A gly^{1} to ala substitution in poliovirus capsid protein VP0 blocks its myristoylation and prevents viral assembly, *J. Gen. Virol.* 72:1151-1157.

Mirzayan, C., Ingraham, R., and Wimmer, E., 1991, Specificity of the polioviral proteinase 3C towards genetically engineered cleavage sites in the viral capsid, *J. Gen. Virol.* 137:1159-1163.

Moscufo, N., Simons, J., and Chow, M., 1991, Myristoylation is important at multiple stages in poliovirus assembly, *J. Virol.* 65:2372-2380.

Nicklin, M. J. H., Toyoda, H., Murray, M. G., and Wimmer, E. , 1986, Proteolytic processing in the replication of polio and related viruses, *Biotechnology* 4:33-42.

Nicklin, M. J. H., Kraussiich, H. G. , Toyoda, H. , Dunn, J. J. , and Wimmer, E. , 1987, Poliovirus polypeptide precursors: expression *in vitro* and processing by exogenous 3C and 2A proteinases, *Proc. Natl. Acad. Sci. USA* 84:4002-4006.

Nicklin, M. J. H., Harris, K. S., Pallai, P. V., and Wimmer, E. , 1988, Poliovirus proteinase 3C: Large-scale expression, purification, and specific cleavage activity on natural and synthetic substrates *in vitro*, *J. Virol.* 62:4586-4593.

Pallai, P. V., Burkhardt, F., Skoog, M., Schreiner, K., Baxt, P., Cohen, K. A., Hansen, G., Palladino, D. E. H., Harris, K. S., Nicklin, M. J., and Wimmer, E. , 1989, Cleavage of synthetic peptides by purified poliovirus 3C proteinase, *J. Biol. Chem.* 264:9738-9741.

Parks, G. D., Duke, G. M. , and Palmenberg, A. C. , 1986, Encephalomyocarditis virus 3C protease: efficient cell-free expression from clones which link viral 5' noncoding sequences to the P3 region, *J. Virol.* 60:376-384.

Parks, G. D., and Palmenberg, A. C., 1987, Site specific mutations at a picornavirus VP3/VP1 cleavage site disrupt *in vitro* processing and assembly of capsid precursors, *J. Virol.* 61:3680-3687.

Parks, G. D., J. C. Baker, and Palmenberg, A. C. , 1989, Proteolytic cleavage of encephalomyocarditis virus capsid region substrates by precursors to the 3C enzyme, *J. Virol.* 63:1054-1058.

Paul, A., Schultz, A., Pincus, S. E., Oroszlan, S., Wimmer, E., 1987, Capsid protein VP4 of poliovirus is N-myristoylated, *Proc. Natl. Acad. Sci. USA* 84:7827-7831.

Pelletier, J., Kaplan, G., Racaniello, V. R., and Sonenberg, N. , 1988, Cap-independent translation of poliovirus mRNA is conferred by sequence elements within the 5' noncoding region, *Mol. Cell. Biol.* 8:1103-1112.

Putnak, J. R., and Phillips, B. A., 1981, Picornaviral structure and assembly, *Microbiol. Rev.* 45:287-315.

Reynolds, C., Birnby, D., and Chow, M., 1992, Folding and processing of the capsid protein precursor P1 is kinetically retarded in neutralization site 3B mutants of poliovirus, *J. Virol.* 66:1641-1648.

Rueckert, R. R., 1985, Picornaviruses and their replication, *In*: "Virology," B. N. Fields, ed., Raven Press, New York.

Semler, B. L., Anderson, C. W., Kitamura, N., Rothberg, P. G., Wishart, W. L. and Wimmer, E. , 1981, Poliovirus replication proteins: RNA sequence encoding P3-1b and the sites of proteolytic processing, *Proc. Natl. Acad. Sci. USA* 78:3464-3468.

Semler, B. L., Johnson, V. J., Dewalt, P. G., and Ypma-Wong, M. F. , 1987, Site specific mutagenesis of cDNA clones expressing a poliovirus proteinase, *J. Cell. Biochem.* 33:39-51.

Sun, X. H. and Baltimore, D., 1989, Human immunodeficiency virus tat-activated expression of poliovirus protein 2A inhibits mRNA translation, *Proc. Natl. Acad. Sci. USA* 86:2143-2146.

Toyoda, H., Nicklin, M. J. H., Murray, M. G., Anderson, C. W., Dunn, J. J., Studier, F. W., and Wimmer, E. , 1986, A second virus-encoded proteinase involved in proteolytic processing of poliovirus polyprotein, *Cell* 45:761-770.

Trono, D., Pelletier, J., Sonenberg, N., and Baltimore, D. , 1988, Translation in mammalian cells of a gene linked to the poliovirus 5' noncoding region, *Science* 241:445-448.

Vakharia, V. N., Devaney, M. A., Moore, D. M., Dunn, J. J., and Grubman, M. J., 1987, Proteolytic processing of foot-and-mouth disease virus polyproteins expressed in a cell-free system from clone-derived transcripts, *J. Virol.* 61:3199-3207.

Ypma-Wong M. F., and Semler, B. L. , 1987a, *In vitro* molecular genetics as a tool for determining the differential cleavage specificities of the poliovirus 3C proteinase. *Nucleic Acids Res.* 15:2069-2088.

Ypma-Wong, M. F., and Semler, B. L. , 1987b, Processing determinants required for *in vitro* cleavage of the poliovirus P1 precursor to capsid proteins, *J. Virol.* 61:3181-3189.

Ypma-Wong, M. F., Dewalt, P. G., Johnson, V. H., Lamb, J. G., and Semler, B. L. , 1988a, Protein 3CD is the major poliovirus proteinase responsible for cleavage of the P1 precursor, *Virology* 166:265-270.

Ypma-Wong, M. F., Filman, D. J., Hogle, J. M., and Semler, B. L. , 1988b, Structural domains of the poliovirus polyprotein are major determinants for proteolytic cleavage at gln-gly pairs, *J. Biol. Chem.* 63:17846-17856.

STUDIES ON THE MECHANISM OF INTERNAL INITIATION OF TRANSLATION ON POLIOVIRUS RNA

Karen Meerovitch and Nahum Sonenberg

Department of Biochemistry
McGill University
Montreal, Quebec
Canada H3G 1Y6

INTRODUCTION

The genome of the picornaviruses, of which poliovirus is a member, is a plus-sense single stranded RNA molecule. Unlike cellular mRNAs, the 5' end of picornavirus mRNAs is not capped, but instead is terminated with pUp (Hewlett et al., 1976; Nomoto et al., 1976). The 5' untranslated region (UTR) is unusually long (approximately 750 nucleotides for poliovirus) and highly structured (for recent review see Agol, 1991). In addition, the 5' UTR contains multiple upstream AUGs (6-8 for poliovirus), none of which appear to be used as initiators of translation (Pelletier et al., 1988a). Because poliovirus RNA is naturally uncapped and translates in vivo and in vitro under conditions where cap-dependent translation is impaired, it must translate by a cap-independent mechanism. Consistent with this, an internal sequence in the poliovirus mRNA 5' UTR (spanning approximately 450 nucleotides) is sufficient for cap-independent translation (Pelletier et. al., 1988b), and this sequence can also confer cap-independent translation to heterologous mRNAs (Pelletier et. al., 1988b; Trono et al., 1988). This region serves as an internal binding site for ribosomes and is termed Ribosome Landing Pad (RLP) (Pelletier and Sonenberg, 1988c). Internal initiation of translation also occurs for other members of the picornavirus group (e.g. Jang et al., 1988, for review see Jackson et al., 1990), plant viral mRNAs (Thomas et al., 1991) and for several cellular mRNAs (Macejak and Sarnow, 1991; Oh et al., 1992).

Regulation of Gene Expression in Animal Viruses, Edited by
L. Carrasco *et al.*, Plenum Press, New York, 1993

The picornavirus family can be divided into two classes based on the comparison of secondary structure of their mRNA 5' UTRs. Secondary structure predictions of picornavirus RNAs, based on phylogenetic comparisons as well as enzymatic and chemical probing, revealed two different patterns of conserved structures, one for entero- and rhinoviruses (Le and Zuker, 1990; Pilipenko et al., 1989a; Rivera et al., 1988; Skinner et al., 1989) and the other for cardio- and aphthoviruses (Pilipenko et al., 1989b). These structural differences may underlie the differences in the translation characteristics between cardio/aphthovirus and entero/rhinovirus RNAs (Jackson, 1989). RNAs from EMCV and FMDV are translated with high efficiency in reticulocyte lysates, whereas poliovirus and rhinovirus RNAs are translated poorly in the reticulocyte lysate, but translation can be augmented by the addition of HeLa cell extracts (Dorner et al., 1984). Not surprisingly, the mechanism of internal initiation appears to be different in certain aspects for these two classes of RNAs (Jackson et al., 1990). Our work is aimed at understanding the mechanism of internal initiation on poliovirus mRNA.

CHARACTERIZATION OF THE RLP

To define the boundaries of the poliovirus RLP, progressive deletions were made from the 5' and 3' ends in the poliovirus type 2 5' UTR and tested in the translation of a bicistronic mRNA in vivo (Nicholson et al., 1991). The results have shown that the 5' boundary of the RLP lies between nucleotides 134 and 155 in stem loop III (Fig. 1). The experiments placed the 3' boundary of the core RLP (containing sequences critical for translation) between nucleotides 556 and 585 (within a polypyrimidine stretch) and an extended border (containing sequences that augment translation) between nucleotides 585 and 612 (Nicholson et al., 1991). Not all of the sequences within the core RLP are critical for translation, for example; stem loop IV (nucleotides 189-223) can be deleted without deleterious effects on virus replication (Dildine and Semler, 1989) or to internal initiation of translation (Nicholson et al., 1991; Percy et al., 1992).

The sequences essential for translation initiation on poliovirus RNA (nucleotides 556-585) comprise a pyrimidine stretch (UUUCCUUUU) that was first noted for foot and mouth disease virus RNA (Beck et al., 1983), and is conserved in all picornaviruses. This conservation is striking in light of the lack of other conserved sequences in the rest of the 5' UTR of picornavirus mRNAs. We have demonstrated by point mutagenesis that only the 5' half of the pyrimidine stretch, UUUCC (nucleotides 560-564 of poliovirus RNA Lansing type 2), is critical for poliovirus translation in vitro and in vivo (Nicholson et al., 1991; Meerovitch et al., 1991). Similar results were reported by Pestova et al. (1991). These results are consistent with the effects of deletion of these nucleotides on poliovirus viability (Iizuka et al.,1989). Therefore, this sequence plays a critical role that is common for all picornaviruses. Another striking finding is the presence of a conserved AUG triplet ca. 15 to 20 nucleotides downstream from the pyrimidine tract. In some genera (cardio- and

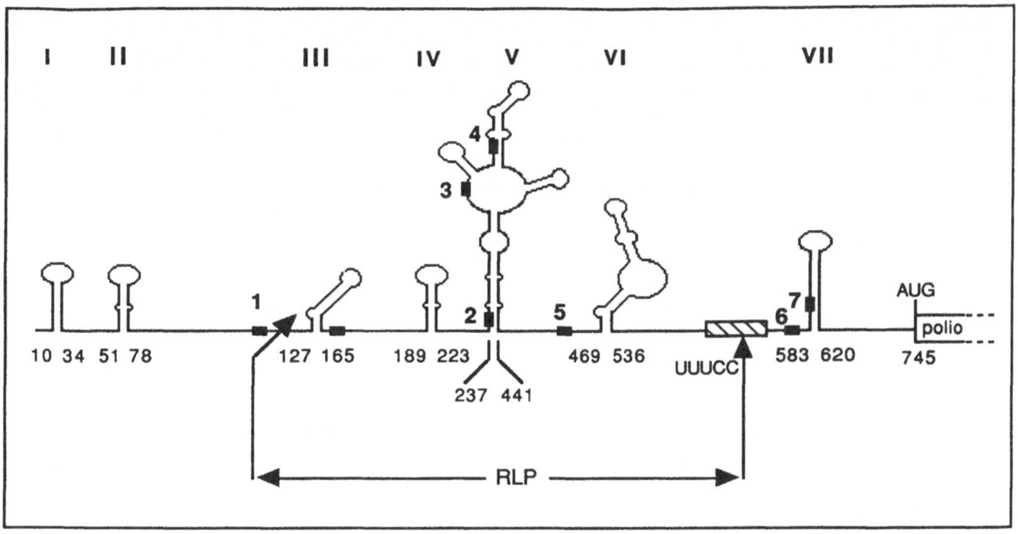

Figure 1. Predicted secondary structure of the 5' UTR of poliovirus type 2 RNA. The figure has been adapted from Skinner et al. and is not to scale. The nucleotide positions for start and end of the stem-loop structures are shown. The region required for cap-independent translation is indicated as RLP for 'ribosome landing pad'. It starts between nucleotides 134 and 155 in loop III and ends at nucleotide 564. Black boxes indicate the upstream AUGs. The sequence UUUCC is part of a polypyrimidine stretch (hatched box) that is essential for ribosome binding.

aphthoviruses) the polypyrimidine tract is located at a conserved distance upstream from the initiator AUG, but in the case of two other genera (entero- and rhinovirus), it resides upstream of the AUG preceeding the initiator AUG (Jackson et al., 1990). Our results from translations of point mutants in the 5' UTR also concluded that the upstream AUG (number 7, figure 1), and not the stem loop structure (stem loop VII, figure 1) in which it resides, is an important feature for modulation of poliovirus translation in vitro (Meerovitch et al., 1991). Furthermore, it has recently been confirmed that stem loop VII, but not AUG 7 and surrounding nucleotides, can be deleted without impairing viral growth (Pilipenko et al., 1992) or translation in vivo (Percy et al., 1992).

Our results demonstrate that the conserved pyrimidine stretch in poliovirus is functionally equivalent to that of FMDV and EMCV. The 3' border of the internal ribosomal entry site in the EMCV 5' UTR has been mapped within the pyrimidine stretch (Jang and Wimmer, 1990). Deletion of this pyrimidine stretch in EMCV, and point mutations in this stretch in FMDV and EMCV decreased dramatically translational efficiency (Jang and Wimmer, 1990; Kuhn et al., 1990; Pestova et al., 1991). The conserved distance between the pyrimidine tract and an AUG was shown to be important in vitro and in vivo (Jang and Wimmer, 1990; Pilipenko et al., 1992). These findings indicate that a unique arrangement of an AUG and the preceeding polypyrimidine tract is crucial. In poliovirus, this AUG lies approximately 150 nucleotides upstream of the initiator AUG, whereas in EMCV it serves as the initiation codon. The conserved polypyrimidine tract/AUG motif was postulated to function in 40S ribosome binding via base-pairing with 18S rRNA in an analogous manner to the Shine-Delgarno sequence in the prokaryotic system (Nicholson et al., 1991; Pestova et al., 1991; Pilipenko et al., 1992). For poliovirus RNA, ribosomes that would presumably bind to this motif would have to translocate to the initiator AUG. This translocation is impeded when extra sequences containing AUGs or secondary structure are introduced between AUG 7 and the initiator AUG (Kuge et al., 1989; Pelletier et al., 1988c).

PROTEINS INTERACTING WITH THE RLP

The finding that a large segment of the 5' UTR is required for internal initiation of translation suggests that tertiary structure is likely to play a key role in ribosome binding. The 3-dimensional structure of the 5' UTR is not known. It is conceivable that proteins interact with different elements of the 5' UTR to promote and stabilize secondary and tertiary structures and to facilitate ribosome binding. A HeLa 52 kDa protein, p52, was shown to bind to a poliovirus type 2 RNA fragment encompassing nucleotides 559-624, which contains the polypyrimidine tract/AUG motif (Meerovitch et al., 1989). p52 is abundant in HeLa cell extracts, but not in rabbit reticulocyte lysates (Meerovitch et al., 1989). This finding might be of significance, because translation of poliovirus RNA in rabbit reticulocyte lysates is very inefficient and yields aberrant polypeptides (Brown and Ehrenfeld, 1989, Dorner et al., 1984; Phillips and Emmert, 1986). The synthesis of the aberrant polypeptides does not result from cleavage of the RNA in the reticulocyte lysate (Dorner et al., 1984). Translation initiation is dramatically enhanced, and aberrant initiation is corrected by supplementing the reticulocyte lysate with HeLa or Krebs-2 cell extracts (Brown and Ehrenfeld, 1989, Dorner et al., 1984; Phillips and Emmert, 1986; Svitkin et al., 1988). It is conceivable that p52 might function in poliovirus RNA internal translation and also might correspond to the "initiation correction factor" (Svitkin et al., 1988).

To characterize p52, we purified the protein. HeLa S100 was used as starting material for column chromatography using sequentially heparin-sepharose and poly U-sepharose. We used a radiolabeled RNA spanning nucleotides 559-624 for the mobility gel shift assay to monitor the purification of p52. p52 was purified to homogeneity and the sequence from one tryptic peptide was determined. Comparison of the amino acid sequence of the p52 peptide with sequences in the Genbank database showed absolute identity with human La autoantigen (Meerovitch, K. and Sonenberg, N., unpublished observations). The La protein is a human antigen that is recognized by antibodies from patients with autoimmune disorders such as systemic lupus erythematosus and Sjogren's syndrome (for review see Tan, 1989). La protein possesses RNA binding activity primarily, but not exclusively, to RNA polymerase III transcripts bearing 3' oligouridylates (for review see Keene et al., 1987). The role of La in RNA metabolism, although reported to play a role in polymerase III transcription termination (Gottlieb and Stietz, 1989a; Gottlieb and Steitz, 1989b), is not definitive.

Our striking and unexpected result that p52 is La antigen was confirmed by comparison of biochemical, immunological and functional properties of p52 and La (Meerovitch, K., and

Sonenberg, N., unpublished observations). We have shown that p52 (La) can bind to several poliovirus RNA fragments including the polypyrimidine stretch. The binding affinity of p52 to poliovirus RNAs containing only two 3' terminal uridylates was approximately 4 nM (Meerovitch, K., and Sonenberg, N., unpublished observations). However, our results suggest that p52 (La) can also bind to internal stretches of uridines, because an RNA that contains the polypyrimidine stretch and possesses 3' terminal guanosines binds with almost the same affinity (8 nM). We showed that the La protein stimulates translation of poliovirus RNA, but not BMV, TMV and AMV-4 RNA in a reticulocyte lysate (Svitkin, Y., unpublished observations). As evidence that La might function in poliovirus translation is the finding that La stimulates the synthesis of authentic poliovirus protein and suppresses the synthesis of several aberrant products in a rabbit reticulocyte lysate, where this factor is limiting. Furthermore, antibodies against La antigen inhibit translation of poliovirus RNA, but not other RNAs in HeLa cell extracts (Svitkin, Y., unpublished observations). We conclude that La protein is involved in poliovirus internal initiation of translation, and might function through a similar mechanism in the translation of cellular mRNAs.

Other factors recognizing picornavirus 5' untranslated regions

Several proteins interact with picornavirus 5' UTRs. Svitkin and colleagues have purified an "initiation correction factor" (ICF) from Krebs-2 cells that in addition to stimulating and correcting translation of poliovirus RNA in reticulocyte lysates, is more active on neurovirulent rather than attenuated viral RNAs (Svitkin et al., 1988). Because the attenuating mutations reside in the mRNA 5' UTR, ICF would presumably bind to this region of the mRNA. For poliovirus, protein-RNA complexes with 5' UTR RNA fragments encompassing nucleotides 98 to 182 and 510 to 629 were reported (Del Angel et al., 1989). These complexes contained among other proteins the eukaryotic initiation factor 2. In another study four proteins were observed to crosslink to the poliovirus 5' UTR (Gebhard and Ehrenfeld, 1992). One of these, a 54 kDa protein, was shown to bind specifically between nucleotides 457 and 626. However, this RNA competed for translation of other mRNAs and the authors concluded that this protein might be a general translation factor. This protein might be a subunit of eIF-2, as eIF-2 has been proposed to bind RNA, although its specificity of interaction is moot. The other three proteins in this study were observed to bind between nucleotides 286 and 456 and were specific for poliovirus translation (Gebhard and Ehrenfeld, 1992). A 50 kDa protein from HeLa cells was shown to interact covalently with nucleotides 186 to 221 (Najita and Sarnow, 1990). In another study, HeLa cellular protein(s) have been shown to interact with stem loop structures IV (nucleotides 186 to 220) and V (nucleotides 220 to 460) (Dildine and Semler, 1992). The significance and mode of action of these proteins in the binding of ribosomes is not understood.

The UV crosslinking technique has been used to detect cellular proteins that interact specifically with the 5' UTRs of EMCV (Jang and Wimmer, 1990; Borovjagin et al., 1990) and FMDV (Luz and Beck, 1991). Specific binding of p57 to a stem loop structure (E) in the 5' UTR of EMCV was found to correlate with efficient translation in vitro (Jang and

Wimmer, 1990). In a more recent study, deletion of or linker insertion in this stem loop structure (E) did not significantly affect translational activity (Duke et al., 1992). Instead, two stem loop elements downstream of E were suggested to play a more critical role in translation initiation of EMCV RNA (Duke et al., 1992), and one of the stem loops was reported to be an efficient binding site for eIF-2/2B (Scheper at al., 1991). The reason for the difference in these results is not immediately clear. However, it is possible that various 5' deletions result in different folding of tertiary structure which determines translational efficiency.

p57 binds also to the 5' UTR of FMDV RNA, although in this case two distinct binding sites were identified (Luz and Beck, 1991). One site corresponds to Jang and Wimmer's EMCV p57 binding site, and the other is close to the initiator AUG. The sequence of the p57 binding site includes a UUUC motif which occurs in the conserved polypyrimidine tract that is important for translation (Kuhn et al., 1990). Crosslinking of p57 to the polypyrimidine tract is reduced when the RNA is mutated in this region. p57 has also been proposed to bind to the 5' UTR of poliovirus RNA (Pestova et al., 1991; Luz and Beck, 1991), although the binding affinity is weaker and the binding site(s) remains unidentified. In addition, p57 binding to poliovirus RNA can be outcompeted by FMDV RNA (Luz and Beck, 1991).

The above results imply that cellular factors play a role in internal translation and would presumably function also in the internal initiation of cellular mRNAs. In this respect, the mechanism for internal initiation should be the same for cellular and picornavirus mRNAs. However, as discussed above, there are distinct mechanisms of translation within the picornavirus family; one for entero/rhinoviruses and the other for cardio/aphthoviruses (Jackson, 1990). This notion is consistent with the apparent specificty of cellular factors binding to either group (see above), with the exception of eIF-2/2B which seems to be a common factor. Clearly, much work needs to be done in order to define the mechanism of internal translation, especially in regard to factor requirements for internal translation of cellular mRNAs. Understanding of the mechanism of internal translation is further complicated by the recent finding that in poliovirus-infected cells, there is a reduced requirement for structural motifs in the 5' UTR (Percy et al., 1992). This work implies that in poliovirus-infected cells there may be an efficient translational mechanism that includes a viral transactivator protein that supports internal initiation of translation.

MODEL FOR INTERNAL INITIATION ON POLIOVIRUS RNA

We propose a model for the mechanism of internal initiation of translation on poliovirus RNA (Fig. 2). The La protein binds to an internal region in the 5' UTR that is functionally important in vitro and in vivo for poliovirus translation (Meerovitch et al., 1991; Nicholson et al., 1991; Pelletier et al., 1988b; Pilipenko et al., 1992). This region contains the polypyrimidine tract which lies upstream of the 3' boundary in the RLP. A polypyrimidine tract constitutes an important translational domain for internal initiation on picornavirus RNAs (Jackson et al., 1990). It was proposed that the polypyrimidine tract presents an

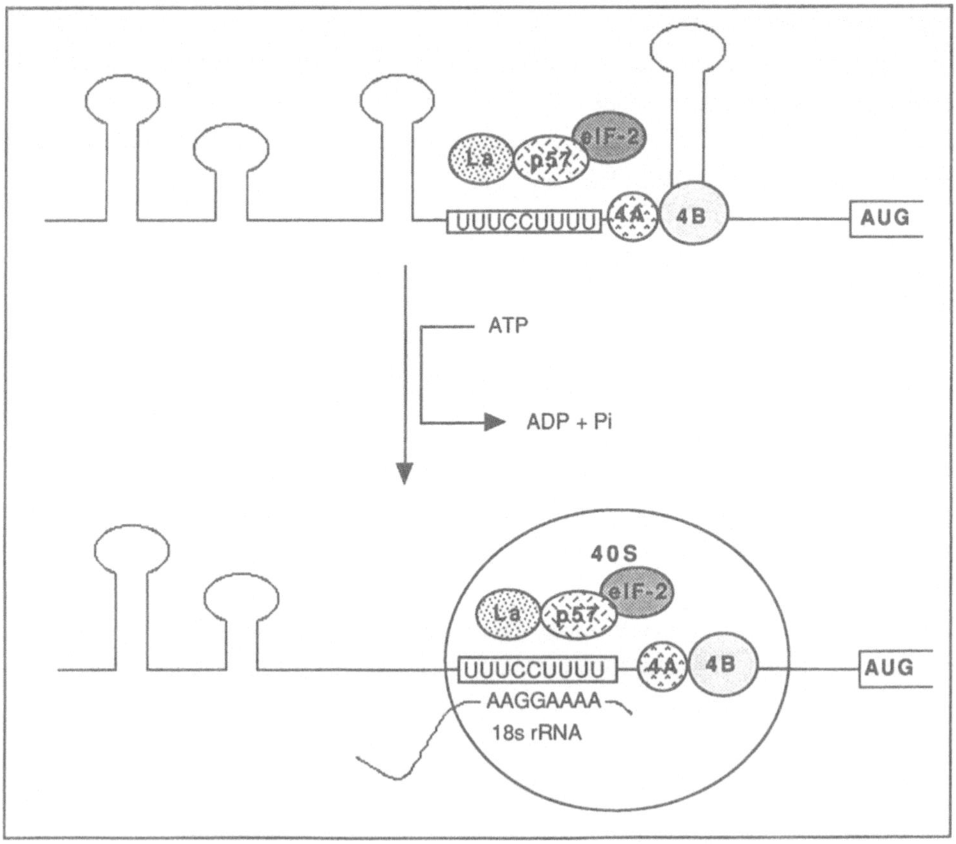

Figure 2. Model of factor interaction with the 5' UTR of poliovirus RNA. La binding to the pyrimidine tract and binding of other factors such as eIF-2 and p57 is postulated to be a prerequisite for RNA secondary structure unwinding mediated by eIF-4A and 4B and ATP hydroysis to facilitate internal 40S ribosome binding. Interactions between the indicated factors are speculative and have not been demonstrated. The ribosome is illustrated as a large circle binding to an unstructured site in the mRNA. This interaction could be facilitated by base pairing of 18S rRNA to the pyrimidine tract.

unstructured site which enables ribosomes to bind directly to the 5' UTR through an interaction with 18S rRNA (Nicholson et al., 1991; Pilipenko et al., 1992). La could, for example, stabilize such an interaction. Alternatively, La may function by assembling sequences in the 5' UTR that are needed for internal initiation (the polypyrimidine tract by itself is not sufficient) into higher order structures that are recognized by initiation factors and/or ribosomes. La interaction with poliovirus RNA may be a prerequisite for binding of other initiation factors, such as eIF-4A and eIF-4B, which promote unwinding of the mRNA 5' secondary structure (Rozen et al., 1990). Other factors such as eIF-2/2B (Svitkin et al., 1988; Scheper et al., 1991), or p57 (Borovjagin et al., 1991; Luz and Beck, 1991; Pestova et al., 1991) may also be required to promote internal ribosome binding.

ACKNOWLEDGEMENTS

We thank our coworkers Yuri V. Svitkin and Han Lee for their valuable contributions to this work. We also thank our collaborators Daniel J. Kenan, Jack D. Keene, Edward K. L. Chan and Vadim I. Agol for their reagents and advice. This work was supported by a grant from the Medical Research Council of Canada to N. S.

REFERENCES

Agol, V.I., 1991, The 5' untranslated region of picornaviral genomes, *Adv. Virus Res.*40:103-180.

Beck, E., Forss, S., Strebel, K., Cattaneo, R., and Feil, G., 1983, Structure of the FMDV translation initiation site and of the structural protein, *Nucl. Acids. Res.* 11:7873-7885.

Borovjagin, A.V., Evstafieva, A.G., Ugarova, T.Y., and Shatsky, I.N., 1990, A factor that specifically binds to the 5'-untranslated region of encephalomyocarditis virus RNA, *FEBS Lett.* 261:237-240.

Brown, B.A., and Ehrenfeld, E., 1979, Translation of poliovirus RNA in vitro: changes in cleavage pattern and initiation sites by ribosomal salt wash, *Virology* 97:396-405.

Del Angel, R.M., Papavassiliou, A.G., Fernandez-Tomas, C.S., Silverstein, J., and Racaniello, V.R., 1989, Cell proteins bind to multiple sites within the 5' untranslated region of poliovirus RNA, *Proc. Natl. Acad. Sci. USA* 86:8299-8303.

Dildine, S.L., and Semler, B.L., 1989, The deletion of 41 proximal nucleotides reverts a poliovirus mutant containing a temperature-sensitive lesion in the 5' noncoding region of genomic RNA, *J. Virol.* 63:847-862.

Dildine, S.L., and Semler, B.L., 1992, Conservation of RNA-protein interactions among picornaviruses, *J. Virol.* 66:4364-4376.

Dorner, A.J., Semler, B.L., Jackson, R.J., Hanecak, R., Duprey, E., and Wimmer, E., 1984, In vitro translation of poliovirus RNA: utilization of internal initiation sites in reticulocyte lysate, *J. Virol.* 50:507-514.

Duke, G.M., Hoffman, M.A., and Palmenberg, A.C., 1992, Sequence and structural elements that contribute to efficient encephalomyocarditis virus RNA translation, *J. Virol.* 66:1602-1609.

Gebhard, J.R., and Ehrenfeld, E., 1992, Specific interactions of HeLa cell proteins with proposed translation domains of the poliovirus 5' noncoding region, *J. Virol.* 66:3101-3109.

Gottlieb, E., and Steitz, J.A., 1989a, The RNA binding protein La influences both the accuracy and the efficiency of RNA polymerase III transcription in vitro, *EMBO J.* 8:841-850.

Gottlieb, E., and Steitz, J.A., 1989b, Function of the mammalian La protein: evidence for its action in transcription termination by RNA polymerase III, *EMBO J.* 8:851-861.

Iizuka, N., Kohara, M., Hagino-Yamagishi, K., Abe, S., Komatsu, T., Tago, K., Arita, M., and Nomoto, A., 1989, Construction of less neurovirulent polioviruses by introducing deletions into the 5' noncoding sequence of the genome, *J. Virol.* 63:5354-5363.

Hewlett, M.J., Rose, J.K., and Baltimore, D., 1976, 5'-terminal structure of poliovirus polyribosomal RNA is pUp, *Proc. Natl. Acad. Sci. USA* 73:327-330.

Jackson, R.J., 1989, Comparison of encephalomyocarditis virus and poliovirus with respect to translation initiation and processing in vitro, in: "Molecular Aspects of Picornavirus Infection and Detection," B.L. Semler and E. Ehrenfeld, ed., A.S.M., Washington, D.C.

Jackson, R.J., Howell, M.T., and Kaminski, A., 1990, The novel mechanism of initiation of picornavirus RNA translation, *TIBS* 15:477-483.

Jang, S.K., Kräusslich, H.-G., Nicklin, M.J.H., Duke, G.M., Palmenberg, A.C., and Wimmer, E., 1988, A segment of the 5' nontranslated region of encephalomyocarditis virus RNA directs internal entry of ribosomes during in vitro translation, *J. Virol.* 62:2636-2643.

Jang, S.K., and Wimmer, E., 1990, Cap- independent translation of encephalomyocarditis virus RNA: structural elements of the internal ribosomal entry site and involvement of a cellular 57-kD RNA binding protein, *Genes Dev.* 4:1560-1572.

Keene, J.D., Deutscher, S.L., Kenan, D., and Kelekar, A, 1987, Nature of the La and Ro RNPs, *Mol. Biol. Reports* 12:235-238.

Kuge, S., Kawamura, N., and Nomoto, A., 1989, Genetic variation occuring on the genome of an in vitro insertion mutant of poliovirus type 1, *J. Virol.* 63:1069-1075.

Kühn, R., Luz, N., and Beck, E., 1990, Functional analysis of the internal translation initiation site of foot-and -mouth disease virus, *J. Virol.* 64:4625-4631.

Le, S.-Y., and Zuker, M., 1990, Common structures of the 5' non-coding RNA in enteroviruses and rhinoviruses. Thermodynamical stability and statistical significance, *J. Mol. Biol.* 216:729-741.

Luz, N., and Beck, E., 1990, A cellular 57 kDa protein binds to two regions of the internal translation initiation site of foot-and-mouth disease virus, *FEBS Lett.* 269:311- 314.

Macejak, D.G., and Sarnow, P., 1991, Internal initiation of translation mediated by the 5' leader of a cellular mRNA, *Nature* 353:90-94.

Meerovitch, K., Pelletier, J., and Sonenberg, N., 1989, A cellular protein that binds to the 5'-noncoding region of poliovirus RNA: implications for internal translation initiation, *Genes Dev.* 3:1026-1034.

Meerovitch, K., Nicholson, R., and Sonenberg, N., 1991, In vitro mutational analysis of cis-acting RNA translational elements within the poliovirus type 2 5' untranslated region, *J. Virol.* 65:5895-5901.

Najita, L., and Sarnow, P., 1990, Interaction of a cellular 50kDa protein with an RNA hairpin in the 5' noncoding region of the poliovirus genome: evidence for a transient covalent RNA-protein bond, *Proc. Natl. Acad. Sci. USA* 87:5846-5850.

Nicholson, R., Pelletier, J., Le, S.-Y., and Sonenberg, N., 1991, Structural and functional analysis of the ribosome landing pad of poliovirus: in vivo translation studies, *J. Virol.* 65:5886-5894.

Nomoto, A., Lee, Y.F., and Wimmer, E., 1976, The 5' end of poliovirus mRNA is not capped with m7G(5')ppp(5')Np, *Proc. Natl. Acad. Sci. USA* 73:375-380.

Oh, S.-K., Scott, M.P., and Sarnow, P., personal communication.

Pelletier, J., Flynn, M.E., Kaplan, G., Racaniello, V., and Sonenberg, N., 1988a, Mutational analysis of upstream AUG codons of poliovirus RNA, *J. Virol.* 62:4486-4492.

Pelletier, J., Kaplan, G., Racaniello, V.R., and Sonenberg, N., 1988b, Cap-independent translation of poliovirus mRNA is conferred by sequence elements within the 5' noncoding region, *Mol. Cell. Biol.* 8:1103-1112.

Pelletier, J., and Sonenberg, N., 1988c, Internal initiation of eukaryotic mRNA directed by a sequence derived from poliovirus RNA,. *Nature* 334:320-325.

Percy, N., Belsham, G.J., Brangwyn, J.K., Sullivan, M., Stone, D.M., and Almond, J.W., 1992, Intracellular modifications induced by poliovirus reduce the requirement for structural motifs in the 5' noncoding region of the genome involved in internal initiation of protein synthesis, *J. Virol.* 66:1695-1701.

Pestova, T.V., Hellen, C.U.T., and Wimmer, E., 1991, Translation of poliovirus RNA: role of an essential cis-acting oligopyrimidine element within the 5' nontranslated region and involvement of a cellular 57-kilodalton protein, *J. Virol.* 65:6194-6204.

Phillips, B.A., and Emmert, A., 1986, Modulation of the expression of poliovirus proteins in reticulocyte lysates, *Virology* 148:255-267.

Pilipenko, E.V., Blinov, V.M., Romanova, L.I., Sinyakov, A.N., Maslova, S.V., and Agol, V.I., 1989a, Conserved structural domains in the 5'-untranslated region of picornaviral genomes: an analysis of the segment controlling translation and neurovirulence, *Virology* 168:201-209.

Pilipenko, E.V., Blinov, V.M., Chernov, B.K., Dmitrieva, T.M., and Agol, V.I., 1989b, Conservation of the secondary structure elements of the 5'-untranslated region of cardio-and aphthovirus RNAs, *Nucl. Acid Res.* 17:5701-5711.

Pilipenko, E.V., Gmyl, A.P., Maslova, S.V., Svitkin, Y.V., Sinyakov, A.N., and Agol, V.I., 1992, Prokaryotic-like cis elements in the cap-independent internal initiation of translation on picornavirus RNA, *Cell* 68:119-131.

Rivera, V.M., Welsh, J.D., and Maizel, J.V. Jr., 1988, Comparative sequence analysis of the 5' noncoding region of enteroviruses and rhinoviruses, *Virology* 165:42-50.

Rozen, F., Edery, I., Meerovitch, K., Dever, T.E., Merrick, W.C., and Sonenberg, N., 1990, Bidirectional RNA helicase activity of eucaryotic translation initiation factors 4A and 4F, *Mol. Cell. Biol.* 10:1134-1144.

Scheper, G.C., Thomas, A.A.M., and Voorma, H.O., 1991, The 5' untranslated region of encephalomyocarditis virus contains a sequence for very efficient binding of eukaryotic factor eIF-2/2B, *Biochem. Biophys. Acta.* 1089:220-226.

Skinner, M.A., Racaniello, V.R., Dunn, G., Cooper, J., Minor, P.D., and Almond, J.W., 1989, New model for the secondary structure of the 5' non-coding RNA of poliovirus is supported by biochemical and genetic data that also show that RNA secondary structure is important in neurovirulence, *J. Mol. Biol.* 207:379-392.

Svitkin, Y.V., Pestova, T.V., Maslova, S.V., and Agol, V.I., 1988, Point mutations modify the response of poliovirus RNA to a translation initiation factor: a comparison of neurovirulent and attenuated strains, *Virology* 166:394-404.

Tan, E.M., 1989, Antinuclear antibodies: diagnostic markers for autoimmune diseases and probes for cell biology, *Adv. Immunol.* 44:93-151.

Thomas, A.A.M., ter Haar, E., Wellink, J., and Voorma, H.O., 1991, Cowpea mosaic virus M-RNA contains a sequence that allows internal binding of ribosomes and that requires eIF-4F for optimal translation, *J. Virol.* 65:2953-2959.

Trono, D., Pelletier, J., Sonenberg, N., and Baltimore, D., 1988, Translation in mammalian cells of a gene linked to the poliovirus 5' noncoding region, *Science* 241:445-448.

PICORNAVIRUS VARIATION

Esteban Domingo, Cristina Escarmís, Encarnación Martínez-Salas, Ana M. Martín Hernández, Mauricio G. Mateu, and Miguel A. Martínez

Centro de Biología Molecular "Severo Ochoa"
(CSIC - UAM), Universidad Autónoma de Madrid
Canto Blanco, 28049 Madrid, Spain

INTRODUCTION

Variation of viruses has multiple theoretical and practical implications. Because mutations are stochastic events, the extensive variability of RNA viruses and retroviruses has manifested, perhaps more than ever before, the indetermination and unpredictability that may be inherent to biological phenomena. RNA virus populations do not have a precisely defined genome, but consist of indeterminate mixtures of related genomes termed *viral quasispecies* (Fig. 1), an application to viruses of a theoretical concept first proposed by M. Eigen and colleagues to describe replicons in early life on earth (Eigen and Schuster, 1979; Eigen and Biebricher, 1988; Domingo *et al.*, 1993). Initially, the quasispecies concept was ignored by many virologists as being purely "theoretical". However, as pointed out by Holland and colleagues (1992) in a recent volume which summarizes the state of the art in virus variation: "Quasispecies populations are not hypothetical abstractions. They exist in animals and humans infected by RNA viruses such as vesicular stomatitis virus, foot-and-mouth disease virus, poliovirus, human immunodeficiency virus, influenza viruses, etc.". Obviously, if comparative nucleotide sequencing of genomes from one population is restricted to a few hundred nucleotides several "identical genomes" may be found for simple statistical reasons (Fig. 1). The mutual influence between theoretical studies on quasispecies and experimental molecular virology has been determinant in providing the basis of what has been termed "RNA genetics", a number of distinctive features associated with the error-prone replication of RNA genomes (Holland *et al.*, 1982, 1992). Here we review evidence that picornaviruses are generally variable and that their populations are best described as quasispecies. The basic principles that influence picornavirus variability and evolution do not differ in any substantial way from the principles guiding variation of other RNA viruses and retroviruses. In spite of the extensive heterogeneity of the human immunodeficiency viruses (HIV) within and between infected individuals, a comparison of the studies carried out with HIV-1 with those for other RNA viruses (including picornaviruses)

Regulation of Gene Expression in Animal Viruses, Edited by
L. Carrasco *et al.*, Plenum Press, New York, 1993

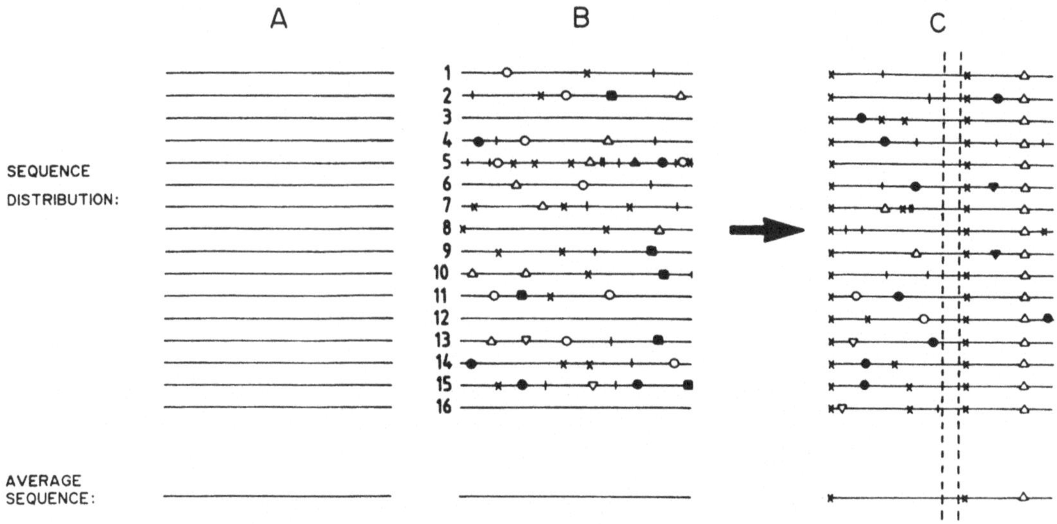

Fig. 1. A schematic representation of a picornavirus quasispecies. A depicts a homogeneous collection of picornavirus genomes, and B a heterogeneous distribution with an average of 4 mutations (symbols on the lines) per genome. In spite of being very heterogeneous, the average or consensus sequence of population B is identical to A. The replication of molecule 8 from the mutant spectrum of B yields a new quasispecies C with a new average sequence. Note that in spite of an extensive heterogeneity in distribution C, the discontinuous vertical lines delimit a genetically homogeneous stretch representing 8% of the genome length (about 600 nucleotides of the picornavirus genomic RNA). Thus, the absence of mutations after repetitive sequencing of a limited genomic region cannot be used as an argument against the quasispecies structure of a genome population. About 10^4 genomes are needed on average to find a mutation at any given nucleotide, but more than 10^{11} infectious genomes can be found in an infected host and about 10^5 in a single plaque on a cell monolayer (see text). Picornavirus populations (even single clones and natural isolates) are as B or C, not A. Mutant distributions are dynamic: they vary with time due to perturbations of equilibrium as the result of input of new variants and/or of changes in the environment. [Adapted from Domingo(1989), with permission from Birkhäuser Verlag, Basel].

led Temin (1989) to conclude that HIV" is not unique but merely different". We hope to persuade virologists that in addition to having important "theoretical" impact, quasispecies and picornavirus variation are intimately linked to "practical" problems, notably the control of picornaviral disease.

TWO APPROACHES TO PICORNAVIRUS VARIATION

Picornaviruses constitute a diverse group of human and animal pathogens which share with other RNA viruses the ability to vary rapidly in response to certain direct selective pressures, and to change more gradually and subtly when facing replication in modified environments. Indeed, picornaviruses able to escape neutralization by monoclonal antibodies (MAbs) or to resist inhibition by antiviral compounds are generally isolated with high frequencies, well above the reciprocal of the number of particles populating an infected host. Those picornaviruses which have been examined in some detail show remarkably high mutation and recombination rates and genetic heterogeneity *in vivo* and in cell culture. Such population microheterogeneity — the quasispecies genetic structure —

constitutes the raw material for long-term evolution and for rapid responses to environmental changes.

Two alternative — albeit complementary — approaches have been used to study RNA virus variation. One approach uses clasical procedures of molecular evolution such as sequence alignments and computations of genetic distances to derive phylogenetic relationships among viral genomes. This approach is very useful to relate different picornaviruses, to trace the possible origin of strains causing disease outbreaks, or to estimate rates of fixation of mutations with time, but it provides limited information on the molecular basis (or driving forces) of picornavirus evolution. The second approach consists on a detailed analysis of individual populations of viruses, including measurements of mutation rates during genome replication, mutant frequencies, and genetic heterogeneity in natural populations and in viruses passaged in a controlled environment. These analyses may alert of the potential of a virus to change in response to selective pressures. It cautions against indiscriminate use of certain antiviral agents, monoclonal antibodies or synthetic vaccines to control virus spread (Domingo and Holland, 1992), but it says little of the relationships among distantly related picornaviruses.

This review is mainly concerned with the general conclusions of these two approaches to study picornavirus variation and long-term evolution. For a broader introduction to picornavirus genetics the reader is referred to recent excellent reviews by Rueckert (1990), and Stanway (1990).

COMPARATIVE MUTATION RATES AND MUTANT FREQUENCIES FOR PICORNAVIRUSES

As it will become more apparent in the following paragraphs, it is important (and even necessary) to distinguish mutation rates from mutant frequencies. A mutation rate expresses the number of misincorporation events per nucleotide site and round of copying during genome replication, irrespective of the fate of the mutant molecules being generated. In contrast, a mutant frequency expresses the proportion of a mutant genome in a virus population. Thus, it is a population number (rather than a biochemical event expressed by the mutation rate), heavily dependent on the fate (increased or decreased stability, or ability to replicate) of the mutant genomes generated. For example, during picornavirus RNA copying into minus strands and subsequent amplification in the replication complex, the proportion of mutants unable to bind the virus-specific replicase would be very low even if they were generated at high rates. The reader is referred to reviews by Domingo and Holland (1988, 1993), Eigen and Biebricher (1988), Domingo (1989) and Domingo *et al.* (1993) for additional examples and comments on this important point.

Table 1 lists a number of determinations of mutant frequencies for picornaviruses. The complexities of RNA genetics become apparent when trying to explain or reconcile the different values that have been reported for poliovirus. Parvin *et al.* (1986) determined the proportion of mutant genomes in progeny virions from a single plaque of poliovirus type 1 by repetitive sequencing of the VP1 gene. Since no mutations were detected in 95,688 nucleotides sequenced, the authors concluded that the mutation rate of poliovirus was less than 2.1×10^{-6} substitutions per nucleotide per infectious cycle. However, any mutant arising in the plaque and showing a slight selective disadvantage had a very good chance of being lost in the subsequent plaque isolations; in consequence it may never have been scored as a mutant

Table 1. Point mutation frequencies, and MAR mutant frequencies for picornaviruses

Virus	Procedure	Value[a]	Reference
Poliovirus	Repetitive nucleotide sequencing	$< 2 \times 10^{-6}$	Parvin *et al.*, 1986
Poliovirus	Reversion of amber mutant	2×10^{-6}	Sedivy *et al.*, 1987
Poliovirus	Reversion of guanidine dependence	2×10^{-4}	de la Torre *et al.*, 1990
Poliovirus	Chemical determination of error oligonucleotides [b]	$3x10^{-3}$ - $5x10^{-3}$	Ward and Flanegan, 1992
Poliovirus	Specific U → C transition	2×10^{-5}	de la Torre *et al.*, 1992
Poliovirus	MAR mutant frequency[c]	10^{-4} - 10^{-5}	Emini *et al.*, 1982; Minor *et al.*, 1983, 1986a
Mengovirus	MAR mutant frequency[c]	$3x10^{-3}$ - $5x10^{-5}$	Boege *et al.*, 1991
Foot-and-mouth disease virus	MAR mutant frequency[c]	10^{-5}	Martínez *et al.*, 1991a
Rhinovirus	MAR mutant frequency[c]	10^{-4} - 10^{-5}	Sherry *et al.*, 1986
Hepatitis A virus	MAR mutant frequency[c]	3×10^{-3}	Stapleton and Lemon, 1987

[a] Expressed generally as substitutions per nucleotide site. However, in some cases one specific base change was measured and in other cases any possible mutation at one or several sites was scored (see specific references for a precise interpretation of each value). MAR mutant frequency is the proportion of mutants resistant to neutralization by a MAb.

[b] Measured at many different sites of the viral genome.

[c] MAR mutant frequencies cannot be directly related to mutation rates or frequencies per site (see text).

— even in the case of its having been generated at a high rate! —. Thus, the value obtained by Parvin *et al.* (1986) was the frequency of neutral mutant genomes generated during early stages of a poliovirus plaque growth. The low value obtained may reflect the good adaptation to HeLa cells of the Mahoney poliovirus 1 used.

Sedivy *et al.* (1987) studied the reversion of an *amber* mutation introduced by site-directed mutagenesis in place of a serine codon at position 28 of the polymerase (3D) molecule of poliovirus. Each of the five revertants that were sequenced had regained the serine codon, even though a single base change in the *amber* codon could have given rise to any of six different amino acids. The specific transversion A→C, which was required to revert the *amber* to a serine codon, occurred at a frecuency of $2.5x10^{-6}$ revertants per *amber* plaque forming unit. In the analysis, however, mutant virus was grown in a suppressor cell line, and well isolated, amber plaques were picked after 3, 4 and 5 days of incubation and analyzed for the presence of revertants. Thus, early-arising revertants (which should produce large plaques) could not have been scored as components of an *amber* plaque. Did this reversion frequency reflect the mutation rate *per site*? Nomoto and associates (Kuge *et al.*, 1989) found a strong bias towards transition mutations at an AUG codon located within a sequence inserted in the non-coding region of Sabin poliovirus type 1. This AUG decreased the plaque size of the virus and permitted identification of revertants originated from independent mutation events. The ratio of transitions to transversions approximated 50 at this deleterious AUG triplet. Thus, if the reversion of the amber mutation studied by Sedivy *et al.* (1987) was dependent on a single transversion (such as A→C, needed to regain the serine in wild type polymerase) and the bias seen by Kuge *et al.* (1989) applied also to codon 28 of the polymerase, the true mutation frequency *per site* would approximate 10^{-4} ($2.5x50x10^{-6}$) mutations per nucleotide (Kuge *et al.*, 1989). The difficulty of extending a mutation rate *per site* from a mutation rate for *a particular base substitution* (a given transition or transversion at one site) will be encountered whenever the reversion of a deleterious mutation is used in the measurement, as discussed in the early work on site directed-mutagenesis of the 3'-extracistronic region of bacteriophage Qß (Domingo *et al.*, 1976; Batschelet *et al.*, 1976; see also Holland *et al.*, 1992).

Not surprisingly, other determinations for poliovirus have yielded high mutant frequencies. Holland and colleagues (de la Torre *et al.*, 1990) measured the frequency of reversion of guanidine-dependent mutants of the Mahoney strain of poliovirus type 1 to guanidine independence. This phenotypic change is due to a single amino acid replacement at position 227 of viral polypeptide 2C. Using several clonal pools of virus they found that the mutation frequency (with all amino acid substitutions occurring at position 227) was $(6.5\pm6.3)x10^{-4}$, and the minimal corrected base substitution frequency per single nucleotide site in the codon analyzed was $(2.1\pm1.9)x10^{-4}$ (de la Torre *et al.*, 1990).

More recently, Ward and Flanegan (1992) have used the error oligonucleotide method developed by Steinhauer and Holland (1986) to measure mutant frequencies at eight G residues of the Mahoney strain of poliovirus type 1. The proportion of molecules devoid of a G residue present in the consensus wild type genome was consistently of $(2.0\pm1.2)x10^{-3}$ to $(4.1\pm0.6)x10^{-3}$ at each of the eight sites analyzed, irrespective of their extent of conservation among related poliovirus genomes isolated in the field. This chemical analysis measures the proportion of mutated residues in viable and non-viable genomes present in the virus population under study. For this reason such chemical methods may yield slightly higher mutant frequencies than procedures which measure mutant frequencies in viable genomes [such as MAb-resistant (MAR) mutant frequencies; see Table 1 and Holland *et al.* (1989)].

A very interesting example of co-substitution of several nucleotides in the poliovirus genome has been recently described by de la Torre *et al.* (1992) in studying the reversion of a *ts* mutant of poliovirus type 1. Using infectious cDNA clones they measured a frequency of $2x10^{-5}$ for a specific U→C transition within the 3AB coding region. However, several revertants had also reverted *four* silent substitutions introduced in the infectious cDNA to provide restriction enzyme recognition sites. Thus, exceedingly high mutation rates may be attained during poliovirus replication, perhaps due to occasional, extremely mutagenic cellular environments or to the presence of error-prone polymerase subsets — polymerase molecules being also subjected to a potential high substitution level —. This finding with poliovirus has a precedent in several observations on hypermutability (and biased hypermutability, with a strong preference for one type of base substitution) made with defective-interfering RNA of vesicular stomatitis virus, measles virus variants, as well as with retroviruses (recent review in Holland *et al.*, 1992, and several chapters of the same book).

Even though studies on variation of picornaviruses are still in their infancy the available evidence suggests that other picornaviruses share with poliovirus high mutation rates and mutant frequencies. In an early study, Eggers and Tamm (1965) estimated a frequency of 10^{-4} mutations per replication for the reversion of coxsackie A9 from dependence on HBB [2(α-hydroxybenzyl)-benzimidazole] to HBB independence. A number of antiviral compounds bind into a hydrophobic pocket in the picornaviral capsid inhibiting virus uncoating. Human rhinovirus 14 mutants with high resistance to one such compounds (WIN 52084) occurred at frequencies of $4x10^{-5}$, and those with low level resistance were found at. $1x10^{-3}$ to $4x10^{-4}$ resistant particles per wild type (Heinz *et al.*, 1989). The ease of selection of viruses resistant to antiviral compounds is shared with other RNA viruses and retroviruses and it represents one of the most important problems for effective antiviral therapy (reviewed by Domingo, 1989, and Domingo and Holland, 1992). In spite of the ease of deriving WIN-resistant rhinovirus variants, Andries *et al.* (1990) could cluster 100 rhinovirus serotypes into two groups, according to their sensitivity to some WIN compounds. Interestingly this grouping correlated not only with sequences lining the pocket at the drug binding site, but also with pathogenicity and with serological relationships (Cooney *et al.*, 1982). It is not clear why such disparate phenotypes correlate so closely in this highly variable virus group.

MAR mutants of picornaviruses can generally be isolated at frequencies of 10^{-3} to 10^{-5} mutants per wild type virus, both in natural isolates and in clonal populations in cell culture (Table 1). These frequencies are only indirectly related to mutation rates since they depend on the number of amino acid substitutions capable of conferring the MAR phenotype while preserving virus viability. Also, MAR mutant frequencies may be underestimated by phenotypic masking of mutant genomes in wild type envelopes or capsids (Valcarcel and Ortín, 1989; Holland *et al.*, 1989; Carrillo *et al.*, 1990) or by the lower fitness of the mutants relative to the parental populations from which they were isolated (Martínez *et al.*, 1991a; González *et al.*, 1991). In spite of these limitations, and since several different neutralizing MAbs have been tested for each picornavirus group, it may be significant that most MAR mutant frequencies are in the range of 10^{-3} to 10^{-5}, irrespectively of the extent of antigenic diversity attained by a picornavirus in the field. For example, no trend towars higher frequencies is observed for rhinoviruses, a group represented by about one hundred serotypes (Table 1). Recently, Domingo and Holland (1993) have reviewed mutation rates and mutant frequencies for several virus groups. The conclusion arising from these comparisons is that picornaviruses do not differ greatly from other RNA viruses and retroviruses in their average mutant frequencies. However, as expected, differences of intrinsic mutability and of tolerance to substitutions may influence mutant frequencies at particular sites on a viral genome.

In a comparative study involving poliovirus and vesicular stomatitis virus, Holland *et al.* (1990) showed that the spontaneous mutation frequencies at defined genomic loci of the two viruses could not be increased by addition of chemical mutagens without loss of virus viability. In keeping with the concept of error threshold for error-prone replication (Eigen and Biebricher, 1988) the results with poliovirus suggest that it replicates with a fidelity very close to what may be compatible with maintaining meaningful genetic information. In other words, viral quasispecies are very adaptable because they replicate close to error catastrophe (Eigen and Biebricher, 1988).

QUASISPECIES. THE "POSITIVE-VERSUS-NEGATIVE SELECTION" GAME

With mutation rates averaging 10^{-3} to 10^{-5} substitutions per nucleotide site per round of copying, any picornavirus-infected cell must generate and contain multiple mutants. It has been estimated that an individual infected with an RNA virus may contain in the order of 10^9 to 10^{12} infectious particles at a given time point of an acute infection (Sellers, 1971; Halstead, 1980; Domingo *et al.*, 1985). Given a genomic complexity of 7kb to 9kb (which embraces picornaviruses) — combined with high population numbers and mutant frequencies (Table 1) — it can easily be calculated that all possible single and double mutants as well as decreasing proportions of triple, quadruple, etc. mutants are potentially present in most virus quasispecies (Domingo *et al.*, 1978, 1985; Temin, 1989). However, during replication, each mutant virus will be rated in competition with all other variants present and (continuously) arising in the population (Fig. 1). The proportion of each particular variant under defined environmental conditions will depend on the rate at which it arises but also on its competitive ability relative to other variants. The dynamic situation in which mutant swarms in continuous modification lead, nevertheless, to the same average or consensus genomic sequence has been termed population equilibrium (Holland *et al.*, 1982; Domingo and Holland, 1988). In general, most components of the mutant spectrum will show a decreased fitness relative to the parental average population (Domingo *et al.*, 1978; Eigen and Biebricher, 1988; Holland *et al.*, 1991; Martínez *et al.*, 1991a). A continuous process of *negative selection* will keep deleterious variants at low levels unless they can replicate by complementation of viral functions from standard genomes. Repeated plaque-to-plaque passage of virus often results in viruses with decreased fitness (Chao, 1990; Duarte *et al.*, 1992; Clarke *et al.*, 1993). This gradual accumulation of deleterious mutations (Muller's ratchet) is an additional proof of the quasispecies structure of RNA virus populations and it may be important to modulate virus virulence in nature (Duarte *et al.*, 1992).

In addition to the sieving effect of negative selection, virus evolution is strongly influenced by positive selection events such as that imposed by immune responses, by antiviral agents, etc. However, a virus variant may also increase in numbers irrespective of its competitive fitness, and without the intervention of positive selection. An example is the amplification of HIV-1 proviruses by antigenic stimulation of lymphoid cells (Wain-Hobson, 1992). It is not possible at present to assess the relative contributions of positive selection, negative selection and random sampling events in long-term evolution of picornaviruses. [This topic has been discussed by Domingo and Holland (1993) in relation to the neutralism-selectionism controversy].

Recombination frequencies can attain values of 10% to 20 % for the entire picornavirus genome (see King, 1988a, and Lai, 1992 for review). Recombinant molecules originate probably from polymerase jumping between genomes replicating in the same cell (Kirkegaard and Baltimore,

1986). The crossover sites do not seem to include a precise nucleotide sequence or a recognition signal. Recombination appears to be more frequent at conserved genomic stretches, and rare at capsid protein-coding regions (King 1988b). Recombination frequencies for FMDV are high in cell culture but only recently Krebs and Marquardt (1992) have provided the first evidence of recombination in nature by suggesting that FMDV 0_1 Burgwedel / 1987 is a recombinant between the related O_1 Kaufbeuren and a C_1 strain.

Since the basic mechanisms involved in RNA recombination appear to be common to most positive strand RNA genomes, recent findings with other systems suggest important evolutionary implications of recombination in the case of picornaviruses. First, as in the case of point mutations, recombination events seem to occur largely at random (Banner and Lai, 1991), and thus, most resulting recombinant genomes must be subjected to very strong negative selection. Occasionally however, a recombinant genome may be sufficiently fit to initiate the exploration of a new ecological niche with the optimizing effects of point mutations generated during replication. If the new replicon is successful, a new virus has come into existence as a result of a large evolutionary jump. Since heterologous recombination may also occur involving disparate RNA molecules of cellular origin (Monroe and Schlesinger, 1983; Meyers *et al.*, 1991), recombination may actively promote reshuffling of viral and cellular domains to produce novel viable genomes (Zimmern, 1988) over a trial and error process lasting hundreds of thousands of years and myriads of recombination events. As will be elaborated further in the last section of this chapter, present day viruses can experience only limited variations and they represent "clouds" in sequence space (Eigen and Biebricher, 1988) which cannot expand beyond certain limits. This is why they possess an identity as viruses and we can classify them, although some times with difficulty. Recombination can be considered a means to rescue viable genomes from highly mutated or defective parental genomes (a conservative activity) but also a way to produce novel genomic combinations (a radical activity) that may mediate long-term evolutionary jumps of viruses.

A significant subclass of defective genomes are the defective interfering (DI) particles able to modulate replication of their standard infectious parental viruses (reviewed by Roux *et al.*, 1991). DI RNAs have been described for a variety of picornaviruses (Cole and Baltimore, 1973; Nomoto *et al.*, 1979; McClure *et al.*, 1980; Radloff and Young, 1983). However, generation of DIs during picornaviral infection is not a frequent event perhaps due to poor complementation of viral functions in systems which use a number of cis-acting activities. The possible role of DIs in natural picornaviral infections remains largely unexplored.

PICORNAVIRUSES VARY WITHIN INFECTED INDIVIDUALS AND DURING DISEASE EPIDEMICS

Poliovirus and foot-and-mouth disease virus (FMDV) have been the picornaviruses most extensively analyzed with regard to genetic heterogeneity and variation *in vivo* and in cell culture. Some interesting points can be made by comparing the two viruses which are quite different regarding host range and the infection process. Humans are the sole natural host of poliovirus, and most (>99%) infections are subclinical. Even for poliovirus type 1— responsible of about 90% of poliomyelitis cases — disease is seen in only about 0.5% of infections by this virus serotype (Minor,

1989). In an outbreak in Finland (1984 - 1985) caused by poliovirus type 3 it was estimated that about 10^5 individuals were infected but only 10 were ill (Hovi *et al.*, 1986). In contrast to the restricted host-range of poliovirus, many different animal species belonging to the artiodactyla are natural hosts for FMDV. Even during one disease outbreak more than one animal species may be afflicted with acute systemic infection; subclinical infections appear to be rare in the common hosts, but a carrier state may be established in ruminants as a sequel of an acute infection. In spite of encountering widely different environments, poliovirus and FMDV show a similar extent of genetic change in nature. By sequencing a VP1/2A-coding region, it was estimated that the genetic heterogeneity during a poliomyelitis epidemic was 1% to 2% of nucleotides (Kew *et al.*, 1990). This is similar to the estimated 0.7% to 2.2% nucleotide heterogeneity over the entire FMDV genome for viruses isolated during an episode of FMD (Domingo *et al.*, 1980). An animal infected with FMDV includes multiple variants (Domingo *et al.*, 1980; Carrillo *et al.*, 1990). Likewise, poliovirus type 2 and 3 vaccine strains can undergo more than 100 mutations during replication in only one or two individual humans (Kew *et al.*, 1981). Thus, not surprisingly, contemporary poliovirus or FMDV virus isolates are also genetically heterogeneous (Sobrino *et al.*, 1986; Rico-Hesse *et al.*, 1987; Villaverde *et al.*, 1988; Kinnunen *et al.*, 1991). Comparison of sequential isolates during an epidemic outbreak suggests a rate of fixation of mutations of 6.9×10^{-3} to, 1.4×10^{-2} substitutions per nucleotide per year for the poliovirus genome (Kew *et al.*, 1990) and of 6.5×10^{-3} for FMDV (Villaverde *et al.*, 1988, 1991). (Table 2).

Examination of poliovirus genomes excreted by primary vaccinees revealed rapid genetic variation by mutation and recombination (Nottay *et al.*, 1981; Kew *et al.*, 1981; Kew and Nottay, 1984; Jameson *et al.*, 1985). A number of mutations have been associated with attenuation of poliovirus for humans. A pyrimidine transition disrupts a base pair at the 5'-untranslated region (5'-UTR) resulting in reduced translation efficiency. In addition, attenuated strains include an amino acid substitution in capsid protein VP1 and VP3 (Westrop *et al.*, 1989; Macadam *et al.*, 1989; Weeks-Levy *et al.*, 1991), the latter being responsible of the *ts* character of the Sabin type 3 vaccine strain. These three mutations are frequently lost within days of replication of poliovirus in vaccinees (Evans *et al.*, 1985; Minor *et al.*, 1986b; Dunn *et al.*, 1990). Reversion of the *ts* character of Sabin type 3 poliovirus occurred by second site mutations (Minor *et al.*, 1989). Mutant and recombinant polioviruses have also been isolated from rare cases of vaccine-associated poliomyelitis. Thus, the efficacy of live attenuated poliovirus vaccines is not due to inability of the virus to change genetically (or even antigenically, at least at some sites), since poliovirus behaves as a quasispecies capable of rapid variation in humans.

Multiple genomic mutations at different loci distinguish attenuated FMDV from their virulent counterparts and no conclusive evidence has been obtained to permit assigning the altered phenotype to a genetic change (Parisi *et al.*, 1985; Cao *et al.*, 1991).

In addition to poliovirus and FMDV, extensive genetic heterogeneity and variations during natural infections have been documented for all picornaviruses examined, including rhinoviruses (Tyrrell, 1990), enterovirus 70 (Takeda *et al.*, 1984), coxsackie A10 virus (Kamahora *et al.*, 1985), coxsackie B4 virus (Prabhakar *et al.*, 1985), hepatitis A virus (HAV) (Siegl *et al.*, 1984; Weitz and Siegl, 1985; Ticehurst *et al.*, 1989; Robertson *et al.*, 1992) and the insect picornavirus Drosophila C virus (Clewley *et al.*, 1983). As complete or partial nucleotide sequences of sufficient length are determined, it becomes clear that each picornavirus isolate is genetically unique (reviews in Domingo and Holland, 1988, 1993; Palmenberg, 1989; Stanway, 1990).

Phenotypic changes can be the result of one or a limited number of mutations in a genome, and variants with few mutations have a good chance of being represented in the quasispecies mutant spectra, provided they do not adversely alter the ability of the virus to replicate. (Domingo *et al.*, 1978, 1985; Domingo and Holland, 1988, 1992, 1993). Thus, it is expected that frequent changes in biological properties occur with picornaviruses within relatively short time periods. Among other examples, a Coxsackie B3 virus with altered receptor apecificity was selected upon passage in rhabdomyosarcoma cells (Reagan *et al.*, 1984). A limited number of mutations distinguish diabetogenic from non-diabetogenic encephalomyocarditis virus (EMCV) (Ray *et al.*, 1983), or fast-growing variants of HAV from slow growing strains (Venuti *et al.*, 1985; review in Ticehurst *et al.*, 1989). Neurological complications have been described for diseases caused by a number of picornaviruses — other than poliovirus — such as enterovirus 70 (Bharucha and Mondkar, 1972; Kono *et al.*, 1973) or FMDV, even though in this case neurotropism is very rare (reviewed by Domingo *et al.*, 1990). It is obvious that a large number of mutations occurring in picornaviruses during natural infections may be without apparent biological relevance. This is unavoidable since mutant spectra must provide a reservoir of variants capable of exploring (and perhaps adapting) to a wide range of environments. Such potential adaptability may require an "excess of variation" relative to what would be strictly optimal for a single lineage completing an evolutionary cycle in a defined environment. An interesting possibility is that a virus replicating in a very specialized site within a restricted number of host species could perform its life cycles with no need of exploring sequence space beyond the one optimized to replicate and assemble infectious and transmissible virions. Such a viral system would not need high mutation rates and its replication machinery could have evolved to possess (or be associated with) some proofreading — repair activity (Holland *et al.*, 1992). However, such a potential virus has yet to be identified. It was suspected that HAV, because of its highly specialized hepatotropic character — and also because of its striking antigenic stability suggested by early results — might be an example of a well adapted, relatively invariant virus. As additional HAV sequences are determined, however, it appears that HAV shows also considerable genetic variability (Ticehurst *et al.*, 1989; Robertson *et al.*, 1992; Brown *et al.*, 1991; see also MAR mutant frequencies in Table 1). Other RNA viruses such as rabies and measles were also considered relatively invariant. However, recent data suggest that they may actually be more heterogeneous and variable than previously suspected (Benmansour *et al.*, 1992; Rota *et al.*, 1992).

In spite of a similar behaviour at the genetic level, poliovirus and FMDV show differences regarding the ability to induce widely protective responses within a serotype. Minor (1990) compared the antigenic structure of representative picornavirus groups. Poliovirus vaccines (live attenuated Sabin vaccines or formalin inactivated Salk vaccines, each of them containing a strain of each of the three poliovirus serotypes) have been effective over many decades (Minor, 1989). In contrast, anti-FMD vaccines must be periodically updated, and tailored according to the predominant viral types and subtypes in a given geographical area (Bachrach, 1968; Brown, 1989; Domingo *et al.*, 1990). A different long-term antigenic stability may result, among other possibilities, of one or several of the following effects. A large proportion of the nucleotide substitutions which are fixed during the evolution of the poliovirus genome in nature is silent (Kew *et al.*, 1990) whereas both silent and non-silent replacements are found during FMDV genome evolution, and, in some cases, non-silent mutations are dominant (Díez *et al.*, 1990a ; Martínez *et al.*, 1992). In spite of a similar overall three-dimensional structure of poliovirus (Hogle *et al.*, 1985) and

FMDV (Acharya *et al.*, 1989) antigenic determinants are located on surface loops which differ in length, composition and flexibility between the two viruses. Some immunodominant antigenic loops may tolerate more (and perhaps more drastic) substitutions in FMDV than in poliovirus. For example, the immunodominat antigenic site 1 of poliovirus type 3 (VP1 residues 89 to 100) — also present but less dominant in type 1 (Minor *et al.*, 1986a) — appears to be more conserved among independent poliovirus isolates (from different years and locations) than the immunodominant site A (VP1 residues 140 to 160) of FMDV [compare Table 2 in Minor (1989) with Fig. 1 in Domingo *et al.* (1992)]. Unlike FMDV in which all antigenic sites are among the most variable capsid regions in primary sequence, some poliovirus sites appear to bear considerable restrictions for variation. No change in antigenic site 1 of poliovirus types 1 or 3 could be evidenced by testing neutralizing MAbs against strains isolated from vaccinees (Minor *et al.*, 1986b, 1987; review in Minor, 1989). Site 3a of poliovirus type 3 appears also as very conserved among recent isolates (Yang *et al.*, 1992). If indeed surface loops of poliovirus are more limited in their tolerance for variation, certain testable predictions can be made. The number of different amino acid substitutions able to confer the MAR phenotype might be, on average, lower for poliovirus than for FMDV. In addition, or alternatively, the relative fitness of the MAR mutants to their parental wild type virus should also be lower for poliovirus than for FMDV. However, no such differences are obvious when comparing MAR mutant frequencies (Table 1). Only a limited number of measurements of relative replication ability have been carried out for picornavirus MAR mutants (Martínez *et al.*, 1991a; González *et al.*, 1991) or for drug-resistant rhinovirus mutants (Ahmad *et al.*, 1987; review in Domingo, 1989), and many more measurements are needed to reveal, in a statistically significant manner, possible differences of tolerance to substitutions among virus groups. Large, flexible loops, structurally independent of other domains, may be prone to variation and capable of inducing large repertoires of antibodies. Some amino acid substitutions at the main antigenic site A (VP1 residues 138 to 150) of FMDV of serotype C abolish many (or all) of the multiple, overlapping epitopes that have been mapped within this site (Mateu *et al.*, 1989, 1990; Martínez *et al.*, 1991b). A critical replacement led to the generation of a subtype-specific epitope in an FMDV of a different subtype (Hernández *et al.*, 1992). Drastic effects of this sort have been described for poliovirus MAR mutants but not for field isolates (Minor *et al.*, 1983). In FMDV the impact of such critical replacements on antigenic specifity may be compensated by additional, non-critical replacements within site A (Mateu *et al.*, 1992). The molecular basis of such compensatory effects is not known. During long-term evolution of FMDV multiple antigenic subtypes and variants are frequently generated (Domingo *et al.*, 1990, 1992; Martínez *et al.*, 1992), whereas intratypic poliovirus variation is generally of limited epidemiological significance in spite of the frequent selection of antigenic variants in cell culture (Table 1) or in the gut (Minor, 1990). HAV shows also considerable conservation of antigenic sites (Lemon and Binn, 1983; Lemon and Ping, 1989) in spite of its genetic diversification (Lemon *et al.*, 1987; Stapleton and Lemon, 1987). Indeed, the nucleotide sequence differences among HAV strains and variants range between 0.3% to 8.5% of nucleotide sites (Ticehurst *et al.*, 1989) values comparable to those for poliovirus or FMDV isolated during disease episodes. In conclusion, picornaviruses share elevated MAR mutant frequencies at most antigenic sites examined (Table 1) but they vary in the extent of long-term antigenic diversification that they can attain (or have attained) in nature. This may be due to different tolerances of antigenic sites to amino acid substitutions, and to the effects of substitutions (and groups of substitutions) on antigenic specificity. Depending on the primary amino acid sequences — which are very divergent among antigenic sites of different picornaviruses — predominance of additive or compensatory effects upon accumulation of amino acid substitutions may result in either substantial or limited antigenic diversification.

PERSISTENT INFECTIONS BY PICORNAVIRUSES AND THE ROLE OF VIRUS VARIATION

The prototype picornavirus capable of establishing a persistent infection in mice is Theiler's murine encephalomyelitis virus (TMEV) — isolated from the central nervous system (CNS) of a paralytic mouse (Theiler, 1934) —. Several variants of TMEV have been described and divided into virulent TMEV (strains GDVII, FA) and persistent TMEV (strains To, BeAn 8386, DA, WW) (reviews in Rodríguez *et al.*, 1987; Welsh *et al.*, 1989; Brahic *et al.*, 1991). Members of the virulent group replicate actively in neurons and cause acute fatal encephalitis. Members of the persistent group (also termed TO for Theiler's original) replicate less efficiently in neuronal cells, and cause an early acute poliomyelitis followed by a persistent infection of the white matter which leads to chronic, demyelinating disease. For this reason, TMEV serves as an adequate model for multiple sclerosis, a human neurological condition of suspected viral etiology with clinical and histological manifestations reminiscent of those caused by TMEV. Pairwise sequence comparisons between TMEV strains (Pevear *et al.*, 1988, and references therein) show about 90% identity at the nucleotide level and 95% at the amino acid level between virulent and persistent strains. Studies using chimaeric viruses constructed with genomic segments originating either from the persistent or the virulent group have led to the conclusion that the capsid polypeptides are the main determinants of efficient viral replication *in vivo*. A virus including the VP1 coding region of GDVII inserted into the genetic context of strain DA was unable to persist in spite of not killing mice. The reciprocal construct, — that is, the VP1 coding region of DA into the genome of GDVII — produced a virus equally attenuated but able to persist and cause demyelination. For the DA strain to regain nearly full virulence it was necessary to include the entire capsid - protein coding region of GDVII. Furthermore, a MAR mutant of TMEV DA with a replacement in protein VP1 showed decreased replication in the CNS and mild demyelination as compared with its parental DA (Zurbriggen *et al.*, 1991). These results (reviewed by Brahic *et al.*, 1991) illustrate how substitutions in the capsid proteins or their corresponding coding RNAs can affect viral multiplication and pathogenesis. The three-dimensional structure of the DA (Grant *et al.*, 1992) and Be An (Luo *et al.*, 1992) particles — both belonging to the persistent group — shows that most differences between the capsids of the two groups map on the surface of a star - shaped plateau, perhaps reflecting differences in receptor recognition or in antigenic behaviour. Genetic variations during persistence of TMEV *in vivo* and in L929 cells persistently infected with the DA strain were detected by Roos *et al.* (1982). In this *in vitro* cell culture system, interferon seems to play a role in persistence, but no extensive phenotypic variations in the virus nor the presence of defective-interfering particles could be shown. Persistence *in vivo* may be favored by the genetic make-up of the virus and host factors which may lead to inefficient clearing by the host immune response. The H-2D genomic region of mice (probably in combination with other genetic factors), contributes to the resistance to TMEV by controling the CTL response to the virus (Rodríguez and David, 1985; Brahic *et al.*, 1991).

HAV can cause subclinical infections mainly in children; chronic disease however, appears to be very rare (Hollinger and Ticehurst, 1990). In cell culture HAV often establishes persistent infections with little perturbation of the host cell metabolism (Provost and Hilleman, 1979; Gauss-Muller and Deinhardt, 1984; Vallbracht *et al.*, 1984). Variants cytolytic for some cell lines but able to persist in other cells have been described (Ticehurst *et al.*, 1989). As is the case also for other picornaviruses, some RNA stretches within the 5'-extracistronic region are prone to variation, and point mutations and deletions often accumulate upon passage of HAV in cell culture (Nüesch *et al.*, 1988; Ticehurst *et al.*, 1989). No detailed analysis of genetic or antigenic variations undergone by HAV during inapparent infections *in vivo* have been reported.

Perhaps the system in which genomic modifications during persistent infections have been examined in greater detail is FMDV. This aphthovirus is able to persist in the esophagous and throat of ruminants either as outcome of an acute infection with incomplete virus clearing or as a result of immunization with some live-attenuated vaccines. This *in vivo* carrier state for FMDV was first described by van Bekkum *et al.* (1959) and it is considered an important reservoir of FMDV in areas where the disease is enzootic (reviewed in Domingo *et al.*, 1990). Persistent infections in cattle established with plaque-purified FMDV have revealed that the virus underwent genetic and antigenic changes in the animals (Costa Giomi *et al.*, 1984, 1988; Gebauer *et al.*, 1988). The rate of accumulation of point substitutions in the VP1 coding region was 9.0×10^{-3} to 7.4×10^{-2} substitutions per nucleotide per year, one of the largest rates of evolution recorded among RNA viruses (Table 2). In contrast, values for several cellular genes are 10^{-8} to 10^{-9} substitutions per nucleotide per year. This million-fold difference was first emphasized by Holland *et al.* (1982) as a hallmark of RNA genetics. The higher rate of fixation of subtitutions in carrier animals as compared with an FMD epizootics (Table 2) can be interpreted as a result of the continuous replication of viral genomes in carriers and the alternation of periods of active replication with others without replication during extended disease episodes (Beck and Strohmaier, 1987; Martínez *et al.*, 1992). Some of the substitutions fixed in carrier cattle were antigenically relevant since they altered FMDV reactivity with MAbs (Gebauer *et al.*, 1988).

Several early experiments (Dinter *et al.*, 1959; Seibold *et al.*, 1964) suggested that FMDV could establish persistent infections in cell culture, but the limited stability of the carrier cells that were derived did not allow detailed studies. A persistent infection with plaque purified FMDV type C was readily established by growing the cloned BHK-21 cells which survived an initial cytolytic infection (de la Torre *et al.*, 1985). As the carrier cells were passaged, genetic and phenotypic changes were observed in the cells and in the resident virus. The virus became increasingly virulent for the parental BHK-21 cells (used to establish persistence) and the cells, in turn, became increasingly resistant to the virus (de la Torre *et al.*, 1988). Such cell resistance was not due to absence or limitation in host cell receptors since a block for virus replication was observed upon transfection with infectious FMDV RNA. Also, cell fusion experiments suggested that the resistance to FMDV was mediated by trans-acting inhibitors rather than by the absence of a factor needed for viral replication (de la Torre *et al.*, 1989a). Surprisingly, the evolution of the cell occurred via extremely heterogeneous cell populations in which many clones — freed of FMDV by treatment with ribavirin (de la Torre *et al.*, 1987) — differed in the degree of cellular transformation, and in resistance to FMDV (de la Torre *et al.*, 1989b). By repetitive nucleotide sequencing of the VP1 gene segment of nine viral clones from passage 59 of the carrier cells the virus population was found to be

Table 2. Rate of accumulation of mutations (evolution rates) of picornaviruses, as compared to immunodeficiency viruses and cellular genes

Virus	Genomic segment	Value[a]	Reference
Foot-and-mouth disease virus	VP1[b]	$9.0x10^{-3}$ - $7.4x10^{-2}$	Gebauer *et al.*, 1988
	VP1	6.5×10^{-3}	Villaverde *et al.*, 1991
	VP1	1.4×10^{-3}	Martínez *et al.*, 1992
Poliovirus	Average over genome	$6.9x10^{-3}$ - $1.4x10^{-2}$	Kew *et al.*, 1990
Human immunodeficiency virus	env	$3.2x10^{-3}$ - $1.6x10^{-2}$	Hahn *et al.*, 1986
	gag	$3.7x10^{-4}$ - $1.8x10^{-3}$	Hahn *et al.*, 1986
Simian immunodeficiency virus	gp120[c]	8.5×10^{-3}	Burns and Desrosiers, 1991
Cellular genes		10^{-8} - 10^{-9}	Britten, 1986; Weissmann and Weber, 1986

[a] Expressed as substitutions per nucleotide per year. Except where indicated (b, c) calculations are based on sequence comparisons of isolates during one or multiple epidemic outbreaks of the indicated virus.

[b] Cattle persistently infected with plaque-purified foot-and-mouth disease virus.

[c] Infections established with a molecular clone of simian immunodeficiency virus.

genetically heterogeneous at a level of $5x10^{-4}$ mutations per nucleotide screened (de la Torre *et al.*, 1988). Thus, persistently infected cell cultures consist of complex distributions of variant cells and viruses. It is likely that this dual heterogeneity endows the system with unique genetic flexibility allowing a coevolution of the cells and the resident virus which permits prolonged persistence.

By passage 58 of the carrier cells, the consensus nucleotide sequence of the resident FMDV had diverged from the parental cloned virus in about 0.3% of genomic nucleotides (de la Torre *et al.*, 1985), and by passage 100 such divergence had attained 1%, with point mutations fairly uniformly distributed along the 5' UTR and the P1 region (Díez *et al.*, 1990a; Escarmís *et al.*, 1992). One of the most striking features of the latter virus (termed R100) is the absence of a cysteine (in position 7 of VP3) involved in disulfide bridges which delimit the narrower part of a pore at the virion five-fold axis (Acharya *et al.*, 1989). This structurally critical cysteine is conserved in all aphthoviruses sequenced to date, and its absence in R100 suggests that even highly conserved residues can be substituted when virus replication takes place in a modified environment (in this case a persistent versus an acute infection) (Díez *et al.*, 1990a).

A second, unexpected feature was the presence in the genome of R100 of the longest polyribocytidylate (poly C) tract so far described, to our knowledge, for a picornavirus (Escarmís *et al.*, 1992). The poly C — whose function in picornaviruses is unknown — of R100 is about 420 residues long and it includes a few uridine residues interspersed at fairly regular intervals at the 3' end region of the homopolymeric tract. It is about 145 residues longer than the poly C present in the genome of the parental virus used to initiate persistence. In spite of its hypervirulence for BHK-21 cells, R100 is highly attenuated for cattle and mice (Díez *et al.*, 1990b). Since attenuated cardiovirus was obtained by engineering a short poly C tract (Duke *et al.*, 1990), the presence of a long homopolymeric tract in R100 was unexpected. It is clear that a long poly C is compatible with attenuation of FMDV, at least for cattle and mice.

It is suspected that persistence of poliovirus *in vivo* may be involved in some degenerative diseases of motor neurons of the brain and spinal cord—such as amyotrophic lateral sclerosis—which may occur many years after acute poliomyelitis (Campbell *et al.*, 1969; Mulder *et al.*, 1972; Dalakas, 1986; Sharief *et al.*, 1991). Early evidence of persistence of poliovirus in cell cultures was provided by Ackermann and Kurtz (1955), and by Vogt and Dulbecco (1958). In the latter study morfologically altered HeLa cells with increased resistance to poliovirus were described. During persistence in lymphoblastoid cell lines, *ts* mutants were identified (Carp, 1981). By cotransfecting poliovirus genomic RNA with a subgenomic RNA —with most of the capsid-coding region deleted— Kaplan *et al.* (1989) established HeLa cell lines persistently infected with poliovirus, which evolved to produce cells highly resistant to poliovirus. In some cells resistance was due to absence of the cellular receptor, and in other cell clones there was an intracellular block to poliovirus replication.

A neuroblastoma cell line persistently infected with poliovirus has been studied by Colbère-Garapin and her colleagues (Colbère-Garapin *et al.*, 1989; Pelletier *et al.*, 1991; Borzakian *et al.*, 1992). Variant polioviruses with altered host range were isolated from these neuroblastoma cells that could establish persistence in nonneuronal cells. In particular, such virus variants, but not wild type or attenuated polioviruses, persisted in HEp-2c cells. Borzakian *et al.* (1992) have shown that establishment of persistence in HEp-2c depends on the virus strain, whereas maintenance of persistence involved selection of variant cells in which multiplication of poliovirus was restricted, a situation parallel to that found during persistence of FMDV in cell culture (see above). A reduced expression of the poliovirus receptor was seen in some cell clones but not in others, and cell fusion experiments suggested that the restriction to poliovirus infection was a dominant trait (Borzakian *et al.*, 1992). These authors suggested the existence of an equilibrium between an abortive and a lytic infectious cycle subjected to alteration by environmental changes. This is a very attractive model — in line with the dynamic nature of picornaviral populations — which deserves further attention when analyzing persistent infections by other highly variable RNA viruses.

Gibson and Righthand (1985) established cloned human WISH cells persistently infected with echovirus 6, a system in which the presence of highly defective virus populations appears to play an important role. In particular, during persistence VPO was present suggesting a defect in its processing to yield VP2 and VP4 in viruses unable to attach to susceptible cells (Righthand and Blackburn, 1989). A persistent infection of HeLa cells with a *ts* rhinovirus type 2 has also been described, in which cell subpopulations resistant to homologous rhinovirus were selected (Gercel *et al.*, 1985).

Thus, virtually for any picornavirus persistent infections have been described *in vivo* or in cell culture, and in many cases virus as well as cell variations appear as important elements in sustaining this type of infection.

PHYLOGENETIC RELATIONHIPS AMONG PICORNAVIRUSES. VIRUSES AS CLOUDS IN A SEQUENCE SPACE

Picornaviruses were initially classified on the basis of biological and physical properties. Rueckert (1990) and Stanway (1990) divided the family Picornaviridae into four genera: *enteroviruses, cardioviruses, rhinoviruses* and *aphthoviruses*. Nucleotide and amino acid sequence comparisons revealed important problems in early classifications. For example, the enteroviruses included not only the related polioviruses and coxsackieviruses, but also hepatitis A virus which shows less than 15% amino acid identity with poliovirus in the capsid proteins. Palmenberg (1989) carried out the first systematic, comparative alignment of picornaviral sequences, and established phylogenetic relationships among picornaviruses. Based on the protein alignments, the corresponding nucleotide sequences were analyzed by application of a maximum parsimony algorithm. For P1 sequences, the percent identities [defined as one hundred times the number of identical nucleotides in a pairwise alignment, divided by the length of the compared sequences (excluding gaps)] ranged from greater than 70% (for different serotypes of the same picornavirus) to less than 30% (for HAV as compared with other picornaviruses). It must be noted however that overall sequence identities averaged over an entire genome (or a specific gene) may be quite at variance with identities for defined genomic segments or the encoded proteins. For example, nucleotide sequence identity for the polymerase (3D) gene is 65% between HRV-14 and poliovirus, and 55% between HRV-14 and HRV-2 (Skern *et al.*, 1984). This difference fades when the capsid protein VP1-coding region is compared, the identities being in the range of 42% to 44% (Andries *et al.*, 1990). Based on comparisons of the P1 genome segment, Palmenberg (1989) proposed four main groups of picornaviruses: I (FMDV), II (Mengo and TMEV), III (HAV), and IV (poliovirus, rhinovirus, coxsackievirus, and bovine enterovirus). Major departures from previous classifications are the recognition that hepatitis A virus constitute a distinct group and that the Theiler's murine encephalomyelitis viruses are related to the cardioviruses. More recently, the ICTV Study Group on picornaviruses has proposed a further division of the Picornaviridae family into five genera: *enterovirus, rhinovirus, hepatovirus, cardiovirus* and *aphthovirus* (Minor *et al.*, submitted 1993). A number of viruses (equine rhinoviruses, Drosophila C virus, etc.) have not been assigned to any genus.

Phylogenetic relationships have also been derived for different serological subtypes and for different isolates within a picornavirus genus. They include, among others, studies with the three serotypes of poliovirus (Rico-Hesse *et al.*, 1987; Kew *et al.*, 1990; Kinnunen *et al.*, 1992), the three European serotypes of FMDV (Dopazo *et al.*, 1988) and isolates from different subtypes of FMDV type C (Martínez *et al.*, 1992). Extremely ramified trees are generally produced reflecting that genetically heterogeneous virus lineages cocirculate during a disease outbreak and that even a single infected individual harbors mixtures of variants (Domingo *et al.*, 1980; Sobrino *et al.*, 1986; Gebauer *et al.*, 1988; Villaverde *et al.*, 1988, 1991; Kinnunen *et al.*, 1991). Viral heterogeneity within infected organisms has generally been disregarded in epidemiological models of virus spread and disease control, and yet it is one of the most relevant parameters to account for vaccine failures and incomplete protection. Furthermore, such heterogeneity is an expected consequence of the quasispecies structure of RNA virus population, and it may define a transient molecular clock when frequent host-to-host transmission occur (Villaverde *et al.*, 1991). A recent model for the development of AIDS places emphasis on antigenic diversity of HIV-1 as the cause rather than a consequence of immunodeficiency (Nowak *et al.*, 1991). Above an antigenic diversity threshold the virus induces, according to this model, a collapse of the host lymphocyte population. For poliovirus type 3, the extent of genetic variation in one infected individual was at times as significant as variation among independent isolates (Kinnunen *et al.*, 1991).

Only in a few cases, the degree of significance of the tree nodes and the standard deviations of branch lengths have been assessed in phylogenetic trees by application of statistical procedures. This practice is particularly important when attempting to define relationships among isolates of a highly variable RNA virus. Martínez *et al.* (1992) applied the bootstrap resampling method (Felsestein, 1985) to assess the significance of a tree obtained by the least-squares method, and which related the VP1-coding region of type C FMDVs isolated in Europe, South America and The Philippines. Six highly significant lineages — one of them corresponding to a virus group which is now probably extinct — were defined, but several sublines were shown to be of questionable significance upon application of the bootstrap resampling method. Obviously, such statistical treatments do not evaluate or correct for problems derived from the sampling of virus genomes which are entered into the tree derivations, and which are due to the fact that the available sequences represent a minority subset of the total population. Interestingly, for FMDV type C, quantification of those mutations that led to an amino acid replacement along well established lineages showed that there was no significant accumulation of amino acid substitutions with time. In contrast, net accumulation of silent replacements at a fairly constant rate was observed. Antigenic variation of FMDV serotype C over six decades was essentially due to combinations among a very limited number (often two) of amino acids per heterogeneous position (Martínez *et al.*, 1992). It will be interesting to extend the type of study of Martínez *et al.* (1992) to other picornaviruses for which representative evolutionary lineages with isolates extending over many decades may be available.

Constraints for variation at the RNA level are also suggested by a number of comparisons of the secondary structure at the extracistronic regions. The internal ribosome entry site (IRES) has a defined secondary structure very similar for cardioviruses and aphtoviruses, but distinct from the corresponding structure for enteroviruses and rhinoviruses (Pilipenko *et al.*, 1989a, b). Perturbation of some key domains of the IRES results in decreased initiation of translation, and natural variations in primary structure often consist of groups of compensatory mutations to ensure conservation of secondary (and probably higher order) structure. A tight secondary structure is also predicted for 5'UTR stretches other than the IRES such as the S fragment (at the 5' side of the poly C) of FMDV (Clarke *et al.*, 1987; Escarmís *et al.*, 1992). In FMDV, block deletions of around 40 or 80 residues are often observed even when comparing viruses of the same serotype. Such deletions (or insertions) correspond to the loss (or gain) of one or two predicted pseudoknot structures (Clarke *et al.*, 1987; Escarmís *et al.*, 1992). In other systems pseudoknots have been implicated in regulation of translation or RNA splicing, but their role in FMDV is not understood. Deletions of this type may be frequent because they involve the complete loss

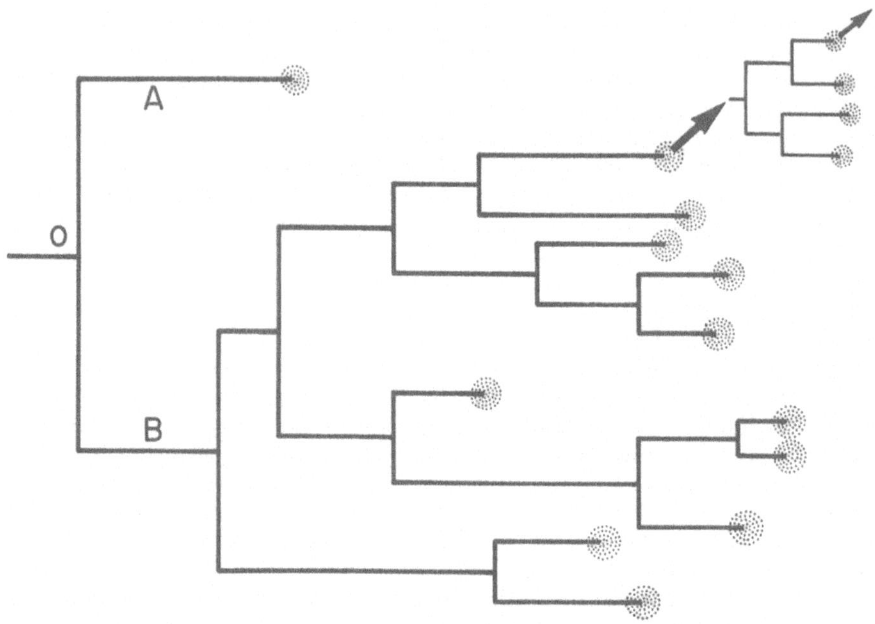

Fig. 2. A schematic phylogenetic tree of the type that has been derived for picornaviruses and other RNA viruses. The tree has been rooted on a theoretical origin O. From O two main branches A and B originate. B gives rise to additional branchings, whereas A represents an evolutionary dead-end. At the tip of each branch a cloud represents the quasispecies distributions of genomes corresponding to each isolate. The arrow at the top right suggests that for a single isolate multiplying in a few hosts, an additional branching and further diversification would be found. The process would repeat in a single host or even in different cells of an infected organ, with decreasing genetic distances. In spite of each quasispecies being restricted to a cloud, diversification in sequence space may be rapid, and large evolutionary jumps may be mediated by homologous and heterologous recombination with optimization by point mutations (see text).

of a non-essential RNA stretch. It has been suggested that RNA insertions or deletions in regulatory regions may affect biological features of picorna viruses (Stanway *et al.*, 1984).

Primary sequence similarity may extend to other virus groups particularly in non-structural proteins such as viral polymerases, helicases, or proteases (Franssen *et al.*, 1984; Haseloff *et al.*, 1984; Kamer and Argos, 1984; Argos *et al.*, 1984; Argos, 1988; Gorbalenya and Koonin, 1989; Koonin *et al.*, 1991; Bruenn, 1991; review in Zimmern, 1988). Such similarities often correspond to catalytic domains of viral enzymes and they could reflect either common ancestry or convergent evolution of viral quasispecies required to use and interact with cellular components. Tenuous relationships can also been identified among structural proteins of viruses as different as FMDV and Sindbis (Fuller and Argos, 1987) or Mengo and hepatitis B (Argos and Fuller, 1988).

RNA viruses, as other more complex life forms, seem to exist under two apparently contradictory forces: variation to ensure adaptability and conservation to ensure functionality. Fig. 2 shows an imaginary phylogenetic tree in which virus isolates —the tips of the branches— are depicted as spherical clouds. Each cloud is the domain of genetic distances in which a given virus with a genetic make-up is allowed to exist. It cannot expand beyond those limits without loosing functionality. The need of certain biochemical functions dictate common features (primary sequences or higher order structures) among members of distant clouds. The tree could represent relationships among different picornaviruses or among representatives of one picornavirus. The arrow in Fig. 2 suggests that if viruses within a narrower space-time coordinates are examined (such as during a disease outbreak, or even within an infected individual) multiple lineages would be found albeit related by shorter genetic distances (branch lengths). Thus, quasispecies is an important step in the molecular evolution of RNA viruses since it is the raw material on which positive selection and random sampling events can act (Domingo and Holland, 1993).

CONCLUDING REMARKS

The field of RNA virus evolution has progressed dramatically during the last decade. Many biological implications of RNA virus variation are becoming apparent. The design of new antiviral compounds and synthetic vaccines and their successful use greatly depend on an understanding of the dynamics of viral populations in infected hosts. The very successful antiviral vaccines (poliomyelitis, measles, mumps) may be the exception rather than the rule, and mankind may have to face other epidemics like AIDS (Temin, 1989; Holland *et al.*, 1992).

Quasispecies is not a mere description of complexity at the population level. Quasispecies represents above all the recognition of the dynamics of mutant generation and competition among variants, a process that cannot be dissociated from genome replication. This creates an important operational difference from mutant generation (and their fixation) in higher organisms. For this reason, quasispecies is conceptually distinct from genetic polymorphisms (Domingo and Holland, 1993). Picornaviruses which have been instrumental in many developments of molecular biology may be so also in the oncoming studies on molecular evolution of RNA viruses.

REFERENCES

Acharya, R., Fry, E., Stuart, D., Fox, G. Rowlands, D., and Brown, F., 1989, The three-dimensional structure of foot-and-mouth disease virus at 2.9 Å resolution, *Nature* 337: 709-715.

Ackermann, W.W., and Kurtz, 1955, Observations concerning a pre-existing infection of HeLa cells with poliomyelitis virus, *J. Exp. Med.* 102: 555-565.

Ahmad, A.L.M., Dowsett, A.B., and Tyrrell, D.A.K., 1987, Studies of rhinovirus resistant to an antiviral chalcone, *Antiviral res.* 8: 27-39.

Andries, K., Dewindt, B., Snoeks, J., Wouters, L., Moereels, H., Lewi, P.J., and Janssen, P.A.J., 1990. Two groups of rhinoviruses revealed by a panel of antiviral compounds present sequence divergence and differential pathogenicity, *J. Virol.* 64: 1117-1123.

Argos, P., 1988, A sequence motif in many polymerases, *Nucleic Acids Res.* 16: 9909-9916.

Argos, P., and Fuller, S.D., 1988, A model for the hepatitis B virus core protein: prediction of antigenic sites and relationships to RNA virus capsid proteins, *EMBO J.* 7: 819-824.

Argos, P., Kamer, G., Nicklin, M.J.H., and Wimmer, E., 1984, Similarity in gene organization and homology between proteins of aminal picornaviruses and a plant comovirus suggest common ancestry of these virus families, *Nucleic Acids Res.* 12: 7251-7267.

Bachrach, H.L., 1968, Foot-and mouth disease virus, *Annu. Rev. Microbiol.* 22: 201-244.

Banner, L.R., and Lai, M.M.C., 1991, Random nature of coronavirus RNA recombination in the absence of selection pressure, *Virology* 185: 441-445.

Batschelet, E., Domingo. E., and Weissmann, C., 1976, The proportion of revertant and mutant phage in a growing population, as a function of mutation and growth rate, *Gene* 1: 27-32.

Beck, E., and Strohmaier, K., 1987, Subtyping of European foot-and-mouth disease virus strains by nucleotide sequence determination, *J. Virol.* 61:1621-1629.

Benmansour, A.M., Brahimi, M., Tuffereau, C., Coulon, P., Lafay, F., and Flamand, A., 1992, Rapid sequence evolution of street rabies glycoprotein is related to the highly heterogeneous nature of the virus population, *Virology* 187: 33-45.

Bharucha, E.P., and Mondkar, V.P., 1972, Neurological complications of a new conjunctivitis, *Lancet ii* 970-971.

Boege, U., Kobasa, D., Onodera, S., Parks, G.D., Palmenberg, A.C., and Seraba, D.G., 1991, Characterization of Mengo virus neutralization epitopes, *Virology* 181: 1-13.

Borzakian, S., Couderc, T., Barbier, Y., Attal, G., Pelletier, I., and Colbère-Garapin, F., 1992, Persistent poliovirus infection: Establishment and maintenance involve distinct mechanisms, *Virology* 186: 398-408.

Brahic, M., Bureau, J.F., and McAllister, A., 1991, Genetic determinants of the demyelinating disease caused by Theiler's virus, *Microbial Pathogenesis* 11: 77-84.

Brown, F., 1989, Towards a molecular vaccine for foot-and-mouth disease, *in*: "Molecular Aspects of Picornavirus Infection and Detection", B.L. Semler, and E. Ehrenfeld, eds., American Society for Microbiology, Washington, D.C.

Brown, E.A., Day, S.P., Jansen, R.W., and Lemon, S.M., 1991, The 5' nontranslated region of hepatitis A virus RNA: Secondary structure and elements required for translation *in vitro*, *J. Virol.* 65: 5828-5838.

Bruenn, J.A., 1991, Relationships among the positive strand and double-strand RNA viruses as viewed througt their RNA-dependent RNA polymerases, *Nucleic Acids Res.* 19: 217-226.

Burns, D.P.W., and Desrosiers, R.C., 1991, Selection of genetic variants of simian immunodeficiency virus in persistently infected rhesus monkeys, *J. Virol.* 65: 1843-1854.

Campbell, A.M.G., Williams, E.R., and Pearce, J., 1969, Late motor neuron degeneration following poliomyelitis, *Neurology* 19: 1101-1106.

Cao, X., Bergmann, I.E., and Beck, E., 1991, Comparison of the 5' and 3' untranslated genomic regions of virulent and attenuated foot-and-mouth disease viruses (strains O_1 Campos and C_3 Resende), *J. Gen. Virol.* 72: 2821-2825.

Carp, R.I., 1981, Persistent infection of human lymphoblastoid cells with poliovirus and development of temperature sensitive mutants, *Intervirology* 15: 49-56.

Carrillo, C., Plana, J., Mascarella, R., Bergadá, J., and Sobrino, F., 1990, Genetic and phenotypic variability during replication of foot-and-mouth disease virus in swine, *Virology* 179: 890-892.

Chao, L., 1990, Fitness of RNA virus decreased by Muller's ratchet, *Nature* 348: 454-455.

Clarke, B.E., Brown, A.L., Currey, K.M., Newton, S.E., Rowlands, D.J., and Carroll, A.R., 1987, Potential secondary and tertiary structure in the genomic RNA of foot-and-mouth disease virus, *Nucleic Acids Res.* 15: 7067-7079.

Clarke, D., Duarte, E., Moya, A., Elena, S.F., Domingo, E., and Holland, J.J., 1993, Genetic bottlenecks and population passages cause profound fitness differences in RNA viruses, *J. Virol.* in press.

Clewley, J.P., Pullin, J.S.K., Avery, R.J., and Moore, N.F., 1983, Oligonucleotide fingerprinting of the RNA species obtained from six Drosophila C virus isolates, *J. Gen. Virol.* 64: 503-506.

Colbère-Garapin, F., Christodoulou, C., Crainic, R., and Pelletier, I., 1989, Persistent poliovirus infection of human neuroblastoma cells, *Proc. Natl. Acad. Sci. Usa* 86: 7590-7594.

Cole, C.N., and Baltimore, D., 1973, Defective interfering particles of poliovirus. II. Nature of the defect, *J. Mol. Biol.* 76: 325-343.

Cooney, M.K., Fox, J.P., and Kenny, G.E., 1982, Antigenic groupings of 90 rhinovirus serotypes, *Infect. Immun.* 37: 642-647.

Costa Giomi, M.P., Bergmann, I.E., Scodeller, E.A., Augé de Mello, P., Gomes, I., and La Torre, J.L., 1984, Heterogeneity of the polyribocytidylic acid tract in aphthovirus: biochemical and biological studies of viruses carrying polyribocytidylic acis tracts of different lengths, *J. Virol.* 51: 799-805.

Costa Giomi, M.P., Gomes, I., Tiraboschi, B., Augé de Mello, P., Bergmann, I.E., Scodeller, E.A., and La Torre, J.L., 1988, Heterogeneity of the poliribocytidylic acid tract in aphthovirus: changes in the size of the poly (C) of viruses recovered from persistently infected cattle, *Virology* 162: 58-64.

Dalakas, M.C., 1986, New neuromuscular symptoms in patients with old poliomyelitis: a three year follow-up study, *Eur. Neurol.* 25: 381-387.

de la Torre, J.C., Alarcón, B., Martínez-Salas, E., Carrasco, L., and Domingo, E., 1987, Ribavirin cures cells of a persistent infection with foot-and-mouth disease virus *in vivo*, *J. Virol.* 61: 233-235.

de la Torre, J.C., Dávila, M., Sobrino, F., Ortín, J., and Domingo, E., 1985, Establishment of cell lines persistently infected with foot-and-mouth disease virus, *Virology* 145: 24-35.

de la Torre, J.C., De la Luna, S., Díez, J., and Domingo, E., 1989a, Resistance to foot-and-mouth disease virus mediated by trans-acting cellular products, *J. Virol.* 63: 2385-2387.

de la Torre, J.C., Giachetti, C., Semler, B.L., and Holland, J.J., 1992, High frequency of single-base transitions and extreme frequency of precise multiple-base reversion mutations in poliovirus, *Proc.Natl. Acad. Sci. USA* 89: 2531-2535.

de la Torre, J.C., Martínez-Salas, E., Díez, J., and Domingo, E., 1989b, Extensive cell heterogeneity during a persistent infection with foot-and-mouth disease virus, *J. Virol.* 63: 59-63.

de la Torre, J.C., Martínez-Salas, E., Díez, J., Villaverde, A., Gebauer, F., Rocha, E., Dávila, M., and Domingo, E., 1988, Coevolution of cells and viruses in a persistent infection foot-and-mouth disease virus in cell culture,*J. Virol.* 62: 2050-2058.

de la Torre, J.C., Wimmer, E., and Holland, J.J., 1990, Very high frequency of reversion to guanidine resistance in clonal pools of guanidine-dependent type 1 poliovirus, *J. Virol.* 64: 664-671.

Díez, J., Dávila, M., Escarmís, C., Mateu, M.G., Domínguez, J., Pérez, J.J., Giralt, E., Melero,J.A., and Domingo, E., 1990a, Unique amino acid substitutions in the capsid proteins of foot-and-mouth disease virus from a persistent infection in cell culture, *J. Virol.* 64: 5519-5528.

Díez, J., Hofner, M., Domingo, E., and Donaldson, A.I., 1990b, Foot-and-mouth disease virus strains isolated from persistently infected cell cultures are attenuated for mice and cattle, *Virus Res.* 18: 3-8.

Dinter, Z., Philipson, L., and Wesslen, T., 1959, Persistent foot-and-mouth disease infections of cells in tissue culture, *Virology* 8: 542-544.

Domingo, E., 1989, RNA virus evolution and the control of viral disease, *Prog. Drug Res.* 33:93-133.

Domingo, E., Dávila, M., and Ortín, J., 1980, Nucleotide sequence heterogeneity of the RNA from a natural population of foot-and- mouth disease virus, *Gene* 11: 333-346.

Domingo, E., Escarmís, C., Martínez, M.A., Martínez-Salas, E., and Mateu, M.G., 1992, Foot-and-mouth disease virus populations are quasispecies, *Curr. Top. Microbiol. Immunol.* 176: 33-47.

Domingo, E., Flavell, R.A., and Weissmann, C., 1976, *In vitro* site-directed mutagenesis: generation and properties of an infectious extracistronic mutant of bacteriophage Qß, *Gene* 1: 3-25.

Domingo, E., and Holland, J.J., 1988, High error rates, population equilibrium, and evolution of RNA replication systems, *in*: "RNA Genetics" vol. 3 "Variability of RNA Genomes", E. Domingo, J.J. Holland, and P. Ahlquist, eds., CRC Press Inc. Florida.

Domingo, E., and Holland, J.J., 1992, Complications of RNA heterogeneity for the engineering of virus vaccines and antiviral agents, *in*:"Genetic Engineering, Principles and Methods". vol. 14. J.K. Setlow, ed., Plenum Press, New York.

Domingo, E., and Holland, J.J., 1993, Mutation rates and rapid evolution of RNA viruses, *in*: "Evolutionary Biology of Viruses", S.S. Morse, ed., Raven Press, New York, in press.

Domingo, E., Holland, J.J., Biebricher, C., and Eigen, M., 1993, Quasispecies: the concept and the word, *in*: "Molecular Evolution of Viruses", A. Gibbs, C. Calisher and F. García-Arenal, eds., Cambridge University Press, in press.

Domingo, E., Martínez-Salas, E., Sobrino, F., de la Torre, J.C., Portela, A., Ortin, J., López-Galíndez, C., Pérez-Breña, P., Villanueva, N., Nájera, R., VandePol, S., Steinhauer, D., DePolo, N., and Holland, J.J., 1985, The quasispecies (extremely heterogeneous) nature of viral RNA genome populations: biological relevance — a review, *Gene* 40: 1-8.

Domingo, E., Mateu, M.G., Martínez, M.A., Dopazo, J., Moya, A., and Sobrimo, F., 1990, Genetic variability and antigenic diversity of foot-and-mouth disease virus, *in*: "Applied Virology Research" vol. 2 "Virus Variation, Epidemiology and Control", E. Kurstak, R.G. Marusyk, F.A. Murphy, and M.H.V. Van Regenmortel, eds. Plenum Publishing Co., New York.

Domingo, E., Sabo, D., Taniguchi, T., and Weissmann, C., 1978, Nucleotide sequence heterogeneity of an RNA phage population, *Cell* 13: 735-744.

Dopazo, J., Sobrino, F., Palma, E.L., Domingo, E., and Moya, A., 1988, Gene encoding capsid protein VP1 of foot-and-mouth disease virus: a quasispecies model of molecular evolution, *Proc. Natl. Acad. Sci. USA* 85: 6811-68915.

Duarte, E., Clarke, D., Moya, A., Domingo, E., and Holland, J.J., 1992, Rapid fitness losses in mammalian RNA virus clones due to Muller's ratchet, *Proc. Natl. Acad. Sci. USA* 89: 6015-6019.

Duke, G.M., Osorio, J.E., and Palmenberg, A.C., 1990, Attenuation of Mengo virus through genetic engineering of the 5' noncoding poly (C) tract, *Nature* 343: 474-476.

Dunn, G., Begg, N.T., Cammack, N., and Minor, P.D., 1990, Virus excretion and mutation by infants following primary vaccination with live oral poliovaccine from two sources, *J. Med. Virol.* 32: 92-95.

Eggers, H.J., and Tamm, I., 1965, Coxsackie A9 virus: mutation from drug dependence to drug resistance, *Science* 148: 97-98.

Eigen, M., and Biebricher, C., 1988, Sequence space and quasispecies distribution, *in*: "RNA Genetics" vol. 3, E. Domingo, J.J. Holland, and P. Ahlquist, eds. CRC Press Inc. Boca Raton, Florida.

Eigen, M., and Schuster, P., 1979, The Hypercycle. A Principle of Natural Self-Organization, Springer-Verlag, Berlin.

Emini, E. A., Jameson, B.A., Lewis, A.J., Larsen G.R., and Wimmer, E., 1982, Poliovirus neutralization epitopes: Analysis and localization with neutralizing monoclonal antibodies, *J. Virol.* 43: 997-1005.

Escarmís, C., Toja, M., Medina, M., and Domingo, E., 1992, Modifications of the 5' untranslated region of foot-and-mouth disease virus after prolonged persistence in cell culture, *Virus Res.* in press.

Evans, D.M.A., Dunn, G., Minor, P.D., Schild, G.D., Cann, A.J., Stanway, G., Almond, J.W., Currey, K., and Maizel, J.V., Jr, 1985, A single nucleotide change in the 5' non-coding region of the genome of the Sabin type 3, poliovaccine is associated with increased neurovirulence, *Nature* 314: 548-550.

Felsestein, J., 1985, Confidence limits on phylogenies: an approach using the bootstrap, *Evolution* 39: 783-791.

Franssen, H., Leunissen, J., Goldbach, R., Lomonosoff, G., and Zimmern, D., 1984, Homologous sequences in non-structural proteins from cowpea mosaic virus and picornaviruses, *EMBO J.* 3: 855-861.

Fuller, S.D., and Argos, P., 1987, Is Sindbis a simple picornavirus with an envelope? *EMBO J.* 6: 1099-1105.

Gauss-Müller, V., and Deinhardt, F., 1984, Effect of hepatitis A virus infection on cell metabolism *in vitro*, *Proc. Soc. Exp. Biol. Med.* 175: 10-15.

Gebauer, F., de la Torre, J.C., Gomes, I., Mateu, M.G., Barahona, H., Tiraboschi, B., Bermann, I., Augé de Mello, P., and Domingo, E., 1988, Rapid selection of genetic and antigenic variants of foot-and-mouth disease virus during persistence in cattle, *J. Virol.* 62: 2041-2049.

Gercel, C., Mahan, K.B., and Hamparian, V.V., 1985, Preliminary characterization of a persistent infection of HeLa cells with human rhinovirus type 2, *J. Gen. Virol.* 66: 131-139.

Gibson, J.P., and Righthand, V.F., 1985, Persistence of echovirus 6 in cloned human cells, *J. Virol.* 54: 219-223.

González, M.J., Sáiz, J.C., Laor, O., and Moore, D.M., 1991, Antigenic stability of foot-and-mouth disease virus variants on serial passage in cell culture, *J. Virol.* 65: 3949-3953.

Gorbalenya, A.E., and Koonin, E.V., 1989, Viral proteins containing the purine NTP-binding sequence pattern, *Nucleic Acids Res.* 17: 8413-8440.

Grant, R.A., Filman, D.J., Fujinami, R.S., Icenogle, J.P., and Hogle, J. M., 1992, Three-dimensional structure of Theiler virus, *Proc. Natl. Acad. Sci. USA* 89: 2061-2065.

Hahn, B.H., Shaw, G.M., Taylor, M.E., Redfield, R.R., Markham, P.D., Salahudin, S.Z., Wong-Staal, F., Gallo, R.C., Parks, E.S., and Parks, W.P., 1986, Genetic variation in HTLV-III/LAV over time in patients with AIDS or at risk for AIDS, *Science* 232: 1548-1553.

Halstead, S.B., 1980, Immunological parameters of Togavirus disease syndromes, *in*: "The Togaviruses. Biology, Structure, Replication", R.W. Schlesinger, ed.,Academic Press, New York.

Haseloff, J., Goelet, P., Zimmern, D., Ahlquist, P., Dasgupta, R., and Kaesberg, P., 1984, Striking similarities in amino acid sequence among nonstructural proteins encoded by RNA viruses that have dissimilar genomic organization, *Proc. Natl. Acad. Sci. USA* 81: 4358-4362.

Heinz, B.A., Rueckert, R.R., Shepard, D.A., Dutko, F.J., McKinlay, M.A., Francher, M., Rossmann, M.G., Badger, J., and Smith, T.J., 1989, Genetic and molecular analyses of spontaneous mutants of human rhinovirus 14 that are resistant to an antiviral compound, *J. Virol.* 63: 2476-2485.

Hernández, J., Martínez, M.A., Rocha, E., Domingo, E., and Mateu, M.G., 1992, Generation of a subtype-specific neutralization epitope in foot-and-mouth disease virus of a different subtype, *J. Gen. Virol.* 73: 213-216.

Hogle, J.M., Chow, M., and Filman, D.J., 1985, Three-dimensional structure of poliovirus at 2.9 Å resolution, *Science*, 229: 1358-1365.

Holland, J.J., de la Torre, J.C., and Steinhauer, D.A., 1992, RNA virus populations as quasispecies, *Curr Top. Microbiol. Immunol.* 176: 1-20.

Holland, J.J., de la Torre, J.C., Clarke, D.K., and Duarte, E., 1991, Quantification of relative fitness and great adaptability of clonal populations of RNA viruses, *J. Virol.* 65:2960-2967.

Holland, J.J., de la Torre, J.C., Steinhauer, D.A., Clarke, D., Duarte, E., and Domingo, E., 1989, Virus mutation frequencies can be greatly underestimated by monoclonal antibody neutralization of virions, *J. Virol.*, 63: 5030-5036.

Holland, J.J., Domingo, E., de la Torre, J.C., and Steinhauer, D.A., 1990, Mutation frequencies at defined single codon sites in vesicular stomatitis virus and poliovirus can be increased only slightly by chemical mutagenesis, *J. Virol.* 64: 3960-3962.

Holland, J.J., Spindler, K., Horodyski, F., Grabau, E., Nichol, S., and VandePol, 1982, Rapid evolution of RNA genomes, *Science* 215: 1577-1585.

Hollinger, F.B., and Ticehurst, J., 1990, Hepatitis A virus,*in*:"Virology", B. N. Fields, D.M. Knipe *et al.*, eds., Raven Press, New York.

Hovi, T., Cantell, K., Huovilainen, A., Kinnunen, L., Kunonen, T., Lapinleimu, K., Pöyry, T., Roivaianen, M., Salama, N., Stenvik, M., Silander, A., Tholen, C.J., Salminen, S., and Weckstrom, P., 1986. Outbreak of paralytic poliomyelitis in Finland: widespread circulation of antigenically altered poliovirus type 3 in a vaccinated population, *Lancet i*: 1427-1432.

Jameson, B.A., Bonin, J., Wimmer, E., and Kew, O.M., 1985, Natural variants of the Sabin type 1 vaccine strain of poliovirus, and correlation with a poliovirus neutralization site, *Virology* 143: 337-341.

Johnson, R.T., 1982, Viral Infections of the Nervous System, Raven Press, New York.

Kamahora, T., Itagaki, A., Hattori, N., Tsuchie, H., and Kurimura, T., 1985, Oligonucleotide fingerprint analysis of coxsackievirus A10 isolated in Japan, *J. Gen. Virol.* 66: 2627-2634.

Kamer, G., and Argos, P., 1984, Primary structural comparison od RNA-dependent polymerases from plant, animal and bacterial viruses, *Nucleic Acids Res.* 12: 7269-7282.

Kaplan, G., Levy, A., and Racaniello, V.R., 1989, Isolation and characterization of HeLa cell lines blocked at different steps in the poliovirus life cycle, *J.Virol.* 63: 43-51.

Kew, O.M., and Nottay, B.K., 1984, Evolution of oral poliovirus vaccine strain in humans occurs by both mutation and intermolecular recombination, *in*: "Modern Approaches to Vaccines", R. Chanock and R. Lerner, eds., Cold Spring Harbor, New York.

Kew, O.M., Nottay, B.K., Hatch, M.N., Nakano, J.H., and Obijeski, J.F., 1981, Multiple genetic changes can occur in the oral poliovaccines upon replication in Humans, *J. Gen. Virol.* 56: 337-347.

Kew, O.M., Nottay, B.K., Rico-Hesse, R., and Pallansch, M.A., 1990, Molecular epidemiology of wild poliovirus transmission, *in*: "Applied Virology Research" vol. 2 "Virus Variability, Epidemiology and Control", E. Kurstak, R.G. Marusyk, F.A. Murphy, and M.H.V. Van Regenmortel, eds., Plenum Publishing Co., New York.

King, A.M.Q., 1988a, Recombination in positive strand RNA viruses, *in*: "RNA Gentics" vol. 2, E. Domingo, J.J. Holland, and P. Ahlquist, eds. CRC Press Inc., Boca Ratón, Florida.

King, A.M.Q., 1988b, Preferred sites of recombination in poliovirus RNA: an analysis of 40 intertypic crossover sequences. *Nucleic Acid Res.* 16: 11705-11723.

Kinnunen, L., Pöyry, T., and Hovi, T., 1991, Generation of virus genetic lineages during an outbreak of poliomyelitis, *J. Gen. Virol.* 72: 2483-2489.

Kinnunen, L., Pöyry, T., and Hovi, T., 1992, Genetic diversity and rapid evolution of poliovirus in human hosts, *Curr. Top. Microbiol. Immunol.* 176: 49-61.
Kirkegaard, K., and Baltimore, D., 1986, The mechanism of RNA recombination in poliovirus, *Cell* 47: 433-443.
Kono, R., Uchida, Y., Sasagawa, A., Akao, Y., Kodama, H., Mukoyama, J., and Fujiwara, T., 1973, Neurovirulence of acute haemorrhagic conjunctivitis virus in monkeys, *Lancet i*: 61-63.
Koonin, E.V., 1991, The phylogeny of RNA-dependent RNA polymerases of positive-strand RNA viruses, *J.Gen. Virol.* 72: 2197-2206.
Koonin, E., Boyko, V.P., and Dolja, V.V., 1991, Small cysteine-rich proteins of different groups of plant RNA viruses are related to different families of nucleic acid-binding proteins, *Virology* 181: 395-398.
Krebs, O., and Marquardt, O., 1992, Identification and characterization of foot-and-mouth disease virus O_1 Burgwedel/1987 as an intertypic recombinant, *J. Gen. Virol.* 73: 613-619.
Kuge, S., Kawamura, N., and Nomoto, A., 1989, Strong inclination toward transition mutation in nucleotide substitutions by poliovirus replicase, *J. Mol. Biol.* 207: 175-182.
Lai, M.M.C., 1992, Genetic recombination in RNA viruses, *Curr. Top. Microbiol. Immunol.* 176: 21-32.
Lemon, S.M., and Binn, L.N., 1983, Antigenic relatedness of two strains of hepatitis A virus determined by cross-neutralization, *Infect. Immun.* 42: 418-420.
Lemon, S.M., Chao, S.-F., Jansen, R.W., Binn, L.N., and LeDuc, J.W., 1987, Genomic heterogeneity among human and nonhuman strains of hepatitis A virus, *J.Virol.* 61: 735:-742.
Lemon, S.M., and Ping, L.-H., 1989, Antigenic structure of hepatitis A virus, *in*: "Molecular Aspects of Picornavirus Infection and Detection", B.L. Semler and E. Ehrenfeld, eds., American Society for Microbiology, Washington, D.C.
Luo, M., He, C., Toth, K.S., Zhang, C.X., and Lipton, H.L., 1992, Three-dimensional structure of Theiler murine encephalomyelitis virus (Be An strain), *Proc. Natl. Acad. Sci. USA* 89: 2409-2413.
Macadam, A.J., Arnold, C., Howlett, J., John, A., Marsden, S., Taffs, F., Reeve, P., Harnada, N., Wareham, K., Almond, J., Cammack, N., and Minor, P.D., 1989, Reversion of the attenuated and temperature-sensitive phenotypes of the Sabin type 3 strain of poliovirus in vaccinees, *Virology* 172: 408-414.
Martínez, M.A., Carrillo, C., González-Candelas, F., Moya, A., Domingo, E., and Sobrino, F., 1991a, Fitness alteration of foot-and-mouth disease virus mutants: measurement of adaptability of viral quasispecies. *J. Virol.* 65: 3954-3957.
Martínez, M.A., Dopazo, J., Hernández, J., Mateu, M.G., Sobrino, F., Domingo, E., and Knowles, N.J., 1992, Evolution of the capsid protein genes of foot-and-mouth disease virus: antigenic variation without accumulation of amino acid substitutions over six decades, *J. Virol.* 66: 3557-3565.
Martínez, M.A., Hernández, J. Piccone, M.E., Palma, E,L., Domingo, E., Knowles, N., and Mateu, M.G., 1991b, Two mechanisms of antigenic diversification of foot-and-mouth disease virus, *Virology* 184: 695-706.
Mateu, M.G., Andreu, D., Carreño, C., Roig, X., Cairo, J.J., Camarero, J.A., Giralt, E., and Domingo, E., 1992, Non-additive effects of multiple amino acid substitutions on antigen-antibody recognition, *Eur. J. Immunol.* 22: 1385-1389.
Mateu, M.G., Martínez, M.A., Capucci, L., Andreu, D., Giralt, E., Sobrino, F., Brocchi, E., and Domingo, E., 1990, A single amino acid substitution affects multiple overlapping epitopes in the major antigenic site of foot-and-mouth disease virus of serotype C, *J. Gen. Virol.*, 71: 629-637.
Mateu, M.G., Martínez, M.A., Rocha, E., Andreu, D., Parejo, J., Giralt, E., Sobrino, F., and Domingo, E., 1989, Implications of a quasispecies genome structure: effects of frequent, naturally occurring amino acid substitutions on the antigenicity of foot-and-mouth disease virus, *Proc. Natl. Acad. Sci. USA* 86: 5883-5887.
McClure, M.A., Holland, J.J., and Perrault, J., 1980, Generation of defective interfering particles in picornaviruses, *Virology* 100: 408-418.
Meyers, G., Tautz, N., Dubovi, E.J., and Thiel, H-J., 1991, Viral cytopathogenicity correlated with integration of ubiquitincoding sequences, *Virology* 180: 602-616.
Minor, P.D., 1989, Humoral immune response to poliovirus, *in*: "Immune Responses, Virus Infections and Disease" N.J. Dimmock and P.D. Minor, eds., Society for General Microbiology, IRL Press, Oxford.

Minor, P.D., 1990, Antigenic structure of picornaviruses, *Curr. Top. Microbiol. Immunol.* 161: 121-154.

Minor, P.D., Dunn, G., Evans, D.M.A., Magrath, D.I., John, A., Howlett, J., Phillips. A., Westrop, G., Wareham, K., Almond, J.W., and Hogle, J.M., 1989, The temperature sensitivity of the sabin type 3 vaccine strain of poliovirus: molecular and structural effects of a mutation in the capsid protein VP3, *J. Gen. Virol.* 70: 1117-1123.

Minor, P.D., Ferguson, M., Evans, D.M.A., Almond, J.W., and Icenogle, J.P., 1986a, Antigenic structure of polioviruses of serotypes 1, 2 and 3, *J. Gen. Virol.* 67: 1283-1291.

Minor, P.D., John, A., Ferguson, M., and Icenogle, J.P., 1986b, Antigenic and molecular evolution of the vaccine strain of type 3 poliovirus during the period of excretion by a primary vaccinee, *J. Gen. Virol.* 67: 693-706.

Minor, P.D., Schild, G.C., Bootman, J., Evans, D.M.A., Ferguson, M., Reeve, P., Spitz, M.,Stanway, F., Cann, A.J., Hauptmann, R., Clarke, L-D., Mountfort, R,C., and Almond, J.W., 1983, Location and primary structure of a major antigenic site for poliovirus neutralization, *Nature* 301: 674-679.

Monroe, S.S., and Schlesinger, S., 1983, RNAs from two independently isolated defective interfering particles of Sindbis virus contain cellular tRNA sequences at their 5'ends, *Proc. Natl. Acad. Sci. USA* 80: 3279-3283.

Mulder, D.W., Rosenbaum, R.A., and Layton, D.D., 1972, Late progression of poliomyelitis or forme fruste amyotrophic lateral sclerosis, *Mayo Clin. Proc.* 47: 756-761.

Nomoto, A., Jacobson, A., Lee, Y.F., Dunn, J., and Wimmer, E., 1979, Defective interfering particles of poliovirus: mapping of deletions and evidence that the deletions in the genomes of DI (1), (2) and (3) are located in the same region, *J. Mol. Biol.* 128: 179-196.

Nottay, B.K., Kew, O.M., Hatch, M.H., Heyward, J.T., and Obijeski, J.F., 1981, Molecular variation of type 1 vaccine-related and wild poliovirus during replication in humans, *Virology* 108: 405-423.

Nowak, M.A., Anderson, R.M., McLean, A.R., Wolfs, T.F.W., Goudsmits, J., and May, R.M., 1991, Antigenic diversity thresholds and the development of AIDS, *Science* 254: 963-969.

Nüesch, J., Krech, S., and Siegl, G., 1988, Detection and characterization of subgenomic RNAs in hepatitis A virus particles, *Virology* 165: 419-427.

Palmenberg, A., 1989, Sequence alignments of picornavirus capsid proteins, *in*: "Molecular Aspects of Picornavirus Infection and Detection", B.L. Semler, and E. Ehrefeld, eds., American Society for Microbiology, Washington, D.C.

Parisi, J.M., Costa Giomi, P., Grigera, P., Augé de Mello, P., Bergmann, I.E., La Torre, J.L., and Scodeller, E.A., 1985, *Virology* 147: 61-71.

Parvin, J.D., Moscona, A., Pan, W.T., Leider, J.M., and Palese, P., 1986, Neasurement of the mutation rates of animal viruses: Influenza A virus and poliovirus type 1, *J. Virol.*, 59: 377-383.

Pelletier, I., Couderc, T., Borzakian, S., Wyckoff, E., Crainic, R., Ehrenfeld, E., and Colbère-Garapin, F., 1991, Characterization of persistent poliovirus mutants selected in human neuroblastoma cells, *Virology* 180: 729-737.

Pevear, D.C., Borkowski, J., Calenoff, M., Oh, C.K., Ostrowski, B., and Lipton, H.L., 1988, Insights into Theiler's virus neurovirulence based on a genomic comparison of the neurovirulent GDVII and less virulent Be An strains, *Virology* 165: 1-12.

Pilipenko, E.V., Blinov, V.M., Chernov, B.K., Dimitrieva, T.M., and Agol, V.I., 1989a, Conservation of the secondary structure elements of the 5'-untranslated region of cardio- and aphthovirus RNAs, *Nucleic Acids Res.* 17: 5701-5711.

Pilipenko, E.V., Blinov, V.M., Romanova, L.I., Sinyakov, A.N., Maslova, S.V., and Agol, V.I., 1989b, Conserved structural domains in the 5'-untranslated region of picornaviral genomes: an analysis of the segment controlling translation and neurovirulence, *Virology* 168: 201-209.

Prabhakar, B.S., Menegus, M.A., and Notkins, A.L., 1985, Detection of conserved and nonconserved epitopes on coxsackie virus B4: frequency of antigenic change, *Virology* 146: 302.

Provost, P.J., and Hilleman, M.R., 1979, Propagation of human hepatitis A virus in cell culture *in vitro, Proc. Soc. Exp. Biol. Med.* 160: 213-221.

Radloff, R.J., and Young, S.A., 1983, Defective interfering particles of encephalomyocarditis virus, *J. Gen Virol.* 64: 1637-1641.

Ray, U.R., Aulakh, G.S., Schubert, M., McClintock, P.R., Yoon, J.W., and Notkins, A.L., 1983, Virus-induced diabetes mellitus. XXV. Difference in the RNA fingerprints of diabetogenic and non-diabetogenic variants of encephalomyocarditis virus, *J. Gen. Virol.* 64: 947-950.

Reagan, K.J., Goldberg, B., and Crowell, R.L., 1984, Altered receptor specifity of coxsackievirus B3 after growth in rhabdomyosarcoma cells, *J. Virol.* 49: 635-640.

Rico-Hesse, R., Pallansch, M.A., Nottay, B.K., and Kew, O.M., 1987, Geographic distribution of wild poliovirus type 1 genotypes, *Virology* 160: 311-322.

Righthand, V.F., and Blackburn, R.V., 1989, Steady-state infection by echovirus 6 associated with nonlytic viral RNA and an unprocessed capsid polypeptide, *J. Virol.* 63: 5268-5275.

Robertson, B.H., Jansen, R.W., Khanna, B., Totsuka, A., Nainan, O.V., Siegl, G., Widell, A., Margolis, H.S., Isomura, S., Ito, K., Ishizu, T., Moritsugu, Y., and Lemon, S.M., 1992, Genetic relatedness of hepatitis A virus strains recovered from different geographical regions, *J. Gen. Virol.* 73: 1365-1377.

Rodríguez, M., and David, C.S., 1985, Demyelination induced by Theiler's virus: influence of the H-2 haplotype, *J. Immunol.* 135: 2145-2148.

Rodríguez, M., Oleszak, E., and Leibowitz, J., 1987, Theiler's murine encephalomyelitis: a model of demyelination and persistence of virus, *Critical Reviews in Immunology* 7: 325-365.

Roos, R.P., Richards, O.C., Green, J., and Ehrenfeld, E., 1982, Characterization of a cell culture persistently infected with the DA strain of Theiler's murine encephalomyelitis virus, *J. Virol.* 43: 1118-1122.

Rota, J.S., Hummel, K.B., Rota, P.A., and Bellini, W.J., 1992, Genetic variability of the glycoprotein genes of current wild type measles isolates, *Virology* 188: 135-142.

Roux, L., Simon, A.E., and Holland, J.J., 1991, Effects of defective interfering viruses on virus replication and pathogenesis, *Advances in Virus Res.* 40: 181-211.

Rueckert, R.R., 1990, Picornaviridae and their replication, *in*: "Virology", B.N. Fields, D.M. Knipe *et al.* eds., Raven Press, New York.

Sedivy, J.M., Capone, J.P., Raj Bhandary, U.L., and Sharp, P.A., 1987, An inducible mammalian amber suppressor: propagation of a poliovirus mutant *Cell* 50: 379-389.

Seibold, H.R., Cottral, G.E., Patty, R.E., and Gailiunas, P., 1964, Apparent modification of foot-and-mouth disease virus after prolonged residence in surviving cells, *Amer. J. Vet. Res.* 25: 806-814.

Sellers, R.F., 1971, Quantitative aspects of the spread of foot-and-mouth disease, *Vet. Bull.* 41: 431-439.

Sharief, M.K., Hentges, R., and Ciardi, M., 1991, Intrathecal immune response in patients with the post-polio syndrome, *N. Engl. J. Med.* 325: 749-755.

Sherry, B., Mosser, A.G., Colonno, R.J., and Rueckert, R.R., 1986, Use of monoclonal antibodies to identify four neutralization immunogens on a common cold picornavirus, human rhinovirus 14, *J. Virol.* 57: 246-257.

Siegl, G. deChastonay, J., and Kronauer, K., 1984, Propagation and assay of hepatitis A virus *in vitro*, *J. Virol. Methods* 9: 53-60.

Skern, T., Sommergruber, W., Blaas, D., Pieler, Ch., and Kuechler, E., 1984, Relationship of human rhinovirus strain 2 and poliovirus as indicated by comparison of the polymerase gene regions, *Virology* 136: 125-132.

Sobrino, F., Palma, E.L., Beck, E., Dávila, M., de la Torre, J.C., Negro, P., Villanueva, N., Ortín, J., and Domingo, E., 1986, Fixation of mutations in the viral genome during an outbreak of foot-and-mouth disease: heterogeneity and rate variations, *Gene* 50: 149-159.

Stanway, G., Hughes, P.J., Mountford, R.C., Minor, P.D., and Almond, J.W., 1984, The complete nucleotide sequence of a common cold virus: human rhinovirus 14, *Nucleic Acids Res.* 12: 7859-7875.

Stanway, G., 1990, Structure, function and evolution of picornaviruses, *J. Gen. Virol.* 71: 2483-2501.

Stapleton, J.T., and Lemon, S.M., 1987, Neutralization escape mutants define a dominant immunogenic neutralization site on hepatitis A virus, 1987, *J. Virol.* 61: 491-498.

Steinhauer, D.A., and Holland, J.J., 1986, Direct method for quantification of extreme polymerase error frequencies at selected single base sites in viral RNA, *J. Virol.* 57: 219-228.

Takeda, N., Miyamura, K., Ogino, T., Natory, K., Yamazaki, S., Sakwiai, N., Nakazono, N., Ishü, K., and Kono, R., 1984, Evolution of enterovirus type 70: oligonucleotide mapping analysis of RNA genome, *Virology* 134: 375-388.

Takemoto, K.K., and Habel, K., 1959, Virus-cell relationship in a carrier culture of HeLa cell and coxsackie A9 virus, *Virology* 7: 28-44.

Temin, H., 1989, Is HIV unique or merely different? *J. AIDS* 2: 1-9.

Theiler, M., 1934, Spontaneous encephalomyelitis of mice — a new virus disease, *Science* 80: 122-123.

Ticehurst, J., Cohen, J.I., Feinstone, S.A., Purcell, R.H., Jansen, R.W., and Lemon, S.M., 1989, Replication of hepatitis A virus: new ideas from studies with cloned cDNA, *in*: "Molecular Aspects of Picornavirus Infection and Detection", B.L. Semler, and E. Ehrenfeld, eds., American Society for Microbiology, Washington, D.C.

Tyrrell, D.A.J., 1990, Virus variation and the epidemiology and control of rhinoviruses, *in*: "Applied Virology Research" vol. 2 "Virus Variability, Epidemiology and Control:, E. Kurstak, R.G. Matrusyk, F.A. Murphy, and M.H.V. Van Regenmortel, eds., Plenum Publishing Co., New York.

Valcarcel, J., and Ortín, J., 1989, Phenotypic hiding: the carry-over of mutations in RNA viruses as shown by detection of *mar* mutants in influenza virus, *J. Virol.* 63: 4107-4109.

Vallbracht, A., Hofmann, L., Wurster, K.G., and Flehmig, B., 1984, Persistent infection of human fibroblasts by hepatitis-A virus, *J. Gen. Virol.* 65: 609-615.

van Bekkum, J.G., Frenkel, H.S., Frederiks, H.H.J., and Frenkel, S., 1959, Observations on the carrier state of cattle exposed to foot-and-mouth disease virus, *Tijdschr. Diergeneesk.* 84: 1159-1164.

Venuti, A., Di Russo, C., del Grosso, N., Patti, A.M., Degener, A.M., Midulla, M., Paña, A., and Pérez-Bercoff, R., 1985, Isolation and molecular cloning of a fast-growing strain of human hepatitis A virus from its double-stranded replicative form, *J. Virol.* 56: 579-588.

Villaverde, A., Martínez, M.A., Sobrino, F., Dopazo, J., Moya, A., and Domingo, E., 1991, Fixation of mutations at the VP1 gene of foot-and-mouth disease virus: can quasispecies define a transient molecular clock? *Gene* 103: 147-153.

Villaverde, A., Martínez-Salas, E., and Domingo, E., 1988, 3D gene of foot-and-mouth disease virus. Conservation by convergence of average sequences, *J. Mol. Biol.* 204: 771-776.

Vogt, M., and Dulbecco, R., 1958, Properties of a HeLa cell culture with increased resistance to poliomyelitis virus, *Virology* 5: 425-434.

Wain-Hobson, S., 1992, Human immunodeficiency virus type 1 quasispecies *in vivo* and *ex vivo*, *Curr. Top. Microbiol. Immunol.* 176: 181-193.

Ward, C.D., and Flanegan, J.B., 1992, Determination of the poliovirus RNA polymerase error frequency at eight sites in the viral genome, *J. Virol.* 66: 3784-3793.

Weeks-Levy, C., Tatem, J.M., DiMichele, S.J., Waterfield, W., Georgiu, A.F., and Mento, S.J., 1991, Identification and characterization of a new base substitution in the vaccine strain of Sabin 3 poliovirus, *Virology* 185: 934-937.

Weitz, M., and Siegl, G., 1985, Variation among hepatitis A virus strains. I. Genomic variation detected by T_1 oligonucleotide mapping, *Virus Res.* 4: 53.

Welsh, C.J.R., Blakemore, W.F., Tonks, P., Borrow, P., and Nash, A.A., 1989, Theiler's murine encephalomyelitis virus infection in mice: a persistent viral infection of the central nervous system which induces demyelination, *in*: "Immune Responses, Virus Infections and Disease", N.J. Dimmock, and P.D. Minor, eds. IRL Press, Oxford.

Westrop, G.D., Wareham, K.A., Evans, D.M.A., Dunn, G., Minor, P.D., Magrath, D.I., Taffs, F., Marsden, S., Skinner, M.A., Schild, G.C., and Almond, J.W., 1989, Genetic basis of attenuation of the Sabin type 3 oral poliovirus vaccine, *J. Virol.* 63: 1338-1344.

Yang, C-F., De, L., Yang, S-J., Gómez, J.R., Cruz, J.R., Holloway, B.P., Pallansch, M.A., and Kew, O.M., 1992, Genotype-specific *in vitro* amplification of sequences of the wild type 3 polioviruses from Mexico and Guatemala, *Virus Research* 24: 277-296.

Zimmern, D., 1988, Evolution of RNA viruses, *in*: "RNA Genetics" vol. 2, E. Domingo, J. Holland, and P. Ahlquist, eds. CRC Press INC., Boca Raton, Florida.

Zurbriggen, A., Thomas, C., Yamada, M., Roos, R.P., and Fujinami, R.S., 1991, Direct evidence of a role for amino acid 101 of VP-1 in central nervous system disease in Theiler's murine encephalomyelitis virus infection, *J.Virol.* 65: 1929-1937.

MODIFICATION OF MEMBRANE PERMEABILITY BY ANIMAL VIRUSES

Luis Carrasco, Luis Pérez, Alicia Irurzun, Juan Lama, Francisco Martínez-Abarca, Pedro Rodríguez, Rosario Guinea, José Luis Castrillo, Miguel Angel Sanz and Mª José Ayala

Centro de Biología Molecular. Universidad Autónoma de Madrid. Cantoblanco 28049, Madrid

1. INTRODUCTION

The infection of susceptible cells by animal viruses leads to profound changes in membrane permeability at two well defined moments of the infection cycle: 1. Early during infection when the virus enters the cells and 2. At the beginning of the late phase, before the onset of viral macromolecular synthesis takes place (see review Carrasco *et al.*, 1989). This late modification of membrane permeability is also known as the late membrane leakiness (Carrasco, 1978; Carrasco and Lacal, 1983; Carrasco *et al.*, 1989). The molecular basis of both phenomena differ. Thus, only virion particles are necessary to accomplish the increase in membrane permeability early during infection (Fernandez-Puentes and Carrasco, 1980; Carrasco 1981), whereas viral gene expression is necessary for the development of the late membrane leakiness (Carrasco, 1978; Contreras and Carrasco, 1979).

Recent progress in the understanding of both phenomena indicates that these modifications of membrane permeability by animal viruses are giving clues to understand two important events of the viral replication cycle, i.e. the entry of animal viruses into cells and the development of the cytopathic effect and cell killing. In addition, the modification of the cellular *internal milieu* by membrane damage modifies the synthesis of cellular macromolecules (see review Carrasco and Castrillo,1987) and regulates a number of viral processes, from viral transcription and translation, to virus assembly (Carrasco *et al.*, 1989). In this article we revise work, mainly from our laboratory, concerning the modification of cellular membranes and membrane permeability by animal virus infection.

Regulation of Gene Expression in Animal Viruses, Edited by
L. Carrasco *et al.*, Plenum Press, New York, 1993

2. MODIFICATION OF MEMBRANE PERMEABILITY EARLY DURING INFECTION

The infection of a cell by an animal virus commences by the recognition and attachment of the virion particle to a specific cellular receptor (Dimmock, 1982; Marsh and Helenius, 1989). This event must be followed by the delivery of the viral genome to the cytoplasm (White, 1990; Marsh and Helenius, 1989). For enveloped animal viruses this is accomplished by fusion of the viral membrane with the plasma membrane (Sendai virus) or with the endosomal membrane (most enveloped animal viruses). The fusion process is triggered by special viral fusion proteins present in the viral envelope, that upon receptor binding induce fusion at neutral pH (Sendai virus), or after the acidification of the endosome by the action of the vacuolar proton-ATPase (Hoekstra and Kok, 1989; Marsh and Helenius, 1989). Much less is known about the delivery of the genome by naked animal viruses, i.e. viruses that lack a lipid envelope in their virion particles (Marsh and Helenius, 1989). Most of these viruses enter cells also by receptor-mediated endocitosis and the virion capsid proteins help the genome to cross the endosomal membrane by an, as yet, unknown process (Marsh and Helenius, 1989). Irrespective of whether the virus directly enters from the plasma membrane, or through an endosome, the viral genome has to cross a membrane. How such a big macromolecule traverses a barrier as a lipid membrane is a fascinating riddle that animal virologists are faced with.

During the entry of animal viruses into cells, not only is the virus able to permeabilize the membrane to its genome, but also to other macromolecules present in the culture medium (for a review see Carrasco *et al.*, 1989). This phenomenon of the co-entry of proteins with viral particles occurs with both enveloped or naked animal viruses. Fernández-Puentes and Carrasco (1980) first observed that viral particles efficiently permeabilized cells to protein toxins that are otherwise unable to cross the membrane because there are no receptors for these toxins. The proteins are efficiently delivered in the cytoplasm shortly after virus addition to the medium, and almost 100% of the cells become permeabilized in a few minutes (Otero and Carrasco, 1987). The majority of animal viruses tested were able to induce this phenomenon, including Semliki Forest virus, vesicular stomatitis virus, adenovirus, poliovirus, etc., (Carrasco, 1981; Carrasco and Esteban, 1982; FitzGerald *et al.*, 1983; Seth *et al.*, 1984; 1985; Otero and Carrasco, 1987; Almela *et al.*, 1991) suggesting that the viral particle contains a component that not only promotes the entry of the viral nucleocapsid into the cell, but also pushes other macromolecules that are not physically bound to the particles towards the cytoplasm. Thus, the simple model indicating that enveloped animal viruses enter cells just by the fusion between two membranes (Marsh and Helenius, 1989) is perhaps more complicate, because that model does not explain how the co-internalization of other macromolecules takes place (Carrasco *et al.*, 1989). Perhaps the proton-motive or the electrical force (membrane potential) generated in endosomes by the activity of the proton-ATPase pump can be used to drive the nucleocapsid through the lipid barrier of the membrane. Therefore, the energy accumulated in endosomes may help the viral nucleocapsid to enter cells. In this model the same route and energy can be used by macromolecules to trespass the membrane and thus, it would account for the co-entry process. The viral protein responsible for fusion will modify its conformation mainly by receptor binding (Flynn *et al.*, 1990; Meyer *et al* .,1992) and perhaps by acidic pH (Marsh and Helenius. 1989). This protein may act by coupling the energy liberated by the movement of protons to the cytoplasm in favor of its gradient to the translocation of the genome or

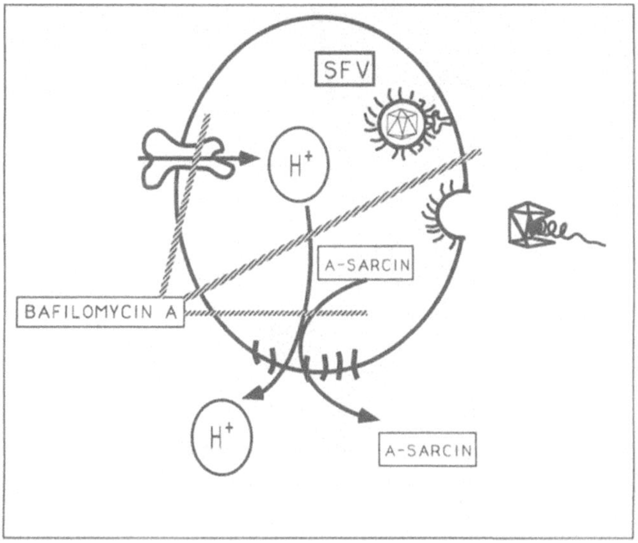

Figure 1. Schematic representation of the co-entry of alpha-sarcin (A-SARCIN) with Semliki Forest virus (SFV). SFV binds to a cellular receptor and is internalized in an endosome. The fusion of the viral membrane with the endosomal membrane releases the viral nucleocapsid in the cytoplasm. This release is inhibited by bafilomycin A1, a selective inhibitor of the vacuolar proton-ATPase, that transports protons into the endosome. The modification of the endosomal membrane by the insertion of the viral glycoproteins allows the co-entry of other macromolecules, such as the toxin alpha-sarcin, that is also liberated in the cytoplasm. The co-entry of this toxin is also inhibited by bafilomycin A1.

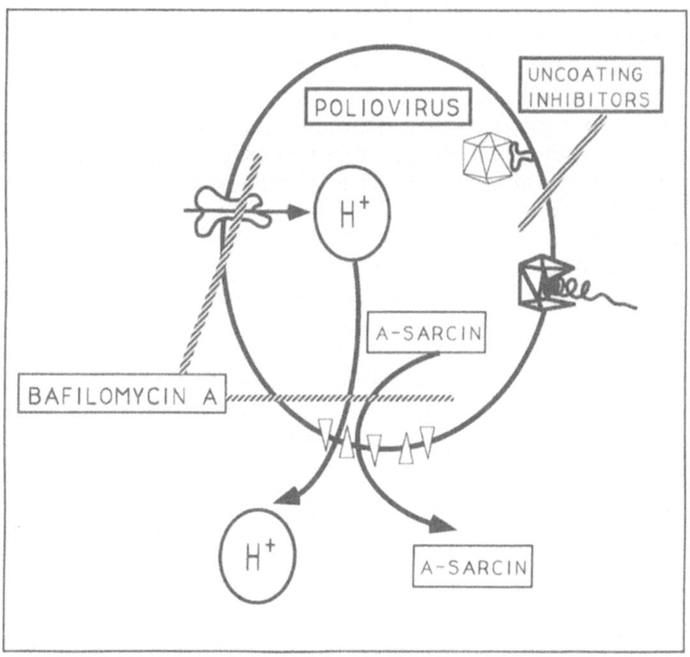

Figure 2. Entry of poliovirus into cells is mediated by endosomes, but low pH is not required because bafilomycin A1 does not block this step. The co-entry of alpha-sarcin is mediated by the insertion of poliovirus structural proteins in the endosomal membrane. The inhibition of the vacuolar proton-ATPase impedes the co-entry of the toxin, but not poliovirus infection.

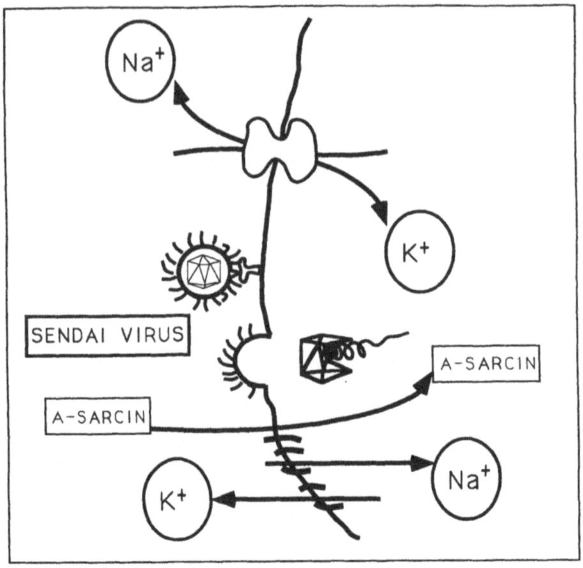

Figure 3. Sendai virus gains access to the cytoplasm by direct fusion of the viral membrane with the plasma membrane. The presence of the viral glycoproteins in the plasma membrane permeabilizes the cell to macromolecules, as alpha-sarcin. Neither Sendai infection, nor alpha-sarcin co-entry are blocked by bafilomycin A1.

protein toxins in the same direction (see Figures 1, 2 and 3). In support of this concept is the fact that virions of animal viruses modify membrane potential in a reversible manner during entry (Fuchs *et al.*, 1980; Seth *et al..*, 1985). In addition, the entry of SFV does not occur when the membrane potential is abolished by modifying the concentration of monovalent cations even under acidic conditions (Helenius *et al.*,1985).

Today we know that not only different protein toxins can trespass the membrane in the presence of animal virus particles, but also other proteins, such as fuciferase, or horse-radish peroxidase (Otero and Carrasco, 1987) and polisaccharides (Gonzalez and Carrasco,1987) are delivered into cells by these particles (Figure 4). No infectious virions are needed for the early membrane permeabilization to occur, because this phenomenon takes place with UV-inactivated (but not heat-inactivated) virions (Carrasco, 1981; Otero,1987).

More recently, the permeabilization of cells by adenovirus particles has been used to introduce DNA into cells. In a series of elegant experiments Birnsteil group has shown that plasmids can bind to cells when they are complexed with transferrin-polylysine molecules (Wagner et al.,1991; Zenke et al.,1990). The entry of these complexes is enhanced by more that 2,000-fold when adenovirus particles are present in the culture medium. In this fashion more than 90% of the cells express the transfected gene (Curiel et al.,1991; Cotten et al.,1992).

2.1. The Early Membrane Permeabilization is Coupled to the Uncoating of Viral Particles

Poliovirus is one of the viruses that efficiently promotes the entry of protein toxins, namely alpha-sarcin, into HeLa cells (Almela *et al.*, 1991).

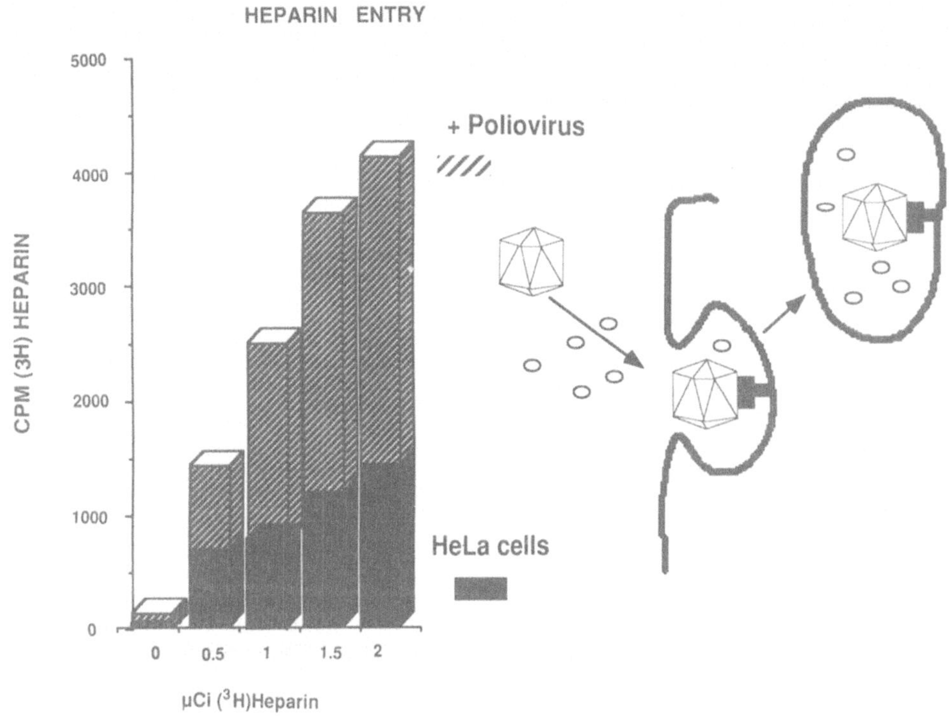

Figure 4. The entry of poliovirus particles into HeLa cells promotes the entry of radioactively labelled heparin (Gonzalez and Carrasco, 1987).

This permeabilization is specific for the interaction of the viral particle with its receptors, because it occurs only in cells that contain the poliovirus receptor, but not in those cells which cannot be infected by poliovirus, such as L cells, or BHK cells (Otero, 1987). In addition, it is a virion component and not a contaminant of the viral preparation because the early permeabilization is blocked by antibodies directed against the poliovirus structural proteins (Otero, 1987).

A number of compounds are now known that interact with the poliovirus particles and inhibit the uncoating process, without interfering with the attachment or the internalization process. These compounds, in addition to block poliovirus uncoating, are also effective inhibitors of the early membrane permeabilization to alpha-sarcin (Almela *et al.*, 1991). The results obtained indicate that the uncoating of the virion particle to deliver the viral genome in the cytoplasm is required to permeabilize the endosomal membrane to the accompanying proteins (Almela *et al.*, 1991).

2.2. Involvement of the $^{+}$H-ATPase in the Early Permeabilization of Membranes by Animal Viruses

As indicated above viruses are internalized in endosomes that are acidified by the activity of a distinct enzyme, the vacuolar proton-ATPase (Mellman *et al.*, 1986). The recent discovery of a potent and selective inhibitor

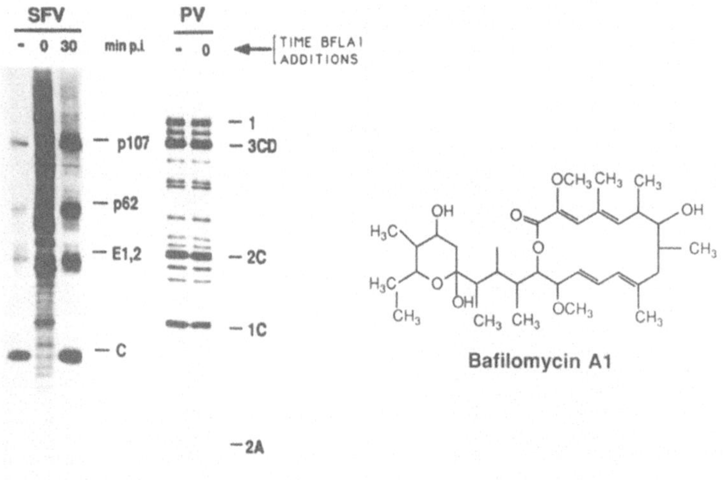

Figure 5. Effect of bafilomycin A1, an inhibitor of the vacuolar proton-ATPase, on the infection of HeLa cells by Semliki Forest virus (SFV) or poliovirus. The addition of this antibiotic from the beginning of infection blocks SFV, but not poliovirus infection.

of this enzyme (bafilomycin A1) (Bowman *et al.*, 1988) has enabled us to analyze the involvement of the vacuolar proton-ATPase not only during animal virus entry, but also in the co-internalization process. Figure 5 shows that concentrations of bafilomycin A (BFLA) as low as 1 μM potently block the infection of HeLa cells by SFV, but only if the compound is present during the beginning of infection, and not if the antibiotic is added thirty minutes after the virus. On the other hand, the infectivity of poliovirus was not blocked by BFLA, indicating that poliovirus uncoats and releases its genome in a pH-independent manner. Since these two viruses had a different behavior in the presence of BFLA, they constituted good models to test the involvement of the proton-ATPase activity in the co-internalization process. Curiously enough, BFLA inhibits the co-entry of alpha-sarcin promoted by SFV, or poliovirus (Perez, L. and Carrasco, L., unpublished results). This result clearly indicates that the exit of alpha-sarcin requires a proton gradient and is coupled to the action of the proton-ATPase. Thus, a virion component is able to couple the proton gradient to the release of macromolecules from the endosome. Therefore, the mechanism of the co-entry of protein toxins is more specific than previously thought (Carrasco *et al.*, 1989) and perhaps does not involve the physical rupture of endosomes by animal virus particles (Carrasco *et al.*, 1989). An updated model of the early membrane permeabilization by viruses is shown in Figures 1, 2 and 3.

3. THE LATE MEMBRANE LEAKINESS

The most drastic modifications of membrane permeability occur in the late phase of the replication cycle of most cytolytic viruses. A detailed review on this subject has already been published (Carrasco *et al.*, 1989). The most salient feature is that a viral gene product is required for the development of the late modifications in membrane permeability. This modification starts at about the third hour in HeLa cells infected with poliovirus (Lacal and Carrasco, 1982; Muñoz and Carrasco, 1983; Lopez-Rivas *et al.*, 1987), or the second hour p.i. in cells infected with SFV (Muñoz *et al*, 1985; Otero and Carrasco, 1987). Thus, the majority of viral macromolecules are synthesized in cells with an altered plasma membrane in a cytoplasm where the concentrations of ions are gradually changing (Carrasco and Smith, 1976; Carrasco, 1977; Lopez-Rivas *et al.*, 1987). The modification of the membrane is such that the gradients of monovalent cations are gradually destroyed. In addition, other metabolites and compounds that are excluded by the membrane leak out from cells and can readily pass to the cytoplasm (Benedetto *et al.*, 1980; Carrasco, 1979; Carrasco and Vazquez, 1983; Carrasco *et al.*, 1989). Thus, inhibitors of translation that are hydrophilic and do not inhibit efficiently protein synthesis in intact cells, are able to selectively block translation in animal virus-infected cells (Lacal *et al.*, 1980; Lacal and Carrasco, 1983; Carrasco and Vazquez, 1983). The structural formulae of some of these "unpermeant translation inhibitors" is shown in Figure 6. Perhaps the most widely used of these agents has been hygromycin

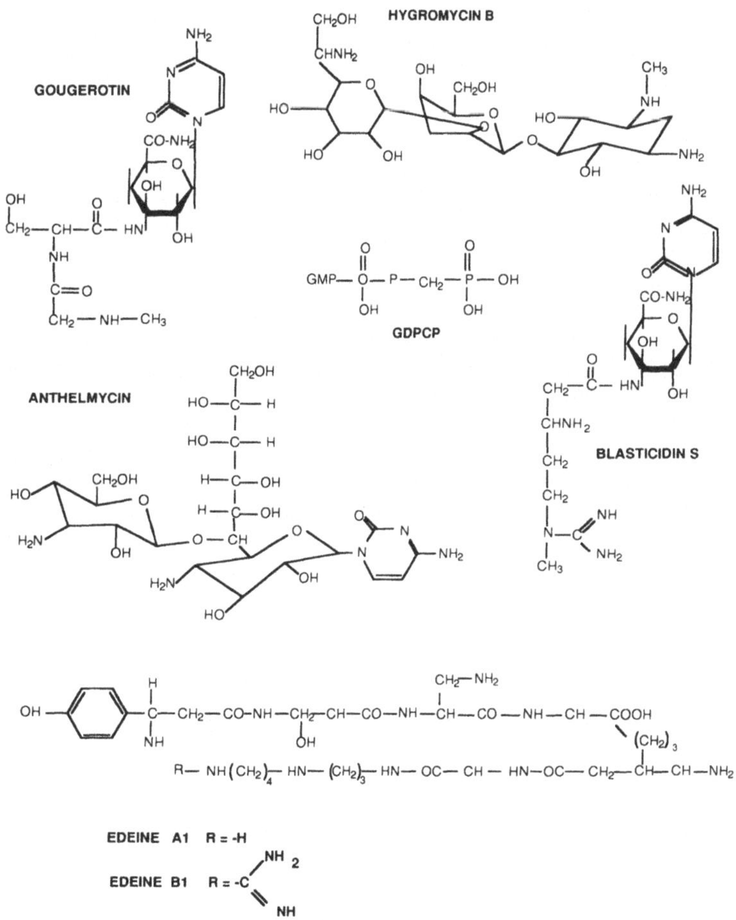

Figure 6. Structural formulae of different "unpermeant translation inhibitors" that selectively enter in animal virus-infected cells (Carrasco and Vazquez,1983).

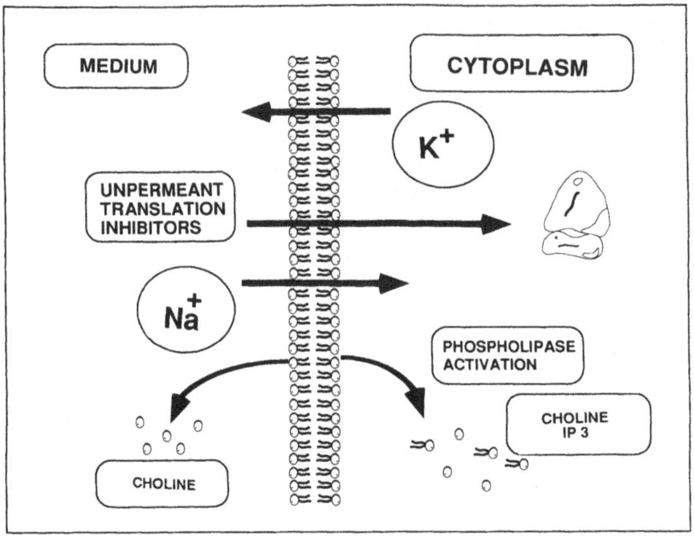

Figure 7. Schematic representation of some of the modifications induced by poliovirus late in infection. The presence in the plasma membrane of a putative viral protein increases membrane permeability to monovalent cations and to a number of translation inhibitors (see figure 6). Phospholipase C is activated as measured by the release of choline, phosphocholine and IP3 (Guinea et al., 1989).

1977). Overwhelming evidence indicates that this may be the case for EMC virus-infected cells. This evidence can be summarized as follows:

1. The *in vitro* and *in vivo* translation of EMC mRNA, is stimulated by sodium ions, as occurs with many viral mRNAs that are translated late during infection, whereas host mRNA translation is inhibited (Carrasco

An updated view of the modifications of membrane permeability by animal viruses is shown in Figure 7. The putative viral protein responsible for membrane modification associates with membranes once it is synthesized and travels through the vesicular system to the plasma membrane. As we have indicated these viral proteins possess an ionophore-like activity (Carrasco, 1977; 1987) in such a way that the pH and ionic content of vesicles will be modified first. Once the proteins reach the plasma membrane they increase its permeability to ions of this membrane leading to the disruption of the membrane potential (Lacal and Carrasco, 1982; 1983). We advanced the suggestion that these changes of ions and pH in the cytoplasm would influence an array of cellular functions, including the translation of host mRNAs (Carrasco, 1977).

3.1. Correlation of the Modification of Membrane Permeability to Monovalent Ions and the Inhibition of Host Protein Synthesis. A Reappraisal

Since animal viruses modify the distribution of monovalent cations between the cytoplasm and the outside medium it seemed feasible that these alterations could affect different cellular and viral functions. In particular, it was initially proposed that the modifications in membrane permeability to monovalent ions in virus-infected cells could be responsible for the virus induced shut-off of host translation (Carrasco and Smith, 1976; Carrasco

B (Contreras and Carrasco, 1979). The inhibition of protein synthesis by this aminoglycoside antibiotic can be used as a test to measure changes in membrane permeability not only in cells infected with animal viruses, but also in any cell type (eukaryotic or prokaryotic) where membrane permeability has been altered (Alonso and Carrasco, 1980; 1981a;1981b;1982a; Otero and Carrasco, 1987; Lama and Carrasco, 1992).
and Smith, 1976; Alonso and Carrasco, 1982b; 1982c; Lacal and Carrasco,1982).
2. The kinetics of monovalent ion fluctuations parallels the inhibition of cellular protein synthesis (Carrasco and Smith, 1976; Lacal and Carrasco, 1982; Muñoz and Carrasco, 1981).
3. A hypotonic medium reverses the shut-off of translation induced by EMC virus (Alonso and Carrasco, 1981c).
4. Ionophores that modify membrane permeability to cations block protein synthesis (Alonso and Carrasco, 1980; 1981a; 1981b). This inhibition is higher on cellular mRNA translation as compared with EMC RNA translation (Alonso and Carrasco, 1981d).

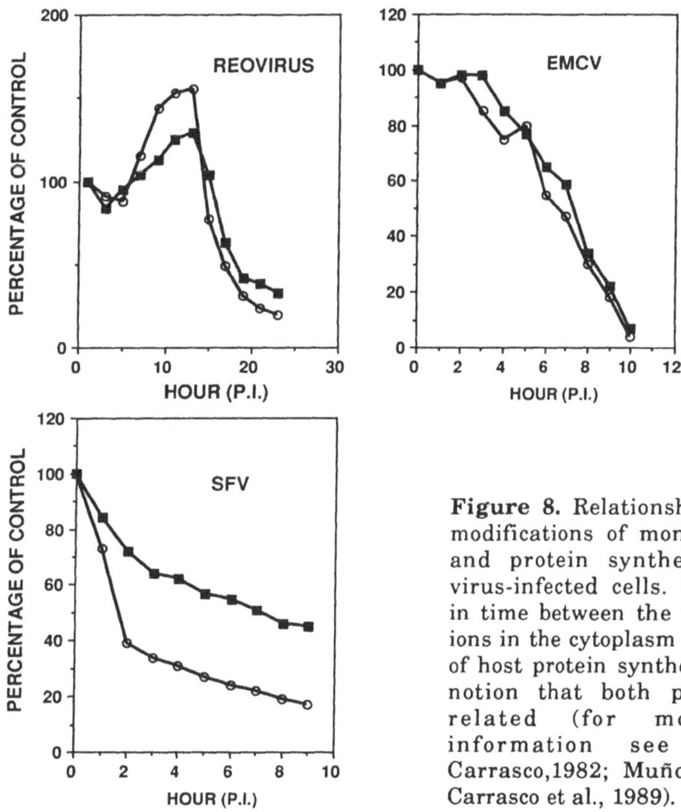

Figure 8. Relationship between the modifications of monovalent cations and protein synthesis in several virus-infected cells. The correlation in time between the modifications of ions in the cytoplasm and the shut-off of host protein synthesis support the notion that both phenomena are related (for more detailed information see Lacal and Carrasco,1982; Muñoz et al.,1986b; Carrasco et al., 1989).

Perhaps, there is no other virus-cell system where more evidence is available to support a model that accounts for the inhibition of host protein synthesis. Togaviruses are thought to block host translation by a similar mechanism (Garry *et al.*, 1979; Ulug *et al.*, 1984; 1987;Muñoz et al.,1985a). In other virus-cell systems analyzed such as VSV (Garry and Waite, 1979) or reoviruses (Muñoz *et al.*, 1985b), there is also a good correlation between the shut-off of translation and the ionic modifications in the infected cells

(Figure 8). It still remains to be tested how general the correlation between the shut-off of host translation and the permeabilization of cellular membranes in other virus-cell systems is.

In the case of poliovirus, it soon became obvious that the alterations in monovalent cations took place one to two hours after the inhibition of host translation had occurred (Lacal and Carrasco, 1982). Thus, both phenomena can be dissociated in time (Lopez-Rivas *et al.*, 1987). A similar argument applies to the current model that the inhibition of host translation by poliovirus is the consequence of p220 cleavage (Etchison *et al.*, 1982; Carrasco and Castrillo, 1987; Sonenberg, 1987;1990). It now seems clear that degradation of p220 occurs prior to the inhibition of host translation (Etchison *et al.*, 1982; Bonneau and Sonenberg, 1987; Perez and Carrasco, 1992). Moreover, the use of inhibitors of poliovirus RNA synthesis still allows the cleavage of p220 under conditions where host translation is taking place at 60-80% of the control for a few hours (Perez and Carrasco, 1992). We believe that the cloning and expression of individual poliovirus proteins in mammalian cells together with site-directed mutagenesis studies could give us clues about the viral proteins responsible for the shut-off of host translation, the modification of membrane permeability, etc.

3.2. Modification of Membrane Permeability by Poliovirus

Although poliovirus does not induce the shut-off of host translation by altering the concentration of monovalent cations, it is still possible that other modifications of the membrane by a viral protein can affect translation as, for instance, changes in pH or divalent cations (Carrasco, 1987; Carrasco and Castrillo, 1987). In fact, we believe that the shut-off of host translation may be a by-product of a viral process, such as viral RNA replication or cell lysis needed to liberate the virus particles made during infection. Thus, if poliovirus needs to put in motion the synthesis of lipids to increase membrane traffic, a process which is necessary to replicate viral genomes (Guinea and Carrasco, 1990), it may cause a harmful effect to cellular metabolism. It must be pointed out, that cytolytic animal viruses lyse the cells by a poorly known mechanism. We have indicated that viruses induce cell lysis by means of ionophore-like proteins (Carrasco, 1977; Carrasco 1987; Carrasco *et al.*, 1989). In principle, poliovirus will synthesize this putative protein from the beginning of infection and the concentration of the protein will augment as infection progresses. The translation of cellular mRNAs will soon be inhibited by the action of the lytic protein, whereas viral RNA translation tolerates two hours of these adverse conditions and it is still translated. However, as the membrane damage caused by this lytic viral protein increases with time, it will also block viral protein synthesis as indeed occurs during poliovirus infection: viral protein synthesis is potently blocked two hours after it has started (Carrasco and Castrillo, 1987). The only means to increase and prolong poliovirus protein synthesis in the infected cells is by changing the concentrations of monovalent ions in the culture medium (Castrillo *et al.*, 1987). Thus far, we know that poliovirus powerfully modifies membrane permeability to monovalent ions from the third hour p.i. and that the majority of viral proteins and genomes are synthesized in a cytoplasm that contains a changing ionic *milieu* (Lopez-Rivas *et al.*, 1987). For this reason many viral mRNAs that are translated late during infection contain special structures in their leader regions that make them more efficient for translation under adverse conditions than cellular mRNAs (Carrasco *et al.*, 1979; Castrillo and Carrasco, 1987; Sonenberg, 1990).

3.3. Modification of Phospholipase Activity by Poliovirus

The studies during the last few years gave us clues about the phenomenology of the modification of membrane permeability by poliovirus, but we need to get more insight about two important questions: 1) The molecular basis of this phenomenon. 2) The viral component involved. To answer the first question we have analyzed the activation of phospholipases during virus infection. Surprisingly, we found that phospholipse A2 activity was inhibited since the early times of poliovirus infection whereas the phospholipase C that hydrolyses phosphatidylinositol is stimulated from the third hour p.i. (Guinea *et al.*, 1989). Thus, in poliovirus infected cells the amount of IP3 present in the cytoplasm is constantly increasing (Guinea *et al.*, 1989). This increase would lead to an augmentation of free calcium. However, we are not aware of any analysis on the modifications of free calcium in cells infected by cytolytic animal viruses. Recently, we have analyzed in detail the activation of the phospholipase C that hydrolyses phosphatidylcholine and have found that both choline and phosphorylcholine drastically increase in the medium of poliovirus-infected cells from the third hour of infection (Irurzun, Perez and Carrasco, unpublished results).

Therefore, a correlation in time exists between the increase in membrane permeability and the activation of phospholipase C in poliovirus-infected cells (Guinea *et al.*, 1989). But, we do not yet know if there is a cause-effect relationship between them. A number of protein toxins that modify membrane permeability also activate phospholipase activity (Durkin and Shier, 1981; Pellkofer *et al.*, 1982). It is tempting to speculate that poliovirus encodes a protein responsible for modifying the membrane (Lama *et al.*, 1992; Lama and Carrasco, 1992). The accumulation of a sufficient number of copies of this protein may lead *per se* to the permeabilization of the membrane. A secondary effect could be the activation of phospholipase C, that will generate diacylglycerol, that in turn, also increases membrane permeability (Felix, 1982; Otero and Carrasco, 1988).

3.4. Phospholipid Synthesis and Membrane Traffic are Required to Replicate Poliovirus Genomes

The replication of poliovirus genomes takes place in close association with newly made membrane vesicles (Semler *et al.*, 1988). During poliovirus infection membrane traffic is profoundly modified, in such a way, that the cytoplasm is loaded with vesicles of various sizes that form a distinct perinuclear structure (Dales *et al.*, 1965; Bienz *et al.*, 1990). This distinctive structure was termed viroplasm (Dales *et al.*, 1965). Little is known about the origin of the membranous vesicles that appear in poliovirus-infected cells or the connection between these vesicles and other membrane compartments. Prevention of lipid synthesis results in an abrupt inhibition of viral RNA synthesis (Guinea and Carrasco, 1990; 1991), suggesting that the replication of viral genomes requires continuous membrane formation. Thus, cerulenin, a potent inhibitor of phospholipid synthesis in prokaryotic and eukaryotic cells, inhibits very quickly the replication of poliovirus RNA when added at any time during infection (Guinea and Carrasco, 1990). Cerulenin does not have a direct inhibitory effect on poliovirus translation, indicating that the synthesis of phospholipids is continuously needed for the synthesis of plus-stranded poliovirus RNA (see Figure 9). Other compounds that block phospholipid synthesis as TOFA or hydroxyacetic acid (see Figure 10) also act in a similar way as cerulenin (Guinea, 1990). These results prompted us

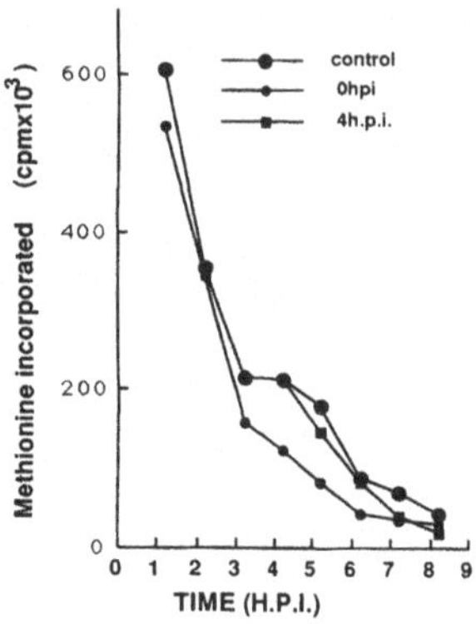

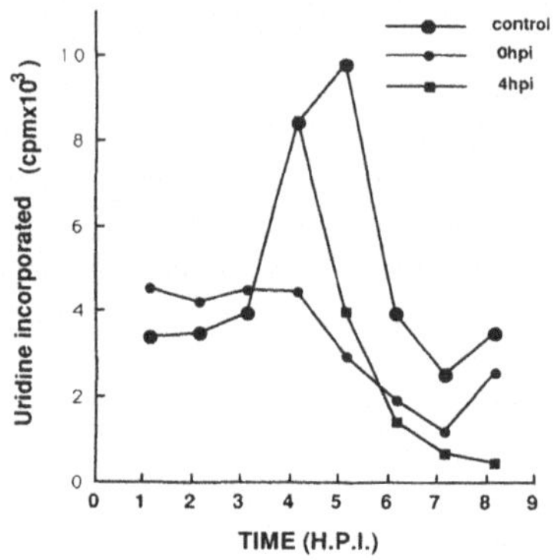

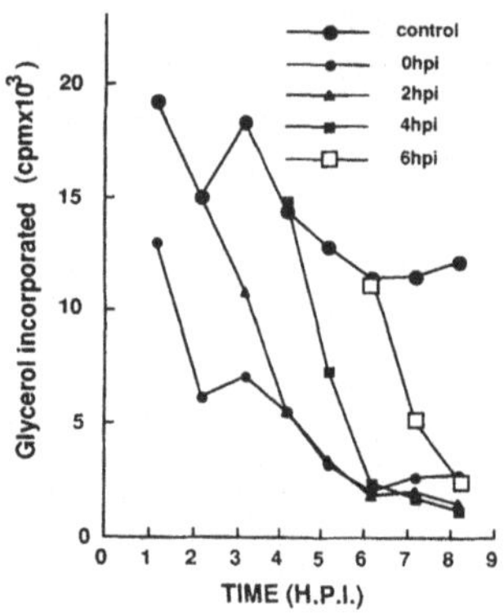

Figure 9. Effects of the inhibitor of phospholipid synthesis cerulenin on the synthesis of proteins (methionine), viral RNA synthesis (uridine) and phospholipid synthesis (glycerol) in poliovirus-infected HeLa cells. Cerulenin was added at the times indicated and the cells were labeled during one hour intervals with the indicated radioactive precursors.

to investigate the connection between membrane traffic and poliovirus RNA replication.

Membranes are continuously moving through several compartments in eukaryotic cells by means of vesicles (Karrenbauer *et al.*, 1990; Morré and Ovtracht, 1977; Tartakoff, 1983; Pfeffer and Rothman,1987). BFA is a fungal antibiotic that blocks protein secretion and vesicular traffic (Misumi *et al* 1986; Fujiwara *et al* 1988; Donaldson *et al.*, 1990; Lippincott-Schwartz *et al.*, 1991). This antibiotic induces a rapid disaggregation of the Golgi complex that redistributes into the ER (Pelham, 1991). The *cis*- and medial Golgi become part of the ER, whereas *trans*-Golgi components do not redistribute to the ER (Pelham, 1991). Therefore, transport of proteins into post-Golgi compartments in the cell is potently blocked by BFA (Doms *et al.*, 1989; Hunziker *et al.*, 1991). An increasing number of animal viruses are reported to be affected by BFA (Tamura *et al.*, 1968; Chen *et al.*, 1991; Cheung *et al.*, 1991; Ulmer and Palade, 1991; Whealy *et al.*, 1991). Since all these viruses possess glycoproteins, BFA, by disrupting the Golgi apparatus, inhibits the processing and transport of viral glycoproteins. In this manner, enveloped viruses that mature from the plasma membrane or intracellular are inhibited by BFA (Chen *et al.*, 1991; Whealy *et al.*, 1991; Cheung *et al.*, 1991; Ulmer and Palade, 1991).

We and others have provided evidence that the macrolide antibiotic BFA is not only active against animal viruses that contain glycoproteins, but also shows antiviral activity against poliovirus (Maynell *et al.*, 1992; Irurzun *et al.*, 1992), a virus lacking a lipid envelope and which does not encode glycoproteins (Urzainqui and Carrasco, 1988). Concentrations of BFA as low as 1 μg/ml prevented the synthesis of poliovirus proteins when the antibiotic was present from the beginning of infection (but after poliovirus entry), whereas the addition of BFA at later stages of infection had no effect on the synthesis of poliovirus polypeptides (Irurzun *et al.*, 1992). These results suggest that this macrolide antibiotic does not inhibit poliovirus translation directly, but blocks an early step required for the synthesis of late proteins. BFA potently inhibited RNA synthesis not only when it was present from the beginning of infection, but also when it was added at the fourth hour p.i., indicating that it had a rapid inhibitory action on poliovirus genome replication. High concentrations of BFA have no effect on cellular RNA synthesis even after prolonged incubation times, i.e. cellular RNA synthesis was 92% of control after eight hours of incubation in the presence of 5 μg/ml BFA. Addition of BFA at the beginning of infection inhibits the appearance of the poliovirus-induced vesicles, whereas, if the antibiotic is added at the fifth hour, the vesicles are resistant to the antibiotic and do not disintegrate. Therefore, BFA does not induce the disintegration of the replication complexes already formed (Irurzun *et al.*, 1992).

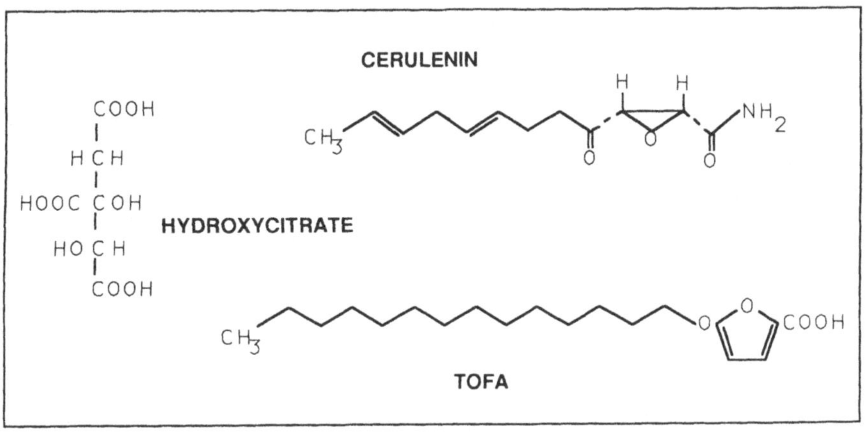

Figure 10. Structural formulae of some inhibitors of phospholipid synthesis that block poliovirus genome replication (Guinea,1990).

The results obtained with BFA indicate that the poliovirus replication complexes need to mature through the Golgi apparatus in order to be functional (Irurzun *et al.*, 1992). If so, the poliovirus proteins once synthesized will bind to membranes in the RER and then they will migrate through the Golgi complex, following the same route as glycoproteins do. Once the poliovirus proteins reach the *trans*-Golgi network they will be sorted to the specialized structures formed by the membranous vesicles and the RNA replication complexes (Semler *et al.*, 1988; Bienz *et al.*, 1990). The finding that BFA immediately blocks poliovirus RNA synthesis without affecting translation suggests that RNA synthesis takes place in association with newly-made vesicles and pre-existing vesicles do not support the formation of new genomes (Guinea and Carrasco, 1990; Irurzun *et al.*, 1992). These findings will help to clarify the action of a number of antiviral agents that interfere with the replication of genomes without affecting the viral polymerase itself (Castrillo *et al.*, 1986; Gonzalez *et al.*, 1991; Rodriguez and Carrasco, 1992). In addition, they can be exploited to devise more selective antiviral agents because, at least theoretically, we could interfere with the action of the viral proteins involved in the coupling between the membranous vesicles and the RNA replication complexes.

3.5. Summary of the Modifications of Cellular Membranes by Animal Viruses: The Plasma Membrane and the Vesicular System

We have indicated in the previous sections that most cytolytic animal viruses, and in particular poliovirus, modify the permeability of the plasma membrane. We now know that the integrity of the membrane in the case of poliovirus is changed by the selective activation of phospholipase C (Guinea *et al.*, 1989). Activation of phospholipases is common to agents that modify membranes such as ionophores and membrane-active peptides (Carrasco *et al.*, 1989). Thus, we believe that activation of phospholipase C is the consequence of membrane modification by a putative poliovirus protein. This toxin-like protein would act as inophores or as membrane active peptides do, as we have repeatedly suggested (Carrasco, 1977; 1987; Carrasco *et al.*, 1989). A general term for this kind of proteins could be "viroporins" i.e. viral proteins with ionophore-like properties, that make pores in cellular membranes. We also have evidence that other animal viruses such SFV or VSV are able to stimulate phospholipases. Both the action of phospholipase C and A_2 are increased by those two viruses (Perez, L., Irurzun, A. and Carrasco, L., unpublished results).

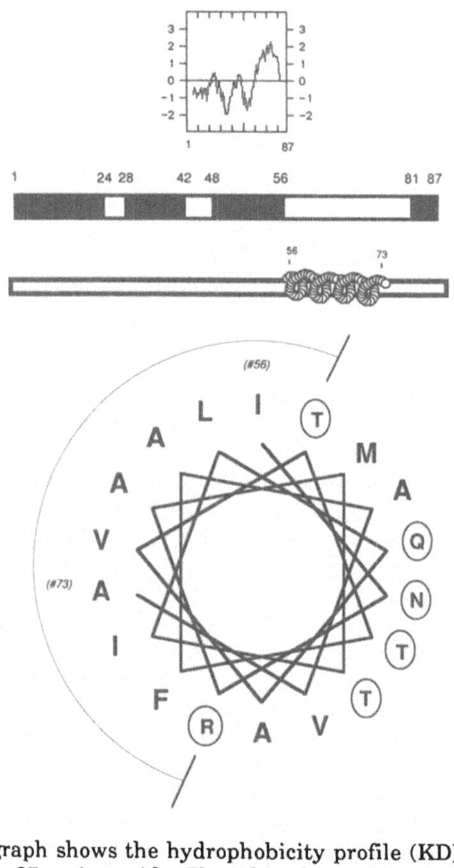

Figure 11. The upper graph shows the hydrophobicity profile (KD) of the poliovirus protein 3A. The protein contains 87 amino acids. The white boxes indicate the hidrophobic regions. Computer analysis shows that amino acids 56-73 can form an amphipathic helix.

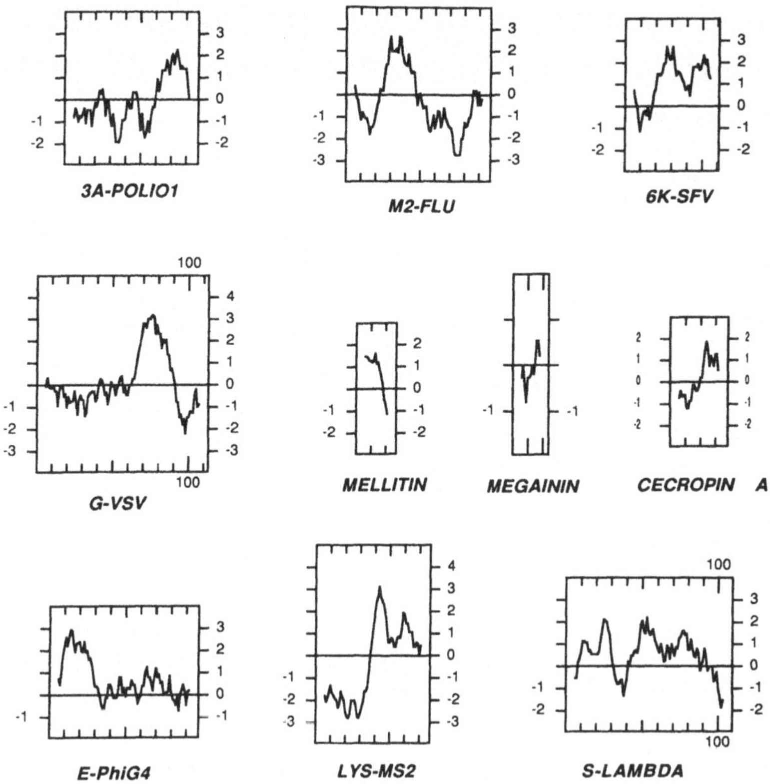

Figure 12. Hydrophobic profiles (KD) of several animal virus proteins : Poliovirus 3A (3A-POLIO1), influenza virus M2 (M2-FLU), Semliki Forest virus 6K (6K-SFV) and the glycoprotein of vesicular stomatis virus (G-VSV). The lytic proteins mellitin, megainin and cecropin. The bacteriophage proteins from G4 (E-PhiG4), MS2 (LYS-MS2) and the S protein of l phage (S-LAMBDA).

The conclusion is that membrane damage has consequences not only for the physical integrity of the membrane (activation of lipases), but also for membrane function. One of the major functions of membranes is to keep gradients of divalent and monovalent cations, including protons. These gradients are generated by the action of special pumps located in cellular membranes. Thus the Na^+/K^+ ATPase creates a gradient of these ions, whereas proton pumps generate a ΔpH. Since these pumps are also electrogenic they create a $\Delta\Psi$. The damage of membranes by animal viruses leads to the disruption of these ionic gradients (Carrasco *et al.*, 1989), with the consequent drop in $\Delta\Psi$ and most probably ΔpH.

As already indicated in section 3.4, some animal viruses as poliovirus not only use the vesicular system to transport and sort their proteins, but also to replicate their genomes. In addition, there are profound modifications in the physical characteristics of the vesicular system during poliovirus infection, though the changes in the functional characteristic of these vesicles remain to be analyzed. If poliovirus and other animal viruses encode for proteins that damage the plasma membrane (viroporins) the first target of their action, once they are synthesized, could be the vesicular system.

4. CHARACTERIZATION OF THE POLIOVIRUS COMPONENT RESPONSIBLE FOR THE MODIFICATION OF MEMBRANE PERMEABILITY

4.1. Cloning and Expression of Individual Poliovirus Proteins

In an effort towards the elucidation of the poliovirus protein involved in modifying membrane permeability, the non-structural poliovirus proteins have been cloned and expressed in *E. coli* using an inducible system (Studier *et al.*, 1990; Studier, 1991). The poliovirus proteins 2A, 2B, 2BC, 3A, 2C3A, 3AB and 2C3AB have been expressed at high levels in bacteria (Lama and Carrasco, 1992; Lama *et al.*, 1992). The most remarkable feature of the expression of these proteins is that two of them 2B and 3A are very toxic for *E. coli* cells and rapidly induce the lysis of bacteria (Lama and Carrasco, 1992). In particular, proteins 3A or 3AB modify the membrane of *E. coli* in a way akin to that observed during the infection of mammalian cells by poliovirus (Lopez-Rivas *et al.*, 1987; Carrasco *et al.*, 1989). Thus, cells expressing 3A or 3AB alter membrane permeability in such a way that compounds unpermeant for normal cells can exit or enter the cells readily (Lama and Carrasco, 1992). The expression of protein 3C or 3D in that system does not modify the membrane. Therefore, there are at least two non-structural proteins encoded by poliovirus i.e. 2B and 3A as potential candidates to account for the modifications of cellular membranes.

4.2. Structural Characteristics of Poliovirus Protein 3A

Since the poliovirus protein 3A is a membrane associated protein (Takegami *et al.*, 1983; Giachetti and Semler,1991), we have concentrated our efforts in the analysis of its mode of action. As part of those studies we have analyzed some of the structural characteristics of the protein. Figure 11 shows that this protein contains a hydrophobic region in the carboxi terminus that can form an amphipathic helix. Deletion mutants of this protein indicate that the presence of this amphipathic helix is not only necessary, but sufficient to induce cell lysis when expressed in *E. coli* (Castrillo, J. L. and Carrasco, L., unpublished results).

Computer search of proteins with a similar structure in other RNA-containing animal viruses has indicated that at least two more proteins share some similarities with 3A as regards to their hydrophobic profile. One of these proteins is the SFV 6K protein (Welch and Sefton, 1980) (Figure 12). We have already cloned and expressed this protein in *E. coli* cells and our results indicate that the SFV 6K behaves in a very similar way as poliovirus 3A does (Sanz, M. A., Perez, L. and Carrasco, L., unpublished results). The other protein with a structure similar to 3A is influenza M2 (Holsinger and Lamb, 1991). This protein has been recently cloned and expressed in Xenopus oocytes (Pinto *et al.*, 1992). Remarkably, recent elegant studies indicate that M2 behaves as an ionic channel (Pinto *et al.*, 1992) and could thus be classified as an inophore-like protein (Carrasco, 1977;1987). It still remains to be investigated if M2 is the protein responsible for cell lysis and the modifications in membrane permeability that take place during influenza virus infection or if the effects of M2 are restricted to the vesicular system (Ciampor *et al.*, 1992).

Future efforts on the cloning of individual viral proteins and the study of their mode of action will give us clues not only about the genes and proteins responsible for causing membrane damage but also about their involvement in modifying other cellular functions.

ACKNOWLEDGEMENTS: Plan Nacional project number (BIO 92-0715), DGICYT project number PB90-0177 and Fundación Ramón Areces are acknowledged for financial support.

REFERENCES

Almela, M.J., Gonzalez, M.E., and Carrasco L., 1991, Inhibitors of poliovirus uncoating efficiently block the early membrane permeabilization induced by virus particles, J Virol. 65:2572-2577.

Alonso, M.A. and Carrasco, L., 1980, Action of membrane-active compounds on mammalian cells. Permeabilization of human cells by ionophores to inhibitors of translation and transcription, Eur J Biochem. 109:535-540.

Alonso, M.A. and Carrasco, L., 1981a, Relationship between membrane permeability and the translation capacity of human HeLa cells studied by means of the ionophore nigericin, Eur J Biochem. 118:289-294.

Alonso, M.A. and Carrasco L., 1981b, Permeabilization of mammalian cells to proteins by the ionophore nigericin, FEBS Lett. 127:112-114.

Alonso, M.A. and Carrasco, L., 1981c, Reversion by hypotonic medium of the shutoff of protein synthesis induced by encephalomyocarditis virus, J Virol. 37:535-540.

Alonso, M A. and Carrasco, L., 1981d, Selective inhibition of cellular protein synthesis by amphotericin B in EMC virus-infected cells, Virology. 114:247-251.

Alonso, M.A. and Carrasco, L., 1982a, Molecular basis of the permeabilization of mammalian cells by ionophores, Eur J Biochem. 127:567-569.

Alonso, M.A. and Carrasco, L., 1982b, Translation of capped virus mRNA in encephalomyocarditis virus-infected cells, J Gen Virol. 60:315-325.

Alonso, M.A. and Carrasco, L., 1982c, Translation of capped viral mRNAs in poliovirus-infected HeLa cells, EMBO J. 1:913-917.

Benedetto, A., Rossi, G.B., Amici, C., Belardelli, F., Cioe, L., Carruba, G., and Carrasco, L., 1980, Inhibition of animal virus production by means of translation inhibitors unable to penetrate normal cells, Virology. 106:123-132.

Bienz, K., Egger, D., Troxler, M. and Pasamontes, L., 1990, Structural organization of poliovirus RNA replication is mediated by viral proteins of the P2 genomic region, J Virol. 64:1156-1163.

Bonneau, A.M. and Sonenberg, N., 1987, Proteolysis of the p220 component of the capbinding protein conplex is not sufficient for complete inhibition of host cell protein syntheses after poliovirus infection, J Virol. 61:986-991.

Bowman, E.J., Siebers, A. and Altendorf, K., 1988, Bafilomycins: A class of inhibitors of membrane ATPases from microorganisms-animal cells-and plant cells, Proc Natl Acad Sci USA. 85:7972-7976.

Carrasco, L., 1977, The inhibition of cell functions after viral infection: A proposed general mechanism, FEBS Lett. 76:11-15.

Carrasco, L., 1978, Membrane leakiness after viral infection and a new approach to the development of antiviral agents, Nature. 272:694-699.

Carrasco, L., 1979, The development of new antiviral agents based on virus-mediated cell modification, in: "Chandra," ed., Pub, City

Carrasco, L., 1981, Modification of membrane permeability induced by animal viruses early in infection, Virology. 113:623-629.

Carrasco, L., 1987, Animal virus-induced cytotoxicity: An overview, in: "Mechanisms of Viral Toxicity in Animal Cells," L. Carrasco, ed., CRC Press, Florida.

Carrasco, L. and Castrillo, J.L., 1987, The regulation of translation in picornavirus infected cells. In Mechanisms of viral toxicity in animal cells, Carrasco, L., ed., CRC Press, Boca-Raton. 115-146.

Carrasco, L. and Esteban, M., 1982, Modification of membrane permeability in vaccinia virus-infected cells, Virology. 117:62-69.

Carrasco, L. and Lacal, J.C., 1983, Permeabilization of cells during animal virus infection, Pharmacol Ther. 23:109-145.

Carrasco, L., Otero, M.J. and Castrillo, J.L., 1989, Modification of membrane permeability by animal viruses, Pharmacol Ther. 40:171-212.

Carrasco, L. and Smith, A.E., 1976, Sodium ions and the shut-off of host cell protein synthesis by picornaviruses, Nature. 264:807-809.

Carrasco, L. and Vazquez, D., 1983, Viral translation inhibitors, in "Antibiotics," vol.6. D. Vazquez, ed., Springuer-Verlag, Berlin. 1980-1990.

Castrillo, J. L. and Carrasco, L., 1987, Adenovirus late protein synthesis is resistant to the inhibition of translation induced by poliovirus, J Biol Chem. 262:7328-7334.

Castrillo, J.L., López-Rivas, A., and Carrasco, L., 1987, Effects of extracellular cations on translation in poliovirus infected cells, J Gen Virol. 68:325-333.

Castrillo, J.L., Van den Berghe, D. and Carrasco, L., 1986, 3-Methylquercetin is a potent and selective inhibitor of poliovirus RNA synthesis, Virology. 152:219-227.

Ciampor, F., Bayley, P.M., Nermut, M.V., Hirst, E.M.A., Sugrue, R.J. and Hay, A.J., 1992, Evidence that the amantadine-induced, M2-mediated conversion of influenza A virus hemagglutinin to the low pH conformation occurs in an acidic trans Golgi compartment, Virology. 188:14-24.

Chen, S-Y., Matsuoka, Y. and Compans, R.W., 1991, Assembly and polarized release of punta toro virus and effects of brefeldin A, J Virol. 65:1427-1439.

Cheung, P., Banfield, B.W. and Tufaro, F., 1991, Brefeldin A arrests the maduration and egress of herpes simplex virus during infection, J Virol. 65:1893-1904.

Contreras, A. and Carrasco, L., 1979, Selective inhibition of protein synthesis in virus-infected mammalian cells, J Virol. 29:114-122.

Cotten, M., Wagner, E., Zatloukl, K., Phillips, S., Curiel, D.T. and Birnstiel, M.L., 1992, High-efficiency receptor-mediated delivery of small and large(48 kilobase gene constructs using the endosome-disruption activity of defective or chemically inactivated adenovirus particles, Proc Natl Acad Sci USA. 89:6094-6098.

Curiel, D.T., Agarwal, S., Wagner, E., and Cotten, M., 1991, Adenovirus enhancement of transferrin--polylysine-mediated gene delivery, Proc Natl Acad Sci USA. 88:8850-8854.

Dales, S., Eggers, H.J., Tamm, I. and Palade, G.E., 1965, Electron microscopic study of the formation of poliovirus, Virology. 26:379-389.

Dimmock, N.J., 1982, Review article inital stages in infection with animal viruses, J Gen Virol. 59:1-22.

Doms, R.W., Russ, B. and Yewdell, J.W., 1989, Brefeldin A redistributes resident and itinerant Golgi proteins to the endoplasmic reticulum, J Cell Biol. 109:61-72.

Donaldson, J.G., Lippincott-Swartx, J., Bloom, B.S., Dreis, T.E., and Krausner, R.D., 1990, Dissociation of a 110 kD peripheral membrane protein from the Golgi apparatus is an early event in brefeldin A action, J Cell Biol. 111:2295-2306.

Durkin, J.P. and Shier, W.T., 1981. Staphylococcal delta toxin stimulates endogenous phospholipase A2 activity and prostaglandin synthesis in fibroblasts, Biochim Biophys Acta. 814:319-326.

Etchison, D., Milburn, S.C., Edery, I., Conenberg, M. and Hershey, J.W.B., 1982, Inhibition of HeLa cell protein synthesis following poliovirus infection correlates with the proteolysis of a 220,000-dalton polypeptide associated with eukaryotic initiation factor 3 and a cap-binding protein complex, J Biol Chem. 257:14806-14810.

Felix, H., 1982, Permeabilized cells, Anal Biochem. 120:211-234.

Fernández-Puentes, C. and Carrasco L., 1980, Viral infection permeabilizes mammalian cells to protein toxins, Cell. 20:769-775.

FitzGerald, D.J., Padmanabhan, R., Pastan, I. and Willingham, M.C., 1983, Adenovirus-induced release of epidermal growth factor and pseudomonas toxin into the cytosol of KB cells during receptor-mediated endocytosis, Cell. 32:607-617.

Flynn, D.C., Meyer, W.J., Mackenzie, J.M.J., and Johnston, R.E., 1990, A conformational change in Sindbis virus glycoproteins E1 and E2 is detected at the plasma membrane as a consequence of early virus-cell interaction, J Virol. 64:3643-3653.

Fuchs, P., Gruder, E., Gitelman, J. and Kohn, A., 1980, Nature of permeability changes in membrane of HeLa cells adsorbing Sendai virus, J Cell Physiol. 103:271-278.

Fujiwara, T., Oda, K., Yokota, S., Takatsuki, A. and Ikehara, Y., 1988, Brefeldin A causes disassembly of the Golgi complex and accumulation of secretory proteins in the endoplasmic reticulum, J Biol Chem. 263:18545-18552.

Garry, R.F., Bishop, J.M., Parker, S., Westbrook, K., Lewis, G. and Waite, M.R., 1979, Na^+ and K^+ concentrations and the regulation of protein synthesis in Sindbis virus-infected chick cells, Virolgy. 96:108-120.

Garry, R.F. and Waite, M.R.F., 1979, Na^+ and K^+ concentrations and the regulation of the interferon system in chick cells, Virology. 96:121-128.

Giachetti, C. and Semler, B.L., 1991, Role of a viral membrane polypeptide in strand-specific initiation of poliovirus RNA synthesis, J Virol. 65:2647-2654.

González, M.E. and Carrasco, L., 1987, Animal viruses promote the entry of polysaccharides with antiviral activity into cells, Biochem Biophys Res Commun. 146:1303-1310.

González, M.E., Martínez-Abarca, F., and Carrasco, L., 1990, Potent inhibitors of poliovirus RNA synthesis, Ant Chem Chemoth. 1,203-209.

Guinea, 1990, Ph D Thesis. Autonomous University Madrid.

Guinea, R. and Carrasco, L., 1990, Phospholipid biosynthesis and poliovirus genome replication, two coupled phenomena, EMBO J. 9:2011-2016.

Guinea, R. and Carrasco, L., 1991, Effects of fatty acids on liid synthesis and viral RNA replication in poliovirus-infected cells, Virology. 185:473-476.

Guinea, R., Lopez Rivas, A. and Carrasco, L., 1989, Modification of phospholipase C and phospholipase A2 activities during poliovirus infection, J Biol Chem. 264:21923-21927.

Helenius, A., Kielan, M., Wellsteed, J., Mellman, I., and Rudnick, G., 1985, Effects of monovalent cations on Semliki Forest virus entry into BHK-21 cells, J Biol Chem. 260:5691-5697.

Hoekstra, D. and Kok, J.W., 1989, Entry mechanisms of enveloped viruses. Implications for fusion of intracellular membranes, Biosci Rep. 9:273-305.

Holsinger, L. and Lamb, R., 1991, Influenza virus M2 integral membrane protein is a homotetramer stabilized by formation of disulfide bonds, Virology. 183:32-43.

Hunziker, W., Whitney, J.A. and Mellman, I., 1991, Selective inhibition of transcytosis by brefeldin A in MDCK cells, Cell. 67:617-627.

Irurzun, A., Perez, L. and Carrasco, 1992, Involvememt of membrane traffic in the replication of poliovirus genomes: Effects of brefedin A,Virology. In Press.

Karrenbauer, A., Jeckel, D., Just, W., Birk, R., Schmidt, R.R., Rothman, J.E. and Wieland, F.T., 1990, The rate of bulk flow from the Golgi to the plasma membrane, Cell 63:259-267.

Lacal, J.C. and Carrasco, L., 1982, Relationship between membrane integrity and the inhibition of host translation in virus-infected mammalian cells. Comparative studies between encephalomyocarditis virus and poliovirus, Eur J Biochem. 127:359-366.

Lacal, J.C. and Carrasco, L., 1983, Antiviral effects of hygromycin B, a translation inhibitor nonpermeant to uninfected cells, Antimicrob Agents Chemother. 24:273-275.

Lacal, J.C., Vazquez, D., Fernandez Sousa, J.M. and Carrasco, L.,1980, Antibiotics that specifically block translation in virus-infected cells, J Antibiot. 33:441-446.

Lama, J. and Carrasco, L., 1992, Expression of poliovirus nonstructural proteins in Escherichia coli cells. Modification of membrane permeability induced by 2B and 3A, J Biol Chem. 267:15932-15937.

Lama, J., Guinea, R., Martinez-Abarca, F. and Carrasco, L., 1992, Cloning and inducible synthesis of poliovirus nonstructural proteins, Gene. 117:185-192.

Lippincott-Schwartz, J., Yuan, L., Tipper, C., Amherdt, M., Orci, L. and Klausner, R.D., 1991, Brefeldin A's effects on endosomes, lysosomes, and the TGN suggest a general mechanism for regulating organelle structure and membrane traffic, Cell. 67:601-616.

López-Rivas, A., Castrillo, J.L., and Carrasco, L., 1987, Cation content in poliovirus-infected HeLa cells, J Gen Virol. 68:335-342.

Marsh and Helenius, 1989, Virus entry into animal cells, Adv Virus Res. 36:107-151.

Maynell, L.A., Kirkegaard, K. and Klymkowsky, M.W., 1992, Inhibition of poliovirus RNA synthesis by brefeldin A, J Virol. 66:1985-1994.

Meyer, W.J., Gidwitz, S., Ayers, V.K. Schoepp, R.J., and Johnston, R.E., 1992, Conformational alteration of Sindbis virion glycoproteins induced by heat, reducing agents, or low pH, J Virol. 66:3504-3513.

Mellman, I., Ruchs, R. and Helenius, A., 1986, Acidification of the endocytic and exocytic pathways, Annu Rev Biochem. 55:663-700.

Misumi, Y., Misumi, Y., Miki, K., Takatsuki, A., TAmura, B. and Ikehara, Y., 1986, Novel blockade by brefeldin A of intracellular transport of secretory proteins in cultured rat hepatocytes, J Biol Chem. 261:11398-11403.

Morré, J. and Ovtracht, L., 1977, Dynamics of the Golgi Apparatus: membrane differentiation and membrane flow, Int Rev Cytol. 5:61-189.

Muñoz, A. and Carrasco, L., 1981, Protein synthesis and membrane integrity in interferon-treated HeLa cells infected with encephalomyocarditis virus, J Gen Virol. 56:153-162.

Muñoz, A. and Carrasco, L., 1983, Effect of interferon treatment on blockade of protein synthesis induced by poliovirus infection, Eur J Biochem. 137:623-629.

Muñoz, A., Castrillo, J.L. and Carrasco, L., 1985a, Modification of membrane permeability during Semliki Forest virus infection, Virology. 146:203-212.

Muñoz, A., Alonso, M.A. and Carrasco, L., 1985b, The regulation of translation in reovirus-infected cells. J. Ge. Virol. 66: 2161-2179.

Otero, M.H., 1987, Ph. Thesis. Autonomous University Madrid.

Otero, M.J. and Carrasco, L., 1987, Proteins are cointernalized with virion particles during early infection, Virology. 160:75-80.

Otero, M.J. and Carrasco, L., 1988, Exogenous phospholipase C permeabilizes mammalian cells to proteins, Exp Cell Res. 177:154-161.

Pelham, H.R.B., 1991, Multiple targets for brefeldin A, Cell. 67:449-451.

Pellkofer, R., Marsh, D., Hoffmann-Bleihauer, P. and Sandhoff, K., 1982, Melittin stimulates incorporation and degradation of sphyngomyelin in synaptosomal plasma membranes, J Neurochem. 38:1230-1235.

Pérez, L. and Carrasco, L., 1992, Lack of direct correlation between p220 cleavage and the shut-off of host translation after poliovirus infection, Virology. 189:178-186.

Pfeffer, S.R. and Rothman, J.E., 1987, Biosynthetic protein transport and sorting by the endoplasmic reticulum and Golgi, Annu Rev Biochem. 56:829-852.

Pinto, L.H., Holsinger, L.J. and Lamb, R.A., 1992, Influenza virus M_2 protein has ion channel activity, Cell. 69:517-528.

Rodriguez, P.L. and Carrasco, L., 1992, Gliotoxin: Inhibitor of poliovirus RNA synthesis that blocks the viral RNA polymerase 3Dpol, J. Virol. 66:1971-1976.

Semler, B.L., Kuhn, R.J. and Wimmer, E., 1988, in "RNA Genetics," E. Domingo, J.J. Holland and P. Ahlquist, eds., CRC Press, Florida. pp23-48.

Seth, P., Willingham, M.C., and Pastan, I., 1984, Adenovirus-dependent release of ^{51}Cr from KB cells at an acidic pH, J Biol Chem. 23:14350-14353.

Seth, P., Pastan, I. and Willingham, M.C., 1985, Adenovirus-dependent increase in cell membrane permeability, J Biol Chem. 260:9598-9602.

Sonenberg, N., 1987, Regulation of translation by poliovirus, Adv Virus Res. 33:175-204.

Sonenberg, N., 1990, Poliovirus translation, Cur Top Microbiol Immunol. 161:23-47.

Studier, F.W., 1991, Use of bacteriophage T7 lysozyme to improve an inducible T7 expression system, J. Mol. Biol. 219:37-44.

Studier, F.W., Rosenberg, A.H. and Dunn, J.J., 1990, Use of T7 RNA polymerase to direct expression of cloned genes. Methods in Enzymology. 185:60-89.

Takegami, T., Semler, B.L., Anderson, C.W., and Wimmer, E., 1983, Membrane fractions active in poliovirus RNA replication contain VPg precursor polypeptides, Virology. 128:33-47.

Tamura, G., Ando, K., Suzuki, S., Takatsuki, A. and Arima, K., 1968, Antiviral activity of brefeldin A and verrucarin A, J Antibiot. 21:160-161.

Tartakoff, A.M., 1983, The confined function model of the Golgi complex; center for ordered processing of biosynthetic products of the rough endoplasmic reticulum, Int Rev Cytol. 85:221-252.

Ulmer, J.B. and Palade, G.E., 1991, Effects of brefeldin A on the processing of viral envelope glycoproteins in murine erythroleukemia cells, J Biol Chem. 266:9173-9179.

Ulug, E.T., Garry, R.F., and Bose, H.R., Jr, 1987, Cell killing by enveloped RNA viruses. In: "Mechanisms of viral toxicity in animal cells," pp. 91-113, Carrasco L., ed., CRC Press. Boca Ratón.

Ulug, E.T., Garry, R.F., Waite, M.R. and Bose, H.R.J., 1984, Alterations in monovalent cation transport in Sindbis virus-infected chick cells, Virology. 132:118-130.

Urzáinqui, A. and Carrasco, L., 1989, Post-translational modifications of poliovirus proteins, Biochem Biophys Res Commun. 158:263-271.

Wagner, E., Cotten, M., Foisner, R., and Birnstiel, M.L., 1991, Transferrin-polycation-DNA complexes: The effect of polycations on the structure of the complex and DNA delivery to cells, Proc Natl Acad Sci USA. 88:4255-4259.

Welch, W.J. and Sefton, B.M., 1980, Characterization of a small-nonestructural viral polypeptide present late during infection of BHK cells by Semliki Forest virus, J Virol. 33:230-237.

Whealy, M.E., Card, J.P., Meade, R.P., Robbins, A.K. and Enquist, L.W., 1991, Effect of brefeldin A on alphaherpesvirus membrane protein glycosylation and virus egress, J Virol. 65:1066-1081.

White, J.M., 1990, Viral and cellular membrane fusion proteins, Annu Rev Physiol. 52:675-697.

Zenke, M., Steinlein, P., Wagner, E., Cotten, M., Beug, H., and Birnstiel, M.L., 1990, Receptor-mediated endocytosis of transferrin-polycation conjugates: An efficient way to introduce DNA into hematopoietic cells, Proc Natl Acad Sci USA. 87:3655-3659.

POLIOVIRUS NEUROVIRULENCE AND ITS ATTENUATION

Vadim I. Agol

Institute of Poliomyelitis &
Viral Encephalitides
Russian Academy of Medical Sciences
Moscow Region 142782, and
Moscow State University
Moscow, 119899
Russia

INTRODUCTION

The Family *Picornaviridae* includes a huge number of viruses. Many of its members have been named after the syndromes they cause: poliomyelitis, encephalomyelitis, myocarditis, hepatitis, foot-and-mouth disease etc. Although not reflected in the virus names, a number of other pathologies may also arise as a consequence of the picornavirus infection, such as meningitis, herpangina, myositis, diabetes, and so on. In some cases, the same syndrome can accompany the infection caused by different viruses, and, on the other hand, a given virus can sometime induce different syndromes. Furthermore, close relatives of some pathogenic agents can be used as efficient live vaccines, the best known example being the Sabin Oral Polio Vaccine. Despite all this diversity, these viruses share structural and genomic organization and follow similar reproduction strategies. Now that the RNA primary structure of the representatives of all the major picornavirus groups has been elucidated, how can we explain the peculiarities of their biological behavior? This question is a big challenge, and we are making only the very first and sometimes awkward steps toward answering it. Here, I am going to discuss just some aspects of this problem; namely, I will focus on the poliovirus and the disease it causes.

OVERVIEW OF THE POLIOMYELITIS PATHOGENESIS

Human beings are the only natural host of the virus, and the infection is transmitted via the fecal/oral route. The initial reproduction takes place in lymphoid and perhaps mucosal cells of oropharynx and, to a greater extent, of lower intestine. Being acid-stable, the virus safely passes through the stomach. From the gastro-intestinal tract, the viral

Regulation of Gene Expression in Animal Viruses, Edited by
L. Carrasco *et al.*, Plenum Press, New York, 1993

progeny can go into the blood, and from there, it may enter the central nervous system. To do that the virus should overcome the blood/brain barrier. How this is achieved is not known for certain, but limited multiplication in endothelial cells or direct intraneuronal movement are likely, and not necessarily mutually exclusive, possibilities. Certain neurons, primarily in the anterior horns of the spinal cord, are the final target. Their destruction leads to the flaccid paralysis. Such an outcome is, however, characteristic of no more than 1% of the infected non-immune human beings; asymptomatic infection is the prevalent form.

The above description refers to the so called wild-type, virulent poliovirus strains. As already mentioned, variants with a significantly reduced neurovirulence have been selected. They elicit a strong and lasting immune response, while exhibiting limited reproduction in the human body. That is why we have an excellent live oral poliomyelitis vaccine.

This overview reveals at least two very important points. Firstly, the host range of poliovirus, even of its virulent strains, is highly restricted in the human body; in addition to neural and lymphoid cells, only endothelial and mucosal cells are known to serve as hosts for the virus growth. Secondly, the preponderance of asymptomatic infections indicates that the damage to the neural cells should also be relatively limited; such a mildness can theoretically be due to either difficulties in reaching neural cells (e. g., in crossing the blood/brain barrier), or the low intrinsic sensitivity of these target cells to poliovirus, or the existence of strong defense mechanisms. It is likely that all these possibilities can be realized, but their individual contributions are yet to be ascertained.

The outcome of the infection depends on the interplay between the virus-encoded macromolecules and cellular factors controlling the efficiency and consequences of poliovirus reproduction, especially in neural cells. Thus, the problem is two-fold. First, we should define the properties of the viral macromolecules that affect the development of the pathology. The scientific paradigm dictates to us to have at least two agents to compare: an experimental and a control ones. That is why poliovirus neurovirulence and attenuation are Siamese twins. Second, by the same token, to understand the specific contribution of the neural or other natural target cells to the pathogenesis, it would be appropriate to compare poliovirus reproduction in these cells, on the one hand, and in some other control cells, on the other. Thus, ideally, experiments should be conducted with at least two kinds of viruses in at least two kinds of host cells. In addition, the contribution of the macro-organism should be evaluated. And most importantly, we should put the right questions to the experimental systems. Unfortunately, these requirements are not always fulfilled.

ATTENUATION AND ATTENUATING MUTATIONS

In an attempt to generate live vaccine against poliomyelitis, Albert Sabin carried out, in the 50s, multiple *in vitro* and *in vivo* passages of poliovirus strains belonging to the 3 existing serotypes; by the multistep selection of variants with progressively lowered virulence, he eventually obtained what is known as the Sabin attenuated strains. The most important distinctive property of these strains is that they fail to cause a paralytic disease upon intracerebral or intraspinal inoculation of monkeys. These strains are successfully used as the live vaccine all over the world for more than 30 years (Sabin, 1985).

The first primary structure of the genome of a Sabin strain, it happened to be type 1, was determined by Nomoto *et al.* (1982). Given the RNA sequence of its virulent predecessor, strain Mahoney, had already been known (Kitamura *et al.*, 1981; Racaniello and Baltimore, 1981), it turned out that the RNA of Sabin 1 did acquire, during the selection, over 50 point mutations; as a consequence, 21 amino acid residues were changed in the viral polyprotein. The mutations were more or less evenly distributed all over the genome. When a similar comparison of the type 3 attenuated and virulent strains was performed by Almond and coworkers, differences in only 10 nucleotides and 3 amino acids were found (Stanway *et al.*, 1984). It was highly unlikely, however, that all the mutations found in the Sabin strains were directly connected with the decreased pathogenicity.

Mapping of the attenuating mutations was achieved primarily by the recombination analysis: segments of the genome of attenuated strains were replaced by the homologous segments derived from their virulent counterparts, and *vice versa*; then the neurovirulence of these chimeras was appropriately assayed. The first, though rather crude, analysis of this sort was performed in our Institute (Agol *et al.*, 1984; 1985a; 1985b). We coinfected cells with pairs of virulent and attenuated strains, and then selected recombinants with the crossover points in the middle of the genome. By choosing appropriate pairs of viruses and convenient selectable markers, a set of recombinants was obtained with all possible arrangements of the genomic halves: attenuated/virulent, virulent/attenuated and so on. The neurovirulence of these chimeras was assayed in monkeys. By using this approach, we mapped the major determinants of attenuation of the Sabin 1 and Sabin 3 strains to the 5' half of the RNA molecule, and also found additional attenuating mutations in the 3' half of the Sabin 1 RNA.

Several laboratories performed a similar, but much more refined, analysis by engineering recombinants with the aid of the recombinant DNA technology rather than by selecting them *in vivo* (reviewed in Nomoto and Wimmer, 1987; Almond, 1987; Racaniello, 1988; see also, Kawamura *et al.*, 1989; Moss *et al.*, 1989; Westrop *et al.*, 1989; Macadam *et al.*, 1991). Remarkably, the attenuated phenotype of the Sabin strains turned out to be due to a few, but only to a few, point mutations. According to the genomic localization [5'-noncoding region (5NCR) and regions encoding capsid or replication proteins], they can be classified into three groups. The first encompasses attenuating mutations within a narrow segment of the 5NCR: in the Sabin type 1, type 2 and type 3 strains, positions 480, 481 and 472 are altered. Attenuating mutations of the second group modify capsid proteins; they also appear to be present in all the three Sabin strains, though in different polypeptides and at different locations. The exact localization of attenuating mutations in the capsid proteins are known for serotypes 2 and 3, and the best studied example concerns a mutation in capsid polypeptide VP3 of Sabin 3. A serine residue at position 91 of the virulent Leon strain is replaced by phenylalanine in Sabin 3. This residue confers thermosensitivity to the virus growth. This phenotypic property seems to correlate with neurovirulence, because the loss of the *ts* character, due either to the direct reversion or second-site suppression, is accompanied by an increase in the level of neurovirulence (Macadam *et al.*, 1989). In Sabin 2, as can be judged from the experiments with its close relative P712, a significant attenuating effect depends on an isoleucine residue at position 143 of VP1 (Ren *et al.*, 1991). It is impossible to name an amino acid residue characteristic of the predecessors of Sabin 2, because such predecessors are unknown.

Less clear is the nature of attenuating mutations in the proteins of Sabin 1. Evidence derived from the experiments with recombinants between the Sabin 1 and Mahoney strains suggests that there are such mutations in both the capsid and at least two replication polypeptides (Nomoto and Wimmer, 1987), but the precise locations of these mutations are yet to be determined. There is no evidence that sufficiently strong attenuating mutations affecting non-capsid, replication proteins are present in the RNAs of Sabin strains other than type 1.

A few specifications perhaps should be made at this point. It is obvious that functions of the poliovirus can be damaged in hundreds of ways. The ability of such crippled viruses to grow in neural cells will, in most cases, be also diminished. This, however, should hardly be termed attenuation. From the practical point of view, necessary amounts of the poorly growing virus simply could not have been prepared for vaccination; moreover, such virus could likely be a poor immunogen. Here, we shall use the term "attenuation" operationally, implying that, being inoculated in monkeys in equal doses (in terms of plaque forming units), attenuated strains elicit milder, if any, symptoms as compared to neurovirulent ones.

Generally speaking, two different mechanisms can be responsible for the attenuated phenotype. Firstly, certain alterations can *specifically* jeopardize the interaction of poliovirus with the disease-related target cells in the human body. Thus, the Sabin-specific mutations in the 5'-noncoding region should definitely be regarded as truly attenuating in this strict sense: they are much more harmful for the virus growth in neural than, say, in monkey kidney or HeLa cells (see below). On the other hand, such an *unspecific* property as thermosensitivity of viral reproduction is for long, and rightly, considered as an attenuation-related quality (Lvoff, 1959); the simplest, if not simplistic, explanation of the attenuating effect of *ts* mutations consists in the fact that the temperature in human and monkey CNS is, especially in the course of the infection-induced fever, above the standard 37°. Interestingly, certain poliovirus mutations confer thermosensitivity in a host-dependent manner; for example, attenuating mutations in the 5NCR may or may not exhibit a *ts* phenotype depending on the type of cultured cells used for the assay (Macadam *et al.*, 1991). Thus, the distinction between specific and unspecific mechanisms of attenuation may be complicated.

It should also be noted that our discussion will be largely limited to the attenuating mutations just in the Sabin strains. It is very likely that there could exist truly attenuating mutations other than those obtained by Sabin. For example, an attenuated strain of serotype 2 (W-2), selected by Koprowski also in the 50s, lacks the known Sabin-specific mutations in the 472-481 area and appears to possess a distinct set of attenuating mutations (Pevear *et al.*, 1990).

Now let us consider the major factors controlling the poliovirus infection and pathogenicity.

VIRION/RECEPTOR COMPATIBILITY: A MAJOR FACTOR OF CELLULAR SUSCEPTIBILITY

Significance of the Receptor

The first of the factors controlling the very possibility of poliovirus infection is obvious. It is the availability of appropriate receptors on the cellular membranes. The molecular nature of the receptors for poliovirus has been elucidated recently (Mendelsohn *et al.*, 1989; Koike *et al.*, 1990; 1991b; for a review, see Racaniello, 1990). The receptor is represented by a three-domain immunoglobulin-like transmembrane protein molecule. Its presumed bulge(s) appear to be able to interact with characteristic cavities known to be present on the surface of poliovirus particles (Hogle *et al.*, 1985) in a key/lock manner. This is not merely a trivial metaphor. The receptor-driven conformational changes in the protein shell of the virions trigger the intracellular release of the genomic RNA.

For unknown reasons, the receptor gene is expressed only in a minor subset of tissues of the human organism, and this fact is a major factor controlling the viral tropism. Understandably, the locked door cannot be opened without a key, and if there are no receptors, no infection with intact virions is possible. But the opposite notion is not necessarily true. The existence of the receptors is a condition, although necessary, but not sufficient to make the cell susceptible to infection with poliovirus.

In particular, this follows from the experiments in transgenic mice. By introducing, into the zygote, the gene encoding the human receptor for poliovirus, mice carrying this gene in the chromosomes of each cell were obtained (Ren *et al.*, 1990; Koike *et al.*, 1991a). The receptor gene is transcribed in many, if not all, tissues of the transgenic animals, though at markedly different levels; the efficient gene expression results in the appearance of the receptor protein on, and the poliovirus binding activity of, the cellular surface. As a result, some cells of the transgenic animals, for example, neurons in the CNS, can be infected with the poliovirus strains that are normally nonpathogenic for mice. Nevertheless, some of the mouse cells expressing the receptor protein fairly well, for example, kidney epithelial cells or lung alveolar cells, failed to support poliovirus growth (Ren and Racaniello, 1992). Similarly, expression of the receptor gene in human cells and susceptibility of appropriate cells to poliovirus do not necessarily coincide with one another (Freistadt *et al.*, 1990). This phenomenon remains unexplained at the moment. It suggests, however, that the susceptibility to poliovirus is also controlled by cellular factor(s) other than receptors. There are other locks.

Virion Adjustments to Improper Receptors

Nevertheless, overcoming of the receptor barrier often results in the productive poliovirus infection of nonsusceptible cells. The well known general method permitting to do so consists in using viral RNA rather than virions. There is, however, another possibility, more relevant to the scope of this Chapter. Mice could be infected, upon the intracerebral inoculation, with certain strains of poliovirus type 2, but they are resistant to infection with type 1 or type 3 strains. This fact appears to be due to inadequacy of the mice analog of the human poliovirus receptor. The polioviruses naturally noninfectious for mice could, however, be made infectious by modifying a short peptide in VP1; several amino acid residues in this peptide should be replaced by appropriate residues from a mice-pathogenic type 2 strain (Murray *et al.*, 1988; Martin *et al.*, 1988; 1991). Unexpectedly, this region of VP1, being critically important for infectivity for mice, does not seem to be directly involved in the interaction with the human receptor (Yeates *et al.*, 1991).

The possibility of infecting mice with poliovirus prompted attempts to investigate, in this animal, the effect of mutations known to attenuate the virus for monkeys. This has been done by using either appropriate viruses [e. g., strains infectious for mice naturally or as a result of engineered mutations (La Monica *et al.*, 1987; Ren *et al.*, 1991; Martin *et al.*, 1991)] or modified, transgenic, hosts (Ren *et al.*,

1990; Koike *et al.*, 1991). These studies demonstrated that mice and monkeys exhibited qualitatively similar, though quantitatively different, capacities to discriminate between attenuating and neurovirulent poliovirus strains. This important observation testifies that neural cells of the two hosts share certain "sensors" of attenuating mutations. Most likely, these sensors are located intracellularly.

The above notion should not, however, be interpreted as ruling out the possibility that the attenuating mutations can exist that selectively impair early steps of the poliovirus interaction with neural cells, not interfering with the infection of other hosts (like HeLa or monkey kidney cells). No compelling evidence is currently available that such a trick is indeed exploited by the Sabin strains, but the location of an attenuating mutation in the Sabin 2 capsid just in the vicinity of the locus that reacts with the receptor is well compatible with such a possibility (Ren *et al.*, 1991).

TRANSLATIONAL COMPATIBILITY: ANOTHER MAJOR REPRODUCTION-CONTROLLING FACTOR

Efficiency of Translation and Attenuation of Neurovirulence

The next major step in the reproduction of poliovirus after its genome enters the cell, is translation. To start protein synthesis, the picornavirus RNA templates take advantage of an unusual mechanism of internal (rather than canonical, i. e., 5' cap-dependent) binding of ribosomes, as discussed in detail by Sonenberg in another Chapter of this volume. Evidence is accumulating that this peculiarity of translation initiation is intimately related to poliovirus pathogenicity and attenuation. The original observation was made by Svitkin *et al.* (1985) who showed that the attenuating mutations in the 5NCR affected the efficiency of translation. Indeed, RNAs from all the 3 Sabin strains exhibited a markedly lower *in vitro* template activity compared to that from their virulent predecessors (see also, Svitkin *et al.*, 1990; Muzychenko *et al.*, 1991). The phenomenon was due to a decrease in the translation initiation rather than elongation. These mutations affect the structure of the *cis*-acting control element which ensures the cap-independent internal binding of ribosomes to the picornaviral RNAs. In type 1 and type 3, these mutations seem to somewhat destabilize, not significantly altering, the secondary structure of the element, whereas the type 2 mutation changes a looped-out nucleotide (Blinov *et al.*, 1988; Pilipenko *et al.*, 1989).

Our laboratory obtained some evidence that the attenuating mutations in the 5NCR of the Sabin strains may affect the template interaction with certain initiation factors. It has been long known that poliovirus RNA is a very poor template in rabbit reticulocyte lysates, and that the translation efficiency can be improved by adding of some proteins from nucleated cells. Svitkin *et al.* (1988) partially purified the initiation correcting factor (ICF) and demonstrated that the RNAs from virulent strains responded much better to the addition of this factor compared to the RNAs from their attenuated relatives. The results obtained in collaboration with the Sonenberg's group suggest that a significant part of the ICF activity could be reproduced with a polypeptide known as p52 (Meerovitch *et al.*, 1989), recently identified as protein La. This latter protein appears to serve as a novel translation initiation factor specifically required for the poliovirus RNA template (Meerovitch *et al.*, in preparation) and able to sense the attenuating mutations in the 5NCR (Svitkin *et al.*, unpublished observations); somewhat smaller correcting effect appeared to be observed also in the presence of some other translation initiation factors. It can well be that ICF, having an apparent molecular mass of over 400 kDa (Svitkin *et al.*, 1988), is actually a complex of different polypeptides.

Thus, modifications of the *cis*-element by the attenuating mutations resulted in certain impairment of its response to specific host cell proteins serving as translation initiation factors. Several years ago we proposed that there should be a tissue-specific component of the translation machinery involved in the expression of the attenuating mutations: we argued that the effect of those mutations should be particularly prominent in neural cells (Svitkin *et al.*, 1985; Agol, 1988). Indeed, experiments carried out in our Institute demonstrated that cultured cells of a human neuroblastoma failed to support efficient reproduction of Sabin 1 and Sabin 2 (Agol *et al.*, 1989), whereas the results reported by La Monica and Racaniello (1989) indicated that the failure was partly due to the deteriorating effect of the attenuating mutation in the 5NCR on translation in neural cells.

There is another line of evidence suggesting that the 5NCR attenuating mutations may affect interaction of the viral RNA with host-specific factors. Indeed, the mutation at position 481 in the Sabin 2 RNA exhibits a host cell-dependent *ts* phenotype (Macadam *et al.*, 1991).

Remarkably, some extended deletions in downstream portions of the poliovirus 5NCR are accompanied by a significant decrease in the level of neurovirulence, exerting much lesser effect on the ability of the mutants to grow in monkey kidney or HeLa cells (Iizuka *et al.*, 1989; unpublished data from this laboratory). These observations support the notion that the translational machinery of human neural cells appears to be particularly discriminative with regard to the structure of the 5NCR *cis*-acting control element; they also suggest that some component(s) of this element behave in a tissue-specific manner, being only required in a subset of host cells. Experiments are underway in our laboratory to check the validity of this proposal and, if it is correct, to define the boundaries and the structure of the neuro-specific component of the viral RNA. Obviously, the existence of such tissue-specific elements in translational templates should have broad biological implications, suggesting that cellular differentiation may involve modifications of the protein-synthesizing apparatus aimed at ensuring the recognition (or non-recognition) of particular mRNA species.

Poliovirus Is a Mild Pathogen

In neuroblastoma cells, even the virulent poliovirus is not particularly vigorous; it takes much longer time to complete the infectious cycle in these cells than in HeLa or monkey kidney cells (cf., Agol *et al.*, 1989). Similarly, poliovirus growth is relatively inefficient in the blood human cells and this inefficiency is again related to the low compatibility of poliovirus RNA and the translation machinery of the cells (López-Guerrero *et al.*, 1989). Thus, we are arriving at a very important conclusion. Poliovirus, in its natural host, is a rather mild pathogen. A major reason for this mildness is incomplete compatibility of the viral template with the host translation machinery.

Accepting this notion, it is easier to explain why, as already mentioned, less than 1% of the non-immune human beings infected with wild type poliovirus develop the paralytic disease. Furthermore, the critical dependence of the poliovirus RNA template activity on host factors involved in the control of translation, may suggest that the rare cases when the disease did develop were actually due, at least in part, to the physiological or pathological alterations of the activity of some of these host factors (Agol, 1990).

Attenuating mutations in the 5NCR exacerbate the already low compatibility between the viral RNA templates and the translational machinery of the natural target cells, contributing significantly to the conversion of the mild pathogen into a safe vaccine. Again, at least some of extremely rare cases of the vaccine-associated poliomyelitis might be due to quantitative or qualitative changes in the activity of translational host factors allowing to overcome the intrinsic template inefficiency of the genome of Sabin strains.

ADDITIONAL WAYS TO MODULATE INFECTION: ATTENUATING MUTATIONS IN THE CODING REGION

The mutations affecting translation, being a major factor contributing to the attenuated phenotype of the Sabin strains, cannot, however, fully explain the extremely low level of their virulence. Alterations in the capsid proteins are also physiologically significant. The mechanisms of expression of these mutations, though poorly understood, is likely to vary among the Sabin serotypes (at least, among serotypes 2 and 3). This follows primarily from quite different localizations of the mutations. As already mentioned, the attenuating mutation in serotype 2, being located near the region involved in the interaction with receptors, could be assumed to restrict the entry of the respective genome into the target cells, but this conclusion should be supported by direct evidence. In Sabin 3, the Ser → Phe substitution at position VP3-91 leads to the capsid instability at elevated temperatures; as a result, it may interfere with different steps of the virus growth (Filman *et al.*, 1989).

Little can be said about the attenuating mutations in the replication proteins except that they do exist in Sabin 1 (Nomoto and Wimmer, 1987). The neuroblastoma cell experiments with recombinants between attenuated and virulent polioviruses of type 1 suggest that the Sabin 1-specific mutation(s) in replication proteins may selectively restrict the viral growth in neural cells as compared to HeLa or monkey kidney cells: the recombinant with the 5' half of the genome derived from the virulent Mahoney and the 3' half from the attenuated Sabin 1 strains generated less infectious progeny than the virulent Mahoney did (Agol *et al.*, 1989).

Despite the scarcity of information, the very existence of true attenuating mutations in replication proteins is illuminating; it strongly suggests the involvement of host factors in the replication of the viral genome; moreover, it suggests that this involvement varies, at least quantitatively, in different types of host cells.

SELF-DEFENSE OF THE HOST

Non-canonical Effect of Antibodies

Disregarding the fact that poliovirus is a rather mild pathogen, the infected humans turn on different defense measures to safeguard themselves against possible adverse consequences. Some of these measures are of the standard type, but others appear to be unique. Poliovirus infection elicits a prompt antibody response; serum neutralizing antibodies appear a few days after infection and, in the case of disease, even before clinical signs (Melnick, 1985). The immunology of poliovirus is outside the scope of this Chapter, but one of its recently discovered aspects appears to be directly related to the neuropathogenicity of the virus.

Rather unexpectedly, it turned out that the neuroblastoma cells undergoing productive poliovirus infection could be cleared of the virus by the addition of external antiviral antibodies (Tolskaya *et al.*, 1992). Neuroblastoma cultures were infected with virulent poliovirus at a multiplicity ensuring the infection of the overwhelming majority of the cells. Two hours after, antiserum against poliovirus was added. This did not prevent the infection, and the viral antigen could be revealed in nearly all the cells. Moreover, there was a significant accumulation of infectious progeny within the antiserum-treated cells; by 6 hours post-infection, each cell, treated with the antiserum or not, contained on the average more than 300 PFU. Nevertheless, the infected cells did not die; instead they appeared to recover from the infection, began to divide, and, after a few passages in the presence of antiserum, they exhibited no signs of the infection experienced. Remarkably, the preventive effect could be reproduced with monoclonal antibodies known to react with the capsid polypeptides; at least one of such active antibodies should recognize a conformational epitope formed by the adjacent pentamers.

Thus, the productive infection of neural cells with virulent poliovirus can be uncoupled from the fatal cytopathic effect by externally present antibodies, and the infected cells can be "cured". The mechanism of this curing remains obscure, however. We propose that the permeability of cellular membranes to macromolecules is increased upon poliovirus infection. In fact, this idea has got a significant support from many studies done years ago in the laboratory of Carrasco (Carrasco and Lacal, 1983). What we are proposing is, firstly, that these permeability changes allow antibodies to enter the cell; secondly, that the changes are completely reversible, if antiviral antibodies are added to infected neural cells; and thirdly, that the interaction of antibodies with maturing virions prevents the cellular death. How this could be achieved remains a mystery. One may speculate that trapping of virus-specific products by the antibodies diminishes their effective concentration or activity to a level below a certain critical threshold required to trigger an unidentified irreversible cytopathic event.

These ideas, vague as they are, can help to formulate a hypothesis explaining the outcome of the infection of neural cells with attenuated polioviruses. We propose that in these cells, due to the impaired translation initiation, the viral "killer" products are also not accumulated above the critical threshold, in an analogy with the antibody-treated cells infected with the virulent virus.

In any case, the ability of exogenous antiviral antibodies to prevent death of neural cells undergoing productive poliovirus infection can well contribute to the pathogenesis of infection, and in particular, it may help to clear the CNS from the virus. There is enough ground to propose that such "curing" is a more or less general phenomenon responsible for the antibody-mediated clearance from neural cells also of Theiler's murine encephalomyelitis virus (Fujinami *et al.*, 1989), reovirus (Virgin *et al.*, 1988; Tyler *et al.*, 1989), Sindbis virus (Levine *et al.*, 1991), and perhaps some other viruses.

Resistance to Superinfection

There is another peculiarity in the poliovirus interaction with the neuroblastoma cells. The cells that survive the infection with virulent and attenuated strains (either as a result of the antiserum treatment or "spontaneously") acquire, after a few passages in the absence of antibodies, resistance to super-infection with heterotypic or homotypic poliovirus strains (Colbère-Garapin *et al.*, 1989; Tolskaya *et al.*, 1992). The nature of this phenomenon remains unknown, but a reversible impairment of the function of the receptor system is a possibility (Tolskaya *et al.*, unpublished observations). Again, it is tempting to speculate that this type of cellular "post-infection immunity" may contribute to the poliomyelitis pathogenesis, protecting the cells against superinfection with more virulent variants which are likely to evolve during the viral reproduction in the CNS (see below).

Despite the general resistance to superinfection, the neuroblastoma cultures that have experienced, and survived after, the primary infection with poliovirus are not completely cured, if passaged in the absence of the antiserum; instead the state of persistent infection is readily developed (Colbère-Garapin *et al.*, 1989; Tolskaya *et al.*, unpublished observations). The relevance of this phenomenon to the pathogenesis of poliomyelitis is obscure, but there is a suggestion that it could be associated with chronic forms of poliomyelitis (Colbère-Garapin *et al.*, 1989). If so, it is a semantic problem whether such chronization could be regarded as a kind of self-defense measure.

Are There Additional Mechanisms Limiting Poliovirus Spread over the Central Nervous System?

The lesions in the CNS of poliomyelitis victims are relatively local. Even in monkeys inoculated with the wild type poliovirus intracerebrally or intraspinally, only a portion of the potentially susceptible cells are injured. The important question is whether the relatively local character of injuries and the eventual clearance of the CNS from the virus are due to the standard immunological response in conjunction with the special self-defense measures outlined above. Or are there some additional uncovered mechanisms? The question is the more appropriate as poliovirus may long persist and multiply in patients with severe immunodeficiencies (Davis *et al.*, 1977). The answer to this question is yet to come.

INSTABILITY OF ATTENUATING MUTATIONS

Being remarkably safe, the Sabin strains are nevertheless known to cause very rare cases of so-called vaccine-associated poliomyelitis (Nkowane *et al.*, 1987; Lipskaya *et al.*, 1987; Kutitova *et al.*, 1990). The disease is more often observed among contacts than among the primary recipients of the vaccine. The instability of attenuating mutations, and, in particular, those in the 5NCR can be listed as a possible reason for at least some of these cases. The rapid loss of the 5NCR attenuating mutations in vaccinated children was originally described by Evans *et al.* (1985) and since then documented in numerous studies (Minor and Dunn, 1988; Muzychenko *et al.*, 1991). Obviously, the Sabin strains experience, during the reproduction in human gut, a considerable selective pressure toward the elimination of these attenuating mutations. This fact can be interpreted to mean that the target cells (whatever their nature) in the gastro-intestinal tract resemble neural cells in their "distaste" for the improper structure of the translational *cis*-element in the 5NCR. Indeed, although the selection against these attenuating mutation could be observed upon the virus reproduction in tissue culture, especially at supraoptimal temperatures (Christodoulou *et al.*, 1990; Chumakov *et al.*, 1992), the rate of reversions of the attenuating mutations occurring in the human gut seems to be especially high.

Such an instability is not particularly surprising. The intrinsic fidelity of the poliovirus RNA replication is moderate, to say the least. If there is some incompatibility between the host cell and the genomes of certain mutants, but the possibility does exist to generate more fit variants by point mutations, deletions, or insertions, this possibility will be remarkably readily realized (Pilipenko *et al.*, 1992; Agol, 1992). If the selection of such revertants is carried out in the standard tissue cultures (e. g., in monkey kidney cells) rather than in the gut, some of the front runners, well adapted to their particular hosts, exhibit a diminished, or lost altogether, ability to infect neural cells in culture and to cause the paralytic disease upon the intraspinal inoculation of monkeys (unpublished observations). These experiments involving modern genetic engineering and analytical techniques reproduce, in fact, some steps of the selection procedure used by Sabin to obtain his famous strains some 40 years ago.

The instability of the attenuating mutations in the capsid-coding region of the genome is less spectacular, but it could also be readily demonstrated with the *ts* mutation in VP3 of Sabin 3 upon *in vitro* cultivation at supraoptimal temperatures; both the true reversion and second-site suppression of the *ts* phenotype resulted in an increase in the level of neurovirulence (Macadam *et al.*, 1989).

On the basis of the above observations, one may formulate the following "rules". The more a given attenuating mutation is specific (i. e., impairing viral growth preferentially in neural cells), the more it will be stable upon the virus cultivation in standard cell cultures; it is expected, however, to be unstable in the infecting organism. On the other hand, if an attenuating mutation is based on a relatively general invalidation of a viral function, it would hardly be genetically stable under any cultivation conditions. These notions may look somewhat pessimistic with regards to the prospects of a significant improvement in the Sabin vaccine strains. The prospects are, however, not as dim as it may appear. This point will be elaborated in the next section.

Before switching to the vaccine problem, it would be appropriate, however, to briefly discuss another point related to the plasticity of the poliovirus genome. It was already emphasized that wild-type poliovirus strains are rather mild pathogens. One may then ask why a highly pathogenic variant did not evolve during the long history of multiple passages of the virus in humans. Questions of such kind can rarely, if ever, be definitively answered. An attractive explanation may however be suggested. One may envision that the translational *cis*-acting control element in the poliovirus 5NCR could not be converted, by a reasonable number of mutational steps, into a form able to promote efficient internal binding of ribosomes in natural target (e. g., neural) cells. A remarkably high conservation of the secondary, and partly primary, structures of this element among all entero- and rhinoviruses (Pilipenko *et al.*, 1989) testify to the existence of severe constraints imposed on its organization.

PROSPECTS FOR VACCINE IMPROVEMENT

Obviously, one may wish to deprive the vaccine from any, even extremely rare, untoward effects. In the framework of this discussion, the worst consequence is the vaccination-associated paralytic disease. Evidently, two kind of causes may underlie such mishaps. Either the vaccine virus is loosing or weakening its attenuated phenotype, or the victim is exhibiting an extreme susceptibility to the residual (very low) virulence of the Sabin strains. But can the cases of vaccine-associated poliomyelitis be totally eliminated? I believe the answer is "Yes". To achieve this goal, one should completely uncouple the antigenicity of the virus from the concomitant pathogenicity. This can perhaps be accomplished in two different ways. Indeed, the overwhelming majority of such cases were due to Sabin 2 and Sabin 3, whereas Sabin 1 appears to be nearly completely innocuous (Nkowane *et al.*, 1987; Lipskaya *et al.*, 1987;

Kutitova *et al.*, 1990). Such a reliability of Sabin 1 is obviously due to the quantity as well as the quality of the attenuating mutations it possesses. Therefore the simplest way to make other vaccine serotypes safer may consist in combining appropriate capsid protein-coding regions with the noncoding and replication protein-coding regions from Sabin 1 RNA (cf., Nomoto *et al.*, 1988; Kohara *et al.*, 1988).

There is perhaps another way to design completely safe poliovirus mutants, making use of the fact, discussed above, that the protein-synthesizing machinery of human neural cells appears to be particularly discriminative with regard to the structure of the translational *cis*-acting control element (cf., Agol, 1992). It can be imagined that some configurations of this element will be totally unacceptable for neural cells, but will permit its more or less normal function in other hosts. The lack of neurotropism of many picornaviruses can be taken as an indirect argument in favor of such an assumption. Although no obvious blueprint is available to improve the vaccine according to this line, it is known that some gross alterations in the poliovirus 5NCR are accompanied by a significant decrease in the level of neurovirulence (Iizuka *et al.*, 1989; unpublished observations from this laboratory). Ideally, the new generation of attenuated strains will not only be devoid of residual neurovirulence but will not be able at all to acquire pathogenic potential by reversion (because of negligible chances to restore the original 5NCR structure). The creatures generated in this way, while inheriting the parental name of "poliovirus", would possess the genomes altered to such an extent that the above pessimistic rules on the instability of attenuating mutations could no longer be applicable to them.

Thus, it seems quite feasible to engineer highly attenuated poliovirus strains by working along these lines. There is a real hazard, however, to make the virus overattenuated. To induce an adequate immune response, the mutants should retain their capacity to grow in human gut. The key point is to what extent the translation machineries of poliovirus-susceptible cells in the intestines and central nervous system share tastes and prejudices with regard to the template structure. Nevertheless, the existence of the Sabin strains with a nearly perfect combination of attenuation and intestinal "takes" testifies to the feasibility of an acceptable balance between these opposite requirements (Agol, 1992).

CONCLUDING REMARKS

A good deal of valuable data on the molecular basis of poliovirus virulence and attenuation have accumulated during the last decade or so. The locations of the majority of strong attenuating mutations in the genomes of Sabin strains have been established, and the mechanisms of phenotypic expression of some of these mutations have been outlined; in particular, a close link between the virulence and the translational potential of the viral RNA has been uncovered. Important peculiarities of the poliovirus interaction with neural cells have been noticed. Nevertheless, many key questions remain unanswered. It is perhaps appropriate to list some of them. Why are poliovirus reproduction and concomitant cell injury in the human organism are largely confined to motoneurons of the anterior horns of the spinal cord? What are the mechanisms restricting poliovirus reproduction in the susceptible cells of the infected host? What are pecualirities of the translational machinery of natural target cells? What is the best way to improve the oral poliovaccine?

It seems that in-depth answers to some of these questions are largely a matter of good work along established directions. It may turn out, however, that the solution of some other problems will be difficult to find unless fresh ideas and/or methodology are put forward. Moreover, if the goal set by the WHO to eradicate poliomyelitis by the year 2000 is achieved and any work with virulent polioviruses is therefore prohibited, it can happen that some of the problems will never be solved. So, we should be in a hurry.

REFERENCES

Agol, V. I. (1988). Genetic determinants of neurovirulence and attenuation of poliovirus. *Mol. Genet.* No 1, 3-9.

Agol, V. I. (1990). Current approaches to the problem of poliovirus attenuation. *In* "New Aspects of Positive-Strand RNA Viruses" (M. Brinton and F. Heinz, Eds), pp. 311-318. ASM, Washington.

Agol, V. I. (1992). Genetic stability and instability of the cis-acting control element of the 5' untranslated region of the poliovirus RNA. *Dev. Biol. Stand.* (in press).

Agol, V. I., Grachev, V. P., Drosdov, S. G., Kolesnikova, M. S., Kozlov, V. G., Ralph, N. M., Romanova, L. I., Tolskaya, E. A., Tyufanov, A. V., and Viktorova, E. G (1984). Construction and properties of intertypic poliovirus recombinants: first approximation of the major determinants of neurovirulence. *Virology* **136**, 41-55.

Agol, V. I., Drozdov, S. G., Frolova, M. P., Grachev, V. P., Kolesnikova, M. S., Kozlov, V. G., Ralph, N. M., Romanova, L. I., Tolskaya, E. A., and Viktorova, E. G. (1985a). Neurovirulence of the intertypic poliovirus recombinant v3/a1-25: characterization of strains isolated from the spinal cord of diseased monkeys and evaluation of the contribution of the 3' half of the genome. *J. Gen. Virol.* **66**, 309-316.

Agol, V. I., Drozdov, S. G., Grachev, V. P., Kolesnikova, M. S., Kozlov, V. G., Ralph, N. M., Tolskaya, E. A., Tyufanov, A. V., and Viktorova, E. G. (1985b). Recombinants between attenuated and virulent strains of poliovirus type 1: Derivation and characterization of recombinants with centrally located crossover points. *Virology* **143**, 467-477.

Agol, V. I., Drozdov, S. G., Ivannikova, T. A., Kolesnikova, M. S., Korolev, M. B., and Tolskaya, E. A. (1989). Restricted growth of attenuated poliovirus strains in cultured cells of a human neuroblastoma. *J. Virol.* **63**, 4034-4038.

Almond, J. W. (1987). The attenuation of poliovirus neurovirulence. *Annu. Rev. Microbiol.* **41**, 153-180.

Blinov, V. M., Pilipenko, E. V., Romanova, L. I., Sinyakov, A. N., Maslova, S. V., and Agol, V. I. (1988). A comparison of the secondary structures of the 5'-untranslated segment of neurovirulent and attenuated poliovirus strains. *Dokl. AN SSSR* **298**, 1004-1006.

Carrasco, L., and Lacal, J. C. (1983). Permeabilization of cells during animal virus infection. *Pharmac. Ther.* **23**, 109-145.

Christodoulou, C., Colbère-Garapin, F., Macadam, A., Taffs, L. F., Marsden, S., Minor, P., and Horaud, F. (1990). Mapping of mutations associated with neurovirulence in monkeys infected with Sabin 1 poliovirus revertants selected at high temperature. *J. Virol.* **64**, 4922-4929.

Chumakov, K. M., Norwood, L. P., Parker, M. L., Dragunsky, E. M., Ran, Y., and Levenbook, I. S. (1992). RNA sequence variants in live poliovirus vaccine and their relation to neurovirulence. *J. Virol.* **66**, 966-970.

Colbère-Garapin, F., Christodoulou, C., Crainic, R., and Pelletier, I. (1989). Persistent poliovirus infection of human neuroblastoma cells. *Proc. Natl. Acad. Sci. USA* **86**, 7590-7594.

Davis, L. E., Bodian, D., Price, D., Butler, I., and Vickers, J. H. (1977). Chronic progressive poliomyelitis secondary to vaccination of an immunodeficient child. *N. Engl. J. Med.* **297**, 241-245.

Evans, D.H.A., Dunn, G., Minor, P.D., Schild, G.C., Cann, A.J., Stanway, G., Almond, J.W., Currey, K., and Maizel, J.V., Jr. (1985). Increased neurovirulence associated with a single nucleotide change in a non-coding region of the Sabin type 3 poliovaccine genome. *Nature (London)* 314, 548-550.

Filman, D. J., Syed, R., Chow, M., Macadam, A. J., Minor, P. D. and Hogle, J. M. (1989). Structural factors that control conformational transitions and serotype specificity in type 3 poliovirus. *EMBO J.* 8, 1567-1579.

Freistadt, M. S., Kaplan, G., and Racaniello, V. R. (1990). Heterogeneous expression of poliovirus receptor-related proteins in human cells and tissues. *Mol. Cell. Biol.* 10, 5700-5706.

Fujinami, R. S., Rosenthal, A., Lampert, P. W., Zurbriggen, A., and Yamada, M. (1989). Survival of athymic (*nu/nu*) mice after Theiler's murine encephalomyelitis virus infection by passive administration of neutralizing monoclonal antibody. *J. Virol.* 63, 2081-2087.

Hogle, J. M., Chow, M., and Filman, D. J. (1985). Three-dimensional structure of poliovirus at 2.9 A resolution. *Science* 229, 1358-1365.

Iizuka, N., Kohara, M., Hagino-Yamagishi, K., Abe, S., Komatsu, T., Tago, K., Arita, M., and Nomoto, A. (1989). Construction of less neuro-virulent polioviruses by introducing deletions into the 5' noncoding sequence of the genome. *J. Virol.* 63, 5354-5363.

Kawamura, N., Kohara, M., Abe, S., Komatsu, T., Tago, K., Arita, M., and Nomoto, A. (1989). Determinants in the 5' noncoding region of polio-virus Sabin 1 RNA that influence the attenuation phenotype. *J. Virol.* 63, 1302-1309.

Kitamura, N., Semler, B. L., Rothberg, P. G., Larsen, G. R., Adler, C. J., Corner, A. J., Emini, E. A., Hanecak, R., Lee, J. J., van der Werf, S., Anderson, C. W., and Wimmer, E. (1981). Primary structure, gene organization and polypeptide expression of poliovirus RNA. *Nature (London)* 291, 547-553.

Kohara, M., Abe, S., Komatsu, T., Tago, K., Arita, M., Nomoto, A. (1988). A recombinant virus between the Sabin 1 and Sabin 3 vaccine strains of poliovirus as a possible candidate for a new type 3 poliovirus live vaccine strain. *J. Virol.* 62, 2828-2835.

Koike, S., Horie, H., Ise, I., Okitsu, A., Yoshida, M., Iizuka, N., Takeuchi, K., Takegami, T., and Nomoto, A. (1990). The poliovirus receptor protein is produced both as membrane-bound and secreted forms. *EMBO J.* 9, 3217-3224.

Koike, S., Taya, C., Kurata, T., Abe, S., Ise, I., Yonekawa, H., and Nomoto, A. (1991a). Transgenic mice susceptible to poliovirus. *Proc. Natl. Acad. Sci. USA* 88, 951-955.

Koike, S., Ise, I., and Nomoto, A. (1991b). Functional domains of the poliovirus receptor. *Proc. Natl. Acad. Sci. USA* 88, 4104-4108.

Kutitova, O. K., Lipskaya, G. Y., Maslova, S. V., Leshchinskaya, E. V., Martynenko, I. N., Koreshkova, G. V., Drozdov, S. G., and Agol, V. I. (1990). Molecular epidemiology of poliomyelitis: characterization of strains isolated from patients in Moscow in 1975-1986. *Zh. Mikrobiol.* 6, 43-49.

La Monica, N., Almond, J. W., and Racaniello, V. R. (1987). A mouse model for poliovirus neurovirulence identifies mutations that attenuate the virus for humans. *J. Virol.* 61, 2917-2920.

La Monica, N., and Racaniello, V. R.. (1989). Differences in replication of attenuated and neurovirulent polioviruses in human neuroblastoma cell line SH-SY5Y. *J. Virol.* 63, 2357-2360.

Levine, B., Hardwick, J. M., Trapp, B. D., Crawford, T. O., Bollinger, R. C., and Griffin, D. E. (1991). Antibody-mediated clearance of alpha-virus infection from neurons. *Science* 254, 856-860.

Lipskaya, G. Y., Kutitova, O. K., Feldman, E. V., Votyakov, V. I., Maslova, S. V., Gracheva, L. A., Drozdov, S. G., and Agol, V. I. (1987)

The nature of strains isolated from polio cases in Byelorussia in 1967-1985. *Mol. Genet.* 10, 30-37.

López-Guerrero, J. A., Carrasco, L., Martinez-Abarca, F., Fresno, M., and Alonso, M. A. (1989). Restriction of poliovirus RNA translation in a human monocytic cell line. *Eur. J. Biochem.* 186, 577-582.

Lwoff, A. (1959). Factors influencing the evolution of viral disease at the cellular level and in the organism. *Bacteriol. Rev.* 23, 109-124.

Macadam, A. J., Arnold, C., Howlett, J., John, A., Marsden, S., Taffs, F., Reeve, P., Hamada, N., Wareham, K., Almond, J., Cammack, N. and Minor, P. D. (1989). Reversion of the attenuated and temperature-sensitive phenotypes of the Sabin type 3 strain of poliovirus in vaccinees. *Virology* 172, 408-414.

Macadam, A. J., Pollard, S. R., Ferguson, G., Dunn, G., Skuce, R., Almond, J. W., and Minor, P. D. (1991). The 5' noncoding region of the type 2 poliovirus vaccine strain contains determinants of attenuation and temperature sensitivity. *Virology* 181, 451-458.

Martin, A., Benichou, D., Couderc, T., Hogle, J. M., Wychowski, C., van der Werf, S., and Girard, M. (1991). Use of type 1/type 2 chimeric polioviruses to study determinants of poliovirus type 1 neurovirulence in a mouse model. *Virology* 180, 648-658.

Martin, A., Wychowski, C., Couderc, T., Crainic, R., Hogle, J., and Girard, M. (1988). Engineering a poliovirus type 2 antigenic site on a type 1 capsid results in a chimaeric virus which is neurovirulent for mice. *EMBO J.* 7, 2839-2847.

Meerovitch, K., Pelletier, J., and Sonenberg, N. (1989). A cellular protein that binds to the 5'-noncoding region of poliovirus RNA: Implications for internal translation initiation. *Genes Dev.* 3, 1026-1034.

Mendelsohn, C. M., Wimmer, E., and Racaniello, V. R. (1989). Cellular receptor for poliovirus: molecular cloning, nucleotide sequence, and expression of a new member of the immunoglobulin superfamily. *Cell* 56, 855-865.

Melnick, J. L. (1985). Enteroviruses: polioviruses, coxsackieviruses, echoviruses, and newer enteroviruses. *In* "Virology" (B.N. Fields, Ed.), pp. 739-794. Raven Press, NY.

Minor, P. D., and Dunn, G. (1988). The effect of sequences in the 5' non-coding region on the replication of polioviruses in the human gut. *J. Gen. Virol.* 69, 1091-1096.

Moss, E. G., O'Neill, R. E., and Racaniello, V. R. (1989). Mapping of attenuating sequences of an avirulent poliovirus type 2 strain. *J. Virol.* 63, 1884-1890.

Murray, M. G., Bradley, J., Yang, X.-F., Wimmer, E., Moss, E. G., and Racaniello, V. R. (1988). Poliovirus host range is determined by a short amino acid sequence in neutralization antigenic site I. *Science* 241, 213-215.

Muzychenko, A. R., Lipskaya, G. Y., Maslova, S. V., Svitkin, Y. V., Pilipenko, E. V. Nottay, B. K., Kew, O. M, and Agol, V. I. (1991). Coupled mutations in the 5'-untranslated region of the Sabin poliovirus strains during in vivo passages: structural and functional implications. *Virus Res.* 21, 111-122.

Nkowane, B. M., Wassilak, S. G. F., Orenstein, W. A., Bart, K. J. B., Schonberder, L. B., Hinman, A. R., and Kew, O. M. (1987) Vaccine-associated paralytic poliomyelitis. United States: 1973 through 1984. *J. Am. Med. Assoc.* 257, 1335-1340.

Nomoto, A., Omata, T., Toyoda, H., Kuge, S., Horie, H., Kataoka, Y., Genba, Y., Nakano, Y., and Imura, N. (1982). Complete nucleotide sequence of the attenuated poliovirus Sabin 1 strain genome. *Proc. Natl. Acad. Sci. USA* 79, 5793-5797.

Nomoto, A., and Wimmer, E. (1987). Genetic studies of the antigenicity and the attenuation phenotype of poliovirus. *In* "Molecular Basis of

Virus Disease" (M.C. Russel and J.W. Almond, Eds.), Soc. Gen. Microbiol. Symp., Vol. 40, pp. 107-134. Cambridge Univ. Press, Cambridge.

Nomoto, A., Iizuka, N., Kohara, M., and Arita, M. (1988). Strategy for construction of live picornavirus vaccines. *Vaccine* 6, 134-137.

Pevear, D. C., Oh, C. K., Cunningham, L. L., Calenoff, M., and Jubelt, B. (1990). Localization of genomic region specific for the attenuated, mouse-adapted poliovirus type 2 strain W-2. *J. Gen. Virol.* 71, 43-52.

Pilipenko, E. V., Blinov, V. M., Romanova, L. I., Sinyakov, A. N., Maslova, S. V., and Agol, V. I. (1989). Conserved structural domains in the 5'-untranslated region of picornaviral genomes: an analysis of the segment controlling translation and neurovirulence. *Virology* 168, 201-209.

Pilipenko, E. V., Gmyl, A. P., Maslova, S. V., Svitkin, Y. V., Sinyakov, A. N., and Agol, V. I. (1992). Prokaryotic-like cis element in the cap-independent internal initiation of translation on picornavirus RNA. *Cell* 68, 119-131.

Racaniello, V. R. (1988). Poliovirus neurovirulence. *Adv. Virus Res.* 34, 217-246.

Racaniello, V. R. (1990). Cell receptors for picornaviruses. *Curr. Top. Microbiol. Immunol.* 161, 1-22.

Racaniello, V. R., and Baltimore, D. (1981). Molecular cloning of poliovirus DNA and determination of the complete nucleotide sequence of the viral genome. *Proc. Natl. Acad. Sci USA* 78, 4887-4891.

Ren, R., Constantini, F., Gorgacz, E, J., Lee, J. J., and Racaniello, V. R. (1990). Transgenic mice expressing a human poliovirus receptor: a new model for poliomyelitis. *Cell* 63, 353-362.

Ren, R. R., Moss, E. G., and Racaniello, V. R. (1991). Identification of two determinants that attenuate vaccine-related type 2 poliovirus. *J. Virol.* 65, 1377-1382.

Ren, R., and Racaniello, V. R. (1992). Human poliovirus receptor gene expression and poliovirus tissue tropism in transgenic mice. *J. Virol.* 66, 296-304.

Sabin, A.B. (1985). Oral poliovirus vaccine: history of its development and use, and current challenge to eliminate poliomyelitis from the world. *J. Infect. Dis.* 151, 420-436.

Stanway, G., Hughes, P. J., Mountford, R. C., Reeve, P., Minor, P., Schild, G. C., and Almond, J. W. (1984). Comparison of the complete nucleotide sequences of the genomes of the neurovirulent poliovirus P3/Leon/37 and its attenuated Sabin vaccine derivative P3/Leon $12a_1b$. *Proc. Natl. Acad. Sci. USA* 81, 1539-1543.

Svitkin, Y. V., Maslova, S. V., and Agol, V. I. (1985). The genomes of attenuated and virulent poliovirus strains differ in their in vitro translation efficiencies. *Virology* 147, 243-252.

Svitkin, Y. V., Pestova, T. V., Maslova, S. V., and Agol, V. I. (1988). Point mutations modify the response of poliovirus RNA to a translation initiation factor: a comparison of neurovirulent and attenuated strains. *Virology* 166, 394-404.

Svitkin, Y. V., Cammack, N., Minor, P. D., and Almond, J. W. (1990). Translation deficiency of the Sabin type 3 poliovirus genome: association with an attenuating mutation $C_{472} \rightarrow U$. *Virology* 175, 103-109.

Tolskaya, E. A., Ivannikova, T. A., Kolesnikova, M. S., Drozdov, S. G., and Agol. V. I. (1992). Postinfection treatment with antiviral serum results in survival of neural cells productively infected with virulent poliovirus. *J. Virol.* (in press).

Tyler, K. L., Virgin, H. V.,IV, Bassel-Duby, R., and Fields, B. N. (1989). Antibody inhibits defined stages in the pathogenesis of reovirus serotype 3 infection of the central nervous system. *J. Exp. Med.* 170, 887-900.

Virgin, H. V., IV, Bassel-Duby, R., Fields, B. N., and Tyler, K. L. (1988). Antibody protects against lethal infection with the neurally spreading reovirus type 3 (Dearing). *J. Virol.* 62, 4594-4604.
Westrop, G. D., Wareham K. A., Evans, D. M. A., Dunn, G., Minor, P. D., Magrath, D. I., Taffs, F., Marsden, S., Skinner, M. A., Schild, G. C., and Almond, J. W. (1989). Genetic basis of attenuation of the Sabin type 3 oral poliovirus vaccine. *J. Virol.* 63, 1338-1344.
Yeates, T. O., Jacobson, D. H., Martin, A., Wychowski, C., Girard, M., Filman, D. J., and Hogle, J. M. (1991). Three-dimensional structure of a mouse-adapted type 2/type 1 poliovirus chimera. *EMBO J.* 10, 2331-2341.

INDEX

The manufacturer's authorised representative in the EU is Springer Nature Customer Service Centre GmbH, Europaplatz 3, 69115 Heidelberg, Germany. If you have any concerns regarding our products, please contact ProductSafety@springernature.com

Printed and bound by CPI Group (UK) Ltd, Croydon, CR0 4YY
28/07/2026
02179215-0008